THE CELL

Volume V

THE CELL

Biochemistry, Physiology, Morphology

Edited by

JEAN BRACHET
Faculté des Sciences, Université libre de Bruxelles
Bruxelles, Belgique

ALFRED E. MIRSKY
The Rockefeller Institute
New York, New York

VOLUME V
Specialized Cells: Part 2

1961

ACADEMIC PRESS, New York and London

ACADEMIC PRESS, INC.
111 Fifth Avenue, New York, New York 10003

United Kingdom Edition published by
ACADEMIC PRESS, INC. (LONDON) LTD.
Berkeley Square House, London W1X6BA

LIBRARY OF CONGRESS CATALOG CARD NUMBER: 59–7677

Third Printing, 1970

PRINTED IN THE UNITED STATES OF AMERICA

LIST OF CONTRIBUTORS

L. ARVY, *Laboratoire de Physiologie du Centre National de la Recherche Zootechnique, Jouy-en-Josar (Seine-et-Oise), Paris, France*

W. BERNHARD, *Institut de Recherches sur le Cancer, Villejuif (Seine), France*

MARCEL BESSIS, *Centre National de Transfusion Sanguine, Paris, France*

ROY P. FORSTER, *Department of Zoology, Dartmouth College, Hanover, New Hampshire*

M. GABE, *Laboratoire d'Evolution des Etres organisés, Paris, France*

P. LACROIX, *Institut d'Anatomie, Louvain, Belgium*

ELIANE LE BRETON, *Centre de Recherches de Physiologie et Biochimie Cellulaires du C.M.R.S., Villejuif (Seine), France*

PHILIP D. MCMASTER, *The Rockefeller Institute, New York*

WILLIAM MONTAGNA, *Department of Biology, Brown University, Providence, Rhode Island*

YVONNE MOULÉ, *Centre de Recherches de Physiologie et Biochimie Cellulaires du C.M.R.S., Villejuif (Seine), France*

CH. OBERLING, *Institut de Recherches sur le Cancer, Villejuif (Seine), France*

PREFACE

The excellent reception accorded the first volume has been very encouraging. It is most rewarding for the contributors to realize that this treatise is filling a real need for a synthesis of up-to-date knowledge in the many active fields of research on the cell.

The chapters in these volumes deal with many different types of cells. Of course what one would desire is to have presented cells of the whole animate creation. "We ought not to hesitate nor to be abashed," said Aristotle, "but boldly to enter upon our researches concerning animals of every sort and kind, knowing that in not one of them is Nature or Beauty lacking." Selection, however, was necessary so that not all areas could be covered.

J. BRACHET

A. E. MIRSKY

July, 1960

CONTENTS

CONTENTS

THE CELL: *Biochemistry, Physiology, Morphology*
COMPLETE IN 5 VOLUMES

xiii

CHAPTER 1

Gland Cells

By M. GABE and L. ARVY

I. INTRODUCTION

Glandular function is a special case of secretion, which is a phenomenon common to all cells.

The secretory process is usually defined as the result of a cellular activity proceeding in three stages: (1) absorption by the cell of certain compounds present in the surrounding medium; this passage of materials

1

through the cell membrane is usually called *ingestion*; (2) formation inside the cell of the product(s) of secretion; this stage of the cell activity is called *synthesis* by Junqueira and Hirsch (1956); (3) elimination of the product(s) of secretion into the external or internal environment; this last stage is called *extrusion*.

The secretory process thus defined is evidently a very widespread phenomenon; it would be hard to find in any multicellular organism, a single type of cell not capable of absorbing various materials from its environment, and, after their more or less complete transformation eliminating them. Therefore is it possible to state with Prenant *et al.* (1904) that "secretion is a general property, but not a function of the cell." Consequently, glandular activity must be more closely defined.

Among the substances rejected by the cell in the course of its activity, some have merely passed through it without undergoing any chemical change. These are the "recrements" (Frey-Wyssling's *Rekrete*, 1945), such as water and certain ions, for instance, and the extrusion of these substances is obviously not the result of secretory activity.

Other substances are produced by catabolic phenomena, i.e., the splitting of complex molecules into simpler ones; their elimination is called excretion, and this activity of the cell must be distinguished from secretion proper.

Still other substances are the end products of the synthetic work of the cell; they differ chemically from the compounds absorbed from the environment and are, to a certain extent, specific for the type of cells which produce them; their rejection into the extracellular environment is often preceded by the storing of either the finished secretory product or its precursor; the cell may store these in particulate form and the storage represents the final stage of intracellular synthesis. The secretory product, no longer an integral part of the protoplasm after its accumulation in the cell, can be expelled either in particulate form or after redissolution in the cytoplasm. Only these substances are truly the products of secretion.

The notion of glandular cell implies functional specialization. Indeed, one of the criteria of the glandular nature of a cell is the primary function of periodical or continuous manufacture and extrusion of a secretory product. Another criterion may be the nature of the substance synthesized. The term glandular activity cannot legitimately be used to designate the formation of the common products of cell metabolism; the substance formed must possess a certain specificity for its process of formation properly to be termed glandular activity. There is unquestionable glandular activity in unicellular organisms, but it is obvious that only multicellular organisms can possess glandular cells.

II. Criteria of Glandular Activity in a Cell

The glandular nature of a cell is determined by means of morphological, histophysiological, physiological, and biochemical data.

A. Morphological Criteria

The general morphology of a glandular cell may vary considerably, and it seems impossible at present to find a pattern applicable to all cases. Descriptions based on the structure of the mammalian salivary glands or pancreas are no more than crude approximations. There is no doubt that a well-developed ergastoplasm, an abundance of mitochondria arranged in definite patterns, and grouped Golgi bodies are indications of the possible glandular nature of a cell, but there are innumerable exceptions to this rule. In fact, the general appearance of a cell and the shape and arrangement of its principal organelles are merely clues leading the experienced histologist to suspect, but not to affirm, that he is dealing with a gland cell.

One morphological criterion of the glandular nature of a cell deserves special mention: this is the accumulation, at some stages of the functional cycle, of a secretory product in particulate form. Whenever the presence of such a product can be shown by means of histological techniques, the argument in favor of the glandular nature of the cell is considerably strengthened. The opposite, however, is not true: in a great many gland cells there is no accumulation of particulate material to indicate their activity. A typical example of this is the parathyroid gland cell. It possesses the general characteristics of a gland cell; cytological study of its experimental hyperfunctioning shows (De Robertis, 1940, 1941a) that the fundamental organelles undergo the transformation typical of active gland cells; physiological experimentation and biochemical studies clearly show that parathyroid gland cells do manufacture a specific substance; removal of parathyroid glands results in well-known serious disturbances, which are completely reversed by injections of parathyroid extract. Yet, up to date, no histological technique has enabled us to detect the accumulation of a particulate secretory product in parathyroid cells.

It should be pointed out that these negative results may be merely due to the inadequacy of our means of investigation. In the thyroid cell, for instance, all attempts at detecting the "intracellular colloid" had failed until Gersh and Caspersson (1940) used ultraviolet microspectrography. A few years later it was discovered that fixation by freezing and drying preserved the secretion product in thyroid cells in a form stained by Heidenhain's azan (De Robertis, 1941b). Today, the presence of this substance can consistently be detected by the periodic acid-Schiff method in preparations fixed by any appropriate chemical method.

Thus, morphological study provides indications of the glandular nature of the cell; this becomes highly probable when accumulation in the cell of particulate material can be observed at some stage of the secretory cycle, followed by its expulsion into the external environment at another stage. The absence of a particulate secretory product is never sufficient justification for formally rejecting the possibility of a glandular function.

B. Histophysiological Criteria

Together with the static study of a cell by histological techniques, valuable evidence of the glandular nature of the cell can be obtained by investigation under well-defined experimental conditions of the various functional stages and by observation of the target area of a secretion, and it becomes possible, in some cases, to determine accurately the site where the synthesis of a secretory product takes place in a given cell category of a complex organ. For example, a parallel investigation of the testis and the target organs of the male hormone shows (Ancel and Bouin, 1904) that the androgenic hormones are actually elaborated by the interstitial cells. Similarly it is possible to locate accurately the site of elaboration of the gonadotropic hormones (Herlant, 1956) by comparative study of the pituitary and the reproductive system in animals with a sexual cycle of sufficient duration. In immature or castrated animals, the epithelial cells of the seminal vesicle or the prostate lack the morphological characteristics of gland cells; injections of appropriate amounts of androgenic hormones bring about the apparition of the morphological characteristics of glandular activity together with the physiological secretory phenomena.

It is thus possible by histophysiological means to ascertain the glandular nature of a cell when morphological techniques alone can, at best, lead to presumptions. It is also possible to determine, at the cellular level, the site of elaboration of a secretory product in complex organs; and, in some cases, to obtain precise information on the functional significance of the product synthesized.

C. Physiological Criteria

The use of physiological methods is obviously of great help in the study of glandular cells. In some cases physiological methods provide the conclusive proof when morphological investigation gives no evidence of the glandular nature of a given cell. It may be appropriate here to recall that the concepts of gland and gland cell originated in Ludwig's investigations (1851) on the effects of electrical stimulation of the glossopharyngeal nerves. The true nature of some vertebrate endocrine glands had of course been suspected because of their structure, but only the results of surgical ablation and injection of extracts provided conclusive evidence of the glandular nature of the thyroid, parathyroids, and pituitary. It is often possible by ligature of the excretory duct of an exocrine gland to determine the functional significance of the product elaborated. Supplemented by histological investigations this method can help to determine the site and mechanism of elaboration of the secretory products. Injections of organ extracts are often very helpful in investigating the nature of a secretory product.

In spite of their importance, physiological criteria are not always adequate. They allow the cellular localization of a secretory process only in organs consisting of a single type of cell. The positive results of injection of extracts confirm, of course, the presence in the organ under investigation of a physiologically active product, but not that this product is formed in this organ, since it could be a substance manufactured somewhere else and simply stored there. The progress of our knowledge of the elaboration of the so-called postpituitary hormones (for references see Bargmann, 1954) is a particularly good case in point.

D. Biochemical Criteria

The investigation of any gland involves as essential steps a biochemical study of the secretory product, the elucidation of its chemical structure, and, if possible, its synthesis. This part of the work makes it possible to give chemical significance to the vague notion of a "secretion product." With respect to physiology, biochemical data often permit a better understanding of the functional significance of the products elaborated. The morphological study itself often benefits from the biochemist's work, since accurate knowledge of the chemical composition of the products secreted by the gland cells can help to detect the product itself, or its precursors, in the cytoplasm of the adenocytes. Thus, apart from its intrinsic value in studying the functions of a gland cell, biochemical study can help to determine and demonstrate the glandular nature of a given category of cells.

To sum up, demonstration of the glandular nature of a given category of cells consists of four steps not necessarily taken in the following order: (1) morphological stage, during which the investigator is looking, on the one hand, for the general morphological characteristics of the cell, and on the other, for the accumulation and extrusion of a particulate secretory product; (2) histophysiological stage, which helps to determine the secretory cycle of the cell; this step often provides information on the significance of the product elaborated; (3) physiological stage, which, in some cases, provides the proof of the glandular nature of the given category of cells and determines the role played by the secretory product; (4) biochemical stage, leading to the replacement of the term "secretion product" by a specific chemical name and a structural formula.

The definition of the gland cell given above obviously implies exclusion of some categories of cells which are, however, undoubtedly secretory in nature. There is no fundamental difference between the elaboration of the secretion granules in the exocrine pancreatic cell on the one hand, and of the vitellus in the oöcyte on the other; and yet, vitellogenesis cannot be accepted as glandular activity because the "secretion product" accumulated in the cytoplasm in particulate form is not extruded in the external or the internal medium. On the other hand, the active participation of the gland cell in manufacturing the secretion product represents one of the essential elements of its definition. This notion, widely accepted since the investigations of R. Heidenhain and his school (for references see Heidenhain, 1880) has unfortunately been forgotten by some more recent authors (Chèvremont, 1956; Verne, 1956) who recognize the existence of "glomerular glands" in which "the cell does not seem to elaborate any product; it merely extracts from the external medium substances which it selects and eliminates without noticeably altering them" (Chèvremont, 1956, p. 320). Such cells, whose very definition corresponds to that of the excretory cells, will be excluded from the present study.

III. STATIC STUDY OF THE GLAND CELL

The static inventory of the constituents of the gland cell is complicated by the extreme diversity of structures encountered as soon as we go beyond the best-known mammalian exocrine glands. Moreover, in order to avoid repetition it will be necessary to retain here only those characteristics which are typical of the gland cell and to refer the reader to the other chapters of this book for the general description of nuclear and cytoplasmic organelles.

A. Size

The dimensions of gland cells vary widely. In mammals, the long axis of most gland cells is not more than 15 or 20 μ; there are, however, exceptions. Thus in the cells of the supraparotid gland of the male albino rat the length of the cell can, at some stages of the secretory cycle, reach 45–50 μ (Guyiesse-Pélissier, 1923; Gabe, 1955) (see Fig. 1). The size of gland cells having identical functional significance and elaborating the same chemical compounds can vary markedly with the species; thus the gland cells of urodele amphibians are much larger than the corresponding

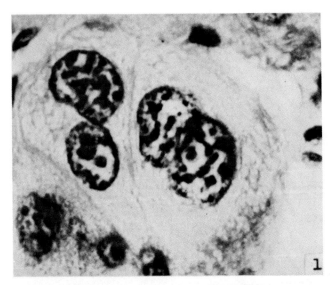

FIG. 1. Cells of the supraparotid gland of a male albino rat, fixed with Bouin's fluid and stained with azan. Note the large, irregular-shaped nuclei. Magnification: ×1200; green filter.

cell categories in mammals (Figs. 2–5). The differences in size mentioned, usually corresponds to the differences in size of the other somatic cells.

Mollusks and arthropods provide instances of giant gland cells. In the poison gland of *Argulus foliaceus,* a branchiuran crustacean, the major axis of the cells is about 150 μ (Fig. 6); the cells of the Verson gland of the lepidopteran larvae are even larger. In the hypobranchial gland of *Thecosomatous pteropods* the long axis of the cells is frequently more than 200 μ, and some mucocytes in the mantle of *Aplysia depilans* reach a height of 300 μ.

The dimensions of a given category of cells may vary greatly during the secretory cycle; some specific cases will be discussed below.

B. Topography

The location of the different types of gland cells varies so much that any attempt at classification would at best be sketchy.

It is necessary to differentiate the isolated gland cells, so-called unicellular glands, from the organized gland cells forming multicellular glands.

In some cases, isolated gland cells are wedged between the cells of a surface epithelium or between the cells lining the lumen of a duct; the integuments of mollusks, teleosts, and cyclostomes provide numerous

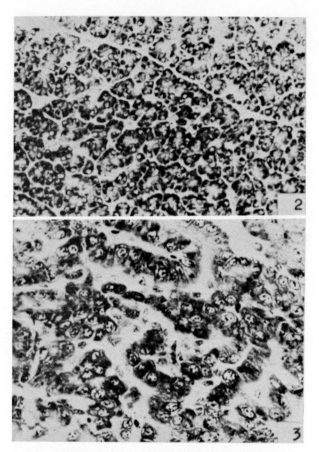

FIGS. 2 and 3.   Exocrine pancreas of the guinea pig (2) and the salamander (3) fixed with Bouin's fluid and stained with Mann-Dominici's technique. Note the size difference of cells. Magnification: × 200; orange filter.

examples of gland cells encased between epidermal cells, and it is unnecessary to insist here on the well-known morphology of the goblet cells in the trachea of mammals and the intestine of vertebrates (Fig. 7).

In other cases, isolated gland cells are located under the integument, communicating with it by means of either a protoplasmic process or a differentiated area representing a true excretory duct. A great many subtegumental unicellular glands of mollusks and most unicellular glands of the skin of arthropods belong to this category.

Clusters of cells aggregated into glands may be wedged between the cells of a surface epithelium; typical instances are the glandular crypts in the integument of Polyplacophora and Sipunculoidea (Fig. 8).

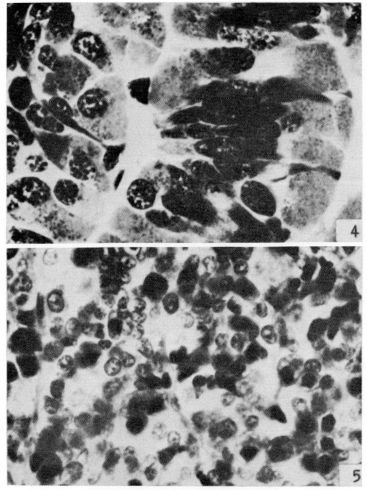

FIGS. 4 and 5.   Adenohypophysis of *Amblystoma mexicanum* (4) and *Bos taurus* (5). Fixation with Halmi's fluid, staining with azan. Note the size difference of cells. Magnifiation: ×800; green filter.

In other cases the glands are subepithelial and communicate with the epithelial surface through very short excretory ducts (Fig. 9). Cutaneous glands of amphibians are examples of associated gland cells located so close to the epidermis bathed by their secretory product that their excretory ducts need be very short (Fig. 10); the salivary glands of the tongue of mammals are embedded in the muscular masses of the tongue, and their relation to the lingual epithelium is similar to that of the skin glands.

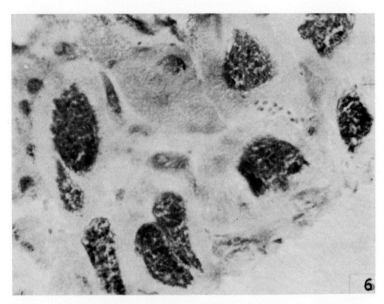

FIG. 6.   Cells of the poison gland in *Argulus foliaceus,* fixed with Carnoy's fluid and stained with methyl green and pyronin. Note the large size of cells and the polymorphism of nuclei. Magnification: × 250; red filter.

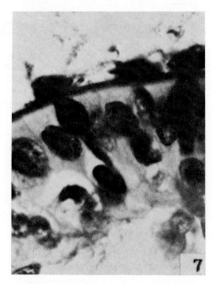

FIG. 7.   Duodenal goblet cells in *Amblystoma mexicanum,* fixed with Helly's fluid and stained with PAS-hematoxylin. Note the extrusion of the secretion product. Magnification: × 1500; green filter.

Fig. 8. Cutaneous gland crypt in *Phascolion strombi*. Fixation with Bouin's fluid, staining with azan. Note the gland cells and their secretion product. Magnification: × 1000; green filter.

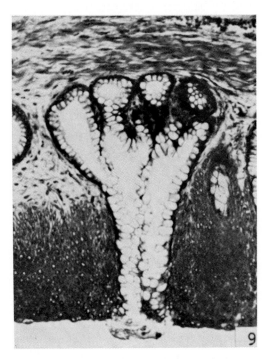

Fig. 9. Palatine salivary gland of a turtle, fixed with Bouin's fluid and stained with Prenant's triple stain. Note the subepithelial glandular acini and their excretory ducts. Magnification: × 150; green filter.

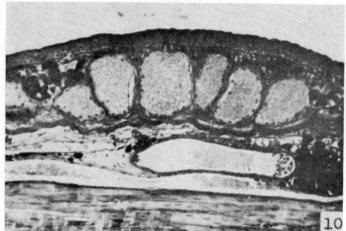

FIG. 10. Subcutaneous glands of a young *Rana temporaria* fixed with Bouin's fluid and stained with azan. Magnification: × 130; green filter.

More numerous are the cases where exocrine gland cells are organized into an anatomically differentiated organ, clearly separated from the neighboring tissues by a connective sheath and communicating with the external surface by means of an excretory duct of appreciable length. This type comprises most of the mammalian exocrine glands. The main anatomical types briefly reviewed below are distinguished by the shape of the groups of gland cells and their relation to the excretory ducts to which they are attached.

In *tubular glands* the secreting portion of the organ forms a tube of roughly the diameter of the excretory duct; the whole can be straight or coiled. Tubular glands can be simple (sudorific glands of mammals, and salivary glands of opistobranch gastropods), branched (gastric glands of mammals), or compound (liver of urodele amphibians, digestive gland of mollusks).

In *acinous glands,* the secretory portion is larger in diameter than the excretory duct, and groups of gland cells form alveoli suspended from the ducts; they too can be subdivided into simple (sebaceous glands), branched (preputial glands of rodents), or compound (main salivary glands of mammals, exocrine pancreatic glands).

In *tubuloacinous* glands, glandular *acini* are associated with excretory tubes, some portions of which are lined with gland cells (submaxillary glands of Muridae, and salivary glands of the Cryptocerata of the order Hemiptera are examples of this type of gland).

A group of special importance is that of the endocrine glands in which there is no excretory duct and whose secretory products are directly emptied into the internal environment. In these organs, the cells may be grouped in islets of joining cells (corpora allata of insects), in cords separated by capillaries (pituitary, endocrine pancreas, parathyroids), in follicles lining small, completely closed vesicles (thyroid), in small masses separated from each other by a tissue of a different nature (interstitial cells of testis). The ductless glands of all animals with a closed circulatory system are richly vascularized.

C. Relation with the Site Where the Secretory Product Is Discharged

The relationship between gland cells and the sites of evacuation of their secretory products varies widely.

First, there are gland cells, isolated or aggregated, embedded in surface epithelium or epithelium lining a natural cavity. In these the secretory products are discharged directly into the external environment or into the lumen of the cavity. Two examples of different types of extrusion are found in the mucous cells of the mammalian digestive tract; the intestinal goblet cells, with open apical pole, from which the mucus flows freely (Fig. 11), differ from the mucous cells of the gastric surface epithelium of the fundus, with a closed mucous apical pole in which the secretory product accumulates near the apex, the extrusion taking place as a kind of dialysis through the apical cell membrane (Fig. 12).

In other cases, subtegumental unicellular glands discharge their secretory product through a cytoplasmic process with no structural peculiarity, but wedged between epithelial cells (Fig. 13); this is the case in the subepithelial mucocytes of pulmonate gastropods.

In arthropods, the unicellular cutaneous glands are usually subepithelial and communicate with the integument by means of very fine chitinous tubes traveling between the epithelial cells and penetrating the chitinous integument.

In the case of multicellular exocrine glands, the secretory products are discharged into the lumen of the tube or the cavity by means of the excretory duct. In some cases myoepithelial cells play a mechanical part in the flow of the secretory product. In other cases the transport is effected by the cilia lining the duct. Besides the multicellular ducts, lined with columnar or cubical epithelium, the unicellular duct of the rosette glands of the malacostracan crustaceans should be mentioned; the whole of the excretory duct consists of one cell, whose nucleus usually lies in the center of the glandular rosette.

In many cases the glandular cells are provided with a network system of very fine intracellular canals such as the hepatic bile canaliculi.

The auxiliary glands of the female reproductive system of prosobranch

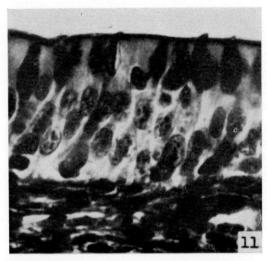

FIG. 11. Intestinal mucosa of *Rana temporaria,* fixed with Helly's fluid and stained with PAS-hematoxylin. Note the mucous goblet cells, with open apical pole. Magnification: ×800; green filter.

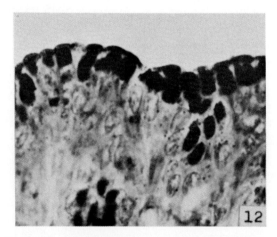

FIG. 12. Gastric mucosa of the guinea pig, fixed with Helly's fluid and stained with PAS-hematoxylin. Note the mucous cells of the surface epithelium with closed apical pole. Magnification: ×800; green filter.

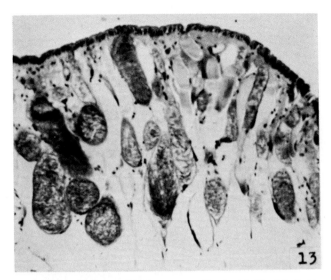

FIG. 13. Pallial epithelium of *Helix aspersa*, fixed with Bouin's fluid and stained with alcian blue and nuclear fast red. Note the large mucous cells wedged between the common epithelial cells. Magnification: × 200; orange filter.

gastropods exhibit a very particular mode of extrusion of the secretory product. The granules leaving the glandular cell in particulate form accumulate in the lumen of the glandular tubes and then pass, still in particulate form and without any morphological or histochemical alteration, through the surface epithelium lining the inside wall of the lumen of the genital tract (Fig. 14). From there the granules are discharged into the lumen of the duct.

The secretory products of the endocrine glands are directly extruded into the internal medium; in vertebrates the communication between these gland cells and the blood capillaries are established, in some cases, by means of the basement membranes.

D. Inventory of Cellular Components

Because of the great morphological diversity of the gland cells, it is particularly difficult to describe all their nuclear and cytoplasmic components. Of course, a very common characteristic is the presence, at some stages of the secretory cycle, of a secretory product in particulate form. But the diversity of appearance of the other organelles is such that it is fallacious to attempt to draw a morphological diagram of the gland cell. In the present state of knowledge, we can only indicate the most frequently occurring types and illustrate them with appropriate examples.

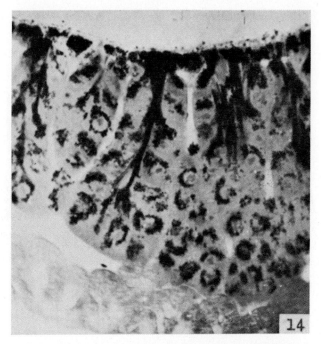

Fig. 14. Shell gland of *Pterotrachea mutica*, fixed with Bouin's fluid and stained with Mann's methyl blue-eosin method. Note the acidophilic secretion granules in the gland cells, in the lumina of glandular tubules, and in the cells of the surface epithelium. Magnification: × 400; green filter.

1. Nucleus

In the gland cell, the *size, shape,* and *structure* of the nucleus may vary within wide limits, as does the location of this organelle in the protoplasm. Besides morphological variations depending on the type of cell under study, very important changes during the secretory cycle may occur.

In most mammalian exocrine gland cells, the nuclei are located in the center or near the base of the cell. Their "chromaticity" varies with the cell type and, in a same category, with the stages of secretion. Descriptions of the various aspects of chromatin in different types of gland cells are given in some classic papers published at the end of the nineteenth and beginning of the twentieth century (Montgomery, 1898; Noll, 1905; Maziarsky, 1910). More recently, the metabolism of nucleic acids during the secretory cycle of the exocrine pancreatic cell was studied by Huber (1949) and Altmann (1952). We shall come back to the main results of these authors with the dynamic study of gland cells.

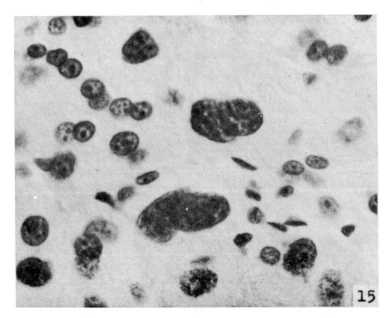

Fig. 15. Supraparotid gland of a male albino rat, fixed with Helly's fluid and stained with Feulgen-Rossenbeck's nucleal reaction and picro-indigo carmine. Note the pronounced nuclear polymorphism. Magnification: ×800; green filter.

Very irregularly shaped nuclei are rare in vertebrate gland cells; a notable exception is the supraparotid gland of the albino rat (Loewenthal gland) (Fig. 15). In invertebrate gland cells on the other hand, truly monstrous, sprouting, distorted nuclei are fairly common (Figs. 16 and 17). In many instances these misshapen nuclei are due to endomitosis with endopolyploidy, which will be described in the appropriate section.

The structure of the *nuclear membrane* is obviously of great interest. Indeed, the problem of exchange of substance between nucleus and cytoplasm has been frequently considered in reference to various objects, including gland cells. Consequently it became necessary to investigate, by means of electron microscopic techniques, the existence in the nuclear membrane of pores through which this exchange of substance could take place. Following the investigations of Callan and Tomlin (1950) and Callan (1955) on the germinative vesicle of the amphibian egg, the presence of solutions of continuity in the external layer of the nuclear membrane was described in the salivary gland cells of *Chironomus* (Bahr and Beermann, 1954) and in the pancreatic cells (Watson, 1954a, b). According to these authors, the diameter of these pores would be sufficient

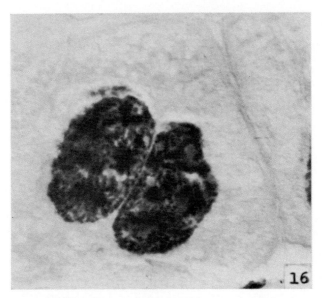

Fig. 16.   A cell of the salivary gland in *Anilocra physodes*, fixed with Regaud's fluid and stained with Feulgen-Rossenbeck's nucleal reaction and picro-indigo carmine. Note the large and bilobated nucleus. Magnification: × 800; green filter.

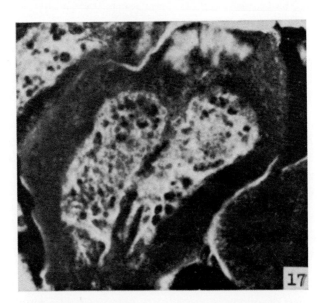

Fig. 17.   Cells of the poison gland in *Argulus foliaceus*, fixed with Carnoy's fluid and stained with methyl green and pyronine. Note the numerous nucleoli. Magnification: × 800; green filter.

to let through large molecules, such as those of ribonucleic acid. However, these pores are clearly too small to explain some nuclear extrusion phenomena described in the literature, and notably the passage into the cytoplasm of nuclear inclusions readily visible under light-field illumination.

The appearance of the *nucleolus* in gland cells has long attracted attention, and descriptions by authors in the early twentieth century include a wealth of accurate information on size, intensity of staining, structural details, and multiplicity of nucleoli in various gland cells.

Improvements in histochemical techniques have permitted a remarkable growth of our knowledge of the nucleoli of gland cells. Studies of this organelle by means of ultraviolet absorption spectrography (Caspersson and Schultz, 1939) and staining with basic dyes, combined with the effect of ribonuclease (Brachet, 1940), have lead to a rational interpretation of nucleolar basophilia. The work of Caspersson *et al.* (1941) has shown that rabbit exocrine pancreatic cells, in which an intensive synthesis of proteins takes place, contain large nucleoli rich in ribonucleic acid (Fig. 18), whereas the Langerhans islet cells, where the synthesis of proteins is less intense, have only small nucleoli containing little ribonucleic acid. Similar observations were made by Caspersson *et al.* (1941) on the chief and parietal cells of the gastric mucosa. These data were confirmed for many other types of cells (Fig. 19). The use of histochemical techniques in the dynamic study of gland cells, has demonstrated that during the secretory cycle the nucleolus undergoes some very distinct changes which will be discussed below.

Recent investigations have confirmed the older descriptions of intranucleolar vacuoles; it is known (for references see Vincent, 1955) that these regions of the nucleolus do not contain nucleic acids or proteins; according to Vincent (1955) supported by Brachet (1957, p. 128), they would be soluble products of nucleolar metabolic activity.

Estable and Sotelo (1951, 1955) describe, in the nucleolus, a more or less coiled filament, *nucleolonema* surrounded by a pars amorpha; this filament, revealed by silver impregnation technique, staining with potassium ferricyanide-pyrogallic acid of sections mordanted with iron alum, observation in living state under phase contrast or dark-field illumination, has been described in a number of gland cells. Corpus luteum of the ovary, exocrine pancreatic cells, intestinal Lieberkühn's glands of the pig, and salivary glands of the *Chironomus* larva are among the objects investigated by Estable and Sotelo. These observations were confirmed by Denues and Motram (1955) and Lettré (1954). In reference to this point it should be mentioned that the electron microscope shows the presence within the nucleolus of granules similar to those described in the ergasto-

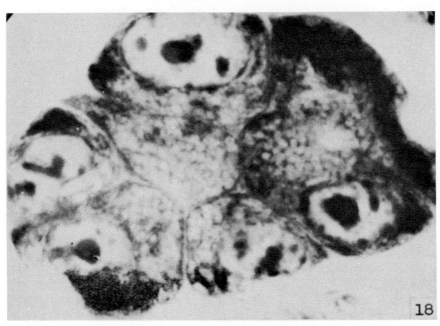

FIG. 18. A pancreatic acinus of the salamander, fixed with Carnoy's fluid and stained with methyl green and pyronine. Note the basal ergastoplasm, the parasomes, and the large nucleoles. Magnification: ×2000; green filter.

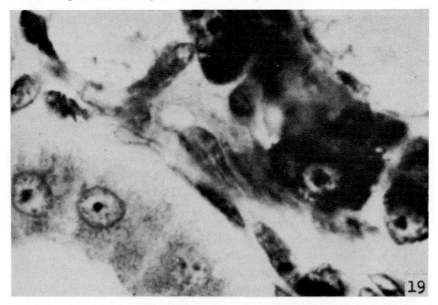

FIG. 19. Demilune cells (right) and cells of an excretory duct (left) of the retrolingual gland in albino rat, fixed with Carnoy's fluid and stained with methyl green and pyronine. Note the large nucleoli of the gland cells and the small nucleoli of the excretory duct cells. Magnification: ×2000; green filter.

plasm and of a filament, possibly corresponding to the nucleolonema (Bernhard *et al.*, 1952a, b; Oberling and Bernhard, 1955).

Intranuclear inclusions different from nucleoli, have been described in various cells, including some gland cells. They are, on the whole, isolated observations, and a classification of these inclusions in definite categories would be very difficult.

In some cases, the inclusions are spherical and their staining affinity is similar to that of the cytoplasmic secretion granules. The enzyme cells of the diverticulate intestine of the araneids (Millot, 1926) are a particularly clear example of this type of intranuclear inclusion. Similar formations have been described in the neurosecretory cell of the preoptic nucleus in some teleosts; Ortmann (1956, 1958) carried out a detailed study of these inclusions and was able to define their cytological and histochemical characteristics in the preoptic nucleus of the carp. This work shows that the properties of these intranuclear inclusions and those of the cytoplasmic secretory product are identical. Ortmann's observations raise the problem of the passage of nuclear material into the cytoplasm. However, only observation in living cells could confirm the hypothesis of such a transfer in particulate form, and the nature of the material makes such a study almost impossible in the living state. On the other hand, these intranuclear inclusions are much larger than the diameter of the pores detected in the nuclear membrane; so much so that the mechanism leading to the extrusion of the secretory product remains obscure. The same remark applies to the observations of Picard and his co-workers (for references see Picard and Stahl, 1956) on certain phenomena of intranuclear synthesis in neurons believed to be neurosecretory.

Next to these inclusions, and having a more or less direct relation to a cytoplasmic secretory product, other intranuclear inclusions must be mentioned which are much less common and more difficult to interpret. These are *crystalloids,* usually octahedric in shape, preserved in most cytological fixations and strongly eosinophilic—hence the name erythrophilic crystalloids given them by the earlier authors. Among the intranuclear crystalloids described in reference to gland cells, it is worth mentioning those of the salivary glands of *Nepa cinerea* (Carnoy, 1884), of the neurosecretory cells of *Delphinapterus leucas* (Scharrer and Scharrer, 1954), and of the digestive gland in heteropods, genus *Pterotrachea* (Gabe, 1952) (Figs. 20 and 21). It seems almost impossible to make any statement on the significance of these formations because their histochemical characteristics are not yet sufficiently known. There are, indeed, signs pointing to an evolution of the crystalloids during the secretory cycle of the cell: in *Pterotrachea* for instance, the number and size of the crystalloids increase with the accumulation of the secretion granules

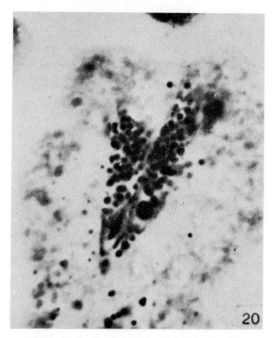

FIG. 20. Epithelium of the digestive gland in *Pterotrachea coronata,* fixed with Helly's fluid and stained with iron, hematoxylin. Note the intranuclear crystalloids. Magnification: × 2000; orange filter.

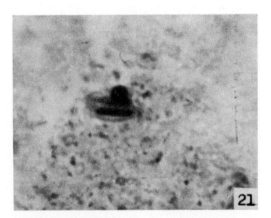

FIG. 21. Intranuclear crystalloids in a cell of the digestive gland in *Pterotrachea mutica.* Fixation with Bouin's fluid, staining with fuchsin and picric acid. Magnification: × 2000; green filter.

in the cytoplasm of the gland cell. However, further investigations are necessary in order to elucidate the role played by these structures. The gland cells are not the only ones that contain intranuclear crystalloids; they have been described in nerve and intestinal cells, in leucocytes and oöcytes, but the results of observations carried out on these objects are not conclusive and it is difficult to give a physiological interpretation to these structures, considered by some authors to be waste products of nuclear metabolism.

2. Cytoplasm

The data relative to the cytoplasm of the gland cells appear to be more homogeneous than the notions concerning the nucleus. This is due to the fact that most of the recent information was obtained through studies of glands of higher vertebrates. Actually, as soon as we go beyond the narrow limits of this group, we find as great a diversity of appearance in the cytoplasm of the gland cells as in the nucleus.

a. Chondriome. The importance of the chondriome of the gland cells is clearly shown by historical considerations. This organelle was discovered in the acinous cells of the parotid gland of the cat by Altmann (1890, 1894). The study of gland cells has long played an essential role in the morphological and physiological study of the chondriome. It is not necessary to insist on the importance of the electron microscopic studies of Sjöstrand and Hanzon (1954a, b) concerning the structure of the exocrine pancreatic cells chondriome. The perusal of recent reviews (Holter, 1952; de Duve and Berthet, 1954; Lindberg and Ernster, 1954; de Duve *et al.*, 1955; Hogeboom and Schneider, 1955) clearly shows the importance of the hepatic mitochondria in biochemical investigations of mitochondria and lysosomes.

Consequently, an analytical survey of the properties of the chondriome of the gland cell would duplicate the subject of another chapter of this treatise; the reader is referred to it for the structural description of the chondriome, the recent data on its chemical composition, its enzymatic apparatus, and its participation in the metabolism of the cell. Only those data indispensable to a static inventory of the gland cell will be mentioned here.

The abundance of chondriomes in gland cells is well known; the number of mitochondria in a liver cell is estimated to be 2500 or more (Allard *et al.*, 1952); such large numbers in gland cells and cells of other types, with very active metabolism, must be related to the enzymatic equipment of the chondriome, especially to the presence of enzymes participating in the Krebs cycle and the aerobic metabolism of cells.

As for the morphology of the chondriome of the gland cells, the

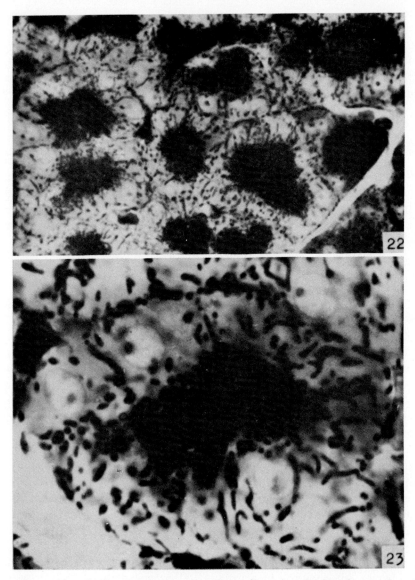

FIGS. 22 and 23. Pancreatic acini of an albino rat, fixed and stained with Nassonov's technique. Note the basal chondriome and the accumulation of secretion granules at the apical pole of the cells. Magnification: ×800 (22) and 2000 (23); green filter.

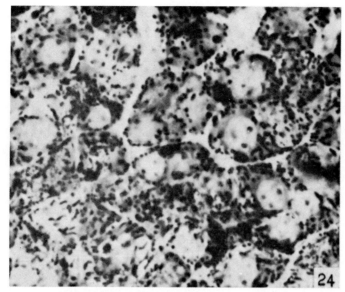

FIG. 24. Parotid gland of a male albino mouse, starved for 12 hours and killed 90 minutes after feeding. Fixation with Regaud's fluid, staining with Altmann's fuchsin and methyl green picrate. Note the well-developed chondriome and the scarcity of secretion granules. Magnification: ×1500; green filter.

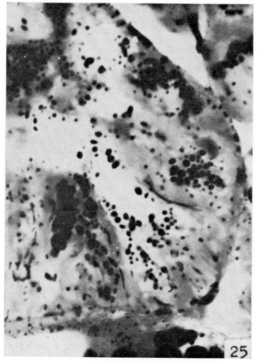

FIG. 25. Shell gland in *Pterotrachea mutica*, fixed with Regaud's fluid and stained with Altmann's fuchsin and methyl green picrate. Note the cytological similarity with exocrine pancreatic cells. Magnification: ×1500; green filter.

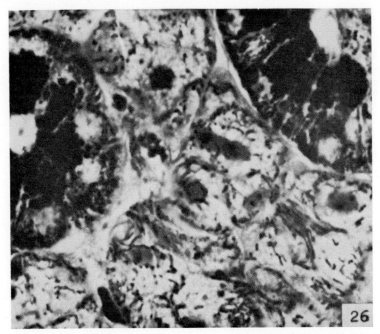

FIG. 26. Submaxillary gland of a male albino rat, fixed with Regaud's fluid and stained with Altmann's fuchsin and methyl green picrate. Note the basal chondriome and the apical secretion granules in the secretory tubules. In the acinous cells, the chondriome is located in the cytoplasm surrounding the secretion granules. Magnification: ×1200; green filter.

standard diagrammatic descriptions, taken up again in some recent publications (Zeiger, 1955), assign to the adenocyte a basal chondriome, made up of chondrioconts of varied length. In fact, in gland cells the chondriome may vary greatly in shape.

In a first morphological type, the nucleus is situated near the base or in the center of the cell, the major portion of the chondriome being made up of chondrioconts lying parallel to the long axis of the cell; these elements thus envelop the nucleus in a kind of discontinuous sheath and extend into the supranuclear region where clustered Golgi bodies are to be found; this region is called "Golgi zone" by many authors. The secretory product collects at the apex of the cell, where very few mitochondria are to be found; during the accumulation of the secretion product, the chondrioconts are, in a way, pushed back toward the base of the cell. The exocrine pancreatic cells, parotid cells, some types of silk-gland cells of arachnids, and cells of the shell gland of heteropods (Gabe, 1951) are of this type (Figs. 22–25).

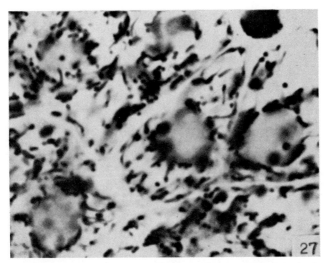

FIG. 27. Hepatic cells of *Triturus cristatus*, fixed with Helly's fluid and stained with Altmann's fuchsin and methyl green picrate. The chondrioconts and mitochondria are located in the cytoplasm surrounding the various secretion products. Magnification: × 1500; green filter.

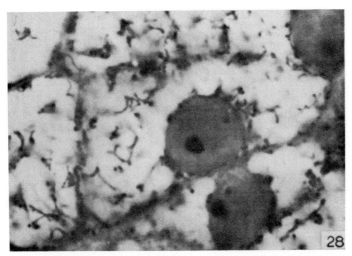

FIG. 28. Hepatic cells of *Amblystoma mexicanum*, fixed with Helly's fluid and stained by Altmann's fuchsin and methyl green picrate. Note the curved form of the chondrioconts located in the cytoplasm surrounding the secretion products. Magnification: × 1500; green filter.

In other gland cells the polarity of the chondriome just referred to is absent; the perinuclear region does not contain more chondriosomes than other cytoplasmic regions; chondrioconts and mitochondria can be located in the cytoplasm surrounding the secretion granules. The chondrioconts are frequently bent and, in some cases, tend to mold themselves around the secretion granules. Such cases are found in the submaxillary gland (Fig. 26), in the sebaceous glands, and in the Harderian glands of mammals.

In other gland cells, the polarity of the chondriome is absent and, moreover, there is no clear relation between the shape and arrangement of these elements on the one hand, and the location and abundance of the secretion product on the other. This is the case for the hepatic cells and the Langerhans islet cells (Figs. 27 and 28).

In all cells which have a rhythmic activity the chondriome undergoes during the secretory cycle important changes which will be discussed later.

b. Ergastoplasm. The gland cell plays as great a part in the history of our knowledge of the ergastoplasm as in that of the investigations devoted to the chondriome. The first descriptions of "basal filaments" (Solger, 1894) and of the "fibrillar basal zone" (Bensley, 1898) and the masterly research on ergastoplasm carried out by the Nancy school (Prenant, 1898, 1899; Garnier, 1899) are primarily based on the study of the gland cell. And since then, it is still to che gland cells that the partisans as well as the detractors of the concept of ergastoplasm have turned for illustrations.

The original notion of ergastoplasm was first extended by histochemical investigations (Caspersson and Schultz, 1939; Brachet, 1940) providing the explanation for the basophilia of this structure and showing, moreover, that basophilia also due to ribonucleic acid occurs in some nonglandular cells. The work of Hydén (1943) confirms in particular the foresight of Prenant (1898, 1899), who compared the Nissl bodies of the nerve cell and the ergastoplasm of the gland cell.

Thanks to the technical developments in electron microscopy, it has been possible to extend further the notion of ergastoplasm. The universality of this organelle such as it appears under the electron microscope is clearly demonstrated by recent authors (Haguenau, 1958). This being the case, the description of the ergastoplasm (Porter's endoplasmic reticulum, Palade's particles) belongs in the chapter treating of the fundamental substance of the cytoplasm, and only properties characteristic of gland cells will be discussed here.

The basal location of the ergastoplasm in gland cells was noticed with great interest by the first investigators; some (Roskin, 1925) even

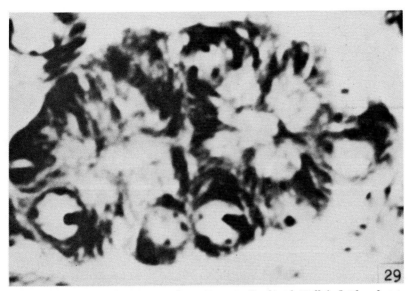

Fig. 29. A pancreatic acinus of a guinea pig, fixed with Helly's fluid and stained by Mann-Dominici's technique. Note the well-developed ergastoplasm. Magnification: × 2000; green filter.

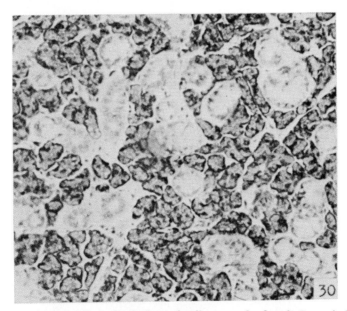

Fig. 30. Submaxillary gland of a male albino rat, fixed with Carnoy's fluid and stained with methyl green and pyronine. Note the basal ergastoplasm of the acinous cells and of the secretory segments of the tubules. Magnification: × 200; green filter.

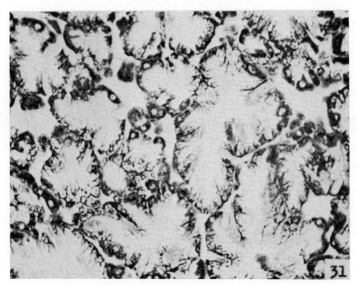

FIG. 31.   Digestive gland of *Pterotrachea mutica,* fixed with Carnoy's fluid and stained with methyl green and pyronine. Note the ergastoplasm at the basal pole of the cells. Magnification : ×1500; green filter.

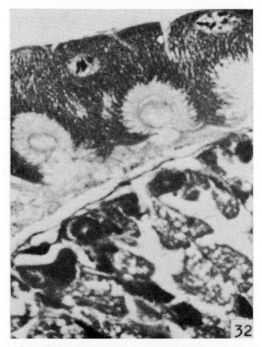

FIG. 32.   Hypobranchial gland (above) and digestive gland (below) in *Euclio pyramidata,* fixed with Bouin's fluid and stained with toluidine blue. Note the well-developed ergastoplasm (Roskin's basoplasm) of the hypobranchial gland cells. Magnification : ×200; orange filter.

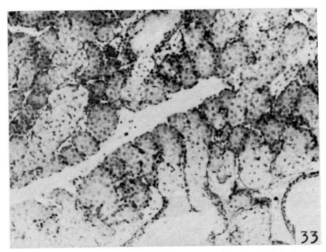

Fig. 33. Preputial gland of an albino mouse, fixed with Carnoy's fluid and stained with gallocyanine. Note the staining differences between the acini at different stages of the secretory cycle. Magnification: ×130; orange filter.

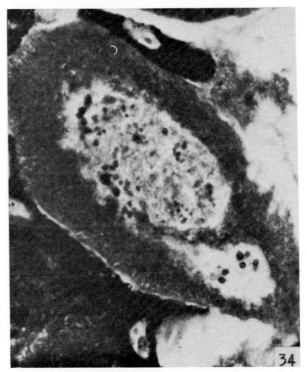

Fig. 34. A cell of the poison gland in *Argulus foliaceus*, fixed with Carnoy's fluid and stained with methyl green and pyronine. Note beginning of delamination of the ergastoplasm. Magnification: ×1200; green filter.

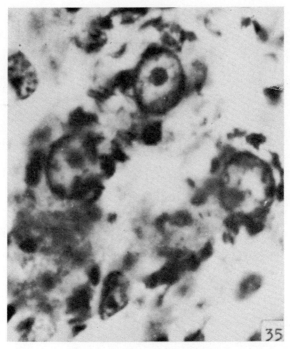

Fig. 35.   Liver cells of an albino rat, fixed with Carnoy's fluid and stained with gallocyanine. Note the blocks of basophilic substance in the cytoplasm. Magnification: × 2000; orange filter.

suggested that the name ergastoplasm should be replaced by basoplasm. This is valid in a great many cases, especially for most of the so-called serous gland cells of higher vertebrates. In these cells, the expanse of the striated basophilic zone, located at the basal pole of the cell, varies in inverse ratio to the abundance of the secretory product (Figs. 29–32). In other gland cells the cytoplasmic basophilia is diffuse; this is the case in the sebaceous gland cells at the beginning of the secretory cycle, for the salivary gland cells of *Chironomus*, for silk-gland cells, and for the poison-gland cells of *Argulus foliaceus* (Figs. 33 and 34). In other preparations, the cytoplasmic ribonucleic acid appears, after chemical fixation, in blocks or lumps lying in the perinuclear region, near the cell membrane, or scattered throughout the cytoplasm (Fig. 35). Another aspect, fairly frequent, is that of strands rich in ribonucleic acid that envelop the product of secretion. It is timely to point out that with the electron microscope it is now possible to detect the morphological elements of ergastoplasm whereas none are visible with the optical microscope. A

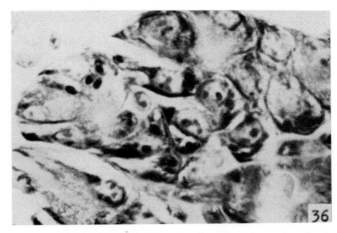

Fig. 36. Exocrine pancreatic cells in *Salamandra salamandra*, fixed with Carnoy's fluid and stained with gallocyanine. Note the numerous parasomes. Magnification: × 800; orange filter.

certain accumulation of basophilic substance is needed for the structures to become visible on sections stained by the usual methods.

The abundance of the basophilic substance and the corresponding ultrastructural elements varies greatly from one cell category to another and, as regards elements of a same type, with the functinonal stage. Generally speaking, the elaboration of protein-rich products is associated with a well-developed ergastoplasm.

The study of the so-called serous exocrine glands had led Garnier (1899) to see the ergastoplasm as being of nuclear origin. Caspersson and his co-workers (for references see Caspersson, 1950) considered the possibility of its nucleolar origin. Today, the data provided by the electron microscope lead one to believe that nucleoli (Bernhard *et al.*, 1955), chromatin (De Robertis, 1954; Watson, 1955), the nuclear membrane (Gay, 1956), mitochondria (Fawcett, 1955; Bernhard and Rouiller, 1956), and the cytoplasmic membrane (Palade, 1955; Bennett, 1956) may all take part in the elaboration of the ergastoplasm. Future investigations will decide between the various hypotheses.

c. Parasomes. Closely connected with the study of ergastoplasm is that of structures described by the early authors under the name of *Nebenkern* (paranucleus); they have been observed in various cells. Through a regrettable confusion in nomenclature, different authors have been led to designate under the same name, structures as dissimilar as the idiosome of the spermatids of the pulmonate gastropods, the mitochondrial body of insect spermatids, the yolk nuclei of oöcytes, and the

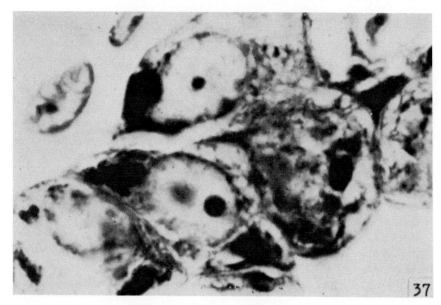

Fig. 37. Exocrine pancreatic cells in *Salamandra salamandra*, fixed with Carnoy's fluid and stained with methyl green and pyronine. Note the well-developed parasomes. Magnification: × 2000; green filter.

parasomes of gland cells. At the present time, the differences between these various structures are well known and the electron microscope has provided, for some of them, diagnostic elements (Grassé *et al.*, 1956; De Robertis and Raffo, 1957). Thus, it seems legitimate to discard the old name of *Nebenkern*, source of confusion, and replace it by the term parasome, suggested by Laguesse (1894, 1905). The general reviews by this author give a list of gland cells containing this organelle, whose principal morphological characteristics are its situation more or less close to the nucleus, its lamellar structure, and a strong affinity for basic dyes. Since then, the nature of parasomes has been discussed by some authors; Benoît (1926), who described them in the cells of the vas deferens of Muridae, considers their relation to the ergastoplasm; Voïnov (1934), who studied them in the salivary gland of the snail, discusses the question of their relation to the Golgi bodies.

Observation under the electron microscope of ultrafine sections shows (for references see Haguenau, 1958) that onion bulb structures, made up of lamellae identical with those of the ergastoplasm and partly filled with Palade's particles, are present in many gland cells. These structures, not always visible under the optical microscope, unquestionably bear a rela-

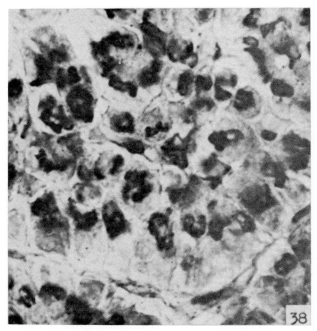

FIG. 38. Exocrine pancreas of an albino rat, fixed and impregnated with osmium tetroxide (Nassonov's technique). Note the Golgi bodies which are located against the apical pole of nuclei. Magnification: ×2000; green filter.

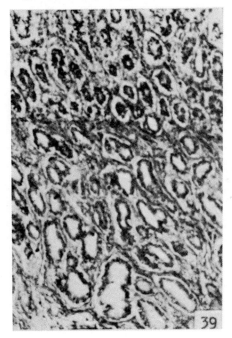

FIG. 39. Lieberkühn's and Brunner's glands of the duodenum in the albino rat. Fixation and silver impregnation with da Fano's method. Note the arrangement of Golgi bodies near the apical pole of cells. Magnification: ×200; orange filter.

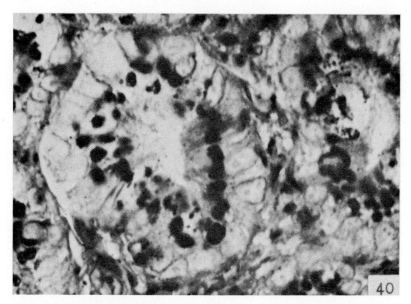

Fig. 40. A Lieberkühn's gland of the duodenum in the albino rat. Fixation and silver impregnation with da Fano's method. Note the typical "Golgi zone." Magnification: × 2000; orange filter.

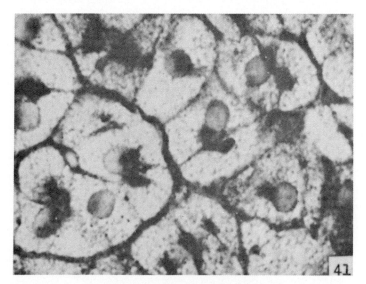

Fig. 41. Cells of the adrenocortical zona fasciculata in the rabbit. Fixation and silver impregnation with da Fano's method, counterstain with nuclear fast red. Note the typical "Golgi zone." Magnification: × 2000; green filter.

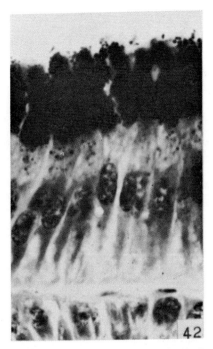

FIG. 42. Epithelium of the glandular segment of *vas deferens* in *Pterotrachea coronata*, fixed with Helly's fluid and stained with Mann-Dominici's technique. The Golgi zone appears in "negative." Magnification: ×800; green filter.

tion to ergastoplasm, and Haguenau (1958) has suggested that they be called "ergastoplasmic *Nebenkern*."

On the other hand, the histochemical study of gland cells, on which is based the usual description of the parasomes, shows (Gabe, 1958) that the basophilia of these structures is due to the presence of ribonucleic acid (Figs. 36 and 37) and that in the parasomes of the exocrine pancreatic cell of the salamander, as well as in those of the salivary gland of the snail, treatment with ribonuclease leaves untouched a residual protein closely related to that of the ergastoplasm. Thus the notion of "ergastoplasmic *Nebenkern*" is histochemically justified.

d. Golgi bodies. Golgi bodies are the most controversial of cellular organelles. They have been the object of innumerable investigations, and the reader is referred for their description to the chapter treating of the Golgi bodies in general.

Most cytologists indicate in the morphological diagram of gland cells a zone, located against the apical pole of the nucleus, containing the "Golgi apparatus" and designated as the "Golgi zone." This view conforms

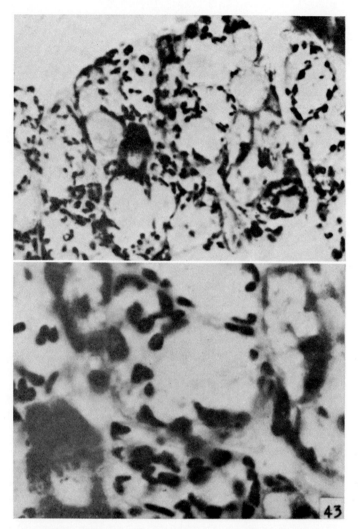

Fig. 43. Salivary gland in *Limnea stagnalis*. Fixation and silver impregnation with da Fano's method. Note the dictyosomes scattered throughout the cytoplasm. Magnification: ×800 (above) and 2000 (below).

to reality in some cases such as that of the exocrine pancreatic cell (Figs. 38–42), but a hasty generalization would be conducive to serious mistakes. Even the cytologists prior to 1930 contrasted vertebrate cells, possessing a "Golgi zone," with invertebrate cells, in which the "Golgi apparatus" consists of scattered dictyosomes. We shall not dwell here on the "Golgi network," nor on the controversies which for a long time divided partisans and adversaries of the "vacuome" theory; it is enough to point out that all these discussions have been rendered void by the developments of electron microscopy. It is, however, worth stressing that even the results of optical microscopy did not justify the interpretation of the "Golgi zone" as an essential attribute of the gland cell. Investigations carried out by means of silver impregnation or osmic acid treatment demonstrated in fact that cells as unquestionably glandular as those of the digestive gland of the crayfish (Jacobs, 1928), of the salivary gland of the snail (Voïnov, 1934), and of the silk gland of the silkworm (Lesperon, 1937) possess a "Golgi apparatus" consisting of dictyosomes, characteristic but scattered throughout the cytoplasm (Fig. 43).

The functional significance of Golgi bodies and their participation in the processes of secretion has also been the subject of innumerable discussions; some of the data derived from them are given in Section IV, which treats of the dynamic study of the gland cell.

e. Tonofibrils. Some gland cells contain, besides chondrioconts and ergastoplasmic filaments, a third fibrillar element. These are fibers running the length of the cell which attach themselves on the basal membrane and play no part in secretion. Heidenhain named these structures tonofibrils. Some categories of nonglandular cells (mesenteron of isopods, some surface epithelia) contain a fibrillar system having the same significance; del Rio Hortego (1917) suggested for these structures the name epitheliofibrils and discussed their relation to the "cytoskeleton" of early authors; Roskin (1925) took up again this notion of cytoskeleton, attributing to the ergastoplasm and the tonofibrils the same supporting function.

The coexistence, in the same cell, of tonofibrils and chondrioconts having the same general orientation has been responsible for a number of errors, the two organelles having been mistaken for one another in many early descriptions. The confusion is facilitated by some morphological peculiarities, such as the threadlike structure, and some staining affinities, such as those for Altmann's fuchsin and iron hematoxylin, common to both. In fact, a distinction between these two organelles is always possible, even by means of optical microscopy alone. The tonofibrils are preserved by fixation at low pH and fixatives having an alcohol base, whereas the chondriosomes are not; staining with crystal violet, after Benda, makes it easy to distinguish the chondriosomes, since the coloration taken by tonofibrils is different from that of the components of the chondriome; moreover tonofibrils differ from chondrioconts by their histochemical characteristics: they stain red by the periodic acid-Schiff method, take up strongly aldehyde-fuchsin, and, according to Baker, do not give the acid hematein reaction. Let us add that vital staining with Janus green shows the chondriosomes, but never the tonofibrils.

f. Secretion products. The intracellular accumulation of a product detectable by histological techniques is certainly not an infallible criterion of the glandular nature of a cell. There are, as remarked earlier, gland cells whose activity is not manifested by elaboration of secretory granules; on the other hand, some gland cells whose normal activity includes accumulation, inside the cytoplasm, of secretion granules, may, under abnormal conditions, cease to elaborate a particulate product while nonetheless continuing to secrete physiologically active compounds. It still remains true, however, that a great many very diverse gland cells do accumulate particulate products in the cytoplasm, at some stages of their functional cycle.

Actually the terms "secretion product" and "secretion granules" constitute only a transitory designation; one of the goals of the investigations on gland cells is the accurate determination of the product(s) elaborated, i.e., the replacing of the morphological notion of "secretion granule" by the statement of its chemical composition. This goal may be reached in some cases. Thus is it possible to define fairly accurately the secretory product of mucous gland cells by means of histochemical techniques on the one hand, chemical analysis on the other. In the same way, the secretion product of the sebaceous glands and the orbital glands, are accessible to histochemical study and chemical analysis. In other cases, we know, at present, only a few properties of the product elaborated by the gland cell. In still other cases, the secretion product may be defined only by its morphology and some staining affinities without chemical significance.

Because of the great diversity of the secretion products they are very difficult to classify.

The form under which the secretion product occurs within the cell varies widely. In some cases, and they are the best known, observation under the microscope reveals, in the apical portion of the cell, more or less voluminous granules, visible without staining thanks to their refractive index, which differs from that of the surrounding hyaloplasm (Figs. 44 and 45); this is the characteristic aspect evoked by the name secretion granules. The size of the granules may vary within fairly wide limits; in some gland cells all the granules have the same size, whereas in others, granules of different sizes may be present at the same stage of the secretory cycle.

In other cases, the secretion product, although still in granular form, accumulates not at the apical pole of the cell, but throughout the cytoplasm, the nucleus being either pushed at the basal pole or hemmed in by the granules and even deformed by their pressure.

In still other cases, instead of spherical granules, we find clumps more or less irregularly shaped, puddles of "intracellular colloids."

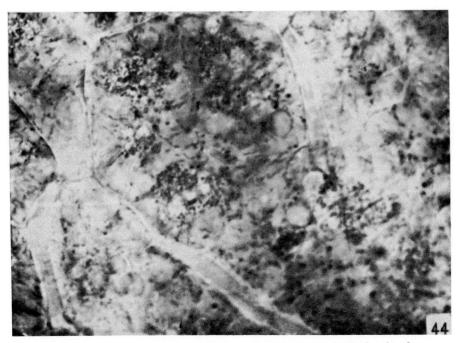

FIG. 44. Pancreatic acini of an albino mouse (Covell's method); the chondriomes are vitally stained with Janus green. Note the capillaries filled with red blood corpuscles, the chondrioconts, and the secretion granules located near the apical poles of the exocrine cells. Magnification: × 1000; red filter.

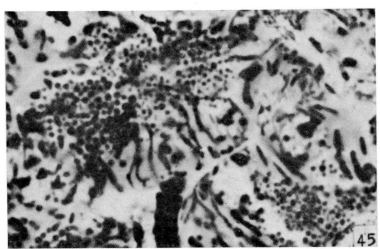

FIG. 45. Exocrine pancreas of an albino mouse, fixed with Regaud's fluid and stained with iron hematoxylin. Note the perinuclear chondrioconts and the apical secretion granules. Magnification: × 2000; green filter.

The different secretion products react in very diverse ways to histological fixatives. Some are preserved by all the usual agents; others resist only the so-called cytological fixatives; the lipid-rich secretory products are not usually preserved in sections in paraffin.

Just as great is the variability in staining properties of the secretion products. There are basophilic secretion granules, their affinity for basic dyes being due to the presence of compounds rich in acidic groups. Other products, on the contrary, take up acid dyes; when methods involving simultaneous or successive action of several acid dyes are used, the diverse secretory granules take up selectively one or another stain—hence, a multiplication of categories of questionable significance. Indeed, it is admitted nowadays that staining affinities toward the standard histological dyes may depend not only on chemical constitution, but also on physical characteristics such as density of structure, degree of hydration, degree of colloidal dispersion. Granulations with varying staining affinities are frequently observed within one cell, but cannot be differentiated by histochemical techniques.

It seems impossible, at present, to state any general histochemical characteristic of secretory products. Aside from the mammalian exocrine glands, the differences from one preparation to another are so great that any attempt at a general description or unification is doomed to failure. Some particularly notable instances will be discussed in the study of the relations between histologically detectable secretory granules of the adenocyte and the substances elaborated by them.

IV. Dynamic Study of the Gland Cell

The structural diversity of the gland cells is accompanied by a corresponding diversity of functional patterns. A review of the activity of the adenocyte is hampered by the lack of information. Only a very few favorable objects familiar to the physiologist, such as the salivary glands and exocrine pancreas of mammals, have been investigated recently. For the great majority of gland cells, the only data available at present do not go beyond morphological study.

A. Activity of the Gland Cell

As has been pointed out in the introduction to this chapter, the activity of gland cells is composed of three parts, namely, ingestion, intracellular elaboration of the secretion product, and extrusion. In cases where the secretory product accumulates in particulate form in the cytoplasm before being discharged into the environment, there is an intermediary step between intracellular elaboration and extrusion: the storage of the product within the cell. The series of steps mentioned here is usually known as the secretory cycle; Hirsch (1929) suggested that the three steps preceding extrusion should be grouped under the name "restoration."

From the manner in which secretion proceeds through the various stages of its cycle, it is possible to distinguish, in the activity of gland cells, several functional patterns.

In some cases the various phases of the cycle occur simultaneously inside one cell, the secretory product, particulate or not, being discharged into the extracellular medium as soon as it is elaborated. This functional pattern is known as *continuous activity*. Since the extrusion of the product, its formation inside the cell, the absorption of material from the environment, all take place at the same time, gland cells of this type exhibit under the microscope no differences from one specimen to another unless there is a change in rhythm. This is the pattern in some mammalian endocrine glands, as long as the hormonal balance is maintained. Parathyroid, thyroid, and adrenocortical cells of the adult male rat, in homeostatic equilibrium, follow the continuous pattern, and there are no morphological changes with time. Among the exocrine gland cells, the mucocytes of the gastric surface epithelium belong in this category; their secretion product is eliminated continuously and restoration takes place at the same rhythm as extrusion, so that the reserves of mucine accumulated at the apical end of the cell remain constant.

In other instances the stages of the secretory cycle are separated in time; a period of ingestion and intracellular elaboration of the product being followed by a period of intracellular accumulation, followed in turn by extrusion. This functional pattern is known as *rhythmic activity*. When the gland cell follows the rhythmic pattern, examination of the cell at various stages of the cycle reveals considerable differences. Apart from metabolic changes—variation in oxygen consumption being their most direct manifestation—the morphological aspect may be totally different. Such structural changes, already conspicuous in the case of most mammalian gland cells following a rhythmic pattern, may be considerable in some gland cells of invertebrates.

Glandular activity, not of a cell but of an organ, may also be continuous or rhythmic, but the differences are not so clear cut as for the individual gland cell. Continuous activity of a gland may occur either because all its cells are continuously active or because, although the activity of its gland cells is rhythmic, their secretory cycles are not synchronous. Thus the salivary glands of rat and mouse are continuously active whereas the functional acini follow a rhythmic pattern. In such cases, examination of sufficiently extensive sections shows, in the gland of any given animal, some acini in which the cells are at different stages of their secretory cycle. This is also the case for some holocrine glands, such as branched sebaceous glands (preputial or clitoridian glands of the mouse for instance). The activity of the organ as a whole is continuous, but examination of a single section may supply gland cells at all stages of their secretory cycle.

There are several possibilities for rhythmic activity. In some cases, the activity of the cells forming the gland is exactly synchronized and it is possible to determine accurately the instant when the gland, considered as a whole, is secreting. In other cases, the gland as a whole is never completely inactive, but there are periods of low secretion separated by periods of great activity. This is the case for the exocrine pancreas of mouse, albino rat, dog, rabbit, and man. Recent investigations (Hirsch *et al.*, 1956) have shown that this fasting secretion (*Hungersekretion*) results from the fact that, throughout the gland itself, extrusion phenomena occur without regularity and in only a few acini at a time. It must be remembered that secretion induced by physiological *stimuli* or pharmacodynamic agents is synchronous in most of the acini throughout the gland. Krijgsman's investigations on the secretory cycle of the salivary gland of the snail, *Helix pomatia,* show instances of the same pattern; the anarchic activity in fasting animal is replaced, after stimulation by ingestion of food, by synchronous activity of the various gland cells, each of them being rhythmically active (Krijgsman, 1925, 1928).

B. Patterns of Activity in Gland Cells. Extrusion of the Secretion Product

The evolution of gland cells during secretion indicates that there are three fundamental patterns: Secretion may be merocine, holocrine, or apocrine.

In *merocrine* or *eccrine* secretion, extrusion of the product elaborated takes place without noticeable alteration of the cellular organelles. The manner in which the secretion product is discharged may vary greatly.

When there is no accumulation of a particulate secretory product in the cell, there is no morphological evidence of extrusion.

When the gland cells are of the continuously active type, elaborating a particulate material, the continuous ejection of this product may be completely masked by the process of resecretion taking place concurrently. Thus, under normal conditions the balance between extrusion and restoration is perfectly maintained in the cell of the surface epithelium of the gastric mucosa.

In other cases, as in rhythmically active gland cells elaborating a histologically identifiable secretory product, extrusion is recognizable under the microscope. Observation of living exocrine pancreatic cells of rabbit (Kühne and Lea, 1882) and of mouse (Covell, 1928; Hirsch, 1931, 1932a, b) provided accurate information on the manner in which the secretion granules leave the cell and gave valuable confirmation of the images seen in preparations after fixation, sectioning, and staining. In numerous cells, masses of secretion granules, individually recognizable, are expelled into the lumen of the acinus, where the granules undergo swelling followed by gradual dissolution and loss of characteristic staining properties. In other cells, the granules undergo, at the apical pole of the cell, a coalescence bringing about formation of large drops of a product which can be discharged into the lumen of the acinus through the apical cytoplasmic region. A last type of extrusion in the exocrine pancreatic cell occurs only after very strong pharmacological stimulation (e.g., injection of pilocarpine). The secretion product is then expelled into the lumen of the acinus with such force that portions of the apical cytoplasm are carried off with the granules (Covell, 1928). This type of extrusion is a transition toward apocrine secretion.

In other rhythmically active merocrine gland cells, the secretory product is collected at the apical end of the cell in a more or less voluminous globule, expelled as a whole at the time of extrusion. Goblet cells of mammalian intestine and trachea and mucocytes of the mantle of pulmonate gastropods operate in this manner. Electron microscopic data seem to indicate a similarity with the apocrine mode of secretion (Palay, 1958).

Finally, in some cases, the secretion product, accumulated in particulate form in the cytoplasm, undergoes at the time of extrusion, intracytoplasmic dissolution accompanied by loss of staining properties. The extrusion itself is not morphologically recognizable and is marked only by disappearance of the stainable product. Neurosecretory cells of *Xiphosurus* (syn. *Limulus*) (Scharrer, 1941) are a particularly good example of this type of extrusion.

In all merocrine gland cells the secretory cycle is repeated a number of times, each extrusion being followed by restoration, itself followed by extrusion. Hirsch (1929) suggested the name "phase" to designate extrusion and restoration taken together; according to this nomenclature the activity of merocrine gland cells is "polyphasic."

The case of *holocrine* gland cells is quite different. Their activity is "monophasic" according to Hirsch's meaning (1929). Ingestion, synthesis, and intracellular storage of the secretory product occur only once during the life of the cell. Here, extrusion takes on a special significance since it coincides with the death of the cell, in which all organelles have undergone changes, some of which will be reviewed below. The entire cell, filled with secretion products, becomes free and falls into the lumen of the gland. Mammalian sebaceous glands and digestive glands of decapod crustaceans are typical examples of holocrine glands.

Holocrine glands are always provided with a reserve of "replacement cells" in the early stages of the secretory cycle; the renewing of cells inherent to this type of activity, explains the abundance of mitoses often observed in these glands.

Apocrine secretion is characterized by the fact that extrusion includes ejection of the apical portion of the cell, where the secretion product is collected, into the lumen of the gland. Axillary sweat glands, Moll's glands of the eyelid, and mammary glands are classic examples. During the secretory cycle, apocrine gland cells undergo important changes in size, shape, and cytological characteristics. According to classical descriptions, the acinous cells of the mammary glands appear cylindrical toward the end of the period of elaboration of the secretory product; there are, in the basal portion, a well-developed chondriome, a well-defined ergastoplasm, and a voluminous nucleus. In the apical region are to be found a nucleus, believed by some authors to be amitotic in origin, and a great many diverse inclusions that represent the products of elaboration. At the time of extrusion, the entire apical portion of the cell is detached and falls into the acinar lumen. At the beginning of the restoration period, the mammary cell remains flattened, the cytoplasm being spread along the basal membrane and protruding as it is pushed up by the nucleus; the cytoplasmic organelles are incompletely developed and not readily visible.

The activity of apocrine gland cells, like that of merocrine gland cells, is most often polyphasic.

In brief, continuously active gland cells are always merocrine; rhythmically active gland cells may be merocrine, apocrine or holocrine.

It should be noted that the description of functional patterns of gland cell activity given above, being entirely based on optical microscopic data, may have to be revised in years to come. In fact, some results obtained with optical, as well as electron microscopy, cast serious doubts on the absolute validity of the hypothesis given in most textbooks of histology concerning the activity of holocrine and apocrine gland cells.

For instance, the description given by Palay (1958), following recent studies with the electron microscope, of the elaboration process in sebaceous glands, is noticeably different from the usual diagram, outlined in this paragraph; according to Palay's results, the "holocrine transformation" of the sebaceous cell and the "physiological degeneration" of the principal organelles should not be accepted without further discussion.

The mechanism of extrusion in the apocrine glands usually chosen as typical of this functional pattern has also been discussed recently.

In the case of the axillary sweat glands, some authors (Zeiger, 1955) accepted and adopted Schaffer's (1927) well-known description. However, it has been pointed out by Montagna (1956) that the characteristic images of amputation of the apical end are absent in frozen sections prepared with sufficient care and may be mere artifacts. Whether these glands are aprocrine is much less evident than in the usual figures.

The extrusion mechanism is even more questionable in the case of mammary gland cells. No agreement has been reached in the interpretation of the images given by the optical microscope; in a recent paper, Dabelow (1958) points out the controversy between authors who state that amputation of the apical portion of the cell takes place (Jeffers, 1935; Weatherford, 1929) and those who consider the possibility of extrusion without rupture of the apical cytoplasm (Dawson, 1935; Grynfeltt, 1937; Richardson, 1947; Howe *et al.*, 1955). Now, electron microscopic study shows (Bargmann and Knoop, 1959) that in the lactating female rat the gland cells contain two sorts of inclusions, namely, large fatty droplets and much smaller granules, probably of proteinic nature. The mechanism of extrusion of the fatty inclusions is very different from that in textbook descriptions; the inclusion, still enveloped by cytoplasm, protrudes and forms at the apex of the cell a salient, attached to the rest of the cell by a cytoplasmic bridge becoming gradually narrower until the "secretory globule," still in its thin cytoplasmic sheath, is separated from the rest of the cell and released into the lumen of the acinus. Thus the cytoplasm is never exposed and there is no rupture of the plasmalemma or the apical cytoplasm. As for the protein granules, they could be extruded, according to Bargmann and Knoop (1959), as the vacuoles containing them reach the apical surface of the cell. Extrusion takes place, whenever the cell is not destroyed, without the organelles, such as mitochondria or Golgi bodies, being discharged into the lumen of the acinus, and this is a strong argument against apocrine secretion in its usual interpretation.

C. Ingestion

Absorption by the gland cell of materials from the environment is necessarily carried out across the cellular membrane. Since its properties and morphological characteristics are described in another chapter of this book, there is no need to enlarge here on the general structure, ultra structure, and cytophysiology of the cellular membrane, and only a few indications relating more particularly to gland cells are given below.

It is known (Sjöstrand and Hanzon, 1954a) that the thickness of the exocrine pancreatic cell membrane is about 60 A. It is through this membrane and the basement membrane, whose average thickness is 150 A., that exchanges with the blood take place through the extracellular fluid.

Permeability of gland cells to various ions has been studied by several authors, most of the investigations having been carried out on the exocrine pancreas. The rapidity of penetration of some ions into the pancreatic cell has been established by the investigations of Ball (1930), confirmed by Montgomery *et al.* (1940); sodium and potassium ions are found in the pancreatic juice 3 minutes after having been injected intravenously. The passage of these cations through the pancreatic cell is not a passive phenomenon, since sodium and potassium concentration of the pancreatic juice remained always lower than the blood concentration (Solomon, 1952). Calcium and magnesium ions pass through the pancreatic cell much more slowly (Ball, 1930).

Solomon's investigation (1952) have shown that the energy expenditure required by the transport of the inorganic constituents present in 1 liter of pancreatic juice is of the order of 329 calories, i.e., less than 8% of the total energy needed for elaboration of 1 liter of pancreatic juice.

Earlier workers had not given much thought to the passage into the pancreatic cell of more complex compounds. Visscher (1942) indicates that out of a total of eight-five dyes injected, only the anionic and amphoteric dyes penetrated into the pancreatic cell, whereas only cationic dyes passed into the gland cells of the gastric mucosa.

Recent investigations, carried out by means of radioactive amino acids, gave more information on the speed of passage of compounds from the external environment through the membrane of the exocrine pancreatic cell. The investigations of Junqueira *et al.* (1955) have shown that glycine, labeled with radioactive carbon, reaches its maximum concentration in pancreatic cells 10 minutes after its intravenous injection. On the other hand, rat plasma proteins labeled by means of previous injection of radioactive glycine to a donor rat are not detectable in the pancreatic juice; consequently, there is reason to believe that they do not penetrate into the pancreatic cell.

In the present concepts of cellular permeability, active participation of the membrane is believed to play an important role. Recent publications (Danielli, 1952, 1954; Dervichian, 1954; Hirsch, 1955; Harris, 1956) emphasize, in the penetration of substances into the cell, the role played by simple diffusion, by diffusion accelerated by immediate utilization of the ingested compounds, and by active transport involving chemical processes in the vicinity of the membrane. These three phenomena participate in the penetration of substances into gland cells. An earlier study by Hirsch (1931) showed that the permeability of the pancreatic cell to Janus green did not change during the secretory cycle, whereas its permeability to other dyes, such as neutral red, underwent considerable changes.

Another phenomenon, instrumental in the passage of substances into the cell, should also be mentioned. This is pinocytosis (Lewis, 1937), which was discovered by optical microscopy. Electron microscopic investigations (Porter, 1955) have shown the existence of folds in the membrane, varying in depth and forming "microvillosities," in a great many cells such as those of parotid glands and of parietal cells of the fundic mucosa in the stomach. It seems likely (Junqueira and Hirsch, 1956; Brachet, 1957) that these structures play an important role in the exchanges between cell and environment.

D. Intracellular Elaboration of Secretory Products

The phase of the activity of gland cells known as intracellular elaboration includes all the phenomena occurring between the penetration into the cell of materials absorbed from the environment and the completion of the secretory product. It is characterized by extensive changes in cellular organelles and a number of chemical reactions, the mechanism, energetics, and sequence of which are still insufficiently known.

1. Morphological Changes in Organelles during Elaboration of the Secretory Product and Their Participation in It

The occurrence of structural changes in gland cells during the secretory cycle was observed by the early investigators of the second half of the nineteenth century. These changes are frequently mentioned in their publications on secretion, particularly in the accounts of Heidenhain (1875, 1880). As cytology progressed, elaboration of the secretory product was attributed in turn to the various organelles as soon as they were discovered. Hence the ergastoplasmic, mitochondrial, Golgian theories of secretion, each of which had a temporary vogue. At present, it is well established that all the organelles cooperate in the synthetic processes occurring in the cell, and only the manner in which this cooperation takes place is still under discussion.

a. Nucleus. The modifications of shape and "chromaticity" of the nucleus of the gland cell during the secretory cycle and its changes of location within the cell have been described a number of times; the early surveys mentioned in the paragraph on static study contain a wealth of morphological information. The older descriptions have been substantiated and supplemented by the more recent data.

Nuclear volume changes are often evident during the secretory cycle and were mentioned in early descriptions. This has been confirmed and extended by karyometric methods.

Round-the-clock observations of the exocrine pancreatic cells of the albino mouse have shown (Nikolaj, 1939) variations of nuclear volume with an average periodicity of 2 hours. Altmann (1952) has observed, in mice treated with atropine to increase storage in the pancreas, two maxima of nuclear volume, in the vicinity of 400 and 800 μ^3. When extrusion is initiated by administration of pilocarpine, the mean nuclear volume decreases at the beginning of the phase of intracellular synthesis and the maxima tend toward smaller values (from 400 to 336 and from 800 to 672 μ^3). As restoration progresses, the nuclear volumes increase and the frequency curve becomes similar to that seen at the beginning of the experiment.

Karyometric study of endocrine glands, during experimental hyper- or hypoactivity, showed that nuclear volume may be a criterion of activity for continuously active gland cells. Kracht (1954, 1955) observed, after injection of glucagon, a statistically significant decrease of nuclear volume in the A cells of the Langerhans islets, as well as in the A cells of the canalicular system of the pancreas in the albino rat. Karyometric studies of B cells showed, in these animals, an increase of mean nuclear volume, probably related to the compensatory hypersecretion of insulin. With a similar experimental technique it is possible to differentiate, in rats, the anterior pituitary cells which secrete the lactogenic and somatotropic hormones—a distinction which is almost impossible with the usual staining methods (Kracht, 1957).

A study by Rohen (1956) of the apocrine axillary sweat glands and the ear-wax glands in man, showed that the nuclear volume first increases linearly and later becomes rhythmic. The normal volume (Jacobj's *Regelgrösse*) is reached when the height of the cell is around 12 μ. When the size of the cell becomes greater, the nuclear growth is no longer linear, but decreases gradually.

In the salivary glands of *Drosophila robusta,* the size of the nuclei increases gradually during the larval stages, this increase being markedly accelerated at the beginning of the third larval stage. Pupation coincides chronologically with a sudden diminution of nuclear volume. The interpretation of these variations must take into account polytenic phenomena which partially explain the enlargement of the nucleus during the larval development. As for the decrease in volume accompanying pupation, it indicates histolysis taking place in the salivary gland during the nymphal stage (Lesher, 1951a, b).

Nuclear evolution comparable to that of salivary glands of *Drosophila* has been described in the salivary glands of the bee; it would be responsible for the elaboration of royal jelly (Painter, 1945).

When nuclear growth occurs without radical change of shape, the phenomenon can be studied quantitatively by karyometric techniques. In other cases, the increase in size of the nucleus is accompanied by changes in shape making mensuration almost impossible. Although this is fairly rare in mammalian exocrine glands, the subparotid gland of the albino rat is a particularly striking example. The nucleus, spherical at some stages of the secretory cycle, becomes at other stages so irregularly shaped that in the course of a karyometric study Collin and Florentin (1930) had to take into account only the spherical nuclei, thus rendering their results valueless. Irregularly shaped, monstrous, sprouting nuclei are fairly frequent in the gland cells of mollusks and arthropods. The relations between nuclear evolution in the salivary gland of some Hemiptera and the silk glands of Lepidoptera, and endomitosis with endopolyploidy are well known and we shall come back to them; moreover, in these two instances, cellular growth and secretion phenomena are intimately linked. Phenomena of nuclear hypertrophy with irregularity of shape, occurring in active mucocytes of the edge of the mantel of the snail, also deserve mention.

Closely connected with these volumetric changes in the nucleus during intracellular elaboration is the problem of *modification of chromosomes and desoxyribonucleic acid*. Altmann (1952) described in detail these changes during the secretory cycle in exocrine pancreatic cells of the albino mouse. At the accumulation stage, the nucleolus, centrally located, dense, and homogeneous, is enveloped in a granular sheath of Feulgen-positive material in which are incorporated two larger clumps, representing "chromocenters" in Heitz's meaning (1933). Very fine filaments connect the perinucleolar granules to Feulgen-positive granules pressed against the nuclear membrane. After injection of pilocarpine and soon after extrusion, these chromocenters lengthen and reach the granules close to the membrane. The nucleolus swells and shows vacuolization affecting all its components, and not only ribonucleic acid, since the intranucleolar "vacuoles" remain visible even after treatment with ribonuclease. The chromocenter undergoes a longitudinal cleavage, thus forming a sort of "canal," through which the nucleolar substance flows. The splitting of the Feulgen-positive granules lying against the nuclear membrane opens a pore through which the nucleolar substance, more or less liquefied, is discharged into the cytoplasm. When the pancreatic cell contains, at the beginning of the cycle, several nucleoli, each of them establishes communications of this type with the nuclear membrane, and the evacuation of nucleolar substances toward the cytoplasm seems to take place always in close connection with the chromosomes, the transfer being effected in well-defined areas of the nuclear membrane. Altmann (1952) emphasized the fact that the transfer of the products of nucleolar origin is in some way "intrachromosomic," the "canal" being made up of portions of a same chromosome, arranged in an anorthospiral. Changes of chromosomic spiralization evidently play an important role in these displacements; Altmann suggested the name "contraction phrase" to designate the sequence of events just described.

In another step, the nuclear volume increases, the chromocenters become less visible, but spectrometric measurements show no diminution in deoxyribonucleic acid content. In fact, this aspect is the result of a lesser spiralization of chromosomes, hence the name "decondensation phase" proposed by Altmann (1952). Later, strongly Feulgen-positive structures appear, mostly touching the nuclear membrane. These structures increase in size, giving images reminiscent of the prophase; some nuclei actually undergo, at this stage of their evolution, mitoses or endomitoses. The last step is characterized by the apparition, inside some Feulgen-positive cords, of "primary nucleoli" having the customary nucleolar staining properties; these primary nucleoli move more and more toward the center of the nucleus, come in contact and merge—thus the picture is the same as at the beginning.

Altmann (1952) emphasizes the fact that this evolution, described on the basis of images seen on fixed and stained preparations, is confirmed by examination of unfixed cells by phase contrast microscopy. Observation in the living cell indicated that 10 minutes is approximately the time necessary for the nucleolar substance to move from the original location of the nucleolus to the nuclear membrane. According to Altmann's hypothesis, chromatin plays an essential role in the nuclear activity of the pancreatic cell during the secretory cycle, the nucleolus being only "an accumulation of products of chromosomic origin, destined to be used up in the cytoplasm in the course of cellular activity."

The lack of variation in deoxyribonucleic acid content during the phenomena described by Altmann (1952) is in agreement with the measurements of other authors (Pollister and Ris, 1947; Bern and Alfert, 1954). Similarly, Altmann's morphological observations tally with other descriptions of the patterns of evacuation, toward the cytoplasm, of the "intranucleolar vacuoles" (Dittus, 1940; Tramezzani and Valeri, 1955). It is nonetheless true that Altmann's hypothesis, and especially his concept of nucleolar origin and behavior, requires further investigations.

Similar observations were made by Pavan and Breuer (1955; Breuer and Pavan, 1955), who observed variations in the nucleic acid content of well-defined areas in the polytenic chromosomes of the salivary glands during the larval development of *Rhynchosciara angelae*. These histochemical changes occur concurrently with morphological changes, such as formation of bulbous swellings originating in some typical euchromatic bands. The authors discuss the relations between this phenomenon and secretion, but interpretation of Breuer and Pavan's observations appears very difficult since, in the case of the cells under study, growth and secretion are intimately linked.

Since their description, at the beginning of the twentieth century, it has become known that the gland cell *nucleoli* undergo more or less extensive changes during the secretory cycle. Most of the studies referred to here were based on examination of preparations stained according to the so-called general methods. The progress of our knowledge has been directly affected by the histochemical investigations which provided a rational interpretation of nucleolar basophilia; Caspersson *et al* (1941) showed that the nucleolus of the exocrine pancreatic cell of the rabbit undergoes during the period of intracellular elaboration of the secretory product a marked increase in size, accompanied by an enrichment in ribonucleic acid. Similar observations were made in the study of a great many gland cells. The results obtained in studies of liver cells (Lagerstedt, 1949; Stowell, 1949; Stenram, 1953), submaxillary gland cells of the rat (Macchi, 1951), and various endocrine glands (Celestino da Costa *et al.*, 1949; Allara, 1952; Sandritter and Hübotter, 1954) should be mentioned as some of the best examples of nucleolar changes during glandular activity in vertebrates. Invertebrate gland cells exhibit during the elaboration period similar phenomena, shown on salivary glands of Diptera Painter, 1945; Caspersson, 1950; Lesher, 1951a, b), salivary glands of the isopod *Anilocra physodes* (Siniscalo, 1951), and cells of pharyngeal glands of the bee (Painter, 1945).

On the whole the observations made on various objects are in agreement that there is a marked increase in nucleolar volume in all protein-elaborating cells during the period of intracellular synthesis. Gland cells, therefore, are only a special case of a more general evolutionary pattern; the nucleolar changes mentioned here occur in nerve cells, oöcytes, and all actively growing cells.

Volume increase may be accompanied by loss of homogeneity of the organelle and formation of "vacuoles," already mentioned in the static study. There are cases where the increase, sometimes considerable, affects a single nucleolus; this occurs in the salivary glands of Chironomidae and certain species of *Drosophila* (Lesher, 1951a). In other cases, e.g., the salivary glands of *Drosophila melanogaster* (Painter, 1945), there is multiplication of nucleoli.

The increase in nucleolar volume during the secretory cycle has been measured in a few cases. Yokoyama and Stowell (1951) gave numerical data on increase in nucleolar volume during intracellular elaboration in the exocrine pancreas of the mouse after injection of pilocarpine. Sandritter and Hübotter (1954) studied, by microplanimetry, variations of nuclear and nucleolar volumes in the cells of the fasiculata of the adrenal cortex of the female albino rat. They observed, in comparing cells rich and poor in lipids, that the nucleolar volume increases considerably during the secretion of proteins, concurrently with the decrease in lipid inclusions of the cytoplasm. In the case studied by Sandritter and Hübotter, the increase in nucleolar volume is far greater than the markedly lesser increase of the nuclear volume.

Schreiber *et al.* (1955) studied, by karyometric techniques, nuclear and nucleolar volumes and their ratio in various gland cells (liver, submaxillary glands, pancreas) of the albino mouse. They showed that the nuclei of the gland cells studied, followed the general laws of nuclear growth with, however, intermediary stages between those of Jacobj's sequence (Schreiber's sesquiphases, 1949). The total nucleolar volume in "nucleolar units," follows, in some cells, the variations of nuclear volume. There are also cells whose nuclei contain a multiple of the number of nucleolar units. Schreiber *et al.* pointed out that the increase in protein synthesis in the cell is accompanied by an increase of the volume of the organelles involved in the synthesis. This increase may take place in two ways: one is multiplication of the whole genome, with subsequent increase in the degree of ploidy of the cell; the other, multiplication of the nucleolar units alone, as well as of the nucleolar organizers, without concomitant multiplication of the whole genome. This second mode of increase of nuclear volume is evidently rhythmic, the number of nucleolar units in a given cell being always a multiple of the original number. It should be noted that studies of nucleolar volume in other cells (nephron of mouse) yield data tallying with this conception (Pellegrino *et al.*, 1955).

The transfer of nucleolar material toward the cytoplasm is not always histologically detectable. Sandritter and Hübotter (1954) believe that morphological indication of transfer of substance from the nucleus to the cytoplasm exists only when protein synthesis is particularly active. The various aspects described by Altmann (1952) in exocrine pancreatic cells of the mouse have been mentioned above; they are in agreement with the earlier descriptions, particularly with that of Huber (1949). This author's work includes a survey of earlier descriptions of the passage of nucleolar into the cytoplasm. It should be emphasized that some descriptions have not been confirmed by observations under vital conditions and that the production of artifacts is particularly likely in this kind of research. It remains true, however, that unquestionable examples of this transfer of substance, morphologically detectable, have been described under condition excluding the possibility of artifacts. Moreover, there is biochemical evidence against the earlier interpretations according to which the nucleolus is an accumulation of reserves or waste products of nuclear metabolism. The nucleolus may be recognized, in agreement with Vincent

(1955), as the site of synthesis of ribonucleic acid, from which the latter is subsequently transferred to the cytoplasm.

b. Cytoplasm. The cytoplasm also shows considerable morphological changes during the intracellular elaboration of the secretory product. The participation in intracytoplasmic syntheses of the ergastoplasm, the chondriome, and the Golgi bodies will be considered in turn.

Relations between the morphology of the *ergastoplasm* and gland cell activity were stated, toward the end of the nineteenth century, by the Nancy school (Prenant, 1898, 1899; Garnier, 1899). According to these authors, the area of the basophilic zone of cytoplasm varies with the stages of the secretory cycle. It is largest, in the serous cells studied, when the cell, after discharging its secretory product, starts to elaborate new granules. At the beginning of the twentieth century, this concept did not receive the acceptance it deserved. We have mentioned in Section III on the static study of gland cells how the concept of ergastoplasm, fallen in disrepute until 1940, has regained favor thanks to progress in histochemistry. Since the first investigations with ultraviolet absorption spectrography or staining with basic dyes controlled by the use of ribonuclease, correlation between the abundance of ribonucleic acid and the intensity of protein synthesis became evident (Caspersson *et al.,* 1941; Brachet, 1942) (Figs. 46 and 47). Numerous histochemical investigations confirmed this hypothesis and, moreover, showed that tissues with very active metabolism but low level of protein synthesis are poor in ribonucleic acid (for references see Brachet, 1957). From then on the work of the Nancy school again received well-deserved attention.

Advances in electron microscopy, particularly improvements in fixation techniques and preparation of ultrafine sections, led to generalization of the concept of ergastoplasm and opened new ways for the study of the relationship of this structure to the secretory process.

Today, Garnier's conclusions (1899) on ergastoplasmic evolution during the secretory cycle of so-called serous glands of mammals, are to a great extent confirmed and are applicable to all gland cells whose activity includes synthesis of notable quantities of proteins. Histochemical techniques made it possible to verify, in all cases, the increase of the basophilic zone during the period of intracellular elaboration of the secretory product. Variations in ribonucleic acid content, as obtained by chemical methods as well as by electron microscopic data, are more difficult to interpret than the results given by histochemistry.

Observation of liver cells shows an appreciable diminution of the extent and stainability of the basophilic areas during fast and their enlargement after a meal rich in proteins (Lagerstedt, 1949). Similarly, liver restoration after partial hepatectomy is accompanied by increased

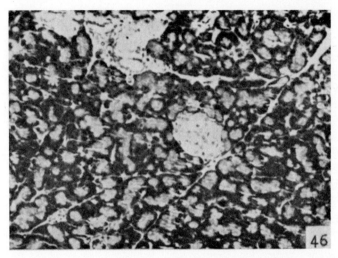

Fig. 46. Pancreas of an albino rat, fixed with Carnoy's fluid and stained with toluidine blue. Note the dark basal ergastoplasm of the exocrine cells and the RNA-less islet of Langerhans. Magnification: × 200; orange filter.

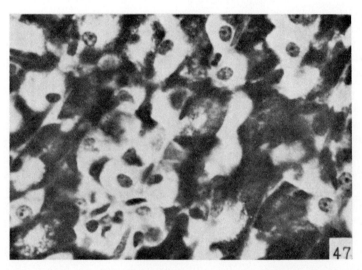

Fig. 47. Gastric mucosa of a guinea pig, fixed with Helly's fluid and stained with toluidine blue. Note the dark, RNA-containing chief cells and the clear, RNA-less parietal cells. Magnification: × 800; orange filter.

cytoplasmic basophilia (Drochmans, 1950). These data were confirmed by Dalton *et al.* (1950) and by Bernhard *et al.* (1951, 1952a), who observed with electron microscopic techniques a notable enlargement of the ergastoplasm in liver cells during active protein synthesis.

The data concerning the exocrine pancreas are more ambiguous. Ultraviolet spectrographic studies carried out by Caspersson *et al.* showed, as early as 1941, a considerable increase of absorption in the basal zones of exocrine pancreatic cells of the rabbit during synthesis consecutive to extrusion. Numerous investigations by means of tracer dyes and subsequent control by ribonuclease led to the same conclusions (for references see Oram, 1955). It should be pointed out that the total quantity per cell of ribonucleic acid is very difficult to evaluate without the integrating microspectrograph of Caspersson *et al.* (1953)—so much so that results of investigations prior to 1953 may be erroneous. Moreover, the results of ribonucleic acid determination during the secretory cycle do not corroborate the histochemical data. Electron microscopic data are still contradictory. Weiss (1953) stated that the ergastoplasm participates directly in the formation of secretion granules; sections of exocrine pancreatic cells from animals just fed after a fasting period show formation of granules from end of elongated and flattened ergastoplasmic sacs and transfer of these granules to the apical region of the cell. In contradiction, Sjöstrand and Hanzon (1954a) found no definite relation between the ergastoplasm and the secretion granules in the exocrine pancreatic cell; they believed rather (Sjöstrand and Hanzon, 1954b) that the secretion granules are formed at the expense of the Golgi bodies. Palay (1958) found no relation between secretion granules and ergastoplasm in the pancreatic cells of mice fasting or having received pilocarpine injections, but described, in agreement with Lacy (1956), the transformation of ergastoplasmic lamellae into vesicles in animals treated with pilocarpine. Palade (1956) described, however, in the ergastoplasm of exocrine pancreatic cells of guinea pigs, granules about 300 mμ in diameter and discussed their relation to the secretion granules.

Observations on thyroid gland cells are equally contradictory. According to Dempsey and Peterson (1955), the ergastoplasm of the thyroid cells elaborate a compact product resembling the colloid; Wissig (1956) declared, on the contrary, that the secretion product of the thyroid cell originated close to the Golgi bodies but that the ergastoplasm become markedly enlarged during stages of hyperactivity.

Similarly, the participation of the ergastoplasm and Golgi bodies in the elaboration of the secretion product of the pituitary gland cells is still controversial (Rinehart and Farquhar, 1953; Farquhar and Rinehart, 1954; Haguenau and Bernhard, 1955; Haguenau and Lacour, 1955).

In other cells, the relation between the ergastoplasm and the secretory product are clearer. According to Hendler *et al.* (1957), the gland cells of the oviduct of the hen contain a well-developed ergastoplasm, the spaces between lamellae being filled with a substance having the same characteristics as the secretion product contained in the lumen of the oviduct. These authors emphasized the fact that the Golgi bodies do not take part in the secretion phenomena.

In the mammary gland the participation of the ergastoplasm in the elaboration of secretion products seems evident also (Bargmann and Knoop, 1959).

In any case, the ergastoplasm undergoes during the period of intracellular elaboration of the secretion product very definite morphological changes. In sections treated by the methods of optical microscopy, the images may be classified into three groups, according to the arrangement of the basophilic substance in the cell, as was mentioned in Section III.

When the ergastoplasm is situated at the base of the cell (exocrine pancreas, parotid gland, etc.) the period of elaboration is characterized by a notable extension of the basophilic zone, followed, during accumulation of the secretion product, by a shrinkage, as if the ergastoplasm were pushed toward the basal pole as the secretion granules increase in number at the apical pole.

When the ergastoplasm is diffuse in the gland cell, at the beginning of the period of elaboration of the secretion product (salivary glands of Diptera, salivary glands of pulmonate gastropods, etc.), accumulation of this product brings about disruption of the ergastoplasm into strands enveloping the secretion products. In some cases, the ergastoplasmic delamination is shown by the appearance of special structures, called "pseudochromosomes" by the early authors, whose significance was established by histochemical detection of ribonucleic acid (Gabe and Prenant, 1948) (Figs. 48 and 49).

When extrusion does not bring about complete ejection of the secretion product, the increase in area of the ergastoplasm during the following elaboration period is indicated by the thickening of the cytoplasmic strands surrounding the secretion granules.

In some cases, especially in holocrine gland cells, the diminution of the ergastoplasm at the end of the elaboration period may be so marked that all histochemically detectable basophilia disappears. Sebaceous glands, for instance, contain an appreciable quantity of pyroninophilic ribonucleoproteins as long as the amount of secretion product remains small; at the end of the elaboration period, the cytoplasm located between the secretion granules has lost its basophilia and takes up avidly the acid dyes (Montagna, 1956).

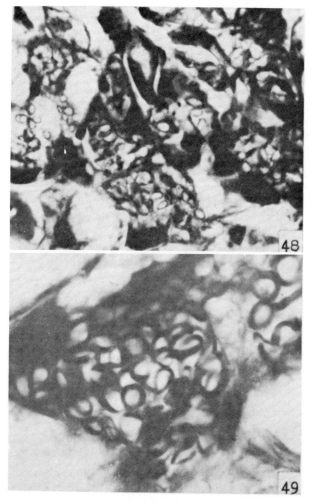

Figs. 48 and 49. Salivary gland in *Limnea stagnalis,* fixed with Carnoy's fluid and stained with toluidine blue. Note the "pseudochromosomes." Magnification: ×350 (48) and 1500 (49); orange filter.

In cases where the ergastoplasm appears, at the beginning of the elaboration phase, in the form of masses scattered throughout the cytoplasm (liver cell), the size and number of these structures increase first, then diminish at the end of the elaboration period.

Participation of the *chondriome* in the elaboration of the secretion products of gland cells was claimed as soon as this organelle was discovered. Altmann (1890, 1894) attributed to the "bioblasts" and "vegetative filaments" an important role in the elaboration of secretion granules. According to him, the vegetative filaments (chondrioconts in recent nomenclature) would become moniliform at the beginning of the elaboration period, then divide into granules, each of which grows into a secretion granule. Zoja and Zoja (1891) and Metzner (1891) agreed with this opinion; this was the origin of one of the theories on the participation of the chondriome in the elaboration of secretion products, the so-called *total transformation theory*.

The investigations of Benda (1899), Michaelis (1900), Bouin (1905), Regaud (1908), Regaud and Mawas (1909), and Policard (1910) added to the knowledge of the morphology and staining properties of chondriosomes and provided a new interpretation of the role of these organelles in secretion. According to these authors, the secretion granules originated in the chondriosomes, as an integral part of them, then became independent during their growth and only part of the chondriosomal substance would be transformed into secretion products. This is the *partial transformation theory*.

Hoven (1912) adopted a similar hypothesis. According to him, the apical portion of the chondrioconts of the serous gland cell would be transformed into secretion product, the remaining basal portion would, after extrusion of the product, reconstitute the whole of the organelle. Hoven admitted also the possibility of a multiplication of the chondrioconts by longitudinal cleavage.

These investigations have been the starting point of many studies in which the role of the chondriome was considered essential for the elaboration of very diverse products. Cowdry (1918) published a list of eighty substances considered to be, more or less directly, of mitochondrial origin.

After 1920, the partisans of the theory of total transformation became less and less numerous. On the other hand, the publications subsequent to this date, although emphasizing that the chondriosomes remained unaltered, described the formation of the secretory product in contact with the chondriome; some of the observations were made in the living state and are therefore reliable. Hirsch (1931, 1932a, b) described, in the pancreas of the mouse, *in vivo* with the technique of Covell (1928), movements of the chondrioconts and formation, in contact with these elements of very fine granules migrating toward the apical portion of the cell. Huber (1949), surveying the descriptions of changes in the chondriome during the secretory cycle of the exocrine pancreatic cell, confirmed Hirsch's data, but declared that the granules originated inside the chondrioconts, not outside.

Morphological study of many objects shows that the secretion product, during the elaboration, remains in close contact with the chondriome. Formation of granules in contact with the chondriosomes is a very frequent occurrence and vital observation shows that not infrequently they are formed inside those elements. This confirms Nassonov's descriptions (1924). On the other hand, the "exhaustion" of the chondriome during the secretory cycle and its transformation into secretory product is not admitted at present.

It should be pointed out that morphological changes in the chondriome during the secretory cycle of gland cells are no longer of great interest. Biochemical studies, starting with isolation of mitochondria (Bensley and Hoerr, 1934), permit today a very satisfactory interpretation of the role played by the chondriome in the elaboration of secretion products. This point will be developed below.

The participation of the *Golgi bodies* in the elaboration of secretory products has been a subject of heated discussion. The descriptions of the morphological changes of these organelles during intracellular synthesis of the granules vary so much, from object to object and, for the same object, from investigator to investigator, that the present situation cannot be summarized. Moreover, Golgi bodies and secretion granules being the subject of another chapter of this treatise, only the principal theories will be mentioned here.

According to partisans of the old theory of the "internal reticular apparatus," the secretion product would be in close relationship with the "Golgi network." Descriptions prior to 1924 mentioned, without, however, emphasizing it, the situation of the secretion granules between the meshes of the net, and Nassonov (1924), who shared this opinion, deserves credit for having clearly stated the problem of the relation between Golgi bodies and secretion product in mammalian gland cells.

After 1925, the concept of "Golgi network" became more and more neglected, but even those authors who considered the silver or osmic acid-impregnated network to be an artifact, believed in a relation between Golgi bodies and the secretion process. Thus Morelle (1927), who rightly opposed the interpretation of the Golgi bodies of the exocrine pancreatic cell as a network, emphasized the increase in number and size of these dictyosomes during the period of intracellular elaboration and their possible participation in secretion. Protagonists of the "vacuome" theory, such as Parat (1928), also maintained that these structures take part in elaborating the secretion granules, emphasizing the grouping of vacuoles in a "Golgi zone," the relation of the chondriome with this zone, and the apparition of products of cellular activity inside the vacuoles. The detractors of the vacuome theory rightly criticized the assimilation of Golgi bodies and vacuoles vitally stainable with neutral red, but agreed in turn that the Golgi bodies take part in intracellular elaboration phenomena (Bowen, 1929; Gatenby, 1931).

Discovery of homogeneous osmiophilic formations in the cells of the digestive gland of the crayfish by Jacobs (1928) was the starting point for a new interpretation of the Golgi bodies, developed at length by Hirsch and his school (for references see Hirsch, 1939). According to these authors, the homogeneous osmiophilic formations, given the name of "presubstance," undergo an evolution with differentiation of an osmiophilic and argyrophilic, crescent-shaped shell enveloping a central argyro- or osmiophobic zone. This would be the classical aspect of the dictyosome, with its external and internal zones. According to Hirsh (1939), it is in these "Golgi systems" that the elaboration of products of intracellular activity take place. For Hirsch, the "Golgi systems" constitute the primary center of elaboration in the cell; elaboration of the secretion granules would take place in the osmiophobic substance (internum). When the granules are formed they become distinct from the osmiophilic substance, afterward named "Golgi remnants," Hirsch's theory represents the extreme expression of the "Golgian concept" of gland cell activity, the other organelles playing a secondary role.

One of the weak points of Hirsch's hypothesis concerns the nature and significance of the "presubstance." In fact, as was shown in Sluiter's pictures (1944) and pointed out by Huber (1949), "presubstance" and "Golgi remnants" coexist in the exocrine pancreatic cell of the mouse at the late stages of the intracellular elaboration period. It is thus impossible to accept the retransformation into "presubstance" of the "Golgi remnants" during the new secretory cycle. Moreover, the identification of the "presubstance" and "Golgi remnants" is based only on selective impregnation with silver or osmic acid, whereas identification of the dictyosomes is based on unquestionable morphological criteria. Furthermore, the dictyosomes are readily visible by vital observation of favorable objects, whereas under such conditions, neither "presubstance" nor "Golgi remnants" are detectable. We may add that, from the histophysiological point of view, Hirsch's theory overlooks all the other organelles of the cell in favor of the Golgi bodies.

A concept worthy of mention, which originated in Hirsch's theory (1939), is that of Worley (1946), according to whom products of nuclear origin, elaborated by the nucleolus and discharged into the cytoplasm, would bring materials to the chondriosomes and Golgi bodies, which would thus develop from a common precursor. For the rest of the evolution of Golgi bodies, Worley's theory rejoins Hirsch's; after a first period of increase of an homogeneous granule, there would be a second period, of "segregation," with formation of an external, osmiophilic, crescent-shaped zone enveloping an internal, osmiophobic zone. In the third period, of discharge, the chromophobic substance, transformed into secretion granules, would be released while the osmiophilic substance (Worley's pycnotic form) re-entered the cycle. Huber (1949) accepted this interpretation and with the support of the data of Caspersson et al. (1941) on the role of the nucleolus in intracellular synthesis of proteins asserted the renewal of the "presubstance" from materials of nucleolar origin.

Another interpretation of the "presubstance" was proposed by Ries (1935) on the basis of studies of the exocrine pancreatic cells of the mouse. According to this author, the "presubstance" would be the result of division of granules known as "lipochondria." As was admitted by Hirsch (1955) this is evidently erroneous, the "lipochondria" of Ries being only a waste metabolic product. As Gresson (1952) and Gatenby (1953) pointed out, the re-use by Baker (1951) of this appellation, with an entirely different meaning, is highly regrettable.

The concepts just mentioned have not been substantiated by electron microscopy. Investigations initiated by Dalton (1951, 1952), Dalton and Felix (1953, 1954), and Sjöstrand and Hanzon (1954a, b) showed that a characteristic ultrastructure corresponding to the dictyosomes of classic cytology existed in all animal and plant cells examined, whereas neither the "presubstance" nor the "Golgi remnants" could be identified.

As for the role of Golgi bodies in intracellular elaboration, Sjöstrand and Hanzon (1954b) discussed the formation, in the exocrine pancreatic cells of the mouse, of secretion granules on the surface of the lamellae (sacculi) of the Golgi apparatus and the possibility of the conversion of the smaller granules embedded in the interlamellar substance into secretion granules. Hyperactivity of the anterior pituitary cells in adenomas induced by estrogens (Haguenau and Lacour, 1955) is evidenced by considerable enlargement of the Golgi bodies. Farquhar and Wellings (1957) described, in the hyperactive acidophilic cells of the adenohypophysis, an increase in the number of granules located between the lamellae and vesicles of the Golgi bodies. They observed the same phenomenon in the exocrine pancreatic cells. Palay (1958) confirmed and completed the description by Sjöstrand and Hanzon of the exocrine pancreatic cell; he also substantiated the role of the Golgi bodies during the elaboration of lipoidal products in the cells of the gland of Meibom in rats. His conclusions tally with those of Bargmann and Knoop (1959) based on studies of protein synthesis in the mammary gland of the rat, i.e., that the Golgi bodies may be responsible for the separation and accumulation—the concentration, so to speak—of products elaborated in the cytoplasm. This would take place before ejection of the completed secretion product into the external environment. This point of view is close to that adopted by Hirsch (1955, 1958) in his review articles.

On the whole, the recent data confirm a histophysiological concept originating in even the most erroneous of the early descriptions, that is, the concept of an increase in the number and volume of the Golgi bodies during intracellular elaboration of the secretory product. Electron microscopy definitely revealed relations between the Golgi bodies and the secretion granules in protein-synthesizing cells, but the role played by the Golgi bodies during this synthesis is as yet unknown. Numerous investigations will have to be carried out before these questions can be adequately answered. Biochemical study of the Golgi bodies is much more difficult than that of chondriosomes, since they have never been satisfactorily isolated by ultracentrifugation. Histochemical data concerning them are very fragmentary and probably have no general validity. Furthermore, progress has been considerably slowed down by two completely opposite tendencies; one, the tendency to hasty and inadequately supported generalizations; the other, rightly opposed by Bensley (1951), the tendency to bypass the difficulties by simply denying the existence of the Golgi bodies and calling them artifacts.

2. Physiological and Biochemical Aspects of Intracellular Elaboration Activity

Glandular activity leads to the ejection into the surrounding medium of secretory products, often in considerable amounts. Numerical data are more or less readily available, depending on the case. It is extremely easy to evaluate the work of the digestive glands in man by comparing Kestner's data (1930) on the quantities of digestive juices elaborated by the various glands with the standard data of descriptive anatomy regarding the weight of these organs. For instance, the weight of the principal salivary glands (parotids, submaxillary, sublingual) in adult man oscillates around 40 gm. and the quantity of saliva secreted in 24 hours varies from 1 to 2 liters, according to the diet. According to Babkin (1944), the adult man's gastric mucosa weighs approximately 100 gm.; the quantity of gastric juice secreted in 24 hours varies from 1 to 2 liters. The pancreas of adult man weighs about 80 gm. and elaborates, in 24 hours, almost 1 liter of pancreatic juice.

These figures illustrate the extent of the work of the gland cells and show the importance of *energy exchanges during the secretory cycle*.

As indicated earlier, ingestion requires little energy (8% of the total energy expenditure in the exocrine pancreas). Intracellular elaboration, on the other hand, is the endergonic process consuming most of the energy required for the secretory cycle. Recent biochemical investigations have shown the essential role played by the chondriome in the functional activity of the cell. This concept is all the more valuable for the gland cell, since most of the data were obtained in the liver cell. It is in the light of these data that the participation of the chondriome in the work of the gland cell should be considered.

It is known that about 70% of the cellular oxidation systems are concentrated in the chondriosomes. Glycolysis takes place in the cytoplasm and without participation of the chondriome, but the oxidation of pyruvate requires the participation of a set of enzyme systems exclusively contained in these organelles. The chondriome is, therefore, one of the principal energy generators of the cell.

It is also in the chondriome that are found the enzymes of the tricarboxylic acid cycle of Krebs, the participation of which in the metabolism of carbohydrates, lipids, and proteins is well known.

Moreover, the chondriosomes contain up to 75% of the total energy-rich phosphates of the cell as well as the enzymes responsible for oxidative phosphorylation. They represent, therefore, one of the greatest reserves of energy of the cell, releasing it at the right time.

These biochemical data explain the mechanisms regulating the increase in oxygen consumption of glands whose activity is initiated by nervous or pharmacological stimuli, as shown in the exocrine pancreas (Barcroft and Starling, 1904; Still *et al.*, 1933; Deutsch and Raper, 1938; Davies *et al.*, 1949). The data also explain the inhibitory action of anoxia on pancreatic secretion.

It is admitted at present that glycolysis does not take any part in the activity of the pancreatic cell as a source of energy. Increased lactic acid production has been described after pharmacological stimulation of pancreatic secretion (Bergonzi and Bolcato, 1930; Ferrari and Höber, 1933), but this has been contradicted by Himwich and Adams (1930) and by Deutsch and Raper (1938). Moreover, Brock *et al.* (1939) pointed out that increased glycolysis after pharmacological stimulation of pancreatic secretion may very well be a nonspecific toxic phenomenon.

Northup (1936), reported increased hydrolysis of phosphocreatine in stimulated salivary glands. This shows the participation of high energy phosphates in the energy metabolism of gland cells and confirms the data on the increase in inorganic phosphorus of the venous blood after stimulation of pancreatic secretion (Camis, 1924; Ferrari and Höber, 1933).

As Junqueira (1955) rightly pointed out, comparison of an unstimulated mammalian digestive gland with a gland under nervous or pharmacological stimulation is open to criticism. Actually, the unstimulated gland is never completely inactive. From this point of view, the experiments of Junqueira and his co-workers (Junqueira, 1951; Fernandes and Junqueira, 1953, 1955; Rothschild and Junqueira, 1951; Junqueira and Rabino-vitch, 1954; Rabinovitch *et al.*, 1951) are much more significant.

These authors studied, from the morphological, histochemical, and biochemical points of view, the submaxillary glands of rat and mouse, comparing normal glands and glands with ligated excretory ducts. Under these conditions, considerable structural changes were observed in the gland cells (flattening of the cells, decrease in the number of chondrio-somes), as well as a marked decrease in weight of the gland. The cyto-plasmic basophilia was decreased in the ligated glands, and ribonucleic acid measurements showed a notable decrease in ribonucleoprotein con-tent. From the biochemical point of view, a sharp decrease in proteolytic activity was also observed; phosphomonoesterase activities, however, did not change markedly. There was no change in glycolysis, but oxygen con-sumption and succinic dehydrogenase activity were greatly reduced after ligature of the excretory ducts. Addition of succinate was followed by only a small increase in succinic dehydrogenase activity. These data suggest the intervention of the Krebs cycle in secretion phenomena. More-over, ligature of the ducts caused a sharp fall in the adenosine tri- and diphosphates, phosphocreatine, and inorganic phosphorus.

Nucleic acid metabolism during gland cell activity was studied by various methods. Besides the histochemical data, given above in the study of the nucleolus and ergastoplasm, the results of chemical determinations, on the one hand, and those obtained by use of labeled compounds on the other, should be mentioned here.

Some biochemical results completely confirm the concept of ribo-nucleic acid participation in protein synthesis (Caspersson *et al.*, 1941; Brachet, 1942). Jeener (1948) showed an increase in the ribonucleoprotein content of the crop of the pigeon during the period of elaboration of the secretory product. Similar results were reported by Rabinovitch *et al.* (1951) in the seminal vesicle and in the submaxillary glands (Rabinovitch *et al.*, 1952a). This view is confirmed also by numerous data relating to mammalian liver. It is known that fasting or a protein-deficient diet in-duces, together with a decrease in cytoplasmic basophilia, a decrease in ribonucleic acid (Brachet *et al.*, 1946; Davidson, 1947; Mandel *et al.*, 1950; Campbell and Kosterlitz, 1952; Daly *et al.*, 1953; Mirsky *et al.*, 1954). Under similar experimental conditions, deoxyribonucleic acid measure-ments, however, showed no significant change.

For the exocrine pancreas, the agreement between histochemical and chemical data is less satisfactory. All histochemical studies (for references see Oram, 1955) show an increase in cytoplasmic basophilia of the exocrine pancreatic cell during the first stage of intracellular elaboration of secre-tion granules. But ribonucleic acid determinations (Rabinovitch *et al.*,

1952b; Daly and Mirsky, 1952) do not show any significant variation of the ribonucleic acid content in the pancreas of the mouse during the secretory cycle experimentally induced by pilocarpine. These results were confirmed by De Deken Grenson (1952) and by Langer and Grassi (1955). It should be pointed out, however, that some of the results of Daly and Mirsky (1952) suggest a participation of ribonucleic acid in the synthesis of pancreatic enzymes; e.g., fetal pancreas, contrary to most embryonic tissues, contains less ribonucleic acid than adult pancreas. Moreover, prolonged fasting induces in mice a notable diminution of pancreatic secretion and a decrease in pancreatic ribonucleic acid.

Actually, the various organs studied present certain differences. In the crop of the pigeon, as well as in the submaxillary glands and the seminal vesicles, cellular growth and secretion phenomena are intimately linked. This is not the case for the hepatic and pancreatic cells.

Besides quantitative evaluations by chemical techniques, autoradiographic studies, after incorporation of isotope-labeled compounds, provide valuable information confirming the histo- and biochemical data of the role of ribonucleic acid in secretion, especially when the activity of the cell leads to intensive protein synthesis.

Comparative study of cytoplasmic basophilia and incorporation of labeled phenylalanine in various organs of the mouse shows (Brachet and Ficq, 1956) a very clear correlation between the intensity of protein synthesis and the degree of basophilia. The incorporation of labeled phenylalanine, detected by autoradiography, is much greater in the exocrine pancreas than in the Langerhans islets, in the Lieberkühn glands than in the rest of the intestinal mucosa. Niklas and Oehlert (1956) obtained similar results; they pointed out the extent of the incorporation of labeled amino acids into gland cells elaborating proteins and emphasized the agreement between their autoradiographic data and the chemical data on ribonucleic acid distribution. Moreover, autoradiographic observation at the cellular level discloses the significant fact of the precocity and intensity of nucleolar incorporation (Taylor, 1953; Ficq, 1955a, b; Ficq and Errera, 1955; Taylor and McMaster, 1955).

As for variations in the incorporation of labeled compounds during the secretory cycle, the first data, obtained with radioactive phosphorus, are unreliable. Some authors declared that after pilocarpine injection incorporation into the pancreas of dogs increases (Guberniev and Il'Ina, 1950), whereas others, working on mice, denied that functional stimulation had any repercussion on incorporation (De Deken Grenson, 1952; Hokin and Hokin, 1954). Actually, studies with radioactive carbon-labeled amino acids proved that functional stimulation is accompanied by increased incorporation (Allfrey *et al.*, 1953, 1955; Fernandes and Junqueira,

1955). The work of Fernandes and Junqueira (1955) on the pancreas of pigeon stimulated by treatment with carbaminoylcholine shows that the incorporation of radioactive glycine into ribonucleic acid is parallel to its incorporation into proteins.

3. Functional Sequence in the Gland Cell

Since the discovery by Bernard (1856) of the pancreatic zymogen granules and the fundamental studies of R. Heidenhain (1875, 1880) on the physiological significance of the secretion granules, the rhythm of activity of the gland cell has been investigated by numerous authors using different methods.

In a first group are found the morphological studies in which the authors were concerned with the interval necessary for the gland cell to regain, after experimentally induced extrusion of the secretion product, its initial appearance.

In some particularly favorable cases, for example in the exocrine cells of the pancreas, vital observation of the gland in action is possible under almost normal conditions and the activity of a gland cell can be studied without having to be interrupted for histological examination. Its dynamic study under vital conditions was begun by Kühne and Lea (1882), who combined, in the rabbit, vital observation of the pancreas with its intact vascular and neural connections, and measurement of glandular activity by quantitative evaluation of the pancreatic juice elaborated during the experiment. With this technique, also used in studies of the pancreas of the mouse by Covell (1928) (Fig. 50), Hirsch (1931, 1932a, b), and others the duration of extrusion was estimated to be about 30 minutes, that of intracellular synthesis, 6–10 hours.

In most cases, the gland cells *in situ* are not accessible to vital observation, and, in order to determine the rhythm of secretion, the investigator must rely on specimens taken from different animals. Variations in the individual responses to stimuli evidently cause some heterogeneity in the results; it is therefore, essential to define certain characteristic aspects, to evaluate their percentages in the preparations corresponding to the various stages of the secretory cycle, and to submit the numerical data to statistical treatment. Comparison of histological preparations of specimens taken from the animals at various stages of the secretory cycle has been a constant practice since the first histophysiological studies of glandular secretion. Hirsch (1915) was the first to number the cells at the various stages of the cycle and to represent graphically his results. This method has been widely used under the name "method of numeration of stages" (*Stufenzählmethode*) by Hirsch and his co-workers (for references see Hirsch, 1929, 1939, 1955, 1958).

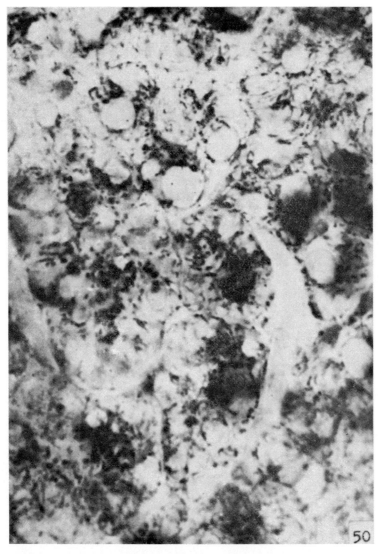

FIG. 50. Pancreatic acini of an albino mouse, examined (Covell's method) 90 minutes after feeding; the chondriomes are vitally stained with Janus green. Note the well-developed chondriomes and the scarcity of the secretion granules (cf. Fig. 44). Magnification: × 1500; red filter.

The results obtained for the exocrine pancreas confirmed those made by vital observation. According to Ries (1935), Sluiter (1944), and Huber (1949), the interval is 30–60 minutes for extrusion, 6–10 hours for return to the starting point. In the parotid gland of the rat, extrusion is complete in 1 hour; restoration takes 10–14 hours (Järvi, 1939). Intracellular synthesis lasts much longer in the salivary glands of the dog; Langstroth *et al.* (1938, 1939) stated that after strong stimulation the cells do not regain their normal appearance before 3–6 days. According to Krijgsman (1925, 1928), the length of the elaboration period in the cells of the salivary and digestive glands of the snail is about 3–4 hours. The secretory cycle of the digestive gland cells of the crayfish lasts 5–6 hours (Hirsch and Jacobs, 1928, 1930).

In the case of large, homogeneous, and easily accessible glands, morphological study of the various stages of the secretory cycle can be carried out on fragments obtained by biopsy. This technique was used by the Babkin school (for references see Babkin, 1944) in histophysiological investigations on histamine-induced gastric secretion. Langer (1957), using the same technique for the study of guinea pig pancreas during pilocarpine-induced secretion, observed that the zymogen granule content of the tissues reaches a minimum during the second hour after injection, then gradually increases.

Physiological and biochemical techniques were used in other investigations; the general procedure was, first, determination of the content in active substances of the gland before application of extrusion-inducing stimulus, and evaluation, by means of determinations done at regular intervals, of the time corresponding to the minimum content in active products and of the return to normal. The first interval indicates the duration of the extrusion, the second, that of intracellular synthesis. Most of the work to be mentioned here was done on the exocrine pancreas. Van Weel and Engel (1938) studied the variations in mice of pancreatic carboxypeptidase during the secretory cycle. They observed a sharp decrease after pilocarpine injection, the minimum being reached during the third hour; from this moment the enzyme content increased, and the return to normal was effected about the ninth hour.

Daly and Mirsky (1952) studied the speed of resynthesis of the most important pancreatic enzymes (amylase, protease, lipase, and carboxypeptidase) in fasting mice receiving subsequently either an abundant meal or pilocarpine injections. These authors observed that the minimum of enzymatic activity in the pancreas was reached during the hour following the stimulation; the return to the initial enzymatic activity occurred in 3–6 hours. Comparative study by Daly and Mirsky (1952) of the variations in total content in trichloroacetic acid-precipitable proteins showed no definite modification of the total proteins during the secretory cycle. From this it may be concluded that protein resynthesis is a very rapid process, probably starting during the extrusion period, and that most of the stage of intracellular elaboration is very likely a period of gradual transformation of these proteins into specific enzymes.

Rabinovitch (1954), studying the variations in ribonuclease content in the pancreas of the mouse, gave for extrusion and resynthesis, data corroborating those of Daly and Mirsky (1952). Similarly, Fernandes and Junqueira (1955) obtained comparable results in the pancreas of the pigeon after stimulation with carbaminoylcholine injections.

According to Langer (1957), amylase in the pancreas of the mouse reaches a minimum about 2 hours after pilocarpine injection, after which the activity increases for about 10 hours after stimulation toward a maximum markedly higher than that observed in the animal after 24 hours of fasting.

The enzymological data of Krijgsman 1925, 1928) on the digestive glands of the snail and those of Hirsch and Jacobs (1930) on the digestive gland of the crayfish indicate that in these glands the periods of intracellular elaboration are, on the whole, comparable in duration to those of the exocrine pancreas and the salivary glands.

A third group of investigations made use of radioactive isotopes to study the rhythm of gland cell activity; the results have been extremely rewarding.

Junqueira et al. (1955) and Rothschild et al. (1957) showed that incorporation of amino acids (glycine, phenylalanine, alanine, histidine) labeled with radioactive carbon is very weak in pancreatic juice proteins during the first hour after injection of the labeled compound. It increases considerably between the second and fourth hours, the intervals necessary to reach maximum incorporation varying slightly with each amino acid.

Other investigations by Junqueira and his co-workers (quoted from Junqueira and Hirsch, 1956) provided more detailed information on the duration of each step: ingestion, intracellular elaboration, and extrusion. Radioactive glycine appears in the pancreatic juice 10 minutes after injection; this confirms the older data of Montgomery et al. (1940) on the speed of passage through the pancreatic cell of radioactive sodium. Radioactive phosphate, injected intravenously, can be detected in the pancreatic juice 3 minutes later; this demonstrates the rapidity of transit in the excretory duct system of the pancreas. From this, it may be assumed that the interval of 2 to 5 hours mentioned above for the duration of intracellular synthesis is probably fairly accurate.

Recent investigations by Junqueira and Rothschild (personal communication by Professor L. C. U. Junqueira) provided valuable additional information on the incorporation of labeled amino acids into the various constituents of the pancreatic cell. These authors experimented on rats receiving an injection of radioactive glycine. Under these conditions, the blood content in radioactive amino acids decreased very rapidly; 1 hour after injection, more than 80% of the radioactive glycine was incorporated into various proteins. The total radioactivity of the pancreas decreased only slightly during the 10 hours following injection, but study of the various fractions showed that the distribution of radioactive glycine among the various constituents of the homogenate varied considerably with time. The content in free radioactive glycine in the pancreatic parenchyma reached its maximum during the first hour, then, until the fourth hour, decreased fairly rapidly owing to the incorporation of this amino acid into proteins. Ten hours after injection, 25% of the maximum radioactivity remained in the amino acids of the pancreatic parenchyma. In the microsomes, radioactivity reached its peak before 1 hour, decreased first rapidly until the third hour, then more slowly, so that 10 hours later about 50% of the maximum radioactivity persisted. In the zymogen granules, radioactivity increased very gradually, the peak being reached about the tenth hour. As for radioactivity in the proteins of the pancreatic juice, it was very weak during the first hour after injection of radioactive glycine, then increased rapidly until the third hour, when it reached its maximum; the latter coincided chronologically with a marked decrease in radioactivity of the fractions corresponding to the microsomes and the free amino acids.

Thus, data obtained with very different techniques from various objects are in agreement that the work of the gland cell takes place rapidly; the longest period, that of intracellular synthesis of the secretion product, in the case of the best-known rhythmically active gland cells lasts only a few hours.

V. Relations between Secretory Granules Detectable by Morphological Techniques and Products of Cellular Activity

As previously mentioned, the presence in a given cell of a secretory product detectable by histological techniques is valuable in indicating the glandular nature of the cell. The name "secretion granules" is somewhat vague; it encompasses very different structures whose relations to the product of cellular activity are variable. There are three main possibilities.

In some gland cells, histological techniques enable the product of cellular activity itself to be detected and characterized more or less accurately. Preparations obtained with appropriate procedures give pre-

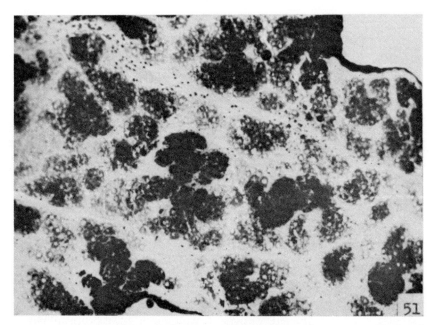

FIG. 51. Frozen section of the preputial gland of an albino mouse, fixed with formaldehyde-calcium and stained with Sudan black B and nuclear fast red. The lipid-containing secretion product shows up in black. Magnification: × 130; orange filter.

cise information on the functional condition of the gland cell under study. This is the case for the ink sac of *Sepia,* which allowed Goodsir to clarify the problem of secretion. Mucous glands of vertebrates are one of the best-known examples. Their secretory product can be chemically analyzed, and in many cases its composition was determined. Moreover, this product can be detected within the cell by various techniques, some of which have precise chemical significance. Thus, it is the presence of compounds identifiable by chemical analysis that enables these structures to be detected by histochemical techniques. Similarly, some constituents of *sebum* are detectable in the sebaceous glands by histochemical techniques, so much so that the functional stage reached by a given cell can be accurately determined by morphological study (Fig. 51). In some cases, identification of the secretory product depends on electron microscopy; e.g., observation of mammary gland cells shows granular constituents which are identifiable also in milk by electron microscopic examination (Bargmann and Knoop, 1959). Generally speaking, the "secretion granules" identifiable in gland cells of this first type retain after extrusion their staining properties, histochemical characteristics, and, quite often, even their shape.

In other cases, morphological techniques can detect in the gland cells compounds more or less directly involved in the metabolism of the secretory product, but not the product itself. For instance, histochemical study shows, in mammalian adrenocortical cells, an abundance of particulate lipids and provides accurate information on the functional state of the organ, but cannot detect the presence of the adrenocortical hormones themselves.

In yet other cases, histological techniques show, in the cytoplasm of the gland cell, "secretion granules" whose staining properties and histochemical characteristics bear no relation whatsoever to the chemical constitution and physiological properties of the products elaborated by the glands. Serous glands are a standard example, particularly certain digestive glands (parotid, chief cells of the gastric mucosa, exocrine pancreas) of higher vertebrates. The name "serous gland" was proposed by the nineteenth-century authors because of the physical characteristics of the secretory product which resembles serous fluid. It was gradually extended to the cells and even to the secretion granules, so much so that the term "serous granules" is frequently used in histology text books. The so-called serous granules do have some characteristics in common, e.g., regular rounded shape on stained sections, accumulation at the apical pole of the cells, affinity for acid dyes and iron hematoxylin lake. Actually, this artificial classification is justified only because it allows the simplification necessary to the teaching of elementary histology. In fact, although there is in the "serous granules" some similitude of shape and staining properties, there is none in the products elaborated by the various categories of so-called serous glands, and the histochemical differences between the various serous granules are considerable. Some of these grains contain substances detectable by the periodic acid-Schiff method, others do not (Junqueira *et al.*, 1951); and even their reactions to the various fixative agents are not identical. The artificiality of a distinction between "mucous" and "serous" gland cells is even better demonstrated by the production, in a same category of gland cells, of secretion alternately "mucous" and "serous." The salivary gland cells of the snail *Helix pomatia,* for instance, during their secretory cycle, pass through a "mucous" phase and a "serous phase" (Krijgsman, 1928). During postnatal histogenesis, the acinous glands of the submaxillary gland of the albino mouse undergo a "serous" phase followed by a "mucous" phase (Gabe, 1956). In some cases, because of its histological characteristics, a gland must be classified as "serous" although the product elaborated is obviously a mucoprotein. The shell gland of certain heteropods, genus *Pterotrachea* (Gabe, 1951), is one of the most typical examples of this. Its cells are similar to those of the exocrine pancreas in their morphological characteristics, they show accumulation, at the apical pole, of strongly eosinophilic and fuchsinophilic secretion granules, and their dictyosomes are grouped in a supranuclear "Golgi zone." Observation at the various stages of the secretory cycle show that the secretion granules are elaborated in exactly the same manner as the zymogen granules of the exocrine pancreas. And yet, the secretion product, released from the cell in particulate form, exhibits all the histochemical properties of mucoproteins, retains these properties in the excretory ducts of the glands, and takes part in the elaboration of the shell of the fertilized eggs.

In any case, in this third category of gland cells, no valid information on the physiological properties and chemical composition of the secretory product can be derived from the morphological characteristics and staining properties of the secretion granules. They seem merely to support the active substances elaborated at the same time, and it is only as such that their detection is of interest from the histophysiological viewpoint. In most cases, elaboration of secretion granules and of active products (enzymes, hormones, venom, etc.) takes place concurrently and both the supporting material and the active substances are extruded together. However, this parallelism, admitted in the early literature, is not absolute; in the case of prolonged, intense stimulation of the gland cells, elaboration of active products takes precedence over granule formation. Even during the normal secretory cycle, the elaboration of physiologically active products and that of their supporting granules is not completely synchronous. Investigations by Langer (1957) have shown that amylase is elaborated more rapidly by exocrine pancreatic cells than are the acidophilic granules with which the precursor of the enzyme is associated.

In some cases, histological techniques can detect, within the structures, either the active substances or granules. Thus, methods of detection of sulfhydryl-containing proteins show, in the mammalian neurohypophysis, a substance whose reactivity very likely corresponds to that of the so-called postpituitary hormones (Barrnett, 1954). Chromium hematoxylin and aldehyde-fuchsin stain the same structures, but this staining property is linked to the presence of a supporting protein whose solubility properties are not the same as those of the hormones (Schiebler, 1951; Acher, 1957). The same situation exists in the B cells of the endocrine pancreas (Barrnett et al., 1955), where the sulfhydryl reagents give the same picture as those obtained by aldehyde-fuchsin or chromium hematoxylin staining (Figs. 52 and 53). The compounds detectable by both techniques follow the same course during various physiological conditions. And yet, study and comparison of the solubility properties with in vitro observations lead to the conclusion that the reagents of protein sulfhydryl indicate the presence of insulin itself, whereas aldehyde-fuchsin or chromium hema-

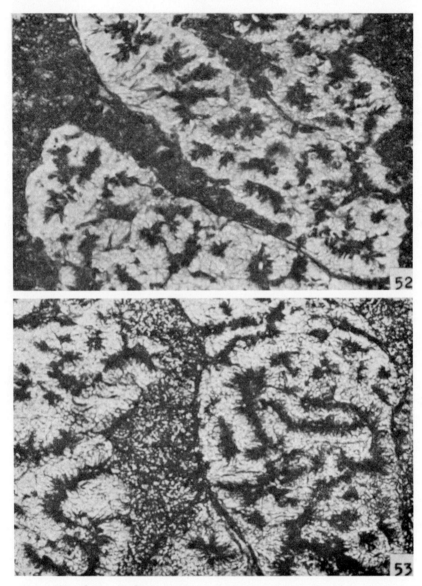

FIGS. 52 and 53.   Endocrine pancreatic islets in *Vipera berus*, fixed with Halmi's fluid and stained with aldehyde-fuchsin (52), with alkaline neotetrazolium (Gomori) (53) for the histochemical detection of protein-bound sulfhydryls. The B cells of the islets are stained in both cases. Magnification: ×200; green filter.

toxylin coloration indicate the presence of the supporting elements alone since insulin itself does not react with these dyes.

To sum up, in some gland cells, it is possible to detect the product of intracellular syntheses and determine the exact stage of functional evolution of the cell by means of morphological techniques. In other cases, the same techniques show, in the cells, products related to the active substances, but not these substances themselves. In yet other cases, they disclose supporting elements elaborated at the same time as the active products but having no direct relationship with them. The parallelism of these two processes gives unquestionable value to the study of this third type of secretion granule as a morphological criterion of cellular activity (Figs. 54–58), but no information as to chemical composition or functional significance of the products associated with the secretion granules can be derived from the morphological or histochemical characteristics of these granules.

VI. Division and Endomitosis in Gland Cells

In gland cells, mitosis follows the same fundamental patterns as in other somatic cells and its description will be found in the appropriate chapter of this treatise. However, the part taken by mitotic and endomitotic phenomena in the functional activity of glands and gland cells will be briefly reviewed here.

Mitotic images are frequently seen in the growing glands of young animals; in adult animals they are present in all glands during regeneration processes, after partial ablation, and during hyperactivity. Under normal conditions they are usually rare in merocrine glands. As Altmann (1952) pointed out in a study of exocrine pancreatic cells, mitosis usually occurs at a definite stage of the secretory cycle.

Cellular multiplication phenomena are obviously essential for the activity of holocrine glands. There is always, in these organs, a reserve of replacement cells taking no part in the secretory process, and whose multiplication insures the replacement of used cells. In continuously active holocrine glands, the number of mitoses occurring in a given gland varies only when the rhythmic pattern of activity changes, e.g., in sebaceous glands. In rhythmically active holocrine glands, the number of mitoses may vary widely according to the phase of the secretory cycle of the gland itself. Investigations by Hirsch and Jacobs (1928, 1930) have shown that the digestive gland of the crayfish (*Astacus leptodactylus*) follows the rhythmic pattern and that the sequence in the secretory cycle after feeding is different in the spring and in the fall. Study of the closed ends of the glandular tubes, where are situated the replacement cells

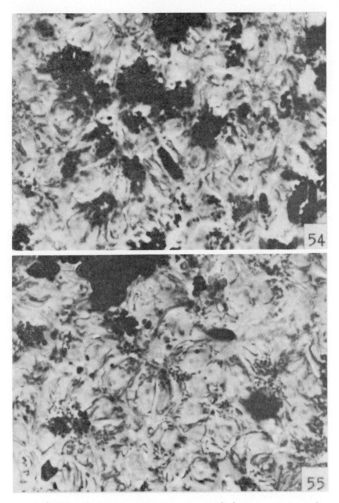

FIGS. 54 and 55.   Exocrine pancreatic acini of the guinea pig after 24 hours
starvation (54), and 120 minutes after feeding (55). Fixation with Regaud's fluid,
staining with Altmann's fuchsin and methyl green picrate. Note the secretion granules
located against the apical pole of the cells in the pancreas of the starved animal (54)
and their reduction after feeding (55). Magnification: × 1500; green filter.

that take no part in the elaboration of the secretion products, shows that
the number of mitoses increases markedly during the secretory cycle, each
wave of mitoses being followed by a corresponding increase in number
of the cell categories responsible for the synthesis of digestive enzymes.
Hirsch and Jacobs concluded that the periodicity of the mitoses is the
prime mover of the rhythmical activity of this digestive gland.

Endomitotic phenomena have been described in many gland cells, and Geitler (1953) published a well-documented survey of this subject; most of the examples in it refer to invertebrate gland cells. Thus, in the salivary glands of Heteroptera, the level of endopolyploidy is high (Geitler, 1937, 1938, 1939). Endomitosis with endopolyploidy (Lesperon, 1937; Geitler, 1940; Barigozzi, 1947) results in morphological peculiarities, especially the large size of the nucleus and the particular structure of its chromatin, in the silk glands of Lepidoptera. Endomitosis occurs also in the digestive gland cells of the isopod crustacean *Anilocra physodes* (Montalenti, 1940) as well as in its salivary glands (Siniscalco, 1952) (Figs. 59 and 60). The polytenic chromosomes in the salivary glands of Diptera larvae have been the subject of many investigations. Painter (1945) emphasized the relation in these organs between secretory activity and polyploidy. According to White (1948), in Itonididae (syn. Cecidomyiidae) endomytosis may result in either endopolyploidy or polyteny within the same salivary gland.

Endomitoses are not less frequent in vertebrate gland cells. In the rat most of the liver cells are tetraploid (Biesele *et al.*, 1942). Polyploidy was pointed out and its functional significance discussed by Schreiber *et al.* (1955) in exocrine pancreatic cells and acinous cells of the submaxillary gland of the rat.

Thus, recent investigations suggest that, in some tissues or organs, there is a relation between activity and endomitosis, but it is difficult to give a functional interpretation of this relation. Geitler (1953) pointed out that, in all probability, endomitosis creates a lesser functional disturbance than mitosis; this would explain the functional significance of endomitosis in especially active gland cells. In illustration of his point of view, he mentions the mesenteron of Heteroptera, in which the cells of the regeneration crypts are diploid, whereas in the functional cells the polyploidy may be as high as 16. A consequence of endomitosis, namely an increase in the number of nucleolar organizers, was emphasized by Schreiber *et al.* (1955), who also mentioned the essential role played by the nucleolus in intracellular synthesis of proteins. According to them, endopolyploidy or polyteny represents one of the two possible ways of increasing, in the cell, the quantity of material directly concerned with intracellular synthesis of proteins; the other way would be the multiplication of nucleolar organizers without concomitant multiplication of the whole genome.

VII. Conclusion

Gland cells, isolated or arranged into glands, are found in all Metazoa. With respect to cellular physiology, their essential characteristic is the

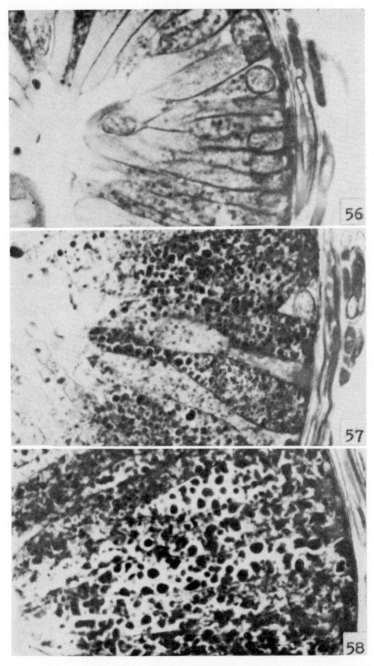

FIGS. 56, 57, and 58.   Epithelial cells of the pelvic gland in *Amblystoma mexi-canum* a short time after extrusion of the secretion product (56), during the intra-cellular synthesis (57), and a short time before extrusion (58). Fixation with Bouin's fluid, staining with azan. Note the size differences of cells and the differences of amount of secretion granules. Magnification: ×800; green filter.

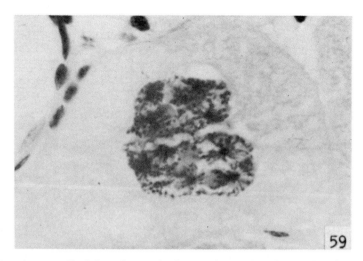

FIG. 59.   A cell of the salivary gland in *Anilocra physodes*, fixed with Regaud's fluid and stained with Feulgen-Rossenbeck's nucleal reaction and picro-indigo carmine. Note the large endomitotic nucleus. Magnification: × 800; green filter.

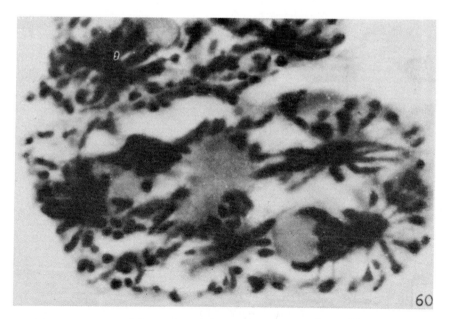

FIG. 60.   A part of the nucleus shown in Fig. 59. Note the starlike arrangement of the chromosome sets and the position of chromocenters (black) and nucleoli (gray). Magnification: × 2000.

predominance over all other functions of the intracellular elaboration of various products which are subsequently released into the environment. The general morphological characteristics of the gland cell are difficult to define because of the infinite variety of the products elaborated and the diversity of the corresponding structures. All the fundamental organelles of animal cells have been found in gland cells, and all, in turn, have been considered as the principal agents of intracellular syntheses.

Actually, recent investigations point more and more to the participation of all organelles in the intracellular elaboration of the secretory product. The instances of intranuclear elaboration are rare, and questionable, but there is no doubt that the nucleus plays a part in the regulation of protein synthesis taking place in the cytoplasm. Biochemical and electron microscopic investigations have clearly confirmed the older concept attributing to the ergastoplasm an active part in intracellular syntheses. The mitochondrial origin of secretory products is no longer accepted, but the essential role of the chondriome in the energy metabolism of cells in general, and gland cells in particular, is firmly established. As for the Golgi bodies, their existence and the unity of their infra-microscopic structures in all animals has been demonstrated by electron microscopy; moreover, the electron microscopic data tend to indicate that the Golgi bodies play a role in the concentration and accumulation, in condensed form, of the secretion products elaborated by the other organelles.

The concept of gland cell originated at a time when the degree of specialization in research had not yet reached its present level and the normal procedure in the study of a biological problem included the use of techniques borrowed from other disciplines. The gradually increased complexity of the techniques and the considerable growth of information have resulted in a very regrettable misunderstanding between morphologists, physiologists, and biochemists who are studying by means of different techniques the various aspects of the secretion phenomenon. It is through the clearing up of this misunderstanding that the advances of the last twenty years were made possible, and this demonstrates the value of the close cooperation of workers in various disciplines in elucidating biological problems.

References

Acher, R. (1957). *Proc. 2nd Intern. Symposium on Neurosecretion, Lund, 1957*, p. 71.
Allara, E. (1952). *Arch. ital. anat. e embriol.* **57**, 186.
Allard, C., Mathieu, R., de Lamirande, G., and Cantero, A. (1952). *Cancer Research* **12**, 407.
Allfrey, V. G., Daly, M. M., and Mirsky, A. E. (1953). *J. Gen. Physiol.* **37**, 1957.
Allfrey, V. G., Daly, M. M., and Mirsky, A. E. (1955). *J. Gen. Physiol.* **38**, 415.

Altmann, H. W. (1952). Z. Zellforsch. u. mikroskop. Anat. 58, 632.

Altmann, R. (1890, 1894). "Die Elementaroganismen und ihre Beziehungen zu den Zellen." Veit, Leipzig.

Ancel, P., and Bouin, P. (1904). Compt. rend. soc. biol. 56, 83, 281.

Babkin, B. P. (1944). "Secretory Mechanism of the Digestive Glands." Hoeber, New York.

Bahr, G. F., and Bermann, W. (1954). Exptl. Cell Research 6, 519.

Baker, J. R. (1951). J. Roy. Microscop. Soc. 71, 94.

Ball, E. G. (1930). J. Biol. Chem. 86, 433.

Barcroft, J., and Starling, E. H. (1904). J. Physiol. (London) 31, 491.

Bargmann, W. (1954). "Das Zwischenhirn-Hypophysensystem." Springer, Berlin.

Bargmann, W., and Knoop, A. (1959). Z. Zellforsch. u. mikroskop. Anat. 49, 344.

Barigozzi, C. (1947). Arch. ital. anat. e embriol. 52, 83.

Barrnett, R. J. (1954). Endocrinology 55, 484.

Barrnett, R. J., Marshall, R. B., and Seligman, A. M. (1955). Endocrinology 57, 419.

Benda, C. (1899). Verhandl. Physiol. Ges. Berlin; cited by Hoven (1912).

Bennett, H. S. (1956). J. Biophys. Biochem. Cytol. 2, Suppl. 99.

Benoit, J. (1926). Arch. Anat. Strasbourg, 5, 173.

Bensley, R. R. (1898). Quart. J. Microscop. Sci. 41, 361.

Bensley, R. R. (1951). Exptl. Cell Research 2, 1.

Bensley, R. R., and Hoerr, N. (1934). Anat. Record 60, 449.

Bergonzi, M., and Bolcato, B. (1930). Arch. sci. biol. (Italy) 14, 573.

Bern, H. A., and Alfert, M. (1954). Rev. brasil. biol. 14, 25.

Bernard, C. (1856). Compt. rend. acad. sci. Suppl. 1, 379.

Bernhard, W., and Rouiller, C. (1956). J. Biophys. Biochem. Cytol. 2, Suppl. 73.

Bernhard, W., Haguenau, F., Gautier, A., and Oberling, C. (1951). Compt. rend. soc. biol. 145, 566.

Bernhard, W., Haguenau, F., Gautier, A., and Oberling, C. (1952a). Z. Zellforsch. u. mikroskop. Anat. 37, 281.

Bernhard, W., Oberling, C., and Haguenau, F. (1952b). Experientia, 8, 58.

Bernhard, W., Bauer, A., Grepp, A., Haguenau, F., and Oberling, C. (1955). Exptl. Cell Research 9, 88.

Biesele, J. J., Poyner, M., and Painter, T. S. (1942). Univ. Texas Publ. No. 4243; cited by Schreiber et al. (1955).

Bouin, P. (1905). Compt. rend. soc. biol. 57, 916.

Bowen, R. H. (1929). Quart. Rev. Biol. 4, 291, 484.

Brachet, J. (1940). Compt. rend. soc. biol. 133, 88.

Brachet, J. (1942). Arch. biol. (Liège) 51, 151.

Brachet, J. (1957). "Biochemical Cytology." Academic Press, New York.

Brachet, J., and Ficq, A. (1956). Arch. biol. (Liège) 67, 431.

Brachet, J., Jeener, R., Rosseels, M., and Thonet, L. (1946).Bull. soc. chim. biol. 28, 460.

Breuer, M. E., and Pavan, C. (1955). Chromosoma 7, 371.

Brock, N., Druckrey, H., and Herkin, H. (1939). Arch. exptl. Pathol. Pharmakol. Naunyn-Schmiedeberg's 191, 687.

Callan, H. G. (1955). In "Symposium on the Fine Structure of Cells: Held at the 8th Congress of Cell Biology, Leiden, Holland, 1954." Interscience, New York.

Callan, H. G., and Tomlin, S. G. (1950). Proc. Roy. Soc. London B137, 365.

Camis, M. (1924). Arch. internat. physiol. 22, 343.

Campbell, R. M., and Kosterlitz, H. W. (1952). Biochim. et Biophys. Acta 8, 664.

Carnoy, J. B. (1884). "La Biologie cellulaire." Lierre, Louvain.

Caspersson, T. (1950). "Cell Growth and Cell Function." Norton, New York.

Caspersson, T., and Schultz, J. (1939). *Nature* **143**, 602, 609.

Caspersson, T., Landstroem-Hyden, H., and Aquilonius, L. (1941). *Chromosoma* **2**, 111.

Caspersson, T., Jacobson, F., Lomakka, G., Svensson, G., and Safström, R. (1953). *Exptl. Cell Research* **5**, 560.

Celestino da Costa, A., Geraldes-Barba, F., and Vasconcelos Frazao, J. (1949). *Compt. rend. Assoc. anat.* **36**, 1.

Chèvremont, M. (1956). "Notions de cytologie et Histologie." Desoer, Liège.

Collin, R., and Florentin, P. (1930). *Compt. rend. soc. biol.* **104**, 1279.

Covell, W. P. (1928). *Anat. Record* **40**, 213.

Cowdrey, E. V. (1918). *Contribs. Embryol. Carnegie* **8**, 39.

Dabelow, A. (1958). *In* "Handbuch der mikroskopischen Anatomie des Menschen" (W. Bargmann, ed.), Vol. III, p. 3. Springer, Berlin.

Dalton, A. J. (1951). *Am. J. Anat.* **89**, 109.

Dalton, A. J. (1952). *Z. Zellforsch. u. mikroskop. Anat.* **36**, 522.

Dalton, A. J., and Felix, M. D. (1953). *J. Appl. Phys.* **24**, 1414.

Dalton, A. J., and Felix, M. D. (1954). *Am. J. Anat.* **94**, 171.

Dalton, A. J., Kahler, H., Striebich, M. J., and Lloyd, B. J., Jr. (1950). *J. Natl. Cancer Inst.* **11**, 439.

Daly, M. M., and Mirsky, A. E. (1952). *J. Gen. Physiol.* **36**, 243.

Daly, M. M., Allfrey, V. G., and Mirsky, A. E. (1953). *J. Gen. Physiol.* **36**, 173.

Danielli, J. F. (1952). *Symposia Soc. Exptl. Biol.* **6**, 1.

Danielli, J. F. (1954). *Proc. Symposium Colston Research Soc.* **7**, 1.

Davidson, J. N. (1947). *Cold Spring Harbor Symposia Quant. Biol.* **12**, 50.

Davies, R. E., Harper, A. A., and Mackay, I. F. S. (1949). *Am. J. Physiol.* **157**, 278.

Dawson, A. B. (1935). Cited by Bargmann and Knoop (1959).

De Deken Grenson, M. (1952). *Biochim. et Biophys. Acta* **8**, 481.

de Duve, C., and Berthet, P. (1954). *Intern. Rev. Cytol.* **3**, 225.

de Duve, C., Pressman, B. C., Gianetto, R. J., Wattiaux, R., and Appelmans, F., (1955). *Biochem. J.* **60**, 604.

del Rio Hortego, P. (1917). Cited by Lesperon (1937).

Dempsey, E. W., and Peterson, R. R. (1955). *Endocrinology* **56**, 46.

Denues, A. R. T., and Motram, F. C. (1955). *J. Biophys. Biochem. Cytol.* **1**, 185.

De Robertis, E. (1940). *Anat. Record* **78**, 473.

De Robertis, E. (1941a). *Anat. Record* **79**, 417.

De Robertis, E. (1941b). *Am. J. Anat.* **68**, 317.

De Robertis, E. (1954). *J. Histochem. and Cytochem.* **2**, 341.

De Robertis, E., and Raffo, H. F. (1957). *Exptl. Cell Research* **12**, 66.

Dervichian, D. G. (1954). *In* "Biocytologia." Masson, Paris.

Deutsch, W., and Raper, H. S. (1938). *J. Physiol. (London)* **92**, 439.

Dittus, P. (1940). *Z. wiss. Zool.* **154**, 40.

Drochmans, P. (1950). *Arch. biol. (Liège)* **61**, 475.

Estable, L., and Sotelo, J. R. (1951). *Inst. invest. cienc. biol. (Montevideo) Publ.* **1**, 105.

Estable, L., and Sotelo, J. R. (1955). *In* "Symposium on the Fine Structure of Cells: Held at the 8th Congress of Cell Biology, Leiden, Holland." Interscience, New York.

Farquhar, M. G., and Rinehart, J. F. (1954). *Endocrinology* **55**, 857.

Farquhar, M. G., and Wellings, S. R. (1957). *J. Biophys. Biochem. Cytol.* **3**, 319.

Fawcett, D. W. (1955). *J. Natl. Cancer Inst.* **15**, Suppl., 1475.

Fernandes, J. F., and Junqueira, L. C. U. (1953). *Exptl. Cell Research* **5**, 329.

Fernandes, J. F., and Junqueira, L. C. U. (1955). *Arch. Biochem. Biophys.* **55**, 54.

Ferrari, R., and Höber, R. (1933). *Arch. ges. Physiol. Pflüger's* **232**, 299.

Ficq, A. (1955a). *Exptl. Cell Research* **9**, 286.

Ficq, A. (1955b). *Arch. biol. (Liège)* **66**, 509.

Ficq, A., and Errera, M. (1955). *Biochim. et Biophys. Acta* **16**, 45.

Frey-Wyssling, A. (1945). "Ernährung und Stoffwechsel der Pflanzen." Gutenberg, Zürich.

Gabe, M. (1951). *Cellule* **54**, 7.

·Gabe, M. (1952). *Cellule* **54**, 363.

Gabe, M. (1955). *Compt. rend. soc. biol.* **149**, 223.

Gabe, M. (1956). *Z. Zellforsch. u. mikroskop. Anat.* **45**, 74.

Gabe, M. (1958). *Compt. rend. acad. sci.* **247**, 1907.

Gabe, M., and Prenant, M. (1948). *Cellule* **52**, 17.

Garnier, C. (1899). *J. Anat. Paris* **35**, 22.

Gatenby, J. B. (1931). *Am. J. Anat.* **48**, 7.

Gatenby, J. B. (1953). *J. Roy. Microscop. Soc.* **73**, 67.

Gay, H. (1956). *J. Biophys. Biochem. Cytol.* **2**, Suppl. 407.

Geitler, L. (1937). *Z. Zellforsch. u. mikroskop. Anat.* **26**, 641.

Geitler, L. (1938). *Biol. Zentr.* **58**, 152.

Geitler, L. (1939). *Chromosoma* **1**, 1.

Geitler, L. (1940). *Naturwissenschaften* **28**, 241.

Geitler, L. (1953). *In* "Protoplasmatologia: Handbuch der Protoplasmaforschung" (L. V. Heilbrunn and F. Weber, eds.), Vol. VI/c. Springer, Wien.

Gersh, I., and Caspersson, T. (1940). *Anat. Record* **78**, 303.

Grassé, P. P., Carasso, N., and Favard, P. (1956). *Ann. sci. nat. Zool. et biol. animale* [11] **18**, 339.

Gresson, R. A. R. (1952). *Cellule* **54**, 399.

Grynfeltt, M. J. (1937). *Arch. anat. microscop.* **33**, 177, 209.

Guberniev, M. A., and Il'Ina, L. T. (1950). *Doklady Akad. Nauk. S.S.S.R.* **71**, 351; cited by Junqueira and Hirsch (1956).

Guyiesse-Pélissier, A. (1923). *Compt. rend. Assoc. anat.* **18**, 243.

Haguenau, F. (1958). *Intern. Rev. Cytol.* **7**, 425.

Haguenau, F., and Bernhard, W. (1955). *Arch. anat. microscop.* **44**, 27.

Haguenau, F., and Lacour, F. (1955). *In* "Symposium on the Fine Structure of Cells: Held at the 8th Congress of Cell Biology, Leiden, Holland, 1954." Interscience, New York.

Harris, E. J. (1956). "Transport and Accumulation in Biological Systems." Butterworths, London.

Heidenhain, R. (1875). *Pflüger's Arch. ges. Physiol.* **10**, 557.

Heidenhain, R. (1880). *In* "Handbuch der Physiologie" (K. Hermann, ed.), p. 173. Vogel, Leipzig.

Heitz, J. (1933). *Z. Zellforsch. u. mikroskop. Anat.* **26**, 239, 473.

Hendler, R. W., Dalton, A. J., and Glenner, G. G. (1957). *J. Biophys. Biochem. Cytol.* **3**, 325.

Herlant, M. (1956). *Arch. biol. (Liège)* **67**, 89.

Himwich, H. E., and Adams, M. A. (1930). *Am. J. Physiol.* **93**, 568.

Hirsch, G. C. (1915). *Zool. Jahrb. Physiol.* **35**, 357.

Hirsch, G. C. (1929). *Wilhelm Roux' Arch. Entwicklungsmech.* **117**, 511.

Hirsch, G. C. (1931).*Biol. Revs. Biol. Proc. Cambridge Phil. Soc.* **6**, 88.

Hirsch, G. C. (1932a). *Z. Zellforsch. u. mikroskop. Anat.* **15**, 290.

Hirsch, G. C. (1932b). *Z. Zellforsch. u. mikroskop. Anat.* **15**, 36.

Hirsch, G. C. (1939). "Form-und Stoffwechsel der Golgi-Körper." Bornträger, Berlin.

Hirsch, G. C. (1955). *In* "Handbuch der allgemeinen Pathologie," Vol. 2. Springer, Berlin.

Hirsch, G. C. (1958). *Naturwissenschaften* **45**, 349.

Hirsch, G. C., and Jacobs, W. (1928). *Z. vergleich. Physiol.* **8**, 102.

Hirsch, G. C., and Jacobs, W. (1930). *Z. vergleich. Physiol.* **12**, 524.

Hirsch, G. C., Junqueira, L. C. U., Rothschild, H. A., and Dohi, S. R. (1956). *Arch. ges. Physiol. Pflüger's* **264**, 78.

Hogeboom, G. H., and Schneider, W. C. (1955). *In* "The Nucleic Acids" (E. Chargaff and J. N. Davidson, eds.), Vol. 2. Academic Press, New York.

Hokin, L. E., and Hokin, M. R. (1954). *Biochim. et Biophys. Acta* **13**, 236, 401.

Holter, H. (1952). *Advances in Enzymol.* **13**, 1.

Hoven, H. (1912). *Arch. Zellforsch.* **8**, 555.

Howe, A., Richardson, K. C., and Birbeck, M. S. C. (1955). *Exptl. Cell Research* **10**, 195.

Huber, P. (1949). *Z. Zellforsch. u. mikroskop. Anat.* **34**, 428.

Hydén, H. (1943). *Acta Physiol. Scand.* **6**, Suppl. 17, 1.

Jacobs, W. (1928). *Z. Zellforsch. u. mikroskop. Anat.* **8**, 1.

Järvi, O. (1939). *Z. Zellforsch. u. mikroskop. Anat.* **30**, 156.

Jeener, R. (1948). *Biochim. et Biophys. Acta* **2**, 439.

Jeffers, K. R. (1935). *Am. J. Anat.* **56**, 257.

Junqueira, L. C. U. (1951). *Exptl. Cell Research* **2**, 327.

Junqueira, L. C. U. (1955). *Rev. Univ. Minas Gerais (Brasil)* **11**, Suppl., 46.

Junqueira, L. C. U., and Hirsch, G. C. (1956). *Intern. Rev. Cytol.* **5**, 323.

Junqueira, L. C. U., and Rabinovitch, M. (1954). *Texas Repts. Biol. and Med.* **12**, 94.

Junqueira, L. C. U., Sesso, A., and Nahas, L. (1951). *Bull. microscop. appl.* **1**, 133.

Junqueira, L. C. U., Hirsch, G. C., and Rothschild, H. A. (1955). *Biochem. J.* **61**, 275.

Kestner, O. (1930). *In* "Handbuch der normalen und Pathologischen Physiologie" Vol. 16, p. 1. Springer, Berlin.

Kracht, J. (1954). *Naturwissenschaften* **41**, 336.

Kracht, J. (1955). *Naturwissenschaften* **42**, 50.

Kracht, J. (1957). *Naturwissenschaften* **44**, 15.

Krijgsman, B. J. (1925). *Z. vergleich. Physiol.* **3**, 264.

Krijgsman, B. J. (1928). *Z. vergleich. Physiol.* **8**, 187.

Kühne, W., and Lea, A. S. (1882). *Untersuch. physiol. Inst. Univ. Heidelberg* **2**, 448.

Lacy, D. (1956). *J. Physiol.* (London) **127**, 26.

Lagerstedt, S. (1949). *Acta Anat.* **9**, Suppl. 7, 1.

Laguesse, E. (1894). *J. Anat.* (Paris) **30**, 591.

Laguesse, E. (1905). *Rev. gén. Histol.* **1**, 543.

Langer, H. (1957). *Z. vergleich. Physiol.* **39**, 241.

Langer, H., and Grassi, A. (1955). *Z. physiol. Chem.* **229**, 139.

Langstroth, G. O., McRae, D. R., and Stavraky, G. W. (1938). *Proc. Roy. Soc. London* **B125**, 335.

Langstroth, G. O., McRae, D. R., and Komarov, S. A. (1939). *Can. J. Research* **D17**, 137.

Lesher, S. (1951a). *Exptl. Cell Research* **2**, 557.

Lesher, S. (1951b). *Exptl. Cell Research* **2**, 586.

Lesperon, L. (1937). *Arch. zool. exptl. et gén.* **79**, 1.

Lettré, R. (1954). *In* "Symposium on the Fine Structure of Cells: Held at the 8th Congress of Cell Biology, Leiden, 1954." Interscience, New York.

Lewis, W. H. (1937). *Am. J. Cancer* **29**, 666.

Lindberg, O., and Ernster, L. (1954). *In* "Protoplasmatologia: Handbuch der Proto-plasmaforschung" (L. V. Heilbrunn and F. Weber, eds.), Vol. III/A/4. Springer, Wien.

Ludwig, O. (1851). *Z. rat. Med.* **1**, 271.

Macchi, G. (1951). *Arch. ital. anat. e embriol.* **56,** 39.

Mandel, H. G., Jacob, M., and Mandel, L. (1950). *Bull. soc. chim. biol.* **32,** 80.

Maziarsky, S. (1910). *Arch. Zellforsch.* **4,** 443.

Metzner, J. (1891). *Arch. Anat. Entwicklungsgeschichte;* cited by Hoven (1912).

Michaelis, L. (1900). *Arch. mikroskop. Anat. u. Entwicklungsmech.* **60,** 558.

Millot, J. (1926). *Bull. biol. France et Belg.* Suppl. **8,** 1.

Mirsky, A. E., Allfrey, V. G., and Daly, M. M. (1954). *J. Histochem. and Cytochem.* **2,** 376.

Montagna, W. (1956). "The Structure and Function of Skin." Academic Press, New York.

Montalenti, G. (1940). *Boll. soc. ital. biol. sper.* **15,** 1108.

Montgomery, M. L., Sheline, G. E., and Chaikoff, I. L. (1940). *Am. J. Physiol.* **131,** 578.

Montgomery, T. H. (1898). *J. Morphol.* **5,** 265.

Morelle, J. (1927). *Cellule* **37,** 178.

Nassonov, D. (1924). *Arch. mikroskop. Anat. u. Entwicklungsmech.* **100,** 433.

Niklas, A., and Oehlert, N. (1956). *Beitr. pathol. Anat. u. allgem. Pathol.* **116,** 92.

Nikolaj, P. (1939). *Boll. soc. ital. biol. sper.* **14,** 23.

Noll, A. (1905). *Ergeb. Physiol.* **4,** 84.

Northup, D. (1936). *Am. J. Physiol.* **114,** 46.

Oberling, C., and Bernhard, W. (1955). *In* "Biocytologia." Masson, Paris.

Oram, V. (1955). *Acta Anat.* **25,** Suppl. **23,** 7.

Ortmann, R. (1956). *Z. Anat. Entwicklungsgeschichte* **119,** 485.

Ortmann, R. (1958). *Z. mikroskop.-anat. Forsch.* **64,** 215.

Painter, T. S. (1945). *J. Exptl. Zool.* **100,** 523.

Palade, G. E. (1955). *J. Biophys. Biochem. Cytol.* **1,** 59.

Palade, G. E. (1956). *J. Biophys. Biochem. Cytol.* **2,** Suppl., 85.

Palay, S. L. (1958). *In* "Frontiers in Cytology." Yale Univ. Press, New Haven, Connecticut.

Parat, M. (1928). *Arch. anat. microscop.* **24,** 73.

Pavan, C., and Breuer, M. E. (1955). *Rev. Univ. Minas Gerais (Brasil)* **11,** Suppl., 90.

Pellegrino, B., Kaiserman Abramof, I.R., and Schreiber, G. (1955). *Rev. Univ. Minas Gerais (Brasil)* **11,** Suppl., 90.

Picard, D., and Stahl, J. (1956). *J. physiol. (Paris)* **48,** 73.

Policard, A. (1910). *Arch. anat microscop.* **12,** 177.

Pollister, A., and Ris, H. (1947). *Cold Spring Harbor Symposia Quant. Biol.* **12,** 147.

Porter, K. (1955). *Federation Proc.* **14,** 673.

Prenant, A. (1898, 1899). *J. Anat. (Paris)* **34,** 657; **35,** 52, 1700, 408, 618.

Prenant, A., Bouin, P., and Maillard, L. (1904). "Traité d'Histologie," 1. Cytologie générale et spéciale. Reinwald, Paris.

Rabinovitch, M. (1954). *Proc. Soc. Exptl. Biol. Med.* **85,** 685.

Rabinovitch, M., Rothschild, H. A., and Junqueira, L. C. U. (1951). *Science* **114,** 551.

Rabinovitch, M., Rothschild, H. A., and Junqueira, L. C. U. (1952a). *J. Biol. Chem.* **194,** 835.

Rabinovitch, M., Valeri, V., Rothschild, H. A., Camara, S., Sesso, A., and Junqueira, L. C. U. (1952b). *J. Biol. Chem.* **198,** 815.

Regaud, C. (1908). *Compt. rend. soc. biol.* **64,** 1145.

Regaud, C., and Mawas, J. (1909). *Compt. rend. soc. biol.* **66,** 461.

Richardson, K. C. (1947). Cited by Dabelow (1957).

Ries, E. (1935). *Z. Zellforsch. u. mikroskop. Anat.* **22,** 523.

Rinehart, J. F., and Farquhar, M. G. (1953). *J. Histochem. and Cytochem.* **1,** 93.

Rohen, J. (1956). *Naturwissenschaften* **43**, 451.

Roskin, G. (1925). *Z. Zellforsch. u. mikroskop. Anat.* **3**, 99.

Rothschild, H., and Junqueira, L. C. U. (1951). *Arch. Biochem. Biophys.* **34**, 453.

Sandritter, W., and Hübotter, F. (1954). *Frankfurt. Z. Pathol.* **65**, 219.

Schaffer, J. (1927). *In* "Handbuch der mikroskopischen Anatomie des Menschen" (W. von Möllendorff, ed.), Vol. 2, p. 1. Springer, Berlin.

Scharrer, B. (1941). *Biol. Bull.* **81**, 96.

Scharrer, E., and Scharrer, B. (1954). *In* "Handbuch der mikroskopischen Anatomie des Menschen" (W. Bargmann, ed.), Vol. 6, p. 5. Springer, Berlin.

Schiebler, T. H. (1951). *Acta Anat.* **13**, 233.

Schreiber, G. (1949). *Biol. Bull.* **97**, 187.

Schreiber, G., Melucci, N., Kaiserman Abramof, I. R., and Pompeu Memoria, J. M. (1955). *Rev. Univ. Minas Gerais (Brasil)* **11**, Suppl., 100.

Siniscalco, M. (1952). *Caryologia* **4**, 1.

Sjöstrand, F. S., and Hanzon, V. (1954a). *Exptl. Cell Research* **7**, 393.

Sjöstrand, F. S., and Hanzon, V. (1954b). *Exptl. Cell Research* **7**, 415.

Sluiter, J. W. (1944). *Z. Zellforsch. u. mikroskop. Anat.* **33**, 87.

Solger, B. (1894). *Anat. Anz.* **9**, 415.

Solomon, A. K. (1952). *Federation Proc.* **11**, 722.

Stenram, N. (1953). *Acta Anat.* **22**, 277.

Still, E. U., Bennett, A. L., and Scott, V. B. (1933). *Am. J. Physiol.* **106**, 509.

Stowell, R. E. (1949). *Cancer* **2**, 121.

Taylor, J. H. (1953). *Science* **118**, 555.

Taylor, J. H., and McMaster, R. D. (1955). *Genetics* **40**, 600.

Tramezzani, J. H., and Valeri, V. (1955). *Rev. Univ. Minas Gerais (Brasil)* **11**, Suppl., 68.

Van Weel, P. B., and Engel, C. (1938). *Z. vergleich. Physiol.* **23**, 214.

Verne, J. (1956). "Précis d'Histologie." Masson, Paris.

Vincent, W. S. (1955). *Intern. Rev. Cytol.* **4**, 269.

Visscher, M. B. (1942). *Federation Proc.* **1**, 246.

Voïnov, D. (1934). *Arch. zool. exptl. et gén.* **76**, 399.

Watson, M. L. (1954a). *Biochim. et Biophys. Acta* **15**, 475.

Watson, M. L. (1954b). *J. Biophys. Biochem. Cytol.* **1**, 257.

Weatherford, A. L. (1929). *Am. J. Anat.* **44**, 199.

Weiss, J. M. (1953). *J. Exptl. Med.* **98**, 607.

White, M. J. D. (1948). *J. Morphol.* **82**, 53.

Wissig, H. (1956). Cited by Haguenau (1958).

Worley, L. G. (1946). *Ann. N.Y. Acad. Sci.* **47**, 1.

Yokoyama, H. O., and Stowell, R. E. (1951). *J. Natl. Cancer Inst.* **11**, 939.

Zeiger, K. (1955). *In* "Handbuch der allgemeinen Pathologie" Vol. 2, p. 1. Springer, Berlin.

Zoja, L., and Zoja, R. (1891). *Mem. reale. ist. Lombardo, Classe lett. sci. (Italy)* **16**, 127.

CHAPTER 2

Kidney Cells

By ROY P. FORSTER

I. Introduction

Vertebrate kidneys, by virtue of their intimate association with the circulatory system, play an important role in keeping constant the internal environment of these complex organisms. This section emphasizes specialized cytological features of renal tubule epithelium which underlie selective transfer processes involved in this regulation of the chemical composition of body fluids. No attempt will be made to deal comprehensively with over-all functions of the kidney as an organ, nor with studies concerned especially with nervous regulation, hemodynamics, or disease. Neither will studies on the myriad excretory devices among invertebrates be considered unless they reveal cytological phenomena of general significance.

With the recognition that form must underlie function, cell structure will be considered first, starting with a description of simple aglomerular nephrons and then proceeding to more specialized tubules which are differentiated regionally and associated with filtering devices. Observations based on cytochemistry and ultrastructure techniques will be examined, as well as studies concerned with alterations in the structure of renal tubules provided by the diversity of nature or occurring during embryological development, senility, and in the course of certain experimentally imposed conditions, to the extent that these have contributed to an understanding of renal function. Special techniques are outlined which apply especially to renal studies, and a brief summary is made of the kidney's chief regulatory functions. Finally, certain active transport systems are considered insofar as they serve to characterize active transport as a general cell process.

Since several thousand papers on renal studies are published each year, references cannot be inclusive. It is hoped, however, that those selected will be helpful as an introduction to the literature.

II. Structure of the Nephron Unit

A. Aglomerular Kidneys

Closed renal tubules, frequently undifferentiated regionally, have been extensively studied, especially in kidneys of certain fishes and insects, because here tubular excretory processes can be observed without the complication of oppositely oriented simultaneous resorptive activities which take place in tubules associated with filtering devices.

1. Invertebrates

It was pointed out by Goodrich (1945) that two ontogenetically distinct excretory organs of independent embryological origin in various invertebrates erroneously are frequently referred to collectively as "nephridia." The true nephridium is primitively excretory in function, is developed centripetally and independent of the coelom. The coelomoduct is primitively reproductive in function, is developed centrifugally as an outgrowth from the coelomic epithelium or wall of the genital follicle, and has a lumen which arises as an extension of the coelomic cavity. These two separate sets of organs can be traced throughout the invertebrates from Platyhelminthes up to the vertebrates. Sometimes in the adult the nephridium is suppressed and excretion carried on entirely by the coelomoduct; sometimes these two originally separate organs are combined into a single compound organ of mixed excretory and genital function (nephromixium) (Fig. 1). This presents a striking example of functional convergence, where morphologically different tubules are pressed into the performance of similar excretory and regulatory functions.

In general, blind-ended tubules, such as the Malpighian tubules of insects, are shorter and structurally simpler than the open-ended tubules of such forms as the annelid worms where coelomic fluid enters the open nephrostome and is modified by both secretory and resorptive processes as it passes along the tubules. The fine anatomy of the excretory system of the stick insect, *Dixippus morosus*, has been studied in detail as a representative of "aglomerular" invertebrate tubules. It is worthy of attention because extensive observations have been made on renal function in the system (Ramsay, 1953a, 1954, 1955a, b) and because of the ease with which these tubules can be studied *in vitro* as isolated tubule preparations. Other invertebrate tubules studied recently, for both structure and function, are *Rhodnius* (Ramsay, 1952), the mosquito larva (Ramsay, 1950, 1951, 1953b), and the earthworm (Ramsay, 1949). The tremendous diversity of renal structures among invertebrates precludes in this chapter any possibiilty of making an inclusive review of these organs. Renal function studies in invertebrates other than the insects have recently been reviewed by Martin (1957).

In the stick insect, as is generally the case in primitive insects, the blind ends of the excretory tubules lie free in the body cavity, where they are surrounded by body fluid. The tubules are attached to and drain into the midgut or at an annulus marking the junction of midgut and hindgut (intestine). Urine draining from the tubules passes into the gut where many of the secreted substances are resorbed, particularly in the rectum. The Malpighian tubules can be recognized as of three types

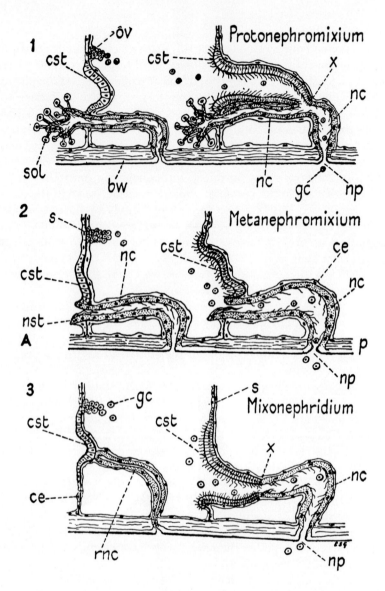

Fig. 1. Formation of three kinds of invertebrate nephromixia by combinations of coelomoduct and nephridium elements. A, anterior early stage before combination; p, posterior, combination completed; bw, body wall; ce, coelomic epithelium; cst, coelomostome; gc, germ cell; nc, nephridial canal; np, nephridiopore; nst, nephridiostome; ov, ovary; rnc, rudiment of nephridium; s, intersegmental septum; sol, solenocyte; x, point where coelomoduct opens into nephridial canal. From Goodrich (1945).

(superior, inferior, and appendices of the midgut) according to their attachments and their position relative to the gut. The cells of superior tubules are binucleate and provided with a well-developed brush border of the *bürstensaum* type (Wigglesworth, 1931). The tubules taper from a diameter of $120\,\mu$ at the proximal end to $80\,\mu$ at the distal tip. The inferior tubules appear the same over their middle and proximal regions as the superior tubules. Distally, however, they dilate to diameters of 250–$300\,\mu$. A brush border is present throughout. The tubules comprising the appendices of the midgut also have a well-developed brush border, but are only $50\,\mu$ in diameter and uniform throughout their length.

The cytochemical and histochemical characteristics of the Malpighian tubule are delineated by Gagnepain (1956 [1957]) for the odonates. Here the tubules consist of five distinct segments: a distal tubule, a median segment, an intermediate section, a mucosal section, and a dilated collecting chamber common to three to six tubules.

2. *Vertebrates*

Among the various classes of vertebrates the nephron unit is a remarkably persistent structure which appears in some of the most primitive fishes in essentially the same degree of development as in mammals (Fig. 2). However, wide variations are encountered within the fishes. A glomerulus may or may not be present, the distal tubule is absent in some fishes, and the proximal convoluted tubule which is composed of two identifiable portions (Edwards, 1929, 1933, 1935), may be represented by only the second portion as in some marine and fresh-water aglomerular fishes. As Grafflin (1937) has pointed out, neither a glomerulus nor a distal convoluted segment is needed for adaptation to fresh water. The simplest teleostean nephron, consisting only of the second portion of a proximal convoluted segment, is capable of sustaining the excretory load imposed by life in either fresh or salt water.

The aglomerular tubules of such marine telecosts as *Lophius, Hippocampus, Opsanus,* etc., except for an intermediate segment possibly homologous to the collecting tubule, are composed entirely of one fundamental cell type (Defrise, 1932; Edwards, 1935; von Möllendorff, 1936; Gabe, 1957). The cells of the toadfish, *Opsanus tau,* for example, resemble in many respects those of the initial portion of the proximal convoluted segment of the mammalian tubule. The nuclei have irregular surfaces and vary in shape and position. The mitochondria are irregularly arranged and distributed throughout the cytoplasm. The apical or luminal surface of the cell as viewed with ordinary microscopy (Defrise, 1932) may be dome-shaped, straight, or cup-shaped. The brush border has the cytological features of brush-bordered epithelium in glomerular tubules. Vesicles

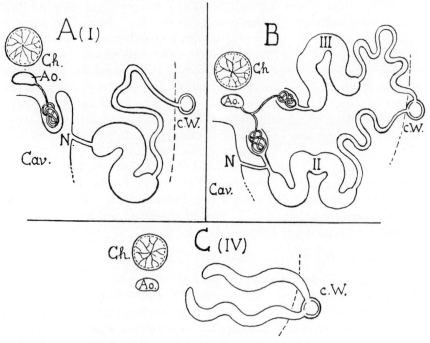

Fig. 2. Four types of vertebrate nephrons. In A(I) the glomus, supplied by a
branch of the aorta, projects into the body cavity opposite the nephridiostome, N.
A broad, brush-bordered segment leads into the narrower distal tubule which
terminates in the Wolffian duct or ureter, c.W. In B, characteristic of the mesonephros
of adult urodeles, the nephridiostome and its duct leading to the body cavity persists
in some tubules (II) and is absent in others (III), but the nephron is always supplied
by an encapsulated glomerulus. In C(IV), characteristic of aglomerular fishes, the
blood supply to the nephron unit is purely venous, and the distal segment has disappeared. From Gerard and Cordier (1934).

are observed in cells of the collecting ducts as well as in the apical part
of brush-border cells. Further structural details of brush-border cells will
be provided in subsequent sections dealing with regional specialization
and with studies relating form and function.

B. The Filtering Surface

The relationship of degrees of glomerular development to the need
for water excretion among various vertebrates was first pointed out by
Marshall and Smith (1930). Morphological studies (Nash, 1931) showed
that kidneys of teleosts living in hypertonic sea water have a smaller
total filtration area than those of fresh-water forms by virtue of both
glomerular size and number. The view that glomeruli are primitive filter-

ing devices pointing to the fresh-water ancestry of vertebrates has been substantiated by certain paleontological evidence which has been re-examined recently by Homer W. Smith (1953). On the other hand, Robertson (1957) has presented the case for a marine habitat as being the more likely environment of early vertebrates. Aside from evolutionary considerations, there is general agreement that glomeruli are effective overflow mechanisms for the elimination of water which enters the body along osmotic gradients conditioned by a hypotonic environment.

Fresh-water invertebrates too have various devices which act as water pumps. These include contractile vacuoles in certain Protozoa, flame cells, solenocytes, nephridiostomes, and a wide array of filtering devices among the Mollusca and Arthropoda (Goodrich, 1945; Martin, 1957). However, it must be pointed out that some marine invertebrates too have kidneys which operate on the general principle of filtration followed by resorption and secretion, with high rates of water turnover, although there is no osmotic gradient between blood and sea water (Webb, 1940; Robertson, 1953). These filtering devices of invertebrates are too diverse to permit detailed consideration here. An introduction to the literature concerning their role in osmotic and ionic regulation can be obtained from Beadle's comprehensive review (1957), which covers the period since the publication of Krogh's classic monograph, "Osmotic Regulation in Aquatic Animals," published in 1939.

The lobulated tuft of capillaries which constitutes the vertebrate glomerulus functions to separate an essentially protein-free filtrate from blood under a head of arterial hydrostatic pressure. The filtrate is received by a thin-walled expansion of the renal tubule which encapsulates the glomerulus. As viewed by ordinary light magnifications, Bowman's capsule consists of flat epithelial cells whose nuclei protrude prominently into the space which exists between the capsule and the filtering surface. The external surface of the capsule is smooth. The following outline of the ultrastructure of the filtering surface is taken largely from the studies of Hall (1953, 1954), who expressed indebtedness to Palade (1952), and Rhodin (1954, 1955) for fixation and sectioning procedures used in electron microscopy, and to Zimmerman (1933), von Möllendorff (1927-1928, 1929), and Bargmann (1933) for interpretations based on their observations made with traditional histological techniques.

The classic notion that Bowman's capsule develops embryologically as the invagination by the vascular tuft of the expanded end of the renal tubule is rejected by Hall and Roth (1956). Electron micrographs support the earlier observation of Huber (1905, 1910) that invagination does not occur. The space between filtering surface and capsule arises in a compact mass of undifferentiated cells by a flattening of cells

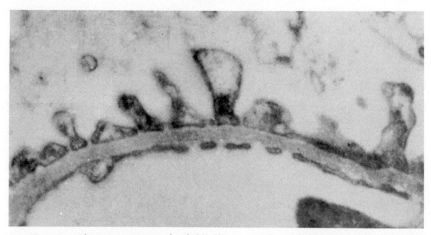

Fɪɢ. 3.   Electron micrograph of the filtering surface of the rat glomerulus showing the continuous homogeneous basement membrane or lamina densa as the middle
layer, with the discontinuous visceral epithelial layer above and the interrupted flattened endothelium below. Magnification: ×30,000. From Hall (1954).

which become the outer capsular layer when this layer subsequently
separates from one which becomes the inner filtering surface. Erythrocytes also differentiate from this mass of undifferentiated cells before
functional connection is made with the general circulation, a conclusion
supported by earlier observations made on explanted embryonic chick
metanephros in tissue culture (Reinhoff, 1922).

The mature cells which comprise the outer layer of the filtering surface
are uniquely specialized cells found only in the glomerulus. Hall has
called them podocytes or foot cells because of the numerous minute
interdigitating pedicils or foot processes which attach to the middle layer,
the definitive basement membrane or lamina densa. Pease and Baker
(1950) and Oberling *et al.* (1951) first saw interdigitating ridges in the
glomerular capillary basement membrane, but it was Dalton (1951) who
identified them as cytoplasmic processes arising from cells of the visceral
layer of Bowman's capsule. The pedicils run together and form large
central ridges or trabeculae on the outer surface. The basement membrane, which under light microscopy with modern staining techniques
appears as a thin homogeneous dark line (McManus, 1948a; Linder, 1949),
is resolved by electron microscopy into a complex of three structures:
(1) the pedicils of the podocytes on the outer surface; (2) a continuous
dense middle layer called the lamina densa or the definitive basement
membrane; and (3) a thin layer of perforated endothelial cytoplasm called
the lining network, lamina fenestrata, or lamina attenuata (Fig. 3).

The numerous large pores of the lining network perforate the endothelial cytoplasmic barrier and provide direct access of plasma to the lamina densa. This layer is the true ultrafilter; because of the channels between the epithelial podocytes it appears to be the only structure in the capillary wall sufficiently continuous to retain particles the size of plasma proteins. This extracellular basement membrane then is the only layer which regulates the composition of glomerular filtrate. Its ultimate structure is not revealed even by the high resolution of electron microscopy.

The fenestration of the endothelial membrane has been viewed as a "honeycomb reinforcements" (Oberling *et al.*, 1951) and as mucoid extensions of epithelial cells which penetrate the basement membrane "at regular intervals in a manner analogous to the cement lines separating hexagonal tile" (Rinehart *et al.*, 1953). Rinehart and Farquhar (1954) suggest that vesicles convey fluid across the glomerular membrane, as had previously been suggested for certain other endothelia (Palade, 1953). It is quite clear that structural details revealed by electron microscopy have not up to now resolved basic problems concerning the precise nature of those physical forces which operate in the elaboration of glomerular fluid.

Electron microscopy has opened up a debate concerning the existence of intercapillary space and whether the space, if it exists, may contain nonendothelial connective tissue cells. Among those who affirm the presence of intercapillary tissue are Zimmerman (1929, 1933), Bensley and Bensley (1930), Kimmelstiel and Wilson (1936), Goormaghtigh (1947), McManus (1948a), Ehrich (1953), Yamada (1955a, b), and Policard *et al.* (1955). However, others claim that the only spaces in the glomerulus are those within capillaries and those between capillaries and the outer layer of Bowman's capsule (Bell, 1950; Pease and Baker, 1950; Allen, 1951; Rinehart *et al.*, 1953; Hall, 1953, 1954; Mueller *et al.*, 1955).

One widely held notion which must be rejected is that the glomerulus is a tuft of capillary loops which branch from the afferent arteriole, and, without anastomoses, terminate separately in the efferent arteriole. Photomontages of electron micrographs and studies on latex-injected kidneys with light microscopy show instead that afferent arterioles divide into several primary branches, and these in turn divide again. These secondary branches give off small lateral capillaries which anastomose with each other and with the second-order vessels. The primary branches usually divide into three or four vessels which seem to afford direct communication, and probably preferential pathways, for the flow of blood from afferent to efferent arterioles (Hall, 1954).

Elias (1957) has recently reviewed current literature dealing with the structure of the glomerulus, and Hall (1957) has summarized his views concerning the protoplasmic basis of glomerular ultrafiltration as revealed by electron microscopy. Bargmann *et al.* (1955) have used histological, cytological and electron microscopic techniques to examine the glomerulus, along with a comprehensive study of the nephron as a whole. Bergstrand (1957) has also described the fine structure of the renal glomerulus with electron microscopic techniques. Pease (1955a) showed that the endothelial sheet of the peritubular capillaries is also a fenestrated membrane.

The glomerulus, as affected by various diseases, has been examined with electron microscopy by Rinehart *et al.* (1953), Piel *et al.* (1955), Simer (1955), Miller and Bohle (1956), Reid (1956), Vernier *et al.* (1956a, b; 1957), and Farquhar *et al.* (1957a, b). The vestigial nature of bird glomeruli is confirmed by electron micrographs which show the central cell mass of indifferent connective tissue to be separate and morphologically different from the functioning filtering surface (Pak Poy and Robertson, 1957). The ultrastructure of the opossum glomerulus is very similar to that of higher mammals (Pak Poy, 1957).

C. Regional Specialization of Tubules

In contrast to the simple aglomerular nephron which consists exclusively of a uniform brush-border cell type comparable to that found in the initial portion of the mammalian proximal segment, most vertebrate nephrons are highly specialized, presumably as a consequence of the division of labor imposed by the processes of infiltration, resorption, and secretion. Comparable specializations are noted in excretory tubules of invertebrates, especially those which are in open communication with the body cavity or are associated with filtering surfaces of one kind or another (Goodrich, 1945). Danielli and Pantin (1950), for example, have used histochemical procedures to point out the analogy of nemertine and turbellarian protonephridia with the differentiated tubule system of the vertebrate kidney. The basic similarity of earthworm and amphibian nephridia was emphasized by Gerard and Cordier (1934), who noted especially the intracellular distribution of absorbed colloidal particles. Other work has been cited earlier concerning the structure of various invertebrate excretory tubules.

The mammalian renal tubule (Fig. 4) can be considered grossly as made up of four morphologically distinct units: (1) the proximal convoluted tubule composed of two closely related parts each possessing a brush order; (2) Henle's loop consisting usually of a thin descending limb of varying length and an ascending limb composed of somewhat thicker cuboidal cells; (3) the distal convoluted tubule whose cells resemble to some extent those of the proximal convoluted segment; and

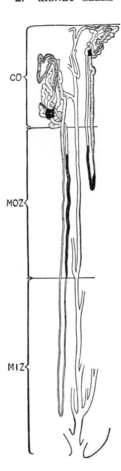

Fig. 4. Diagram of two nephrons in association with a collecting duct. Lengths of the loops of Henle vary widely within individual kidneys. Short nephrons, associated with glomeruli near the outer surface of the kidney, are about seven times more numerous than long ones in the human. *CO*, cortex; *MOZ*, outer zone of medulla; *MIZ*, inner zone of medulla; Malphigian corpuscles, black; proximal convolution, stippled; thin limb of Henle, white; limb of Henle, cross-hatched and then white (to indicate the opacity and clearness seen in macerated preparations, but not in sections); distal convolution, obliquely striated; collecting tubule, white. After Peter *in* Maximov and Bloom (1957).

finally, (4) the excretory ducts or collecting tubules. Except for certain teleosts, vertebrates in general have these same segments, although Henle's thin loop is sometimes replaced by a short, ciliated intermediate segment of small diameter (Marshall, 1934).

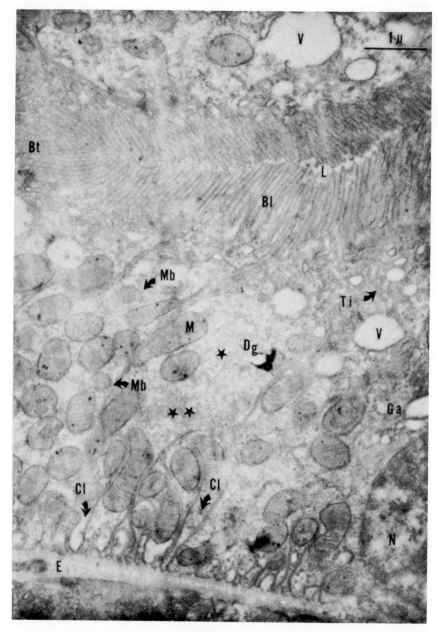

FIG. 5

1. Proximal Tubule

The feature all vertebrate nephrons have in common is a segment resembling the first part of the mammalian proximal tubule. As seen in ordinary light microscopy it is made up of a single layer of pyramidal cells with strongly eosinophilic cytoplasm and with each cell containing a large spherical nucleus. On the basal or peritubular side is a basement membrane, and arranged roughly perpendicular to it are closely packed rod-shaped mitochondria, the basal striations of light microscopy, which fill much of each cell up to the brush border on the luminal side. Numerous infoldings of the basal cell membrane divide the vascular side of the cell into intracellular compartments containing isolated rows of mitochondria distributed in a finely divided ground substance. Electron micrographs show the brush border on the luminal side to be made up of microvilli (Fig. 5), and between the bases of these extensions are invaginations of the surface membrane (vesicles) (Sjöstrand and Rhodin, 1953; Rhodin, 1954; Pease, 1955b, c; Bargmann *et al.*, 1955; Sjöstrand, 1956; Ruska *et al.*, 1957). The cell membrane consists of two osmophilic layers, each about 20 A. thick, separated by a nonosmophilic layer about 20 A. in width. The brush-border processes are enveloped by a membrane containing layers similar to those of the cell membrane. Each mitochondrion is surrounded by a double-walled capsule and crossed internally by a variable number of parallel septa, each again composed of two membranes (Fig. 6). The Golgi apparatus lies close to the nucleus and is composed mainly of pairs of parallel membranes. A tremendously large inner surface area is provided by these membranes of the basal infoldings, the brush border, the vesicles, the mitochondria, and the Golgi apparatus. Sinusoidal interdigitations interlock adjacent cells.

Cells in the second or terminal part of the proximal tubule are lower

FIG. 5. Proximal convoluted tubule cell of the mouse kidney. The tubule is partly collapsed, leaving a small, slit-formed lumen (L) free. The apical part of the cells is provided with an abundance of microvilli, cut longitudinally (Bl) and transversely (Bt). Between the bases of the microvilli are tubular invaginations (Ti) of the surface membrane, some widened to small vacuoles. Large fluorescent granule vacuoles (V) and one small electron-dense granule (Dg) are seen. Several mitochondria (M) and two microbodies (Mb) occupy the greater part of the cytoplasm, the ground substance of which is composed of RNA-granules (☆) with a diameter of 160 A., and areas (☆☆) characterized by a multitude of fine fibrils or short single membranes. The basal part of the tubule cell is clefted by infoldings of the cell membrane, resulting in cytoplasmic lamellae (Cl) which rest on the basement membrane (E). The nucleus (N) is located close to the basement membrane (E), and above is seen part of the Golgi apparatus (Ga), consisting of paired membranes, and vacuoles of varying size. Electron micrograph. Fixation: buffered osmium tetroxide. Magnification: ×17,000. From Rhodin (1958).

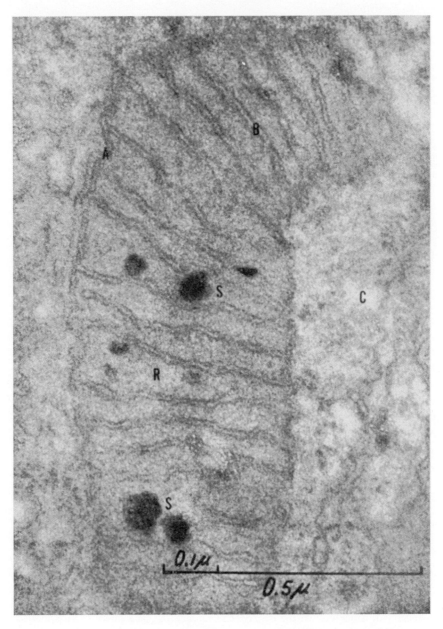

FIG. 6

than in the first part, the individual brush-border processes are more widely separated, mitochondria are large, but less abundant, and infoldings of the basal membrane are scarce or absent. Cell height decreases as the thin part of Henle's loop is approached. Segmental differentiation has also been established in various vertebrates by observed modifications in fat deposition and cell shape (Grafflin and Foote, 1939; Foote and Grafflin, 1938, 1942), the cytochemical demonstration of periodic acid-Schiff staining and phosphatase activity (Longley and Fisher, 1954) and iron-containing pigmentation (Grafflin, 1942), and the fluorescence of mitochondria at different levels of the proximal tubule (Sjöstrand, 1944). The upper segment presents a more uniform appearance among various mammalian species than does the lower segment. This suggests that cellular transport functions which are held in common by all mammals may be located in this segment whereas interspecific functional differences might be sought in the more variable second segment.

Refractometric technique based on phase contrast behavior of unstained sections mounted in media of known refractive index show the brush border to have positive form birefringence and negative intrinsic birefringence with respect to the length of the striations (Kruszynski, 1957). The brush border consists of rodlets 300–600 A. in diameter, with an oriented molecular structure at right angles to their length (Brewer, 1954).

2. Henle's Loop

At the transition between the proximal tubule and the thin loop of Henle, the pyramidal brush-border cells change abruptly to a thin squamous epithelium. The diameter of the lumen, however, narrows gradually. In mammals, the thin loop leads by a marked transition to a second limb with a sudden change in cell shape from squamous to cuboidal and an increase in mitochondrial content. Rod-shaped mitochondria in this thicker limb impart a distinct perpendicular striation to the peritubular part of the cells. Flagella associated with centrioles at the free surface may be present. Wide variations in Henle's loop occur among the various vertebrates and also within the nephron population of an individual kidney. For example, in the human only one tubule in seven

FIG. 6. High-power electron micrograph of one mitochondrion as seen in a cell of the proximal convolution of the mouse kidney. A membrane (A) forms the mitochondrial capsule, and a system of layered membranes or plates (B) occupy the interior of the mitochondrial body. The matrix (R) of the mitochondrion has a higher electron density than the surrounding cytoplasm (C). Freely in the mitochondrial matrix occur highly electron-dense spherical bodies (S). Fixation: buffered osmium tetroxide. Magnification: × 140,000. From Rhodin (1954).

has a long loop which dips down into the pyramid or inner zone of the medulla. The much more numerous short tubules have a thin descending limb which is very short or absent entirely. In the latter the proximal convolution leads directly into the thick ascending limb in the juxta-medullary zone of the cortex (Peter, 1927). These wide variations in the length and structure of this segment must, it seems, be associated with alterations in function, but none has been demonstrated, and physiologists concerned with regional specialization in function along the tubule frequently fail to take these structural variations into account. Oliver (1955) has stressed this heterogeneity of the nephron population from a structural standpoint. Bradley *et al.* (1954) and Childs *et al.* (1955) have presented impressive functional data to support this view.

3. Distal Tubule

The distal segment, when present, appears to be essentially homologous throughout the vertebrates. The cells are lower and less acidophilic and the lumen larger than in the proximal tubule. There is no brush border, but electron microscopy (see Fig. 7) reveals short scattered microvilli. Basal striations are very prominent due to very long and dense mitochondria which sometimes are so tightly packed that in electron micrographs almost no cytoplasm can be discerned between them (Rhodin, 1954). The basal infoldings of the plasma membrane on the peritubular side are very prominent and appear to surround the mitochondria (Pease, 1955b, c).

4. Juxtaglomerular Complex

In the kidneys of many animals, cells of the distal convoluted tubules, where they establish direct contact with the glomerular vascular root of the same nephron, are specialized as the macula densa with the characteristically cuboidal cell converted to a high cuboidal or even columnar

---

Fig. 7. Distal convoluted tubule cell of the mouse kidney. The cell is typical for the cortical part of the thick ascending limb of Henle's loop in the vicinity of the glomerulus. A wide free lumen (*L*) and a cell surface with but few and very short microvilli (*Mv*) characterize the cells of this part of the nephron. In addition, the nucleus (*N*) is located in the apical part of the cell, and the mitochondria (*M*) are elongated, closely packed, and with an electron-dense matrix. The mitochondria are located within narrow cytoplasmic lamellae (*Cl*) which are formed by infoldings of the basal cell membrane and rest on the basement membrane (*E*). The cell border between two cells is seen, consisting of two apposed cell membranes. Near the surface is an apparent firm contact formed by the so-called terminal bars (*T*). A small Golgi apparatus (*Ga*) is seen. The apical part of the cell displays an abundance of small vesicles (arrows), some in free communication with the tubule lumen. Electron micrograph. Fixation: buffered osmium tetroxide. Magnification: × 18,000. From Rhodin (1958).

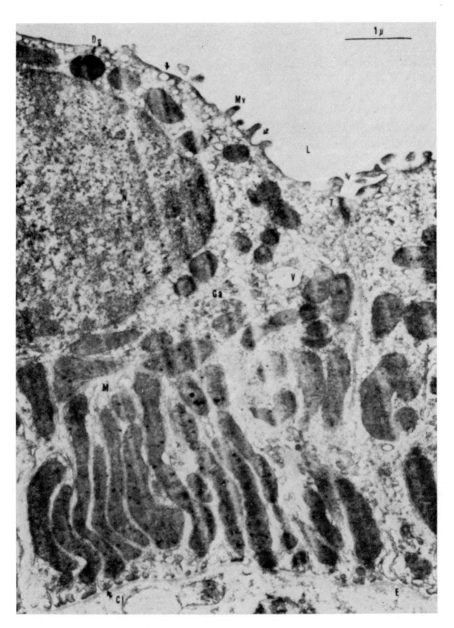

FIG. 7

type recognizable because of the densely crowded and closely aligned nuclei (Zimmerman, 1933). In these cells the Golgi apparatus lies on the side of the nucleus opposite to its position in other distal tubule cells (McManus, 1942, 1944, 1947). Cells of the afferent arteriole where they come into contact with the macula densa possess a characteristically granular cytoplasm. This juxtaglomerular complex of tubular and vascular components is established at a very early stage in embryonic development and appears to be a consistent feature throughout the vertebrates. Goormaghtigh (1940, 1945, 1947) has ascribed an endocrine activity to the apparatus, and he and other earlier workers implicated it in a variety of pathological states involving hypertension and renal function. Oliver (1945) represented another group of investigators who concluded that cellular changes result merely from mechanical stresses where the thin-walled arteriole enters the glomerulus. Electron micrographs, however, disclose in these cells specialized secretory granules which contrast with those of adjacent cells in the arteriolar media (Hartcroft, 1956), and the endocrinelike nature of the cells is further suggested by studies implicating cytological alterations specifically with adrenal activity, salt balance, and hypertension (Dunihue, 1946, 1947, 1949; Dunihue and Robertson, 1953; Hartcroft and Hartcroft, 1953, 1955; Toussaint et al., 1953; Hartcroft, 1957).

5. Collecting Duct

Excretory ducts or collecting tubules are made up of sharply outlined cuboidal cells containing a round nucleus, and clear cytoplasm with relatively few mitochondria. As they pick up branches the ducts enlarge and the cells grow higher, finally acquiring a columnar form. Infoldings of the basal membrane, prominent in the distal tubule, which Rhodin (1956) has associated with water resorption at that site, are few or absent in collecting ducts. Intercalated cells, which Oliver et al. (1957) have implicated in electrolyte imbalance, lie compressed between the predominant clear cells of the collecting tubules. Oliver (1945) describes their appearance as observed in dissected tubules and refers to von Möllendorff's Handbuch (1929) for a historical review of the older literature concerning them.

6. Development

The kidneys of higher vertebrates are traditionally viewed as consisting phylogenetically and ontogenetically of three successive sets of excretory organs separately identified as the pronephros, mesonephros, and metanephros. The pronephros is the functioning excretory organ in the lowest vertebrates and in the larval stages in amphibians; the mesonephros, in fishes and amphibians; and the metanephros is the final and permanent excretory organ in amniotes—reptiles, birds, and mammals. Huber (1932) is an important source of information concerning structural modification in renal tubules of various vertebrates, considered especially from the standpoint of embryological development. With respect to specific cellular functions one is much more impressed by the similarities of these various tubules than by their differences, and indeed the trend today is to stress their fundamental morphological and embryological similarities also (Fraser, 1950). Electron microscopy fails to reveal any significant differences between the fully developed embryonic mesonephros and the adult metanephros in the ultrastructure of glomeruli or tubule cells (Leeson, 1957).

Renal structure in most mammals is incomplete at birth. During the first few weeks after birth in the rat, for example, growth takes place by the formation of new nephrons in the outer zone of the cortex (Kittelson, 1917; Arataki, 1926a), and during this period the cells acquire their specific cytological characteristics. At birth the proximal tubules possess a homogeneous cuticular border on their luminal surface which takes on the appearance of a brush border during the first 2 weeks after birth. At birth mitochondria are arranged randomly, but during the maturing period they line up into rows perpendicular to the peritubular basement membrane and become recognizable as basal striations by light microscopy. During this period the brush border and the basal striations are organized anisotropically, which renders them birefringent (Olivecrona and Hillarp, 1949). The ability of the proximal cells to stain intravitally with trypan blue is acquired coincidentally with the development of the brush border (Baxter and Yoffey, 1948).

Differentiation of the alkaline phosphatase reaction and the periodic acid-Schiff staining properties of the brush border continues for several months after birth in the mouse (Longley and Fisher, 1956). In the human fetus cytogenesis of renal tubules resembles that in the rat, but development is more nearly complete at birth (Policard, 1912).

Electron micrographs show that the brush border develops by the accumulation of numerous microvilli, and basal striations in proximal and distal tubules by an alignment of mitochondria, which increase in number and size during the first 2 weeks in mice. The collecting ducts develop tight pleating of the basal or peritubular cell membrane, and dark cells containing many small cytoplasmic vesicles and microvilli make their appearance during this period. Lipid inclusions and protein droplets in proximal cells disappear a few days after birth, and the cytoplasmic granules of Palade become less apparent as the cells differentiate (Clark, 1957).

III. FORM IN RELATION TO FUNCTION

No attempt will be made here to examine renal hemodynamics or to consider the kidney's over-all function as an organ. These excretory, regulatory, and homeostatic aspects of renal function have been recently reviewed by Smith (1951, 1956), Robinson (1954a), and Winton (1956). Instead, emphasis will be placed on individual cell processes, especially where specific functions may be associated with, or determined by, form. Modern techniques which seek to explore this interdependence will be considered particularly.

A. *Direct Observation and Microdissection*

1. *General*

Gerard and Cordier's monograph (1934) is an important summary of the earlier literature concerning direct observation of tubular activity. Their review of the comparative functional histology of the vertebrate kidney appeared during an important transitional period in the development of techniques used in exploring renal concepts. Clearance techniques were still in the formative stage when the following summary was made by these investigators. The glomerulus was viewed as a filter which allowed not only diffusive substances to pass but also "highly dispersed colloids" which subsequently were resorbed by the cells of the proximal convoluted tubule. They pointed out that this process of atherocytosis of colloids differs from normal resorption of diffusible substances in that it is accompanied by prolonged intracellular retention. The proximal convoluted tubule cells were shown to have in addition a secretory function, which was demonstrated directly by the suppression of glomerular activity in certain amphibians. The distal tubule was implicated in the resorption of water.

Isolation of tubules by microdissection has permitted localization of certain tubular functions and the identification of alterations induced by various pathological conditions. Oliver *et al.* (1951) describe in detail methods for mounting and staining dissected renal tubules for this kind of examination. After maceration in concentrated acid, the formalin-fixed kidneys are washed with water and the nephrons then disentangled with needles under a binocular microscope. The isolated tubules may then be stained or subjected to various histochemical procedures. Considerable cellular detail is revealed in these relatively thick (40–60 μ) uncleared tubules. Mitochondria are altered somewhat in form but still keep their striated orientation, the brush border is clearly identifiable, and the basement membrane appears as a well-defined thin homogeneous band. Supplementally, conventional histological sectioning techniques are usually used to guide interpretation of various cellular alterations revealed in dissected tubules after damage or disease: cast formation, regeneration of atypical cells, hyaline droplets, cloudy swelling, fat accumulation, vacuolar hydropic degeneration, etc.

In confirmation of conclusions reached earlier by Suzuki (1912), microdissection showed that foreign substances such as carmine and trypan blue are absorbed by atherocytosis in the proximal tubules along a gradient with the highest concentrations of particles appearing in cells next to the glomerulus and then decreasing regularly until those cells half-way down the proximal segment may be entirely clear of stored dye (Oliver, 1915). After injection of large amounts of serum or egg white, protein accumulates as intracellular droplets localized in the middle third of the proximal tubule. Microincineration procedures reveal massive accumulation in proximal cells of certain minerals such as calcium salts after the administration of calciferol or parathormone when other segments such as the ascending limb and the distal tubule show so little mineral that cell outlines can barely be traced (Oliver, 1949). In humans with diabetes, glucose resorption is incomplete and glycogen then accumulates preferentially in the terminal part of the proximal segment in very high concentrations. Glucose normally is viewed as being completely resorbed by proximal cells before this region is reached (Oliver, 1945). The ability of proximal tubules to store resorbed substances has been confirmed with the use of various other techniques which improve visualization, such as with the identification of ferrocyanide (Gersh and Stieglitz, 1933) or by labeling protein (e.g., Smetana, 1947; Narahara et al., 1958).

Following potassium depletion there is an impairment in the ability of the tubules to concentrate urinary solute which is associated with specific lesions in the collecting ducts (Oliver et al., 1957). This site is also implicated in water resorption by damage noted here in humans with diabetes insipidus (Darmady, 1954), and other evidence based on comparative studies and micropuncture techniques will be presented later.

2. Protein Resorption

Direct micropuncture observations on various mammals indicate that glomerular capillaries leak protein readily (Walker and Oliver, 1941), and other evidence suggests that cells resorb filtered protein and, in some instances, degrade it (Rather, 1952). Plasma proteins labeled in vivo with the dye T-1824 or tagged with radioactive tracers are later found concentrated in the cells of the proximal convoluted tubules (Dock, 1942; Gilson, 1949; Narahara et al., 1958) and accumulate in the rat at a rate which suggests that 33 % of the circulating plasma proteins are filtered and subsequently resorbed (Sellers et al., 1954). Urinary protein has the electrophoretic and solubility characteristics of serum alpha and beta globulin (Sellers et al., 1952). If these large globulin molecules are filtered it seems likely that the much smaller serum albumin molecules pass through the glomerular membrane even more readily. When proteinuria is induced in rats by the intravenous and intraperitoneal injection of a wide variety of proteins, absorption occurs in proximal cells without droplet formation unless the capacity to absorb protein is exceeded either by the amount or by the nature of the protein. Oliver *et al.* (1954a) view droplet formation as an accessory mechanism characterized by the intracellular combination of mitochondria and the absorbed protein. The droplets react strongly with dilute Janus green, and other histochemical procedures suggest again that the complex contains a mixture of the absorbed protein and mitochondrial elements (Oliver *et al.*, 1954b). Mitochondria in the form of rodlets disappear concomitantly with the formation of droplets, and it is suggested that the mitochondrial enzymes are thus brought into immediate contact with the absorbed protein which is ultimately digested inside the cell. Histochemical procedures disclose that certain specific administered amino acids are selectively concentrated in these droplets, and the mitochondria of the proximal cells were found to contain sulfhydryl in high concentration which correlates with their suggested enzymatic activity (Lee, 1954). Highly purified suspensions of the droplet material reveal that their phospolipid and pentose nucleic acid content is essentially similar to that of the mitochondrial rodlets which disappear as droplets form (Straus, 1954). Twelve hours after intraperitoneal injection of egg white there is a 35–40 % decrease in succinoxidase and cytochrome oxidase activities in the fraction containing the larger particles, i.e., mitochondria and droplets in equal concentration. Thirty hours after injection these enzyme activities have returned to normal (Kretchmer and Dickerman, 1954).

Also after the parenteral injection of amino acids gram-positive droplets formed within 15 minutes, and at this time the free α-amino nitrogen in the kidney cortex had doubled. One hour after injection lysine was found within the particulate protein of the fractions which contained droplets (Kretchmer and Cherot, 1954). Eliasch *et al.* (1955) found significantly higher concentrations of amino acids and polypeptides in the renal vein than in either the aorta or inferior vena cava, which supports the view that the kidney may be an important site of protein catabolism, and may, as a consequence, play a significant role in regulating the metabolism of circulating proteins.

Rhodin (1954) with electron micrographs confirmed Oliver's classic experiment (1948) and described fusion of mitochondria as an early response to protein resorption, followed by a loss of their characteristic internal structure as they combined with the absorbed protein. Eventually these large granules were reconverted into typical mitochondria.

3. Experimental Tubular Lesions

Uric acid nephritis in rabbits induced by injection of moderate doses of lithium monourate affects specifically the first part of the collecting duct (Smith and Lee, 1957), whereas heavier doses produce severe selective damage in the ascending limb and distal convoluted tubule as well (Dunn and Polson, 1926). Experimental phosphate nephritis in rats also affects the lower nephron (McFarlane, 1941), and damage may extend proximally to include also the descending part of the proximal convolution (Oliver, 1945). The crush syndrome involves the nephron segments affected in uric acid nephritis (Dunn *et al.*, 1941), as does severe damage resulting from incompatible blood transfusion (Bywaters and Dible, 1942; Lucké, 1946; Mallory, 1947). However, more recently Oliver *et al.* (1951) have described these lesions associated with acute renal failure as being much more extensive. They demonstrate patchy tubular necrosis and widespread disruption also in the proximal convoluted tubule, its descending limb, the ascending limb of Henle, and the distal convoluted tubule.

Smith and Lee (1957), using Oliver's microdissection techniques, again localize in the lower nephron lesions resulting from uric acid administration. The pH change from alkaline to acid in the distal tubule, and also perhaps water resorption in the initial collecting ducts, facilitates precipitation of urates. An internal hydronephrosis may occur subsequent to obstruction in the collecting duct which is secondarily manifested in dilation of collecting, distal convoluted and ascending Henle segments. This in turn may cause cellular atrophy and ultimately focal regeneration of cells.

Darmady and Stranack (1957) have summarily described spontaneous tubular lesions in human diseases using Oliver's microdissection techniques.

4. Experimental Glomerular Lesions

Masugi (1934) provided a clue to the pathogenesis of glomerulonephritis with the demonstration that kidney tissue of one animal acts as an antigen when introduced into a different species. Antisera collected from the recipient, upon injection into the original species, will produce a form of glomerulonephritis which resembles the spontaneous disease in man. The antigenicity has been demonstrated successively as residing in the renal cortex (Heymann and Lund, 1948), the glomeruli (Greenspon and Krakower, 1950), and finally in the glomerular basement membrane (Krakower and Greenspon, 1951). The antibodies of the nephrotoxic sera have been shown by radioautographic techniques to localize in the glomeruli of the recipient (Pressman *et al.*, 1950). Using the microfluorescence technique of Coons and Kaplan (1950) for the detection of antigens, it was shown that the maximum concentration of nephrotoxic antibodies has a distribution similar to, if not identical with, that of the periodic acid-Schiff stain on the glomerular basement membrane (Mellors *et al.*, 1955a, b; Mellors and Ortega, 1956). These observations suggest an allergic pathogenesis for experimental renal disease and perhaps for some types of human glomerulonephritis. Ellis (1956) gives a detailed description of glomerular lesions induced experimentally in rabbits by the intravenous injection of saccharated iron oxide. In this case glomerular occlusion results from the intravascular precipitation of iron salts and terminates in a condition resembling the nephrotic syndrome.

B. *Cytochemistry and Renal Function*

In his classical paper on the distribution of phosphatase activity in practically every organ of many mammalian species, Gomori (1941) noted that the most uniform picture was obtained in the kidney. In all species, cells of the proximal convoluted segment and, in addition, a portion of varying length of the straight segment of the proximal tubule gave a strongly positive reaction. In man and, much less regularly, in the dog the ascending limb of Henle's loop also exhibited phosphatase activity. In all species, the distal convoluted tubules and the entire system of collecting ducts were negative throughout. Also, in certain invertebrate protonephridia, as in the nemertines and planarians, cells of the proximal ciliated convoluted canal and the ciliated terminal ducts are rich in this enzyme, whereas no alkaline phosphatase activity is found in the flame cells or in the distal glandular canal (Danielli and Pantin, 1950). Modern fixation techniques and studies on unfixed free-floating frozen sections (Wachstein and Meisel, 1953) have resulted in the demonstration of many other enzymatic systems in mammalian kidneys. Wachstein's (1955) paper is a very useful source of information concerning the segmental distribution of enzymatic reactions in the renal tubules of rat, mouse, rabbit, human, dog, and cat. Much of the following is based on his summary and personal observations.

Nonspecific alkaline phosphatase in all species is present in the upper half of the proximal convoluted tubules but absent in the terminal portions in mouse, dog, and cat. The ascending limb shows activity only in the cat; negative reactions are obtained in all other segments irrespective of species. The brush border in fixed sections exhibits the greatest activity and under certain circumstances seems to have a bilaminar distribution (Johnson, 1954). However, in unfixed frozen sections such preferential localization at the brush border does not occur, and a diffuse distribution of stain throughout the cells is noted (Wachstein, 1955). Specific glycerophosphatase at pH 7.2 with substrate concentrations in the range thought to occur in living cells shows the highest activity in the initial proximal segment of most species, but in some (rat and rabbit) the terminal segment shows more. Segments that are negative for nonspecific alkaline phosphatase show activity, and indeed each region of the tubule in some species or other shows specific phosphatase activity under these conditions at pH 7.2. Nonspecific acid phosphatase is widely distributed in the various portions of the tubules including thin loops and collecting ducts in several species, but again, the highest activity generally is found in the first segment of the proximal convolution.

5-Nucleotidase (Gomori, 1949; Newman et al., 1950) in all species is found only in the proximal tubules, with mouse and rabbit exhibiting more activity in the terminal segment, whereas rat and human show an equal distribution between the initial and terminal segments.

Glucose-6-phosphate is of particular interest because it has been implicated in hypothetical schemes to account for glucose resorption by proximal tubule cells. Chiquoine (1953) found free phosphatase in the proximal convoluted tubule of the mouse, as well as some activity in the ascending limb of Henle's loop and in the parietal epithelium of Bowman's capsule. Wachstein's (1955) analysis in six species showed in all the strongest reaction in the proximal tubule, with some activity in the ascending limb of Henle's loop in the rat and mouse. The reaction is not entirely specific because some staining reaction is obtained when glycerophosphate is substituted for glucose-6-phosphate in the substrate mixture. Phlorizin administered to rats in amounts sufficient to inhibit glucose resorption does not affect activity, as previously it had been shown to have no effect on alkaline phosphatase activity (Kritzler and Gutman, 1941). However, the unique method of Lowell et al. (1953) for the separation and isolation of tubule cells and glomeruli from fresh renal cortex of the dog disclosed that phlorizin did inhibit phosphatase activity which was specific for splitting glucose-6-phosphate.

Various other phosphatases for such substrates as creatine phosphate (Maengwyn-Davies et al., 1952), flavine and pyridoxal phosphate (Bourne, 1954), organic phosphate esters (Burgos et al., 1955), and phosphoric acid p-chloroanilide (Eger and Schulte, 1954) have no specific distribution patterns in kidney sections.

Nonspecific esterases can be demonstrated with the use of a variety of substrates including Tweens, α- and β-naphthyl acetate, naphthyl AS acetate, and indoxyl acetate. In a fixative-free preparation and with α-naphthyl acetate as substrate, Wachstein (1955) finds the strongest reaction in the proximal and distal convoluted tubules, and some activity in most species in the ascending limb of Henle's loop and in the collecting ducts. Burstone (1957) with the use of new chromogenic substrates of the naphthol AS series demonstrated spherical bodies which stain intensely and a positive esterase reaction also in the basal portion of the tubules.

β-Glucuronidases are enzymes which split the β-glucoside linkage of glucuronides. With the method of Seligman *et al.* (1954) the distribution of enzyme activity is generally the same as that of the esterases. Concentrations are highest in proximal and distal tubules, and some activity is noted in all the other segments, except the descending limb.

Again, the descending limb shows no succinic acid dehydrogenase activity and the collecting duct none or little. Usually the ascending limb and distal convolution are more positive than even the proximal tubule segment (Wachstein, 1955). Depletion of cytochrome oxidase and succinic dehydrogenase activity occurs in rats when nephrosis is induced by the injection of aminonucleoside (Fisher *et al.*, 1958).

Diphosphopyridine nucleotide (DPN) diaphorase activity, however, is highest in the proximal convoluted tubule, while the ascending loop and distal convoluted tubule stain less intensely in all species but the human. The thin limb of Henle stains heavily in all species except the rat, and collecting tubules stain throughout their entire length. Sternberg *et al.* (1956) used methods for DPN and TPN diaphorase, combined with a succinic dehydrogenase procedure, to delineate effectively the segmental subdivisions of the mammalian nephron. Triphosphopyridine nucleotide diaphorase staining is not as intense as DPN diaphorase, but the pattern of distribution is the same. Amino peptidase activity can be exhibited in the proximal tubule (Gomori, 1954a, b).

The identification of carbonic anhydrase and its regional distribution in the nephron is of great interest because of its role in acid-base regulation and in salt and water excretion by the kidney, but Wachstein (1955) was unsuccessful in his attempts to apply a technique described by Kurata (1953) for its identification in renal tubules of the rat.

Histochemical procedures have been used to test for possible intracellular accumulation of various organic substances undergoing tubular transport. Normal kidneys do not contain histochemically demonstrable quantities of lipid except for the cat and the dog (Foote and Grafflin, 1938; Oliver, 1945; C. Smith and Freeman, 1954). A sharp difference in distribution exists in these species, with lipid in the cat limited to the initial half of the proximal segment, whereas in the dog it is deposited only in the terminal portion of the proximal segment. Lipid is common in glomeruli and various tubular segments in many diseases involving the kidney (Allen, 1951). Glycogen also cannot be demonstrated histochemically in kidneys; however, with diabetes mellitus, deposition occurs in the terminal portion of the proximal segment and in the ascending limb (Oliver, 1945; Polysaccharide identification through use of the periodic acid-

Schiff reaction has been widely applied to renal studies, and its use recommended particularly for the identification of glomerular alterations accompanying various diseases (McManus, 1948a, b, 1950; Jones, 1953). With this procedure the basement membranes of glomeruli and tubules are coloured bright red, as are certain droplets and granules in the arteriolar wall. The brush borders of proximal tubules take the stain markedly, and the cytoplasm only slightly. Species differences exist in the regional distribution of the brush-border staining reaction; the terminal portion of the proximal segment stains more distinctly in the mouse, whereas in the cat the brush border in the upper end is more heavily stained (Longley and Fisher, 1954). As with histochemical staining reactions generally, these species variations cannot be associated with specific differences in function.

Protein-bound sulfhydryl and disulfide groups in various mammalian species are most markedly exhibited in the proximal convoluted tubules. A technique for identifying protein-bound amino groups (Weiss *et al.*, 1954) reveals an intense reaction within the proximal tubule cells and a fainter response in other segments.

McCann (1956) combined precise microchemical methods with microdissection techniques to make quantitative comparisons of water content, lipid content, and the activities of alkaline phosphatase, aldolase, fumarase, phosphohexoisomerase, and hexokinase in isolated tissue samples taken from frozen-dried sections of dog kidney.

To date it is quite clear that not a single enzyme system revealed by these histochemical techniques can be implicated directly in an active transfer process. However, many textbooks advance so forcefully hypothetical schemes involving phosphatases with glucose resorption that students are likely to accept their involvement in phosphorylation-dephosphorylation processes as an established fact. Actually the glucose resorptive process is far from understood, and implication of the hexokinase-phosphatase sequence of reactions is by no means clearly established. The hypothesis that phlorizin blocks glucose resorption by interfering with the esterification of glucose in proximal tubule cells is not supported by experiments on the phosphatase activity of phlorizinized kidneys and on the action of iodoacetic acid, in both amphibians and mammals (Walker and Hudson, 1937b). Cori (1954) pointed out that in glycogen storage disease (von Gierke's) the specific glucose-6-phosphatase is almost completely absent, yet the patients do not usually have glycosuria. The enzyme deficiency offers a satisfactory explanation for glycogen accumulation, but persistence of the ability to resorb glucose normally is irreconcilable with the phosphorylation-dephosphorylation theory of glucose resorption. Alkaline phosphatase activity is similar in both glomerular and aglomerular kidneys, even though glucose is not resorbed in the latter (Browne *et al.*, 1950).

Further evidence that glucose phosphatases may have nothing to do with glucose transport in the kidney is provided by Dratz and Handler (1952), who injected radioactive orthophosphate and measured glucose-6-phosphate and glucose-1-phosphate turnover in the kidney. The turnover rate was found to be very slow, and specific activities of G-1-P and G-6-P were not altered significantly either by inhibiting glucose resorption with phlorizin nor by increasing glucose resorption with direct glucose infusion. Mudge and Taggart (1950) showed that 2,4-dinitrophenal, in concentrations which uncouple aerobic phosphorylation, had no effect on glucose resorption in the dog, though simultaneously *p*-aminohippurate (PAH) secretion was depressed significantly. If glucose resorption is driven by adenosine triphosphate (ATP) via the hexokinase reaction, DNP should interfere since its effect is to diminish the supply of ATP. Furthermore, McCann (1956) showed that in the dog nephron hexokinase activity was four times higher in the terminal segment of the proximal tubule than in any other area. This, again, does not support the notion that hexokinase is implicated in the glucose resorptive process, which all evidence indicates is located in the first half of the proximal tubule (Oliver, 1950). None of these lines of evidence, of course, is conclusive in itself, but taken together they suggest strongly the need for a re-examination of the phosphatase hypothesis as an explanation of glucose resorption by the renal tubule.

C. Development, Growth, and Senescence

The earlier literature on the correlation of structure and function in the developing mesonephros and metanephros was summarized by Gersh (1937), who related onset of glomerular and tubular function in embryos of the rabbit, cat, opossum, chick, and pig to structural differentiation in the developing nephron. Ferrocyanide was used to test glomerular activity and phenol red for secretory activity in the proximal convoluted tubule. These agents were introduced into the fetus in a variety of ways and subsequently identified in sections by histochemical methods.

Gersh's observations are summarized in Table I. It is readily noted that, in all species examined, the mesonephros functions in the elimination of both ferrocyanide and phenol red, the mesonephros functions for a shorter or longer period even after the metanephros has assumed the same functions, water resorption in the metanephros takes place to some extent in the proximal convolution, and, finally, water resorption takes place to a much greater extent in the thin limb of the loop of Henle. As a rule, degeneration of the mesonephros is accompanied by gradual

TABLE I

	RABBIT					CAT					OPOSSUM POUCH-YOUNG						CHICK								PIG					
AGE IN DAYS	19	21	23	29	31	35	38	42	47	52	8	10	13	18	20	24	9	11	11	12	13	13	14	18	29	45	54	61	74	92
PROTOCOL	K	J	H	B	A	D	I	G	H	A	1	3	5	6	8	9	7	11	4	12	15	17	6	19	2	6	1	3	4	5
MESONEPHROS																														
(FERROCYANIDE ELIMINATION)																														
GLOMERULUS	+'	+'					+'	-'	-'		+'	-'	-'				+	+	+	+	+	+	+							
LUMEN OF PROXIMAL PORTION	+'	+'					+'	-'	-'		+'	-'	-'				+	+	+	+	+	+	+							
CONCENTRATION IN DISTAL PORTION	+	+					+					+					+	+	+	+	+	+	+							
WOLFFIAN DUCT	+	-						-			+						+	+	+	+	+	+	+							
METANEPHROS																														
GLOMERULUS	-	+	+	-	+	+	+	+	+	+	+	+	+	+	+		-	-	+	+	+	+	+	-	+	+	+	+	-	-
LUMEN OF PROXIMAL CONVOLUTION	+	+	+	S	S	+	S	S	C	C	+	+	+	S	S		-	-	+	S	+	S	+	+	+	+	S	S	S	S
LUMEN OF LOOP OF HENLE	C	C	C	C	C	C	C	C	C	C	+	+	+	C	C		-	-	+	+	C	C	C	C	+	C	S	C	C	C
LUMEN OF DISTAL CONVOLUTION	+	+	+	+	+	+	+	+	+	+	+	+	+	+	+		-	-	+	+	+	+	+	+	+	+	+	+	+	+
LUMEN OF LARGER DUCTS	+	+	+	+	+	+	+	+	+	+	+	+	C	+	+		-	-	+	S	+	+	+	+	+	+	+	+	-	-
PROTOCOL																														
MESONEPHROS																														
(PHENOL RED ELIMINATION)																														
GLOMERULUS	-						-	-	-		-	-	-				-	-	-	-	-	-	-	-					-	-
CELLS OF PROXIMAL CONVOLUTION	+	+'					+'	+'	+'		+'	+'	+'	+	+		+	+	+	+	+	+	+	+			+	-	+	+
LUMEN OF DISTAL CONVOLUTION	+	+'					+'	+'	+'		+'	+'	+'	+	+		+	+	+	+	+	+	+	+			+	-	+	+
LUMEN OF WOLFFIAN DUCT	+	-?					+				+	+					-	-	-	-	-	-	-	-						
METANEPHROS																														
GLOMERULUS	-						-	-	-		-	-	-	-	-	-	-	-	-	-	-	-	-	-			-	-	-	-
CELLS OF PROXIMAL CONVOLUTION	+	+					+	+	+		+	+	+	+	+	+	-	+	+	+	+	+	+	+		+	+	+	+	+
LUMEN OF DISTAL CONVOLUTION	+	+					+	+	+		+	+	+	+	+	+	-	-	+	+	+	+	+	+		+	-	+	+	+
LUMEN OF LARGER DUCTS	+	+					+	+	+		+	+	+	+	+	+	-	-	+	+	+	+	+	+		-	-	+	+	+

Legend:
+ = POSITIVE
- = NEGATIVE
+' = MOSTLY POSITIVE
-' = MOSTLY NEGATIVE
S = SLIGHT CONCENTRATION
C = MARKED CONCENTRATION

loss in both glomerular and tubular function to the same degree, usually proceeding anteroposteriorly. Shunting or transference of arterial flow is probably the primary cause of structural degeneration and dysfunction.

Direct observations of glomerular filtration rates have been made on sheep fetuses between 61 and 142 days gestation age, using clearance techniques (Alexander *et al.*, 1958). Maximal filtration rates at 61 days, expressed per gram body weight, exceed adult resting values, but by 142 days these drop to only 25% of that of the adult. Sodium, potassium, and chloride are actively resorbed, and hypotonic urine is formed between 61 and 130 days, but between 130 and 140 days tubular resorption of water climbs from 66% of that filtered to 80 or 90% coincidentally with the production of hypertonic urine. The onset of secretory activity in the fetal metanephros of the pig is accompanied by an increase in cytochrome-cytochrome oxidase activity (Nadi test), and the development of a potential difference of the order of 200 millivolts between epithelium and stroma. No such phenomena were observed in the region of the collecting ducts (Flexner, 1939).

Active secretion of phenol red does not take place in the metanephros until structural differentiation of the proximal tubule has occurred. As in the rabbit and chick, this may precede the onset of glomerular function and has no apparent relationship to the growth and deposition of new blood vessels along the outer walls of the tubules. However, the onset of tubular secretory activity in the pronephros has been associated with the apposition of a blood vessel along the outside of the tubule in *Fundulus* (Armstrong, 1932). Water resorption by the descending limb of the loop of Henle first appears sometime after glomerular and tubular secretory activities have been established, and only after its cells become noticeably thinner than the cells of the undifferentiated tubules in the nephrogenic zone and their rounded nuclei extend almost the height of the whole cell.

Gersh (1937) believed that onset of glomerular function was not accompanied by a detectable morphological variation or differentiation in the filtering surface, but depended upon the establishment of favorable extraglomerular hemodynamic factors. On the other hand, Gruenwald and Popper (1940) thought that the high epithelium of the visceral layer of the glomerular capsule forms an inextensible sac which acts as a filtration barrier by reducing blood flow and causing coherence of adjacent loops of the glomerular tuft. At the time of birth this continuous epithelium was thought to rupture and permit onset of urinary function. However, in rats, Wells (1946) noted that the renal pelvis and ureter became distended with fluid above a ligature tied 6–31 hours before birth. The allantoic fluid of fetal rabbits appears to be formed wholly or in part by the mesonephros (Davies and Routh, 1957). Fergusson (1952) confirmed the opinion of Gruenwald and Popper (1940) that the low glomerular filtration rate in infancy is due to the relatively undeveloped state of the vascular tuft in glomeruli of the outer cortex and to the high epithelium of the visceral layer of the glomerular capsule.

Junqueira (1952) has summarized data for the chick embryo, timing the onset of metanephric function and the beginning and end of mesonephric activity with many different methods for evaluating function. In addition, histochemical observations indicated that in the mesonephros acid and alkaline phosphatase activity was evenly distributed throughout the cell, then polarized at the brush-border zone later on.

Davies (1952) concludes on histological grounds and on the distribution of alkaline phosphatases that the sheep mesonephros functions from the eighteenth day; he also used periodic acid-Schiff method to demonstrate protein resorption by proximal tubules cells in both the mesonephros' and metanephros of many species (Davies, 1953).

The metanephros in the rabbit overlaps the mesonephros in capability of function as deduced from histochemical observations, and at birth the metanephric kidney is not fully differentiated, with only the juxtamedullary nephrons being fully formed (Moog and Wenger, 1952; Leeson and Baxter, 1957).

Longley and Fisher (1955) used histochemical procedures in mice to delineate the postnatal development of segmental differentiation within the proximal tubule.

Functional characteristics related to the development and subsequent regression of the pronephros in mammals are not well known (Rossi et al., 1957). The pronephros of the larval frog exhibits the same general secretory characteristics with respect to the active transfer of phenol red previously demonstrated in the adult mesonephros of various cold-blooded vertebrates (Forster, 1948; Forster and Taggart, 1950; Taggart and Forster, 1950; Jaffe, 1954a). Urinary regulation of water and salt does not begin in the pronephros of larval frogs until stage 19 when circulatory and nervous system function has been established (Rappaport, 1955).

During the first few weeks after birth the urea, sodium, and chloride clearances per unit surface area are lower than in human adults, and the urine is always less concentrated in osmotically active materials (McCance and Young, 1941). Reviews of renal function in infants include: McCance (1948, 1950), Smith (1951), Heller (1951), McCance and Widdowson (1952a, b, 1954), and Barnett and Vesterdal (1953). Growth itself in the young, by virtue of selective anaerobic and catabolic activities, relieves the kidney of some regulatory and excretory function imposed upon it in the adult. Dietary experiments in fast-growing animals shows that these general metabolic factors may be more important than the kidney in maintaining homeostasis (McCance and Widdowson, 1956, 1957). These workers stress the need for an integrated view of the kidney's role in the whole organism: "Anyone who would understand the function of the kidney in infancy must think of it in terms of the physiology of the whole animal. We shall only deceive ourselves further if we continue to compare functions isolated in a narrow way, and on some entirely artificial basis, with functions similarly isolated in adults."

Renal hypertrophy may result from various hormonal or dietary factors, or after unilateral loss of kidney function when the contralateral kidney exhibits compensatory overgrowth. Trophic factors include hormones of the anterior pituitary, thyroid, gonads, adrenal cortex, and such dietary factors as vitamins A, B, D, E and, especially, various kinds of high-protein diets. The enormous literature in this field has been reviewed by Smith (1951).

With a high-protein diet the kidneys of young rats undergo hypertrophy with no pathological change (Addis *et al.*, 1926), but no hypertrophy occurs if the diet is begun after the animals are about a year old (MacKay *et al.*, 1926). With intraperitoneal injections, however, renal enlargement is associated with proteinuria and pronounced pathological changes consisting of cellular enlargement and vacuolization, especially in the proximal segment. This hypertrophy is readily reversible, and the kidneys regress to their normal weight within 3–4 days after termination of the injections (Baxter and Cotzias, 1949).

With compensatory renal hypertrophy the glomeruli swell but do not increase in number (Arataki, 1926b; Jackson and Shiels, 1927-1928; Moore, 1929). Hypertrophy begins in rats as soon as the shock of the operation wears off and reaches a maximum in 20 days (Rollason, 1949). Some proximal tubules may increase fivefold in size, although the total kidney mass may only double (Oliver, 1945). Both cortex and medulla increase in size, but the proximal cell mass increases preferentially; this is reflected in corresponding elevation of the maximal secretory rates of diodrast and *p*-aminohippurate. On the second day hyperplasia is indicated by increased mitotic figures in all parts of the tubule except the thin segment, and mitotic activity reaches a peak 72–240 hours after the operation (Sulkin, 1949). Functional studies in the dog 1 week after unilateral nephrectomy show in the remaining kidney an increase in blood flow of nearly 70% and almost a 100% increase in oxygen consumption (Van Slyke *et al.*, 1934).

The interrelationship of compensatory hypertrophy and recovery from injury is revealed in the study of Koletsky (1954), who followed necrotic changes in rat's kidneys after renal circulation had been arrested unilaterally. Damage was always confined to the proximal convoluted tubules and remained indefinitely unless the normal kidney was removed. Complete compensatory hypertrophy and restoration of adequate function resulted if the normal kidney was removed within 3 weeks after injury. Compensatory hypertrophy has been studied in the rabbit, using modern methods of karyometric examination (Fajers, 1957a, b). The possible significance of maturity as a factor in determining the extent of compensatory hyperplasia is noted by Kurnick (1955), who found no increase in DNA content in mature rats following unilateral nephrectomy, despite a great increase in kidney mass. Previously Mandel et al. (1950) had reported increased DNA counts in regenerating kidneys of rats of undisclosed age.

Senile changes in the human kidney have been classified by Binet et al. (1952) and Laroche and Mathé (1955) as follows: congestion of glomeruli and fibrous or hyaline alterations, distension and atrophy of tubules, changes in the small arteries, and sclerosis of connective tissue chiefly in the cortex. It is frequently difficult to distinguish between senescence, as such, and the results of pathological damage to the kidney. Oliver (1942) examined thoroughly the kidneys of seventy-five persons over the age of 70 with no symptoms of renal disease. In every case he found degenerative changes, but these were invariably accompanied by arteriosclerosis. He suggested that the diversion of blood supply which resulted from the narrowed arteries caused the degenerative changes in the nephron, and he drew an analogy to the decline of the mesonephros which Gersh (1937) had shown to be accompanied by a diversion of blood supply to other structures in the developing embryo.

The kidneys of young, "middle-aged," and senile rats were studied extensively by Andrew and Pruett (1957), who noted marked differences in the kidneys of old rats as compared with young or middle-aged ones, but the most predominant features here were not the arteriolar sclerotic changes and glomerular fibrosis mentioned most frequently as the outstanding alterations noted in the senile human kidney. However, the glomeruli of the senile rat show a greater dilation of capillaries with a more pronounced basement membrane, and the glomerular tufts are usually considerably larger than those of young and middle-aged rats. Further histological features are the presence of precipitated colloidlike material in the tubules, and aberrant cells (oncocytes) in various tubular segments. Arterial changes include an accumulation of lymphocytes and plasma cells in the adventitia, clear areas probably caused by deposits of fatty material in the media, and scattered deposits of basophilic material in the adventitia of many arteries.

When hyperphagia was induced in rats by hypothalamic operation, doubling the food intake increased first the size (nitrogen content and

RNA) and then the number (DNA) of cells in the kidney. Obesity so induced caused renal lesions characteristic of aged rats, and other changes
similar to those of regenerative hyperplasia (Kennedy, 1957).

D. Comparative Studies

The aglomerular kidney of such a teleost as *Lophius* with its seg-
mentally undifferentiated "brush border" tubule normally forms hypotonic
urine relatively free of univalent ions. After capture, the rate of urine flow
increases and urine becomes isotonic with blood as univalent ions appear
in concentrations approaching those in plasma. Under laboratory conditions urine volume and composition is very much the same in all marine
teleosts, whether they are totally aglomerular (*Lophius*), partially glomerular (short-horn sculpin, *Myoxocephalus scorpius*), or completely glomerular (long-horn sculpin, *M. octodecimspinosus*). Glomerular fishes in
their normal habitat functionally appear to be essentially aglomerular
until capture, when there is a striking recruitment of glomerular activity
and increases in renal plasma flow accompany markedly increased urine
flows. This "laboratory diuresis" is presumably associated with the drink-
ing of sea water, a falling hematocrit, and alterations in tubular activities
which are identical in glomerular and aglomerular marine teleosts (Forster,
1953).

The ability of proximal convoluted tubule cells to form and excrete
fluid is unequivocally demonstrated in the aglomerular fishes, and under
certain conditions glomerular tubules can also be shown to do so. For
example, the completely glomerular long-horn sculpin after 3 days or
so under the stress of maintenance in diluted sea water will come to have
urine flows more than twice simultaneous glomerular filtration rates
measured as the inulin clearance (Forster, 1953). The ability to secrete
free water is indicated by the hypotonic urine of freshly captured aglomerular fishes where the freezing point or osmotic pressure difference
between urine and plasma may be as high as 20%, equivalent to about
1 atmosphere (Forster, 1953; Forster and Berglund, 1956). This could be
due to selective salt resorption at some distal site, but the complete lack
of regional differentiation makes this explanation rather unlikely.

The presence of a special segment on either side of the proximal
tubule in the elasmobranch nephron has been correlated with the special
ability of such forms as the dogfish to resorb urea actively, but Kempton
(1943) has challenged this suggestion. Cilia in the neck segment and in
the intermediate segment of cold-blooded animals have been viewed as
aids in the propulsion of fluid down the tubules in these animals which
have relatively low filtration pressures. The thin segment of the medullary
loop, characteristic of certain birds and mammals, has been implicated in

their ability to form hypertonic urine (Marshall, 1934), but Sperber (1944) in his exhaustive study of mammalian kidney structure rejects the theory as unfounded.

Structural specializations in the enormously elongated collecting ducts of such arid-living forms as the desert rat have been correlated with their ability to form urine with osmotic urine : plasma concentration ratios as high as 18, compared to maximal values of about 4 in man (Howell and Gersh, 1935; Schmidt-Nielsen and Schmidt-Nielsen, 1952). Their antidiuretic hormone content is also high in urine and the pituitary, compared to the laboratory rat (Ames and Van Dyke, 1950). Localization of the water conservation mechanism in the collecting ducts is suggested by the greatly elongated form characteristic of these species (Sperber, 1944; Vimtrup and Schmidt-Nielsen, 1952). Vimtrup (1949) observed directly that indigo carmine was concentrated progressively as it passed along the collecting duct and dense masses of dye in the terminal portion appeared almost to block the lumen. Vimtrup and Schmidt-Nielsen (1952) described special interstitial cells of the renal papillae in the desert rat, *Dipodymys,* which Sternberg *et al.* (1956) have examined in detail using DPN and TPN diaphorase stains. They point out that, while these cells appear in a variety of mammals, they are most apparent in rodents, and especially in the kangaroo rat, where as end organs for the action of anti-diuretic hormone they may constrict the lumina of collecting ducts, retard flow, and thereby facilitate tubular water resorption.

Adult vertebrate kidneys are traditionally classified as being either pronephric, mesonephric, or metanephric, and they are viewed as having their respective counterparts in homologous structures which appear in sequence during the embryonic development of amniotes. However, no special functions can be ascribed as characteristic of these types. On the contrary, biochemical and physiological characteristics underlying active transport mechanisms appear to be the same, whether the observations on secretory processes held in common are made on fish, amphibians, reptiles, birds, or mammals. Most is known, perhaps, about the organic acid-secreting mechanism whose characteristics with respect to energy dependence and transfer competition are identical in the pronephric embryonic frog renal tubule (Jaffee, 1954a), the mesonephric flounder nephron (Forster and Taggart, 1950), and the metanephric rabbit renal cortex (Forster and Copenhaver, 1956). Similarly, an organic base-secreting mechanism has been shown to have general characteristics which are shared by vertebrate nephrons from fish to mammals (Sperber, 1948a; Beyer *et al.,* 1950b; Forster *et al.,* 1958).

Davies and Davies (1950) have pointed out that the intensive search for homology has led many anatomists to speculations which are unjusti-

fied and misleading. Fraser (1950) minimizes structural differences of the
pro-, meso-, and metanephros and points out that the nephrogenic material
throughout the vertebrates arises embryologically from a single source.

Many active cellular transport systems exist in renal tubules of certain
species and not in others. These unique active transfer mechanisms include urea secretion in the frog (Marshall and Crane, 1924; Marshall,
1932a: Walker and Hudson, 1937a; Forster, 1954; Schmidt-Nielsen and
Forster, 1954) and perhaps in some mammals (Schmidt-Nielsen *et al.*,
1957); uric acid in certain birds (Mayrs, 1924; Gibbs, 1929; Shannon,
1938a; Sperber, 1948b) and reptiles (Marshall, 1932b); trimethylamine
oxide in aglomerular *Lophius* (Grollman, 1929; Forster *et al.*, 1958);
divalent ions in *Lophius* (Marshall and Grafflin, 1928; Forster and
Berglund, 1956; Berglund and Forster, 1958); phosphate in the chicken
(Levinsky and Davidson, 1957); and creatine in fishes (Marshall and
Grafflin, 1928; Pitts, 1936; Forster *et al.*, 1958). Endogenous creatine is
secreted by *Lophius* (Grollman, 1929; Forster *et al.*, 1958) and exogenous
creatinine, by aglomerular and glomerular fishes (Edwards and Condorelli,
1928; Clarke and Smith, 1932; Marshall and Grafflin, 1932; Pitts, 1935–
1936; Shannon, 1933, 1934, 1938b, 1940), the chicken (Shannon, 1938c), the
anthropoid apes (Smith and Clarke, 1938), and man (Shannon, 1935;
Shannon and Ranges, 1941). Urea is actively resorbed in elasmobranchs
(Smith, 1936; Kempton, 1953; Forster and Berglund, 1957) as is trimethylamine oxide (Cohen *et al.*, 1956b). However, none of these secretory
mechanisms can be identified with the presence or absence of specific
intracellular structural features, with any degree of regional specialization
of the tubule or the organ's status as a pronephros, mesonephros, or
metanephros. Whatever form may underly these discrete functions must
be sought for at the molecular, rather than at the cellular or organ, level.

Rhodin (1956) has used electron micrographs to compare structure in
the aglomerular tubules of *Lophius* with tubular cell structure in the
dogfish elasmobranch, *Squalus,* and in Mammalia. The brush-border cell
of the glomerular dogfish appears identical to that of the mammal. However, cells of the aglomerular tubule lack the basal infoldings of the
plasma membrane which Rhodin implicates in the water resorption
process.

IV. Special Techniques

A. *Micropuncture*

The filtration resorption hypothesis advanced by Cushny (1917) was
substantiated by direct micropuncture studies on frogs, necturi, and
snakes (Wearn and Richards, 1924). With exquisite manipulative and

chemical methods glomerular urine in these cold-blooded forms was shown to be an essentially protein-free ultrafiltrate of plasma with concentrations of glucose, chloride, urea, uric acid, phenol red, inulin, hydrogen ions, creatinine, phosphate, and total electrolytes similar to that of their unbound counterparts in plasma (Richards, 1938). The mean glomerular capillary pressure of frogs was found to be 20.0 cm. H_2O or 54% of the average aortic pressure (Hayman, 1927; White, 1929). Later techniques were extended to obtain fluid from known levels of a renal tubule and to perfuse selected tubule segments (Richards and Walker, 1937; Kempton, 1937).

These experiments on amphibians disclosed that no significant change in vapor pressure and chloride concentration occurred until the glomerular filtrate had traversed the entire length of the proximal and intermediate segments. Chloride resorption in the distal convoluted tubule implicated this site in the formation of the characteristically hypotonic amphibian urine (Walker et al., 1937). Acidification did not occur until urine reached the distal tubule. The yellow color of phenol red, indicative of pH 7.0 or lower, did not develop in the proximal tubule, no matter how long the fluid column was retained in it (Montgomery and Pierce, 1937). Normally glucose was resorbed in the first half of the proximal tubule (Fig. 8), but with perfusion experiments involving high glucose levels in tubular urine it was shown that sugar could passively diffuse out of the distal tubule (Walker and Hudson, 1937b).

Micropuncture observations were extended to include mammals, especially the guinea pig and rat, after an examination of many species showed these to be the only forms with glomeruli on the kidney surface. A simplified technique for identifying puncture sites was introduced involving maceration and microdissection of tubules rather than the more laborious examination of serial sections (Walker and Oliver, 1941). Glomerular fluid, as in amphibians, was entirely or almost completely free of protein, and reducing substances and creatinine were in concentrations similar to those in plasma water. All reducing substance was resorbed in the proximal convolution, and at least two-thirds of the fluid volume was resorbed here via an isosmotic process (Walker et al., 1941).

Micropuncture studies on rat kidney show that intratubular pressures in the upper proximal segment average 13.5 mm. Hg and under all experimental conditions are approximately the same as pressures in the peritubular capillaries (Gottschalk and Mylle, 1956). Independently, Wirz (1955) reported that pressures in the proximal tubules of rats averaged 14.5 mm. Hg, when postglomerular capillary pressure was 17.4.

In the rat kidney, as in Necturus, resorption of potassium takes place in the proximal tubule at a site roughly similar to that of active glucose

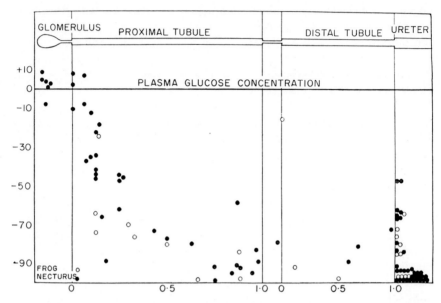

Fig. 8. Diagram showing differences in glucose concentration between plasma and tubular fluid collected by micropuncture from various levels of frog and necturi nephrons. Figures above and below zero on the ordinate represent percentage differences from plasma. Under the circumstances of these experiments the greater part of the glucose resorptive process is accomplished in the upper half of the proximal tubule. From Walker and Hudson (1937b).

resorption (Wirz and Bott, 1954). Osmotic pressures of fluids obtained by micropuncture from rat tubules were measured microcryoscopically by Wirz (1956) while both concentrated and dilute urine were being formed. Proximal tubule fluid is isosmotic with blood even in diuresis when ureteral urine is markedly hypotonic, and urine becomes hypotonic in both concentrating and diluting kidneys before it reaches the distal tubule. In diluting kidneys this hypotonicity is maintained throughout the length of the distal convolution. However, in antidiuresis it becomes isotonic by the middle of the segment and remains so in the rest of the distal tubule. It appears that tubular urine becomes hypotonic before it enters the distal convolution by solute resorption, whereas the epithelium is virtually impermeable to water in both the diuretic and antidiuretic state. Antidiuretic hormone apparently alters permeability in the distal tubule to permit passive resorption of water. Wirz envisages the final concentration of urine as taking place in the collecting ducts by a passive process, with the effective role being played by a countercurrent mechanism involving the loops of Henle. These are viewed as providing a hypertonic environment for the collecting ducts so that urine within the latter is concentrated in solute through loss of water to this hypertonic milieu. The counter-current explanation of osmotic pressure regulation was originally developed in a study involving measurement of freezing point depression in frozen sections by observing the disappearance of ice crystals under a polarizing microscope while temperature was gradually raised (Wirz et al., 1951).

Transtubular electropotential differences measured in rats by insertion of microelectrodes showed that the inside of the tubule is always negative to the outside. Measurements of randomly selected surface tubules resulted in a bimodal population distribution, with one group (probably proximal) ranging from 19 to 39 mv. and another (probably distal) ranging from 34 to 70 mv. Interruption of blood flow was accompanied by a fall in potential difference to low values which were restored to normal with the re-establishment of circulation (Solomon, 1957). Earlier Wilbrandt (1938) found in Necturus also that the inside of the proximal tubule was negative to the outside, but polarity was reversed in the distal tubule. No explanation can be provided at present to account for this species difference.

Micropuncture studies in certain fresh-water invertebrates show, as in amphibians and mammals, that excretory fluid, initially isotonic, is rendered hypotonic in the lower portion of the nephron. Ramsay (1949) demonstrated that the ability to form hypotonic urine in the earthworm nephridium, for example, was definitely present in the terminal "wide" tube, possibly present in the middle tube, and absent in the initial "narrow" tube. Other work with fresh-water invertebrates which suggests a filtration-resorption system of urine formation analogous to that of vertebrates includes that of Picken (1937) and Florkin and Duchâteau (1948) on Anodon, and Peters (1935) on Astacus.

Micropuncture procedures have been used on in vitro preparations of isolated Malpighian tubules of the stick insect to study functions in specific regions designated as proximal, middle, and distal. Electro-potential differences across the tubular epithelium have been measured in each region, as well as freezing point depression, and sodium and potassium concentrations (Ramsay, 1955a).

B. Renal Portal Perfusion

The dual venous and arterial blood supply to kidneys of amphibians and birds has provided methods for studying discrete transfer processes in renal tubules while avoiding the traumatic complications of micropuncture. In the frog, for example, arterial blood is delivered to glomeruli and distal tubules in the ventral half of the kidney, whereas the proximal tubules in the dorsal half are supplied exclusively by blood from the renal portal vein. Because of this twofold blood supply it is possible to deliver different solutions to the two halves and thereby to investigate transfer processes in the proximal tubule separately from glomerular filtration and tubular activity in the distal segment. These advantages were first pointed out by Nussbaum (1878a, b), and Cullis (1906) developed a practical procedure for perfusing the isolated frog kidney with balanced isotonic salt solution. Further refinements by Bainbridge et al. (1914), Atkinson et al. (1921), Barkan et al. (1921), and Richards and Walker (1927) included precautions to ensure differential supplies to the proximal tubules and to the glomeruli and distal epithelia. When normal pressures are maintained in the aortic and portal perfusion systems (24 and 12 cm. H_2O, respectively) these fluids, although having a common venous drainage after passing through the peritubular capillary bed, remain sufficiently separated to allow a distribution to the proximal tubule which is distinct from that to the glomeruli and distal region of the nephron.

As an added precaution, different compounds excreted solely by glomerular filtration may be separately added to the portal and aortic perfusates to measure with precision the individual contributions which the fluids in the two channels may make to the glomerular filtrate (Hogben and Bollman, 1951). With this refinement tubular exchanges of various substances have been studied in the frog: phosphate (Hogben and Bollman, 1951), deuterium oxide (Swanson et al., 1956), Na^{22} and K^{42} (Hoshiko, 1956; Hoshiko et al., 1956), and urea (Love, 1957). Creatine:inulin clearance ratios in these preparations range from 0.95 to 1.50, and average 1.27 (Swanson, 1956), but in intact frogs their simultaneous clearances are identical and are assumed to be valid measurements of filtration rate (Forster, 1938; Sawyer and Sawyer, 1952; Sawyer, 1957). Anesthetization, laparotomy, and extensive traumatization involved in the perfusion experiments might alter tubular activity, but it is difficult to see why these incidental factors should cause the proximal tubule to "secrete" creatinine.

The renal portal circulation of the chicken has been put to use in a method which avoids the traumatization of double perfusion experiments and permits quantitative evaluation of discrete tubular secretory processes. Unilateral peritubular perfusion is attained by the injection of material into a leg vein on one side which bypasses the glomerulus on the first circulation through the kidney, and, only after completing the heart-lung circuit, is filtered equally by both kidneys. Separate urine collections obtained from each ureter immediately after injection permit comparisons of urine composition of the kidney on the perfused side with that of the control kidney on the other side (Sperber, 1948c). By this procedure, various secretory processes have been delineated: organic bases (Sperber, 1948a), sulfuric esters of phenols (Sperber, 1949), glucuronic acid derivatives (Sperber, 1948c, d), histamine (Lindahl and Sperber, 1956), tetraethylammonium ion (Rennick *et al.*, 1954), potassium (Orloff and Davidson, 1956), mercurial diuretics (Campbell, 1957), and phosphate (Levinsky and Davidson, 1957).

C. Clearance Methods, General

Net values for either the tubular excretion or tubular resorption of any given substances can be calculated when glomerular filtration rates are known. To measure the latter, some substance must be used which is freely filterable at the glomerulus and known not to permeate the tubular epithelium. The most widely used substance for this purpose is the nonmetabolizable polysaccharide, inulin. When the total amount of inulin excreted by the kidneys per unit time has been determined, then the milliliters of glomerular filtrate required to deliver that quantity of inulin can be calculated by dividing the amount excreted by the amount of inulin in 1 ml. of glomerular filtrate (protein-free plasma). Measurement of glomerular filtration rate (inulin clearance) requires exact methods for measuring rates of urine flow and precise chemical procedures for determining the relative concentrations of inulin or other test substance in urine and plasma.

When the glomerular filtration rate is known, it is simple to determine for any plasma constituent the net amount secreted by the tubules (amount excreted minus amount filtered) or net resorption (amount filtered minus amount excreted) subsequent to filtration. Likewise, total blood flow through the kidneys can be calculated by dividing the total amount of any given substance removed by the kidneys from the circulation per unit time by the amount extracted from each milliliter of blood as it traverses the renal circulation (concentration in arterial blood minus concentration in renal venous blood). For some substances, such as diodrast and *p*-aminohippurate (PAH), it has been shown that blood is essentially cleared completely with each circuit through the kidney when plasma concentrations of these are low. The value for concentration in renal venous blood emerging from the peritubular capillary bed then approaches zero, and blood flow can be determined simply by dividing the total amount of PAH or diodrast excreted per minute by the amount of either of these substances in 1 ml. of arterial blood.

Clearance techniques provide information on overall renal function and are not by themselves of much value in approaching problems concerned with cellular function in specialized regions of the nephron. However, combined with evidence from micropuncture and comparative renal function studies, some information has been obtained by clearance investigations which serves to localize cell functions in the nephron. These studies will be referred to specifically in subsequent sections when discrete cell processes are discussed. The tremendous literature on the kidney as an organ in health and disease was extensively reviewed by Smith (1937, 1943, 1951) and encapsulated by him more recently in a very useful volume stressing basic principles and techniques (Smith, 1956). Specific techniques used in clearance determinations and in other renal function studies can be found in a section on Methods of Renal Study, edited by Corcoran, in *Methods in Medical Research* (1952).

Clearance determinations of glomerular filtration rates have been correlated with direct microscopic observations of glomerular activity (Richards and Schmidt, 1924; Forster, 1943), with varying degrees of glomerular development in marine teleosts (Forster, 1953), and with sudden alterations in glomerular activity in frogs (Forster, 1942; Schmidt-Nielson and Forster, 1954), in seals (Bradley and Bing, 1942), and in rabbits (Forster and Nyboer, 1955). Localization of renal tubular transport along the nephron is characterized by allowing concentration patterns to develop during ureteral occlusion with stopped filtration in a novel method introduced by Malvin *et al.* (1957). The pattern is then caught in serial urine samples rapidly collected from a ureteral cannula after the brief occlusion is released. This method holds promise of assigning specific secretory and resorptive processes to regions of the tubule relatively proximal or distal to one another.

D. In Vitro Procedures

Isolated renal tubules or slices of kidney tissue when maintained in oxygenated synthetic media may exhibit for several hours high metabolic rates and the capacity to transport or accumulate actively many organic and inorganic substances. The advantages over direct observation *in situ* are: (1) Isolated preparations permit localization of the substance undergoing transport, either by microscopic observation on its accumulation within cell or lumen directly in the case of a colored substance, or indirectly with the aid of radioautographic or fluorescent techniques. (2) Selected inhibitors and antagonists may be used in concentrations which would otherwise have deleterious side effects when administered intravenously to the intact animal. (3) The isolated system is not exposed to extrarenal factors which could affect the transfer system in the intact animal, such as changes in renal blood flow, blood pressure, or alterations in the chemical or hormonal composition of the blood which are difficult to control when the kidney is an integral part of a living organism (Forster, 1948).

Kidney fragments of chick mesonephros cultured at 39° C. in a mixture of chicken plasma and embryonic juice accumulate phenol red intracellularly and within the tubular lumina (Chambers and Cameron, 1932; Chambers and Kempton, 1933). Various physiological conditions associated with this secretory system were examined subsequently by Chambers et al. (1935), Beck and Chambers (1935), and Keosian (1938). Cultured human metanephros taken from a 3½-month fetus has been shown similarly to secrete phenol red and orange G (Cameron and Chambers, 1938).

A simpler system which permits direct observation of phenolsulfonphthalein (PSP) transport in a totally synthetic oxygenated balanced salt solution was devised for isolated tubules and thin slices prepared from kidneys of cold-blooded vertebrates (Forster, 1948). With this method the organic acid transport mechanism was characterized with respect to specific energy dependence, transfer rate maxima, and competitive inhibition (Forster and Taggart, 1950; Taggart and Forster, 1950; Puck et al., 1952), and the energetic relations existing at both cell membranes were examined (Forster and Hong, 1958; Hong and Forster, 1958).

Fish kidneys are best for the isolated tubule preparation because there is little cementing substance binding the nephrons and they are easily disentangled. Marine teleosts such as the common flounder present a very uniform appearance in that the kidney tubule is made up solely of the actively secreting "proximal" brush-border segment. Isolated tubules of fresh-water fishes are equally active, but dye uptake is spottier because of the long, inert distal segments. However, the latter are useful for studying acid-base regulation with the aid of suitable indicators. Kidneys of the other classes of vertebrates are much more compact and are best examined as thin slices prepared as for in vitro respiration studies (Taggart and Forster, 1952).

Cross and Taggart (1950) devised a method which has been widely applied for measuring PAH accumulation in slices of rabbit kidney cortex, and Beyer et al. (1950a) used cortex slices from the guinea pig and other mammals to study phenol red uptake in vitro. Accumulation in mammalian slices appears to be closely related to tubular excretion in intact animals, and the in vitro process shows many of its characteristics; however, uptake here appears to represent only preliminary intracellular accumulation rather than over-all transfer and luminal concentration as occurs in isolated tubules and thin slices of cold-blooded forms (Forster and Copenhaver, 1956).

In vitro methods have been used by Opie (1949) and Robinson (1950) to study the movement of water into and out of kidney slices. Mudge (1951a, b) describes in detail methods for leaching sodium and potassium from slices as preliminary to characterizing the reaccumulation process with respect to energy dependence, competition, etc. Other procedures for observing salt and water balance in mammalian slices *in vitro* include: Robinson (1953), Whittam and Davies (1953), Deyrup (1953a, b, 1957), Schwartz and Opie (1954), Conway and Geoghegan (1955), Whittam (1956), and Leaf (1956). Sulfate is accumulated by slices of kidney cortex and medulla, and S^{35} has been used to measure its uptake (Deyrup, 1955; Deyrup and Ussing, 1955).

Ramsay (1954) describes in detail an *in vitro* procedure for maintaining Malpighian tubules of the stick insect in a droplet of hemolymph. Microtechniques are described for determining osmotic pressure (Ramsay, 1949), chloride (Ramsay *et al.*, 1955), sodium and potassium by flame photometry (Ramsay *et al.*, 1953), and electropotential differences across the tubular wall of the *in vitro* preparation (Ramsay, 1953b). Unlike vertebrate nephrons, this preparation is incapable of carrying on active transport in synthetic media. The contractile elements of the tubule wall, however, maintain normal activity in simple salt solutions long after the secretory cells have completely disintegrated. This preparation actively transports sodium and potassium against an electrochemical gradient. In the majority of cases the urine is slightly but significantly hypotonic to the hemolymph.

V. Excretory and Regulatory Functions

A. *Water and Sodium*

Water accumulates against its own concentration gradient in all the aglomerular renal tubules examined to date. The cysts formed by closure of cut ends of fragments of chick mesonephros proximal convoluted tubule grown *in vitro* elaborate, against a hydrostatic pressure, a fluid which is on the average 10.7% more dilute than the culture medium (Keosian, 1938). The Malpighian tubules of insects *in vitro* form urine which is usually hypotonic to the ambient hemolyph (Ramsay, 1954). Finally, the aglomerular marine fish, *Lophius*, forms urine *in situ* which is very low in chloride and sodium, and 15% more dilute than plasma (Forster, 1953; Forster and Berglund, 1956). The only instances reported where glomerular tubules have been demonstrated to form fluid apart from filtrate are in the frog proximal tubule (Bensley and Steen, 1928; Ellinger and Hirt, 1930, 1931), and in the marine teleost, *Myoxocephalus octodecimspinosus*. The latter, when placed in diluted sea water, will in time have urine flows which significantly exceed simultaneous glomerular filtration rates measured as the inulin clearance (Forster, 1953). Experiments with D_2O indicate that it diffuses from the renal portal blood into the tubular lumen in the perfused bullfrog (Swanson *et al.*, 1956). With sodium also, despite the fact that net movement is from lumen to peritubular capillaries, studies with radiosodium indicate a bidirectional flux. The activation energy for the resorptive sodium flux (24,000 to 34,000 calories per mole) suggests a carrier system in which diffusion of a sodium complex is the rate-limiting step (Hoshiko, 1956; Hoshiko *et al.*, 1956; Thomas, 1956). *In vitro* studies indicate that sodium and potassium are actively resorbed, and chloride passively, along a gradient in the proximal tubule (Whittam, 1956). Krakusin and Jennings (1955) have used radioautographic techniques for the localization of Na^{22} in rat kidney.

The classical filtration-resorption hypothesis assumes that urinary water and salt have their origin in the glomerular filtrate, and hypotonic urine results from the resorption of solute by the tubule. The presence of this "osmotically free water" (Smith, 1952) in tubular fluid means that the tubular epithelium resorbs solute while acting to some extent as a barrier to the passive back diffusion of water. Urine entering the distal tubule is generally assumed to be isotonic with plasma, and free water is formed there (hypotonic urine) by the preferential resorption of solute (Wesson and Anslow, 1952). Considerable evidence indicates that the active concentrating mechanism, which forms hypertonic urine by resorption of solute-free water, is situated in the collecting ducts (Smith, 1956). This operation does not appear to require the antidiuretic neurohypophyseal hormone for its activation (Berliner and Davidson, 1956). Since indications are that the active transport of water by this process is limited by a maximum rate (T_m), its effectiveness in raising the osmolar concentration of the urine is influenced by the volume of tubular fluid presented to it. According to this hypothesis, during water diuresis the distal tubule remains relatively impermeable to water, and free water in excess of the collecting duct T_m passes into the bladder and the urine is hypotonic. When the antidiuretic hormone is present, the distal tubule is permeable to free water so that the volume of isotonic urine entering the collecting duct is relatively small and the active concentrating operation can now remove a sufficient fraction of water to render the urine hypertonic.

Sawyer (1957) suggests that the neurohypophyseal hormones facilitate resorption in the distal tubule by increasing pore size in a manner similar to that proposed by Koefoed-Johnsen and Ussing (1953) and Ussing (1954a, b) for the toad skin.

Amphibians, and all other cold-blooded vertebrates, are incapable of forming hypertonic urine. The effectiveness of antidiuretic hormone in reducing urine flows in toads and frogs indicates that the mechanism of action proposed for mammals, increasing the size of submicroscopic pores in the distal tubule, applies similarly to these forms (Sawyer, 1957).

A vast literature has accumulated concerning the kidney's role in regulating salt and water balance. Smith's review (1957) deals especially with salt and water volume receptors, and Welt's paper (1956) is of particular value as an introduction to literature reporting the influence of disease on the renal excretion of water. The secretion and transport of water in various aquatic and terrestrial animals, as well as by slices *in vitro*, has been reviewed in detail by Robinson (1954b).

B. Acid-Base Regulation

The inadequacy of the over-simplified "filtration-resorption" hypothesis is no more clearly demonstrated than in acid-base and electrolyte regulation. In a complex series of vaguely defined metabolic events, hydrogen ions are generated and transported in processes which are demonstrably linked with the simultaneous excretion of ammonia, bicarbonate, phosphate, sodium, potassium, and chloride. The elimination of these substances is dependent upon equilibrium conditions which are sensitively attuned, on one side of the cell, to the ever-changing composition of blood, and on the other to the composition of tubular urine. Very little is known of the intracellular processes involved in the formation of hydrogen ions, or of the transport mechanism effecting their movement into and subsequent accumulation within the lumen. Matters are further complicated by marked species differences among the various vertebrates.

The aglomerular tubules of marine teleosts, and the glomerular tubules of elasmobranchs and marine fishes invariably form an acid urine. The acidity cannot be altered by administration of sodium bicarbonate, alkaline phosphate, or phlorizin (W. W. Smith, 1939). It appears that this fixation of acidity might be related to the high urinary magnesium content of marine forms which would precipitate as $Mg(OH)_2$ or $MgHPO_4.3H_2O$ and clog tubules and collecting ducts if pH approached neutrality (Pitts, 1934). The mechanism of acidification in marine fishes was explained as a process of exclusion related to exchange of H^+ for cations (Smith, 1937). Later, the H^+ exchange theory was applied to urine formation in amphibians and mammals when sulfanilamide or the sulfonamide derivative Diamox (6063, 2-acetylamine-1,3,4-thiadiazole-5-sulfonamide) was used to inhibit carbonic anydrase implicated in the H^+ exchange mechanism (Höber, 1942; Pitts and Alexander, 1945; Pitts and Lotspeich, 1946; Berliner *et al.*, 1951). A carbonic anhydrase-dependent system is not involved in the acidification of urine in the dogfish (*Squalus acanthias*) nor in the marine sculpins (*Myoxocephalus scorpius* and *M. octodecimspinosus*) (Hodler *et al.*, 1955). In these marine fishes the gills were found to contain carbonic anhydrase which participates in the excretion of sodium bicarbonate at this site. Aside from their role in respiration and acid-base regulation, the gills play an important part in the excretion of univalent electrolytes and ammonia (Smith, 1953). Fresh-water fishes, like amphibians and mammals, regulate the H^+ content of body fluids via a carbonic anhydrase-dependent system in the kidneys (Hodler et al., 1955).

Direct observations on dyes in cultured mesonephric tubules of the embryonic chick disclosed that the intracellular contents of proximal tubule cells have a pH of 6.8 ± 0.2 (Chambers and Cameron, 1932). Phenol red was secreted solely by proximal cells, and when it flowed into the lumen of the distal portion, acidification was indicated by color change from pink to yellow (Chambers and Kempton, 1933). The distal tubule was confirmed as the site of acidification in direct observation on intact amphibian tubules by Montgomery and Pierce (1937) and Giebisch (1956). However, in winter frogs proximal urine may be acidified to pH 6.0, and indicator dye in the transected kidney of the rat discloses pH reactions here of 6.2 to 6.5 with indications of even higher H^+ concentration in acidotic animals (Ellinger, 1940).

No direct micropuncture studies have been made on mammals, but it is inferred that both proximal and distal segments participate in an exchange mechanism involving H^+ and Na^+, with intracellular H_2CO_3 as the source of H^+ (Smith, 1956). According to this hypothesis carbon dioxide diffuses freely from plasma into the cells where it reacts with H_2O to form H_2CO_3. Upon dissociation H^+ passes into urine in exchange for equivalent amounts of resorbed Na^+ to generate acid out of neutral buffer salts. The $NaHCO_3$ thus formed from Na^+ and HCO_3^- is restored to blood. Carbonic anhydrase inhibitors slow down the intracellular formation of H_2CO_3 from CO_2 and H_2O and diminish thereby the supply of H^+ for the $H^+–Na^+$ exchange mechanism. The relatively alkaline urine thus formed would carry with it larger quantities of filtered sodium. A complicating feature of this generally accepted explanation is the paradoxical action of Diamox in the alligator. Some reptiles, such as the turtle, resemble mammals, amphibians, and fresh-water fishes in their response to carbonic anhydrase inhibition (Williams, 1956). In the alligator, however, Diamox administration results in acidification rather than alkalinization, and carbonic anhydrase appears to be involved chiefly in anion excretion, with the result that the output of chloride rather than sodium is increased (Coulson and Hernandez, 1955, 1957).

Potassium is secreted by the distal tubule via a common transport mechanism with H^+, and factors which promote the excretion of one simultaneously inhibit that of the other (Berliner et al., 1945a, b). Final acidification of the urine also involves the formation of ammonia, probably in the second half of the distal tubules and in the collecting ducts, as shown directly by micropuncture studies in the amphibian (Walker, 1940), and inferentially by clearance studies. As urine becomes increasingly acid, diffusible NH_3 is trapped in urine as nondiffusible NH_4^+ and excreted as such. Glutaminase activity has been implicated in the synthesis of ammonia (Davies and Yudkin, 1952; Hines and McCance, 1954; Rector *et al.*, 1954). Pitts (1948) and Smith (1956) have emphasized its role in over-all H^+ and electrolyte regulation. Krebs (1935), Jacobellis *et al.* (1955), Leonard and Orloff (1955), Muntwyler *et al.* (1956), and Richterich-van Baerle *et al.* (1956) have demonstrated important species differences among mammals. Ammonia formation in kidney slices increases with acidity of a buffered medium, is depressed anaerobically and then its formation is independent of pH. Surviving slices of the newborn are not as active as those of the adult with respect to oxygen uptake or ammonia production. Robinson (1954c) concludes that the stimulating factor in ammonia formation by tubular cells is the acidity of fluid within the lumen. Goldstein *et al.* (1956), Richterich-van Baerle and Goldstein (1957), Richterich-van Baerle *et al.* (1957), and Goldstein *et al.* (1957) give evidence for the existence of two glutaminases and a glutamine synthesizing enzyme in guinea pig kidneys, which respond differentially to treatment with acid and alkali. They emphasize the usefulness of this ammonia production system as a model for the study of enzyme adaptation.

C. Mechanism of Active Transport

The most widely studied active transport mechanism is one which moves against high concentration gradients a series of organic acids including phenol red, *p*-aminohippurate, diodrast, and penicillin. The system which operates in many invertebrates and in all classes of vertebrates is restricted to the brush-border ("proximal") cells. Its general features with respect to transfer maxima and competitive inhibition were established in clearance studies on intact animals, but only with the development of *in vitro* techniques was it possible to characterize specifically the individual steps involved in the transcellular movement of these compounds. The main features of active transfer will be delineated as they were revealed in an isolated tubule preparation transporting certain phenolsulfonphthaleins, and in addition reference will be made to other *in vitro* and *in vivo* observations on this and other systems.

1. Energy Dependence

The over-all secretion of phenol red is dependent, at least in part, on the utilization of phosphate bond energy. This was disclosed by the observation that invariably inhibitors of aerobic phosphorylation (2,4-dinitrophenol, 2,4-dinitro-6-methylphenol, 2,4-dinitro-6-phenylphenol, 2,6-dinitro-4-chlorophenol, 2,4-dichloro-4-nitrophenol), in similar concentrations, reversibly inhibited phenol red secretion and stimulated oxygen uptake. Other substituted phenols which were inactive in blocking the uptake of inorganic phosphate in the phosphorylation system (2-nitrophenol, 4-nitrophenol, 2-amino-4-nitrophenol, 2-4-diaminophenol, 2,4-dichlorophenol, 2,4-dinitroanisole) did not affect oxygen uptake or phenol red transport in the isolated tubule preparation. It was suggested that one of the cellular components of the transport system may be raised to a reactive state through interaction with an energy-rich phosphate, as an alternative to the other possibility that phenol red itself was phosphorylated during its passage across the tubule (Forster and Taggart, 1950; Taggart and Forster, 1950). Mudge and Taggart (1950) extended these observations to show by clearance methods that 2,4-dinitrophenol similarly inhibited PAH secretion in the intact dog without simultaneously affecting glucose resorption.

Other general metabolic inhibitors were found also to block dye transport reversibly. These included anaerobiosis, cyanide, cold, and certain more or less specific agents which tie up sulfhydryl compounds or inactivate cytochrome oxidase and succinic dehydrogenase. Several of these agents were examined for their effects on respiration; oxygen uptake was depressed in each instance at the minimal concentration which completely blocked phenol red transport (33% by $M/100$ phlorizin, 60% by $M/100$ arsenite, and 71% by $M/50$ azide) (Forster and Taggart, 1950).

In a preparation using thin slices of rabbit cortex which takes up p-aminohippurate from a synthetic medium, Cross and Taggart (1950) found that the accumulation characteristics resembled those for phenol red in the isolated tubule preparation and for PAH secretion in the intact animal. Intermediate reactions involved specifically in PAH uptake have recently been summarized by Taggart (1956a). Confirmation of PAH and phenol red transport dependence upon Kreb's cycle activity and aerobic phosphorylation was provided by Shideman and Rene (1951) and Beyer et al. (1950a), and a similar dependence has been shown for the secretion of potassium (Mudge, 1951, 1955) and urea in the frog (Forster, 1954).

The temperature coefficient for PAH transport is 2.0 in fishes (Forster, 1953), and similar agreement with van't Hoff's law has been obtained in mammals where E or μ values for tubular maxima of PAH in the order of $+23,500$ cal. per mole, characteristic of first-order biological actions, have been reported (Page, 1955; Blatteis and Horvath, 1958). Inhibition of renal tubular functions by cold has also been noted in the alligator (Hernandez and Coulson, 1957) and in mammals during hypothermia and hibernation (Hong, 1957).

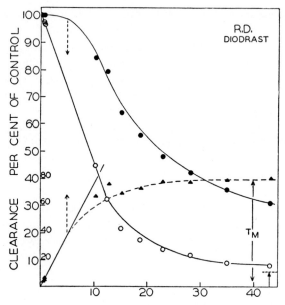

Fig. 9.   Transfer competition of diodrast and phenol red demonstrated in man by elevation of plasma concentrations of diodrast (abscissa). The triangles show the rate of tubular excretion of diodrast, and the solid circles show its clearance. At higher plasma levels the rate of tubular excretion approaches a maximal value, T_m. When the plasma level of diodrast is elevated, the simultaneous phenol red clearance (open circles) is depressed, even though the concentration of phenol red in the plasma is maintained below the level where its self-depression begins. These observations indicate that phenol red and diodrast are excreted by a common mechanism subject to competitive inhibition. Diodrast also displaces phenol red from its combination with plasma protein and thus increases the free and filterable fraction. From Smith *et al.* (1938).

2. *Transfer Competition*

Phenol red transfer is inhibited competitively when other organic acids, as diodrast, penicillin, PAH, and probenecid, are presented simultaneously to the isolated tubule preparation. This reversible competition, presumably for an intracellular or membrane carrier, is very specific (e.g., not involving *p*-aminobenzoic acid or actively transferred organic bases), and inhibition occurs without affecting oxygen uptake by the cells. The interrelationship was first thoroughly established by Smith *et al.* (1938), using clearance techniques on man (Fig. 9), and later extended to include penicillin competition with diodrast by Rammelkamp and Bradley (1943), with PAH by Beyer *et al.* (1944), and with caronamide by Beyer *et al.* (1947). Sperber (1954) has used his chick renal portal preparation to show with the phenolsulfonphthaleins that the more sluggishly transported members of the series are generally the more effective competitive inhibitors.

3. Transfer Maxima

A specific limitation is imposed upon the amount which can be transported across cells per unit time. This transfer maximum (T_m) is probably determined by the fixed amount of membrane or intracellular carrier available for interaction with the transported substances, and also by the amount of free energy available as a source of power to move these substances against an electrochemical gradient. This is in contrast to permeation by filtration or free diffusion where movement across cells is in direct proportion to the electrochemical gradient. The T_m concept has its origins in the study of Marshall and Crane (1924) who compared the relative rates of excretion of phenol red and urea in the dog as plasma concentrations of these substances were progressively increased. At low plasma concentrations the U:P ratio of phenol red was much higher than that of urea, but whereas urea excretion increased in a linear fashion, that of phenol red rose rapidly at first and then very slowly as plasma levels were raised. They suggested that the cells became saturated with phenol red at a critical plasma level, and above that any further increase came only from increased glomerular filtration. The transfer maximum is revealed by a self-depression of clearance values as the plasma concentration is raised, as first demonstrated for phenol red by Goldring et al. (1936) and for diodrast and hippuran in the dog by Elsom et al. (1936). Smith et al. (1938) used the phenomenon to measure active tubular excretory mass. Transfer maxima have been expressed for many renal secretory and resorptive processes throughout the vertebrates (Smith, 1951). The isolated renal tubule preparation and other in vitro systems similarly reveal the T_m phenomenon (Fig. 10) (Forster and Taggart, 1950; Cross and Taggart, 1950; Puck et al., 1952).

4. Cell Membrane Permeation

Actively transported substances are generally assumed to diffuse passively into the cell, and then by some energy-dependent enzymatic or carrier reaction to be actively extruded from the cell into the tubular lumen. However, this must be questioned in the light of direct observations on thin slices and isolated tubules. It appears that both the peritubular and the luminal cell borders are sites of energy-driven transfer mechanisms subject to competitive inhibition.

Normally movement into the cell is the rate-limiting reaction in transcellular excretion, but in two in vitro preparations transcellular movement

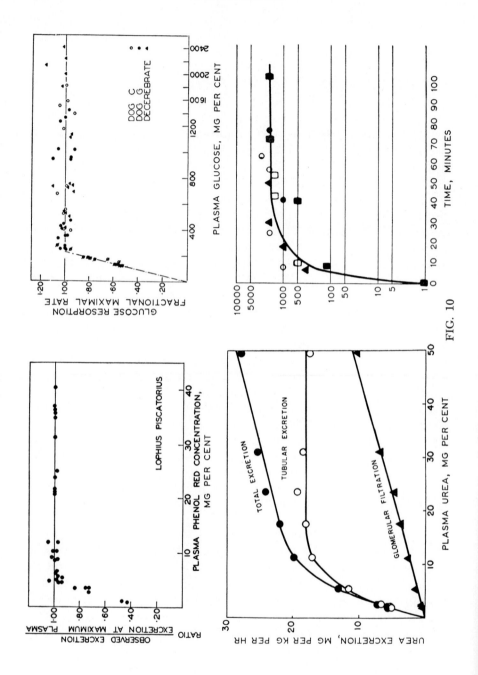

FIG. 10

is blocked at the lumen side of the cell, and under these circumstances phenol red accumulates intracellularly rather than within the lumen. First, in the isolated flounder tubule, where the cell contains no discernible dye during normal transport, by omitting Ca^{++} from the sustaining medium, intracellular:medium concentration ratios as high as 2000:1 can be induced. This intracellular accumulation of phenol red can be inhibited by the simultaneous presence in the medium of such transfer competitors as diodrast and PAH (Puck *et al.*, 1952; Forster and Hong, 1958). Secondly, thin slices of renal cortex taken from rabbits and other mammals accumulate phenol red and chlorphenol red intracellularly without further discernible concentration in the lumen, even when maintained optimally in a balanced isotonic medium. Whether this is due to some obscure experimental deficiency or occurs because only one active step is normally involved in tubular excretion by mammalian kidneys has not been determined. The action of various metabolic inhibitors within this system shows that permeation into cells of proximal tubules is an active process dependent upon aerobic phosphorylation and subject to competitive inhibition involving the phenolsulfonphthaleins, PAH, and probenecid (Forster and Copenhaver, 1956).

That the second step, from cell to lumen, is an active process is evident from simple inspection in the isolated flounder tubule preparation optimally maintained. Here with no discernible dye in the cells, intraluminal concentrations several thousand times that in the ambient medium are achieved. Upon subsequent exposure to cold, dinitrophenol, etc., the luminally concentrated dye diffuses back into the medium without accumulating detectably within the cell during its passage. That the cell membrane on the luminal border is another site of competitive inhibition for transport is revealed when, subsequent to concentration of chlorphenol red within the lumen, tubules are transferred to dye-free medium con-

---

Fig. 10. Transfer maxima in four different active transport systems. Upper left: the tubular excretory rate of phenol red reaches a maximum at plasma concentrations of 6 mg. % in the aglomerular *Lophius piscatorius* (from Shannon, 1938d). Upper right: glucose resorption in the dog is essentially complete when plasma levels are between 200 and 300 mg. %. When delivery by glomerular filtration exceeds the maximal resorptive rate, all the excess glucose passes into the urine (from Shannon and Fisher, 1938). Lower left: in the bullfrog tubular excretion of urea reaches a maximum as plasma levels approach 15 mg. %. Above this level increase in total urea excretion corresponds to that provided solely by glomerular filtration (from Forster, 1954). Lower right: in the flounder isolated tubule preparation phenol red achieves maximal concentrations in the lumen averaging more than 1000 times that in the ambient medium within 30 minutes *in vitro*. Each set of points represents an experiment on a different fish on a different day (from Puck *et al.*, 1952).

taining such competitors as PAH and probenecid, and marked facilitation of chlorphenol red run-out takes place. The facilitating action of competitors on run-out suggests that a steady state prevails when luminal accumulation has occurred with a tendency for dye to diffuse passively into the cell being opposed by an active process which transports dye from the cell back into the lumen. Net luminal accumulation then would be the resultant of these opposing forces. A carrier, in or near the luminal cell membrane, which is capable of combining with several transport competitors could accept a homolog, thereby displacing the original dye and reducing its transfer into the lumen subsequent to back diffusion (Hong and Forster, 1958).

These direct observations on the "active" processes involved in membrane permeation might be interpreted by viewing the cell boundary as a restrictive barrier by virtue of its lipid nature. A high order of activation energy would be required for these organic acids, first, to leave the aqueous peritubular fluid and enter the lipid phase of the basal cell membrane, and, second, to dissociate from the aqueous intracellular fluid and enter the luminal membrane. In studies on the physicochemical characteristics of a series of organic acid dyes, Höber (1940) showed that actively transported members of the series were all dipoles. The "head-and-tail" feature of these molecules was determined by one end which was polar and hydrophilic, and the other nonpolar and lipid soluble. He suggested that the selective accumulation of nonpolar-polar dyes depended upon a preliminary orientation on the cell surface, with the nonpolar end being associated with the lipid phase of the membrane and the hydrophilic half anchored in the aqueous phase.

Tissue culture experiments with proximal convoluted tubules of the chick embryo kidney indicate that a linear arrangement of micellar components in protoplasm may account for active transport in this type of secreting cell. In tissue culture some cells remain in the wall of the tubular segment where they preserve the cylindrical structure, while those which migrate out of the explant form sheets of flattened, polygonal cells having no morphological semblance of what they were in the tubule. Only the former retain their secretory activity. The flattened cells fail to take up ionized phenolsulfonphthalein dyes, but they retain their ability to pick up colloidal materials and accumulate these particulate dyes intracellularly. The loss of secretory activity is not irretrievable because, after the culture has been growing for some time, flattened cells may arrange themselves radially and contribute to cylindroid structures, whereupon they resume their regional secretory activity of passing phenol red and aqueous solutions into a progressively enlarging lumen about which the cells are arranged (Cameron, 1948). Unidirectional transport in this case appears to be associated with the cylindrical organization and an accompanying orientation of intracellular micellae.

5. Intracellular Trapping

In the flounder isolated tubule preparation, and in kidney slices of cold-blooded animals, actively transported compounds such as phenol red and chlorophenol red are moved across proximal tubule cells at a very rapid rate and are concentrated intraluminally without any detectable intracellular accumulation. Smith *et al.* (1938) in clearance studies on man also concluded that at moderate plasma concentrations diodrast and hippuran, as well as phenol red, are not stored in cells of the renal tubule during excretion. However, some evidence from studies on mammals indicates that actively transported compounds may accumulate intracellularly *in vivo* (Marshall and Vickers, 1923; Edwards and Marshall, 1924; Elsom *et al.*, 1937; Rollhäuser, 1957). With high plasma concentrations diodrast accumulates intracellularly in rabbits (Josephson and Kallas, 1953) and its concentration within proximal cells can be detected directly with historadiographs (Engström and Josephson, 1953). The second step, transfer of diodrast from cell to lumen, was viewed in these studies as the rate-limiting reaction in transcellular movement. This is in contrast to observations made on isolated tubules and on thin slices maintained *in vitro*, where the initial step, movement from ambient medium into the cell, is the rate-limiting process for rapidly excreted compounds under optimal conditions.

However, with the more slowly moving members of the phenol-sulfonphthalein series direct visualization in the flounder isolated tubule preparation discloses a tendency to accumulate intracellularly which is inversely related to the speed of their over-all excretory rate (Forster and Hong, 1958). Moreover, these are the most effective competitive inhibitors of others in the homologous series which are more rapidly transported (Sperber, 1954; Forster *et al.*, 1954). These slowly transported compounds, in contrast to the more actively secreted phenol red and chlorophenol red, accumulate intracellularly by passive diffusion even in the presence of metabolic inhibitors; once trapped within the cells they do not run out when tubules with accumulated dye are transferred to dye-free medium in the cold. The intracellular color is distributed uniformly, and in contrast to such cationic stains as neutral red (Weiss, 1955), no selective accumulation is discernible in vacuoles or on particulate matter with the 100× magnification used in these studies. The submicroscopic intracellular carrier appears to be capable of holding transported substances with varying degrees of firmness; whether it acts as a selective solvent or adsorbent, or as an intermediate which combines chemically with the transported compound has not been determined. Evidence for a similar "trapping mechanism" was also obtained from studies on thin slices of mammalian renal cortex (Copenhaver and Forster, 1958).

6. Enzymatic Aspects of Transport

The simplest hypothesis to account for unidirectional movement and establishment of a concentration gradient across restrictive barriers would involve a reaction of an intracellular carrier and the transported substance, followed by diffusion of the complex and subsequent reversal at another site, with the transported compound being deposited in its original form on the far side of a second barrier. At least one of the steps in this scheme would be energy-dependent. Glycine conjugase activity was suggested as an intermediate step in PAH transport because probenecid blocks PAH secretion without affecting cell respiration or phosphorylation, and it also inhibits the conjugation of benzoate and glycine to form hippurate (Beyer et al., 1950a). However, this explanation appears to be unlikely on the basis of tracer studies involving the slow infusion of PAH labeled with C^{14} in the carboxyl group and its subsequent recovery in urine. If hydrolysis and reconjugation were an intermediate step during transport, the carboxyl labeling should be diluted by the incorporation of at least some unlabeled glycine which is present in tubule cells as a consequence of amino acid resorption from glomerular filtrate. However, the specific radioactivity of the PAH originally infused and of the PAH recovered from urine in dogs was found to be identical (Taggart, 1954).

Various metabolic substrates and intermediates markedly stimulate or depress PAH transport without correspondingly affecting oxidative activities (Cross and Taggart, 1950). Taggart (1956a) points out that their actions can be explained in terms of a series of strictly competitive phenomena outside of the transport mechanism per se. The well-known stimulating effects of acetate, for example, may be interpreted as resulting from removal of an endogenous competitor. A long series of careful studies by his group has specifically shown how presumably unrelated cellular metabolic activities may influence the PAH transport mechanism by facilitating either the formation or breakdown of a hypothetical intermediate, p-aminohippuryl-CoA (Taggart, 1954).

Derivatives of PAH were prepared with all reactive groups covered by chemical substitution, and it was shown (Taggart, 1956b) that all are transported except those with the anionic carboxyl group rendered inactive. When O^{18} was used to mark the carboxyl group, and this labeled PAH infused, there was no appreciable loss of O^{18} from the carboxyl group noted in recovered PAH. Reaction with the carrier in the case of PAH probably takes place at the carboxyl end of the molecule, but the specific kind of association and the nature of the carrier is unknown. No information is available concerning specific carrier association for such other actively transported compounds as the organic bases, organic acids with no carboxyl groups, or for such secreted nonelectrolytes as urea in the frog. Metabolic aspects of transport of organic compounds across renal tubular epithelium have recently been summarized and discussed in the Wisconsin Symposium, especially with respect to actions of various substrates and inhibitors, comparison of the acid-secreting and base-secreting mechanisms, and the nature of postulated enzymatic components which may be involved in PAH transport (Shideman, 1957).

D. Other Transport Systems

Actively excreted organic bases do not compete for transport with members of the acid series. Their transport appears to be dependent upon cellular oxidative phosphorylation, but for some, such as tetraethylammonium ion (TEA), transfer does not depend upon Kreb's cycle activity (Farah, 1957). Some weak bases, unlike the actively transported organic acids, are excreted at rates determined by the pH of urine, and their transfer rate is unaffected by simultaneous presence of secreted acids (Orloff and Berliner, 1956). These substances diffuse across the membrane as free base and accumulate within the lumen in the form of the less permeable lipid-insoluble ionic species. At least one is actively secreted and also actively resorbed, with net secretion occurring when the urine is acid and net resorption with alkaline urine (Baer *et al.*, 1956). As with the organic acids (Forster and Hong, 1958; Copenhaver and Forster, 1958), it appears that those basic dyes which are slowly transported themselves tend to accumulate intracellularly and very effectively inhibit transfer of the more actively transported members of the basic series such as TEA and *N*-methylnicotinamide (NMN) (Peters, 1957). Trimethylamine oxide, an endogenous base in the aglomerular *Lophius,* competes in a general base-secreting system of wide occurrence in vertebrates which transfers competitively TEA, NMN, and cyanine dye. Creatine, another endogenous base which is actively secreted in *Lophius,* does not compete in this series (Forster *et al.*, 1958).

Potassium is actively secreted by the Malpighian tubules of insects, and in several species it has been shown to accumulate against an adverse electrochemical gradient (Ramsay, 1953a). It is not actively secreted by the aglomerular kidney of *Lophius* (Forster and Berglund, 1956) nor in glomerular marine teleosts whose nephrons are composed exclusively of cells typical of the proximal segment, even when potassium concentrations in plasma are progressively raised to very high levels (personal observation). In mammals it is inferred that all filtered potassium is resorbed by the proximal tubules, and whatever appears in urine is excreted by the distal segment in response to its accumulation intracellularly (Morel, 1955; Morel and Guinnebault, 1956; Smith, 1956). Its excretion appears to be closely related to mitochondrial activity (Mudge, 1955), and, as discussed earlier, it is competitively involved in the resorption of sodium and in H^+ regulation. The role of renal tubular secretion in potassium homeostasis has recently been studied by Koch *et al.* (1956), and Auditore and Holland (1956) have noted factors affecting the intracellular distribution of potassium in kidney and other tissues. Berliner (1957) and Pitts (1957) have recently reviewed aspects of potassium excretion, ion exchange, and electrolyte transport by the renal epithelium.

Kidney cells detoxify certain substances in the course of their elimination. Several phenolic glucuronides and sulfuric esters of phenols formed in this way are actively secreted by the chicken kidney (Sperber, 1948c, d). Diodrast, and hippuric acid formed by the glycination of benzoic acid in the kidney, decrease excretion rates of these glucuronides and esters. Menthylglucuronide and resorcinol disulfuric ester depress the secretion of phenol red. The excretion of many other conjugated aromatic acids has been studied, many of which have clearances as high as PAH in mammals (see Smith, 1951, for review). Acetylation of PAH does not affect its rate of tubular excretion, but, inasmuch as this interferes with standard methods of its chemical determination by covering the reactive amino group, it can lead to experimental errors in certain species in which the kidney acetylates PAH during transport (Taggart *et al.*, 1953).

Mercurial diuretics are widely used clinically to promote the excretion of water and sodium. They appear to have three sites of action: diminished sulfhydryl activity is noted in the terminal portion of the proximal tubule, in the ascending limb of Henle's loop and in the collecting ducts (Cafruny *et al.*, 1955). Owing to high protein binding little or none of the mercurials is filtered, and what appears in the urine is largely accounted for by tubular secretion (Borghgraef *et al.*, 1956). With the renal portal preparation in the chicken it was noted that diuresis caused by various mercurials was suppressed by the simultaneous presence of probenecid or bromcresol green, and that those diuretics which were themselves slowly eliminated were the most effective, and their diuretic action persisted for the longest time (Campbell, 1957). This is in keeping with the phenomenon in the isolated flounder tubule preparation where a similar inverse relationship was noted for the phenolsulfonphthalein series; those members of the series most effective as competitive inhibitors tended to accumulate intracellularly (Forster and Hong, 1958). However, in the dog no correlation was found between excretory rates, the degree of concentration in renal tissue, and the diuretic activity of various mercurials (Kessler *et al.*, 1957a). Distinct species differences occur also among mammals as noted in the effectiveness of mercurials in depressing the secretion of PAH in man, while having no such action in the dog (Berliner *et al.*, 1948). It appears unlikely that organic mercurials are effective only by virtue of their capacity to dissociate mercuric ions. In the isolated tubule preparation, for example, when organic mercurials with their very low degree of dissociation were compared with inorganic salts in their ability to inhibit phenol red transport, Salyrgan was found to be effective at $M/100,000$ and mercuric chloride at $M/40,000$. Since, on a molar basis, mercuric chloride is only two and one-half times as effective as Salyrgan, one would have to assume that the latter compound was at least 40% dissociated. This is quite unlikely in view of the nature of the carbon-mercury bond present in the organic material which permits an extremely low degree of ionization (Forster and Taggart, 1950). The diuretic effectiveness of organic mercurials is related to their structural configuration. Active agents must have a chain of not less than three carbon atoms, an atom of hydrogen attached to the terminal carbon, and some hydrophilic group not less than three carbons distant from the mercury (Kessler *et al.*, 1957b).

Other tubular secretory systems among vertebrates were discussed previously in the comparative section relating form and function.

Besides the resorptive systems already discussed involving H^+ exchange, glucose, water, and the common univalent ions, other resorptive mechanisms occur for such strong electrolytes as lithium, calcium, magnesium, strontium, iodide, bromide, thiocyanate, nitrate, sulfate, phosphate, and for such organic compounds as various amino acids, acetoacetic acid, uric acid, several vitamins, lactic acid, and certain other metabolic intermediates (see Smith, 1951, 1956, for reviews). It is usually difficult to characterize quantitatively the transport mechanism for these substances because of the strong affinity many have for plasma proteins which makes measurement of filtration rate difficult, or because they are often linked to complex intracellular synthetic and metabolic activities. Moreover, no quantitative *in vitro* technique has been established for studying resorptive mechanisms. However, as with secretory processes, it has been shown that resorptive transfer for some compounds is limited by a maximal rate, common mechanisms involving several related substances exhibit competitive inhibition, and certain resorption systems exhibit a high degree of selectivity and operate entirely independently of others.

Resorptive systems, however, may be interlocked with simultaneous secretory processes and with intracellular metabolic events. Phlorizin, for example, which specifically inhibits sugar resorption, is itself actively secreted in competition with diodrast and phenol red in a probenecid-sensitive transfer system (Braun *et al.*, 1957); bicarbonate appears to be resorbed by two pathways, one involving carbonic anhydrase and via another in competition with phosphate (Malvin and Lotspeich, 1956); sulfate is resorbed at a rate affected competitively by sodium and various amino acids (Berglund and Lotspeich, 1956a, b), but it is taken up actively by rat cortex slices *in vitro* (Berglund and Deyrup, 1956); and glucose competitively inhibits the resorption of acetoacetate, inorganic sulfate, and inorganic phosphate, whereas this inhibition is reversed with resorption levels actually being raised above control values by phlorizin administration (Cohen *et al.*, 1956a). Furthermore, proximal tubule cells, viewed as structurally identical throughout the vertebrates, in certain species actively excrete some substances which are usually resorbed, as is the case with uric acid in birds and arid-living reptiles, and with calcium, magnesium, and sulfate in the marine teleosts (Berglund and Forster, 1958).

Urea is actively secreted in the frog, but it is resorbed against a concentration gradient in the dogfish. Trimethylamine oxide again is actively secreted by aglomerular tubules in teleosts and resorbed actively in the dogfish. Basic differences in the resorptive and secretory mechanisms of these identical compounds are revealed by studies which show that probenecid blocks urea secretion in frogs without affecting its resorption in dogfish when doses are administered which simultaneously inhibit PAH transport (Forster and Berglund, 1957). Also, tetraethylammonium, which competitively depresses secretion of trimethylamine oxide in the aglomerular tubule, has no such inhibiting effect on its resorption in the dogfish (Forster *et al.*, in preparation).

Various hypothetical schemes have been proposed concerning active transport, but none is adequate to account for the multiplicity of specific operations performed by tubule cells in transferring variously secreted and resorbed substances transcellularly in kidneys of different species. Reviews of proposed transport schemes include those of Conway (1954), Danielli (1954), Davies (1954), Davson (1954), Lundegårdh (1954), Robinson (1954a), Rosenberg (1954), Ussing (1954a, b), Wilbrandt (1954), Bayliss (1956), and Harris (1956). For PAH secretion specifically, Chinard (1956) proposes a model postulating the complexing of PAH with some intracellular component, and the relatively slow transit time for secreted substances is viewed as due to the time required for diffusion of the complex across the tubular cell. That the hypothetical PAH-complex may be macromolecular, but not necessarily particulate in size, is inferred from estimation of its diffusion coefficient. This approach, especially when used with the more slowly transported members of the organic acid series, could be of great value in characterizing the separate cell membrane and "trapping" steps which were revealed by direct observation studies on *in vitro* preparations (Forster and Copenhaver, 1956; Forster and Hong, 1958; Hong and Forster, 1958; Copenhaver and Forster, 1958).

VI. Conclusion

Epithelial cells of the regionally differentiated vertebrate nephron carry on special regulatory functions which are shared to some extent by surface membranes of unicellular organisms and by body cells generally in the metazoans. No living cell is in passive equilibrium with its external environment, and "active" processes are perhaps universally involved to insure chemical constancy in the intracellular *milieu*. With the cells of kidney tubules, regulation is complicated by the fact that their cell membranes are exposed to, and in dynamic equilibrium with, *two* external environments—the urine and the extracellular or vascular elements. This two-way orientation of kidney cells is not unique, and, as in marine fishes, for example, kidneys are relieved in great part by the ability of the gills to carry on such selective functions as the elimination of hydrogen ions, certain nitrogenous constituents and univalent inorganic ions. Similar situations prevail in frog skin, choroid plexus, ciliary body, exocrine glands, liver, gut, and perhaps in many other thin-walled and tubular surfaces in intimate association with the circulatory system. The general ability of cells to carry on regulatory functions individually is best revealed in the echinoderms where the entire body surface serves as an overflow mechanism, and every member of this phylum manages to survive without any kidneys at all. Specialized processes in the tubular epithelium of the vertebrate nephron serve in elimination, of course, but to view this surface as being of the nature of a leaky sieve which functions solely for the excretion of noxious wastes is a sterile oversimplification which will only postpone an understanding of the complex regulatory mechanisms involved at this site.

To date, not a single structural feature of the tubular epithelium, at either the microscopic or the submicroscopic level, can be implicated specifically in a transfer process, and, aside from some general synthetic and metabolic reactions, not a single enzyme shown by cytochemical or other procedures to occur in these cells can be associated directly with a discrete secretory or resorptive mechanism. While over-all renal functions at the organ level are well known and can be accurately measured, the need is apparent for more investigation and new approaches directed toward relating intracellular form and function.

REFERENCES

Addis, T., MacKay, E. M., and MacKay, L. L. (1926). *J. Biol. Chem.* **71,** 189.

Alexander, D. P., Nixon, D. A., Widdas, W. F., and Wohlzogen, F. X. (1958). *J. Physiol. (London)* **140,** 14.

Allen, A. C. (1951). "The Kidney—Medical and Surgical Diseases." Grune & Stratton, New York.

Ames, R. G., and Van Dyke, H. B. (1950). *Proc. Soc. Exptl. Biol. Med.* **75,** 417.

Andrew, W., and Pruett, D. (1957). *Am J. Anat.* **100,** 51.

Arataki, M. (1926a). *Am. J. Anat.* **36,** 399.

Arataki, M. (1926b). *Am. J. Anat.* **36,** 437.

Armstrong, P. B. (1932). *Am. J. Anat.* **51,** 157.

Atkinson, M., Clark, G. A., and Menzies, J. A. (1921). *J. Physiol. (London)* **55,** 253.

Auditore, G. V., and Holland, W. C. (1956). *Am. J. Physiol.* **187,** 57.

Baer, J. E., Paulson, S. F., Russo, H. F., and Beyer, K. H. (1956). *Am. J. Physiol.* **186,** 180.

Bainbridge, F. A., Menzies, J. A., and Collins, S. H. (1914). *J. Physiol. (London)* **48,** 233.

Bargmann, W. (1933). *Z. Zellforsch. u. mikroskop. Anat.* **18,** 166.

Bargmann, W., Knoop, A., and Schiebler, T. H. (1955). *Z. Zellforsch. u. mikroskop. Anat.* **42,** 386.

Barkan, G., Broemser, P., and Hahn, A. (1921). *Z. Biol.* **74,** 1.

Barnett, H. L., and Vesterdal, J. (1953). *J. Pediat.* **42,** 99.

Baxter, J. H., and Cotzias, G. C. (1949). *J. Exptl. Med.* **89,** 643.

Baxter, J. S., and Yoffey, J. M. (1948). *J. Anat.* **82,** 189.

Bayliss, L. E. (1956). *In* "Modern Views on the Secretion of Urine" (F. R. Winton, ed.), Chapt. 4. Little, Brown, Boston, Massachusetts.

Beadle, L. C. (1957). *Ann. Rev. Physiol.* **19,** 329.

Beck, L. V., and Chambers, R. (1935). *J. Cellular Comp. Physiol.* **6,** 441.

Bell, E. T. (1950). "Renal Diseases." Lea & Febriger, Philadelphia, Pennsylvania.

Bensley, R. R., and Bensley, R. D. (1930). *Anat. Record* **47,** 147.

Bensley, R. R., and Steen, W. B. (1928). *Am. J. Anat.* **41,** 75.

Berglund, F., and Deyrup, I. J. (1956). *Am. J. Physiol.* **187,** 315.

Berglund, F., and Forster, R. P. (1958). *J. Gen. Physiol.* **41,** 429.

Berglund, F., and Lotspeich, W. D. (1956a). *Am. J. Physiol.* **185,** 533.

Berglund, F., and Lotspeich, W. D. (1956b). *Am. J. Physiol.* **185,** 539.

Bergstrand, A. (1957). *Lab. Invest.* **6,** 191.

Berliner, R. W. (1957). *In* "Metabolic Aspects of Transport Across Cell Membranes" (Q. R. Murphy, ed.), p. 203. Univ. Wisconsin Press, Madison, Wisconsin.

Berliner, R. W., and Davidson, D. G. (1956). *J. Clin. Invest.* **35,** 690.

Berliner, R. W., Kennedy, T. J., Jr., and Hilton, J. G. (1948). *Am. J. Physiol.* **154,** 537.

Berliner, R. W., Kennedy, T. J., Jr., and Orloff, J. (1951). *Am. J. Med.* **11,** 274.

Berliner, R. W., Kennedy, T. J., Jr., and Orloff, J. (1954a). *In* "The Kidney" (A. A. G. Lewis and G. E. W. Wolstenholme, eds.), p. 147. Little, Brown, Boston, Massachusetts.

Berliner, R. W., Kennedy, T. J., Jr., and Orloff, J. (1954b). *Arch. intern. pharmacodynamie* **97,** 299.

Beyer, K. H., Peters, L., Woodward, R., and Verwey, W. F. (1944). *J. Pharmacol. Exptl. Therap.* **82,** 310.

Beyer, K. H., Miller, A. K., Russo, H. F., Patch, E. A., and Verwey, W. F. (1947). *Am. J. Physiol.* **149**, 355.

Beyer, K. H., Painter, R. H., and Wiebelhaus, V. D. (1950a). *Am. J. Physiol.* **161**, 259.

Beyer, K. H., Russo, H. F., Gass, S. R., Wilhoyte, K. M., and Pitt, A. A. (1950b). *Am. J. Physiol.* **160**, 311.

Binet, L. C., Laroche, C., and Mathé, G. (1952). *Presse méd.* **60**, 1211.

Blatteis, C. M., and Horvath, S. M. (1958). *Am. J. Physiol.* **192**, 357.

Borghgraef, R. R. M., Kessler, R. H., and Pitts, R. F. (1956). *J. Clin. Invest,* **35**, 1055.

Bourne, G. H. (1954). *Quart, J. Microscop. Sci.* **95**, 365.

Bradley, S. E., and Bing, R. J. (1942). *J. Cellular Comp. Physiol.* **19**, 229.

Bradley, S. E., Leifer, E., and Nickel, J. F. (1954). *In* "The Kidney" (A. A. G. Lewis and G. E. W. Wolstenholme, eds.), p. 50. Little, Brown, Boston, Massachusetts.

Braun, W., Whittaker, V. P., and Lotspeich, W. D. (1957). *Am. J. Physiol.* **190**, 563.

Brewer, D. B. (1954). *Quart. J. Microscop. Sci.* **95**, 23.

Browne, M. J., Pitts, M. W., and Pitts, R. F. (1950). *Biol. Bull.* **99**, 152.

Burgos, M. H., Deane, H. W., and Karnovsky, M. L. (1955). *J. Histochem. and Cytochem.* **3**, 103.

Burstone, M. S. (1957). *J. Natl. Cancer Inst.* **18**, 167.

Bywaters, E. G. L., and Dible, J. H. (1942). *J. Pathol. Bacteriol.* **54**, 111.

Cafruny, E. J., Farah, A., and DiStefano, H. S. (1955). *J. Pharmacol. Exptl. Therap.* **115**, 390.

Cameron, G. (1948). *Anat. Record* **100**, 646.

Cameron, G., and Chambers, R. (1938). *Am. J. Physiol.* **123**, 482.

Campbell, D. (1957). *Experientia* **13**, 327.

Chambers, R., and Cameron, G. (1932). *J. Cellular Comp. Physiol.* **2**, 99.

Chambers, R., and Kempton, R. T. (1933). *J. Cellular Comp. Physiol.* **3**, 131.

Chambers, R., Beck, L. V., and Belkin, M. (1935). *J. Cellular Comp. Physiol.* **6**, 425.

Childs, A. W., Wheeler, H. O., Cominsky, B., Leifer, E., Wade, O. L., and Bradley, S. E. (1955). *J. Clin. Invest.* **34**, 926.

Chinard, F. P. (1956). *Am. J. Physiol.* **185**, 413.

Chiquoine, A. D. (1953). *J. Histochem. and Cytochem.* **1**, 429.

Clark, S. L. (1957). *J. Biophys. Biochem. Cytol.* **3**, 349.

Clarke, R. W., and Smith, H. W. (1932). *J. Cellular Comp. Physiol.* **1**, 131.

Cohen, J. J., Berglund, F., and Lotspeich, W. D. (1956a). *Am. J. Physiol.* **184**, 91.

Cohen, J. J., Boyarsky, S., and Biggs, A. (1956b). *Bull. Mt. Desert Isl. Biol. Lab.* **4**(2), 52.

Conway, E. J. (1954). *Symposia Soc. Exptl. Biol. No.* **8**, 297.

Conway, E. J., and Geoghegan, H. (1955). *J. Physiol. (London)* **130**, 438.

Coons, A. H., and Kaplan, M. H. (1950). *J. Exptl. Med.* **91**, 1.

Copenhaver, J. H., Jr., and Forster, R. P. (1958). *Am. J. Physiol.* **195**, 327.

Corcoran, A. C., ed. (1952). *Methods in Med. Research* **5**, 134.

Cori, G. T. (1954). *Harvey Lectures* **48**, 145.

Coulson, R. A., and Hernandez, T. (1955). *Proc. Soc. Exptl. Biol. Med.* **88**, 682.

Coulson, R. A., and Hernandez, T. (1957). *Am. J. Physiol.* **188**, 121.

Cross, R. J., and Taggart, J. V. (1950). *Am. J. Physiol.* **161**, 181.

Cullis, W. C. (1906). *J. Physiol. (London)* **34**, 250.

Cushny, A. R. (1917). *In* Monographs on Physiology. "The Secretion of Urine." Longmans, Green, London; 2nd ed., 1926.

Dalton, A. J. (1951). *J. Natl. Cancer Inst.* **11**, 1163.

Danielli, J. F. (1954). *Symposia Soc. Exptl. Biol. No.* **8**, 502.

Danielli, J. F., and Pantin, C. F. A. (1950). *Quart. J. Microscop. Sci.* **91**, 209.

Darmady, E. M. (1954). *In* "The Kidney" (A. A. G. Lewis and G. E. W. Wolsten-holme, eds.), p. 27. Little, Brown, Boston, Massachusetts.

Darmady, E. M., and Stranack, F. (1957). *Brit. Med. Bull.* **13**, 21.

Davies, B. M. A., and Yudkin, J. (1952). *Biochem. J.* **52**, 407.

Davies, J. (1952). *Am. J. Anat.* **91**, 263.

Davies, J. (1953). *Am. J. Anat.* **94**, 45.

Davies, J., and Davies, D. V. (1950). *Proc. Zool. Soc. London* **120**, 73.

Davies, J., and Routh, J. I. (1957). *J. Embryol. Exptl. Morphol.* **5**, 32.

Davies, R. E. (1954). *Symposia Soc. Exptl. Biol. No.* **8**, 453.

Davson, H. (1954). *Symposia Soc. Exptl. Biol. No.* **8**, 16.

Defrise, A. (1932). *Anat. Record* **54**, 185.

Deyrup, I. J. (1953a). *Am. J. Physiol.* **175**, 349.

Deyrup, I. J. (1953b). *J. Gen. Physiol.* **36**, 739.

Deyrup, I. J. (1955). *Am. J. Physiol.* **183**, 609.

Deyrup, I. J. (1957). *Am. J. Physiol.* **188**, 125.

Deyrup, I. J., and Ussing, H. H. (1955). *J. Gen. Physiol.* **38**, 599.

Dock, W. (1942). *New Engl. J. Med.* **227**, 633.

Dratz, A. F., and Handler, P. (1952). *J. Biol. Chem.* **197**, 419.

Dunihue, F. W. (1946). *Anat. Record* **96**, 40.

Dunihue, F. W. (1947). *Am. J .Pathol.* **23**, 906.

Dunihue, F. W. (1949). *Anat. Record* **103**, 442.

Dunihue, F. W., and Robertson, W. van B. (1953). *Anat. Record* **115**, 300.

Dunn, J. S., and Polson, C. J. (1926). *J. Pathol. Bacteriol.* **29**, 337.

Dunn, J. S., Gillespie, M., and Niven, J. S. F. (1941). *Lancet* **ii**, 549.

Edwards, J. G. (1929). *Anat. Record* **44**, 15.

Edwards, J. G. (1933). *Anat. Record* **55**, 343.

Edwards, J. G. (1935). *Anat. Record* **63**, 263.

Edwards, J. G., and Condorelli, L. (1928). *Am. J. Physiol.* **86**, 383.

Edwards, J. G., and Marshall, E. K., Jr. (1924). *Am. J. Physiol.* **70**, 489.

Eger, W., and Schulte, W. (1954). *Acta Histochem.* **1**, 60.

Ehrich, W. (1953). *Proc. 5th Ann. Conf. Nephrotic Syndrome*, p. 117.

Elias, A. H. (1957). *Anat. Anz.* **104**, 26.

Eliasch, H., Sellers, A. L., Rosenfeld, S., and Marmorston, J. (1955). *J. Exptl. Med.* **101**, 129.

Ellinger, P. (1940). *Quart. J. Exptl. Physiol.* **30**, 205.

Ellinger, P., and Hirt, A. (1930). *Arch. exptl. Pathol. Pharmakol. Naunyn-Schmiede-berg's* **150**, 285.

Ellinger, P., and Hirt, A. (1931). *Arch. exptl. Pathol. Pharmakol. Naunyn-Schmiede-berg's* **159**, 111.

Ellis, J. T. (1956). *J. Exptl. Med.* **103**, 127.

Elsom, K. A., Bott, P. A., and Shiels, E. H. (1936). *Am. J. Physiol.* **115**, 548.

Elsom, K. A., Bott, P. A., and Walker, A. M. (1937). *Am. J. Physiol.* **118**, 739.

Engström, A., and Josephson, B. (1953). *Am. J. Physiol.* **174**, 61.

Fajers, C.-M. (1957a). *Acta Pathol. Microbiol. Scand.* **41**, 25.

Fajers, C.-M. (1957b). *Acta Pathol. Microbiol. Scand.* **41**, 34.

Farah, A. E. (1957). *In* "Metabolic Aspects of Transport Across Cell Membranes" (Q. R. Murphy, ed.), p. 257. Univ. Wisconsin Press, Madison, Wisconsin.

Farquhar, M. G., Vernier, R. L., and Good, R. A. (1957a). *Am. J. Pathol.* **33**, 791.

Farquhar, M. G., Vernier, R. L., and Good, R. A. (1957b). *J. Exptl. Med.* **106**, 649.

Fergusson, A. M. (1952). *J. Anat.* **86**, 144.

Fisher, E. R., Tsuji, F. I., and Gruhn, J. (1958). *Proc. Soc. Exptl. Biol. Med.* **97**, 448.

Flexner, L. B. (1939). *J. Biol. Chem.* **131**, 703.

Florkin, M., and Duchâteau, G. (1948). *Physiol. Comparata et Oecol.* **1**, 29.

Foote, J. J., and Grafflin, A. L. (1938). *Anat. Record* **72**, 169.

Foote, J. J., and Grafflin, A. L. (1942). *Am. J. Anat.* **70**, 1.

Forster, R. P. (1938). *J. Cellular Comp. Physiol.* **12**, 213.

Forster, R. P. (1942). *J. Cellular Comp. Physiol.* **20**, 55.

Forster, R. P. (1943). *Am. J. Physiol.* **140**, 221.

Forster, R. P. (1948). *Science* **108**, 65.

Forster, R. P. (1953). *J. Cellular Comp. Physiol.* **42**, 487.

Forster, R. P. (1954). *Am. J. Physiol.* **179**, 372.

Forster, R. P., and Berglund, F. (1956). *J. Gen. Physiol.* **39**, 349.

Forster, R. P., and Berglund, F. (1957). *J. Cellular Comp. Physiol.* **49**, 281.

Forster, R. P., and Copenhaver, J. H. Jr. (1956). *Am. J. Physiol.* **186**, 167.

Forster, R. P., and Hong, S. K. (1958). *J. Cellular Comp. Physiol.* **51**, 259.

Forster, R. P., and Nyboer, J. (1955). *Am. J. Physiol.* **183**, 149.

Forster, R. P., and Taggart, J. V. (1950). *J .Cellular Comp. Physiol.* **36**, 251.

Forster, R. P., Sperber, I., and Taggart, J. V. (1954). *J. Cellular Comp. Physiol.* **44**, 315.

Forster, R. P., Berglund, F., and Rennick, B. R. (1958). *J. Gen. Physiol.* **42**, 319.

Fraser, E. A. (1950). *Biol. Revs. Cambridge Phil. Soc.* **25**, 159.

Gabe, M. (1957). *Arch. Biol. (Liége)* **68**, 29.

Gagnepain, J. (1956 [1957]). *Bull. soc. zool. France* **81**, 395.

Gerard, F., and Cordier, R. (1934). *Biol. Revs. Cambridge Phil. Soc.* **9**, 110.

Gersh, I. (1937). *Contribs. Embryol. Carnegie Inst. Wash. Publ.* **26** (153), 33.

Gersh, I., and Stieglitz, E. J. (1933). *Anat. Record* **58**, 349.

Gibbs, O. S. (1929). *Am. J. Physiol.* **88**, 87.

Giebisch, G. (1956). *Am. J. Physiol.* **185**, 171.

Gilson, G. B. (1949). *Proc. Soc. Exptl. Biol. Med.* **72**, 608.

Goldring, W., Clarke, R. W., and Smith, H. W. (1936). *J. Clin. Invest.* **15**, 221.

Goldstein, L., Richterich-van Baerle, R., and Dearborn, E. H. (1956). *Proc. Soc. Exptl. Biol. Med.* **93**, 284.

Goldstein, L., Richterich-van Baerle, R., and Dearborn, E. H. (1957). *Enzymologia* **18**, 261.

Gomori, G. (1941). *J. Cellular Comp. Physiol.* **17**, 71.

Gomori, G. (1949). *Proc. Soc. Exptl. Biol. Med.* **70**, 7.

Gomori, G. (1954a). *Proc. Soc. Exptl. Biol. Med.* **85**, 570.

Gomori, G. (1954b). *Proc. Soc. Exptl. Biol. Med.* **87**, 559.

Goodrich, E. S. (1945). *Quart. J. Microscop. Sci.* **86**, 113.

Goormaghtigh, N. (1940). *Am. J. Pathol.* **16**, 409.

Goormaghtigh, N. (1945). *J. Pathol. Bacteriol.* **57**, 392.

Goormaghtigh, N. (1947). *Am. J. Pathol.* **23**, 513.

Gottschalk, C. W., and Mylle, M. (1956). *Am. J. Physiol.* **185**, 430.

Grafflin, A. L. (1937). *J. Cellular Comp. Physiol.* **9**, 469.

Grafflin, A. L. (1942). *Am. J. Anat.* **70**, 399.

Grafflin, A. L., and Foote, J. J. (1939). *Am. J. Anat.* **65**, 179.

Greenspon, S. A., and Krakower, C. A. (1950). *A. M. A. Arch. Pathol.* **49**, 291.

Grollman, A. (1929). *J. Biol. Chem.* **81**, 267.

Gruenwald, P., and Popper, H. (1940). *J. Urol.* **43**, 452.

Hall, B. V. (1953). *Proc. 5th Ann. Conf. Nephrotic Syndrome* p. 1.

Hall, B. V. (1954). *Proc. 6th Ann. Conf. Nephrotic Syndrome* p. 1.

Hall, B. V. (1957). *Am. Heart J.* **54**, 1.

Hall, B. V., and Roth, L. E. (1956). *In* "Electron Microscopy," p. 176. Academic Press, New York.

Harris, E. J. (1956). "Transport and Accumulation in Biological Systems," Chapt. 9. Academic Press, New York.

Hartcroft, P. M. (1956). *Anat. Record* **124**, 458.

Hartcroft, P. M. (1957). *J. Exptl. Med.* **105**, 501.

Hartcroft, P. M., and Hartcroft, W. S. (1953). *J. Exptl. Med.* **97**, 415.

Hartcroft, P. M., and Hartcroft, W. S. (1955). *J. Exptl. Med.* **102**, 205.

Hayman, J. M. (1927). *Am. J. Physiol.* **79**, 389.

Heller, H. (1951). *Arch. Disease Childhood* **26**, 195.

Hernandez, T., and Coulson, R. A. (1957). *Am. J. Physiol.* **188**, 485.

Heymann, W., and Lund, H. Z. (1948). *Science* **108**, 448.

Hines, B. E., and McCance, R. A. (1954). *J. Physiol. (London)* **124**, 8.

Höber, R. (1940). *Cold Spring Harbor Symposia Quant. Biol.* **8**, 40.

Höber, R. (1942). *Proc. Soc. Exptl. Biol. Med.* **49**, 87.

Hodler, J., Heinemann, H. O., Fishman, A. P., and Smith, H. W. (1955). *Am. J. Physiol.* **183**, 155.

Hogben, C. A. M., and Bollman, J. L. (1951). *Am. J. Physiol.* **164**, 662.

Hong, S. K. (1957). *Am. J. Physiol.* **188**, 137.

Hong, S. K., and Forster, R. P. (1958). *J. Cellular Comp. Physiol.* **51**, 241.

Hoshiko, T. (1956). *Am. J. Physiol.* **185**, 545.

Hoshiko, T., Swanson, R. E., and Visscher, M. B. (1956). *Am. J. Physiol.* **184**, 542.

Howell, A. B., and Gersh, I. (1935). *J. Mammal.* **16**, 1.

Huber, G. C. (1905). *Am. J. Anat.* **4**, Suppl. **1**.

Huber, G. C. (1910). *Harvey Lectures* **5**, 100.

Huber, G. C. (1932). *In* "Special Cytology" (E. V. Cowdry, ed.), 2nd ed., Chapt. 2. Hoeber, New York.

Jackson, C. M., and Shiels, M. (1927–1928). *Anat. Record* **36**, 221.

Jacobellis, M., Muntwyler, E., and Griffin, G. E. (1955). *Am. J. Physiol.* **183**, 395.

Jaffee, O. C. (1954a). *J. Cellular Comp. Physiol.* **44**, 347.

Jaffee, O. C. (1954b). *J. Morphol.* **95**, 109.

Johnson, F. R. (1954). *Proc. Roy. Soc.* **B142**, 169.

Jones, D. B. (1953). *Am. J. Pathol.* **29**, 33.

Josephson, B., and Kallas, J. (1953). *Am. J. Physiol.* **174**, 65.

Junqueira, L. C. U. (1952). *Quart. J. Microscop. Sci.* **93**, 247.

Kempton, R. T. (1937). *J. Morphol.* **61**, 51.

Kempton, R. T. (1943). *J. Morphol.* **73**, 247.

Kempton, R. T. (1953). *Biol. Bull.* **104**, 45.

Kennedy, G. C. (1957). *Brit. Med. Bull.* **13**, 67.

Keosian, J. (1938). *J. Cellular Comp. Physiol.* **12**, 23.

Kessler, R. H., Lozano, R., and Pitts, R. F. (1957a). *J. Pharmacol. Exptl. Therap.* **121**, 432.

Kessler, R. H., Lozano, R., and Pitts, R. F. (1957b). *J. Clin. Invest.* **36**, 656.

Kimmelstiel, P., and Wilson, C. (1936). *Am. J. Pathol.* **12**, 83.

Kittelson, J. A. (1917). *Anat. Record* **13**, 385.

Koch, A. R., Brazeau, P., and Gilman, A. (1956). *Am. J. Physiol.* **186**, 350.

Koefoed-Johnson, V., and Ussing, H. H. (1953). *Acta Physiol. Scand.* **28**, 60.

Koletsky, S. (1954). *A. M. A. Arch. Pathol.* **58**, 592.

Krakower, C. A., and Greenspon, S. A. (1951). *A. M. A. Arch. Pathol.* **51**, 629.

Krakusin, J. S., and Jennings, R. B. (1955). *A. M. A. Arch. Pathol.* **59**, 471.

Krebs, H. A. (1935). *Biochem. J.* **29**, 1951.

Kretchmer, N., and Cherot, F. J. (1954). *J. Exptl. Med.* **99**, 637.

Kretchmer, N., and Dickerman, H. W. (1954). *J. Exptl. Med.* **99**, 629.

Kritzler, R. A., and Gutman, A. B. (1941). *Am. J. Physiol.* **134**, 94.

Krogh, A. (1939). "Osmotic Regulation in Aquatic Animals." Cambridge Univ. Press, London and New York.

Kruszynski, J. (1957). *Exptl. Cell Research* **12**, 108.

Kurata, Y. (1953). *Stain Technol.* **28**, 231.

Kurnick, N. B. (1955). *J. Histochem. and Cytochem.* **3**, 290.

Laroche, C., and Mathé, G. (1955). *In* "Précis de Gérontologie" (L. R. Binet, ed.), Chapt. 11. Masson, Paris.

Leaf, A. (1956). *Biochem. J.* **62**, 241.

Lee, Y. C. (1954). *J. Exptl. Med.* **99**, 621.

Leeson, T. S. (1957). *Exptl. Cell Research* **12**, 670.

Leeson, T. S., and Baxter, J. S. (1957). *J. Anat.* **91**, 383.

Leonard, E., and Orloff, J. (1955). *Am. J. Physiol.* **182**, 131.

Levinsky, N. G., and Davidson, D. G. (1957). *Am. J. Physiol.* **191**, 530.

Lindahl, K. M., and Sperber, I. (1956). *Acta Physiol. Scand.* **36**, 13.

Linder, J. E. (1949). *Quart. J. Microscop. Sci.* **90**, 427.

Longley, J. B., and Fisher, E. R. (1954). *Anat. Record* **120**, 1.

Longley, J. B., and Fisher, E. R. (1955). *J. Histochem. and Cytochem.* **3**, 392.

Longley, J. B., and Fisher, E. R. (1956). *Quart. J. Microscop. Sci.* **97**, 187.

Love, J. K. (1957). *Dissertation Abstr.* **17**, 2045.

Lowell, D. J., Greenspon, S. A., Krakower, C. A., and Bain, J. A. (1953). *Am. J. Physiol.* **172**, 709.

Lucké, B. (1946). *Military Med.* **99**, 371.

Lundegårdh, H. (1954). *Symposia Soc. Exptl. Biol. No.* **8**, 262.

McCance, R. A. (1948). *Physiol. Revs.* **28**, 331.

McCance, R. A. (1950). *Am. J. Med.* **9**, 229.

McCance, R. A., and Widdowson, E. M. (1952a). *Ber. phys.-med. Ges. Wurzburg* **66**, 115.

McCance, R. A., and Widdowson, E. M. (1952b). *Lancet* **ii**, 860.

McCance, R. A., and Widdowson, E. M. (1954). *Rec. progr. med.* **16**, 62.

McCance, R. A., and Widdowson, E. M. (1956). *J. Physiol. (London)* **133**, 373.

McCance, R. A., and Widdowson, E. M. (1957). *Brit. Med. Bull.* **13**, 3.

McCance, R. A., and Young, W. F. (1941). *J. Physiol. (London)* **99**, 265.

McCann, W. P. (1956). *Am. J. Physiol.* **185**, 372.

McFarlane, D. (1941). *J. Pathol. Bacteriol.* **52**, 17.

MacKay, L. L., MacKay, E. M., and Addis, T. (1926). *Proc. Soc. Exptl. Biol. Med.* **24**, 335.

McManus, J. F. A. (1942). *Lancet* **ii**, 394.

McManus, J. F. A. (1944). *Quart. J. Microscop. Sci.* **85**, 97.

McManus, J. F. A. (1947). *Quart. J. Microscop. Sci.* **88**, 39.

McManus, J. F. A. (1948a). *Am. J. Pathol.* **24**, 1259.

McManus, J. F. A. (1948b). *Am. J. Pathol.* **24**, 643.

McManus, J. F. A. (1950). "Medical Diseases of the Kidney." Lea & Febriger, Philadelphia, Pennsylvania.

Maengwyn-Davies, G. D., Friedenwald, J. S., and White, R. T. (1952). *J. Cellular Comp. Physiol.* **39**, 395.

Mallory, T. B. (1947). *Am. J. Clin. Pathol.* **17**, 427.

Malvin, R. L., and Lotspeich, W. D. (1956). *Am. J. Physiol.* **187**, 51.

Malvin, R. L., Sullivan, L. P., and Wilde, W. S. (1957). *The Physiologist* **1**, 58.

Mandel, P., Jacob, M., and Mandel, L. (1950). *Compt. rend.* **230**, 786.

Marshall, E. K., Jr. (1932a). *J. Cellular Comp. Physiol.* **2**, 349.

Marshall, E. K., Jr. (1932b). *Proc. Soc. Exptl. Biol. Med.* **29**, 971.

Marshall, E. K., Jr. (1934). *Physiol. Revs.* **14**, 133.

Marshall, E. K., Jr., and Crane, M. R. (1924). *Am. J. Physiol.* **70**, 465.

Marshall, E. K., Jr., and Grafflin, A. L. (1928). *Bull. Johns Hopkins Hosp.* **43**, 205.

Marshall, E. K., Jr., and Grafflin, A. L. (1932). *J. Cellular Comp. Physiol.* **1**, 161.

Marshall, E. K., Jr., and Smith, H. W. (1930). *Biol. Bull.* **59**, 135.

Marshall, E. K., Jr., and Vickers, J. L. (1923). *Bull. Johns Hopkins Hosp.* **34**, 1.

Martin, A. W. (1957). *In* "Invertebrate Physiology" (B. T. Scheer, ed.), p. 247. Univ. Oregon Publ., Eugene, Oregon.

Masugi, M. (1934). *Beitr. pathol. Anat. u. allgem. Pathol.* **92**, 429.

Maximow, A. A., and Bloom, W. (1957). "Textbook of Histology," 7th ed. p. 452. Saunders, Philadelphia and London.

Mayrs, E. B. (1924). *J. Physiol. (London)* **58**, 276.

Mellors, R. C., and Ortega, L. G. (1956). *Am. J. Pathol.* **32**, 455.

Mellors, R. C., Siegel, M., and Pressman, D. (1955a). *Lab. Invest.* **4**, 69.

Mellors, R. C., Arias-Stella, J., Siegel, M., and Pressman, D. (1955b). *Am. J. Pathol.* **31**, 687.

Miller, F., and Bohle, A. (1956). *Klin. Wochschr.* **34**, 1204.

Montgomery, H., and Pierce, J. A. (1937). *Am. J. Physiol.* **118**, 144.

Moog, F., and Wenger, E. L. (1952). *Am. J. Anat.* **90**, 339.

Moore, R. A. (1929). *J. Exptl. Med.* **50**, 709.

Morel, F. (1955). *Helv. Physiol. et Pharmacol. Acta* **13**, 276.

Morel, F., and Guinnebault, M. (1956). *Helv. Physiol. et Pharmacol. Acta* **14**, 255.

Mudge, G. H. (1951a). *Am. J. Physiol.* **165**, 113.

Mudge, G. H. (1951b). *Am. J. Physiol.* **167**, 206.

Mudge, G. H. (1955). *In* "Electrolytes in Biological Systems" (A. M. Shanes, ed.), p. 112. Am. Physiol. Soc. Washington, D. C.

Mudge, G. H., and Taggart, J. V. (1950). *Am. J. Physiol.* **161**, 173.

Mueller, C. B., Mason, A. D., Jr., and Stout, D. G. (1955). *Am. J. Med.* **18**, 267.

Muntwyler, E., Jacobellis, M., and Griffin, G. E. (1956). *Am. J. Physiol.* **184**, 83.

Narahara, H. T., Everett, N. B., Simmons, B. S., and Williams, R. H. (1958). *Am. J. Physiol.* **192**, 227.

Nash, J. (1931). *Am. J. Anat.* **47**, 425.

Newman, W., Feigin, I., Wolf, A., and Kabat, E. A. (1950). *Am. J. Pathol.* **26**, 257.

Nussbaum, M. (1878a). *Arch. ges. Physiol. Pflüger's* **16**, 139.

Nussbaum, M. (1878b). *Arch. ges. Physiol. Pflüger's* **17**, 580.

Oberling, C., Gantier, A., and Bernhard, W. (1951). *Presse méd.* **59**, 938.

Olivecrona, H., and Hillarp, N. (1949). *Acta Anat.* **8**, 281.

Oliver, J. (1915). *J. Exptl. Med.* **21**, 425.

Oliver, J. (1942). *In* "Problems of Aging" (E. V. Cowdry, ed.), 2nd ed., p. 302. Williams and Wilkins, Baltimore, Maryland.

Oliver, J. (1945). *Harvey Lectures* **40**, 102.

Oliver, J. (1948). *J. Mt. Sinai Hosp. N.Y.* **15**, 175.

Oliver, J. (1949). *Conf. on Renal Function Trans. 1st Conf.* p. 15.

Oliver, J. (1950). *Am. J. Med.* **9**, 88.

Oliver, J. (1955). *Livre jubilaire Prof. Paul Govaerts* **1955**, 1.

Oliver, J., MacDowell, M., and Tracy, A. (1951). *J. Clin. Invest.* **30**, 1307.

Oliver, J., MacDowell, M., and Lee, Y. C. (1954a). *J. Exptl. Med.* **99**, 589.

Oliver, J., Moses, M. J., MacDowell, M. C., and Lee, Y. C. (1954b). *J. Exptl. Med.* **99**, 605.

Oliver, J., MacDowell, M., Welt, L. G., Holliday, M. A., Hollander, W., Jr., Winters, R. W., Williams, T. F., and Segar, W. E. (1957). *J. Exptl. Med.* **106**, 563.
Opie, E. L. (1949). *J. Exptl. Med.* **89**, 185.
Orloff, J., and Berliner, R. W. (1956). *J. Clin. Invest.* **35**, 223.
Orloff, J., and Davidson, D. G. (1956). *Federation Proc.* **15**, 139.
Page, L. B. (1955). *Am. J. Physiol.* **181**, 171.
Pak Poy, R. K. F. (1957). *Australian J. Exptl. Biol. Med. Sci.* **35**, 437.
Pak Roy, R. K. F., and Robertson, J. S. (1957). *J. Biophys. Biochem. Cytol.* **3**, 183.
Palade, G. E. (1952). *J. Exptl. Med.* **95**, 285.
Palade, G. E. (1953). *J. Appl. Phys.* **24**, 1424.
Pease, D. C. (1955a). *Anat. Record* **121**, 701.
Pease, D. C. (1955b). *Anat. Record* **121**, 723.
Pease, D. C. (1955c). *J. Histochem. and Cytochem.* **3**, 295.
Pease, D. C., and Baker, R. F. (1950). *Am. J. Anat.* **87**, 349.
Peter, K. (1927). "Untersuchungen über Ban und Entwickelung der Niere." Gustav Fischer, Jena.
Peters, H. (1935). *Z. Morphol. Ökol. Tiere* **30**, 355.
Peters, L. (1957). *In* "Metabolic Aspects of Transport Across Cell Membranes" (Q. R. Murphy, ed.), p. 271. Univ. Wisconsin Press, Madison, Wisconsin.
Picken, L. E. R. (1937). *J. Exptl. Biol.* **14**, 20.
Piel, C. F., Dong, L., Modern, F. W. S., Goodman, J. R., and Moore, R. (1955). *J. Exptl. Med.* **102**, 573.
Pitts, R. F. (1934). *J. Cellular Comp. Physiol.* **4**, 389.
Pitts, R. F. (1936). *Carnegie Inst. Wash., Yearbook* **35**, p. 90.
Pitts, R. F. (1939). *J. Cellular Comp. Physiol.* **13**, 151.
Pitts, R. F. (1948). *Federation Proc.* **7**, 418.
Pitts, R. F. (1957). *In* "Metabolic Aspects of Transport Across Cell Membranes" (Q. R. Murphy, ed.), p. 221. Univ. Wisconsin Press, Madison, Wisconsin.
Pitts, R. F., and Alexander, R. C. (1945). *Am. J. Physiol.* **144**, 239.
Pitts, R. F., and Lotspeich, W. D. (1946). *Am. J. Physiol.* **147**, 481.
Policard, A., Collet, A., and Giltraire-Ralyte, L. (1955). *Bull. microscop. appl.* **5**, 5.
Policard, M. A. (1912). *Arch. anat. microscop. morphol. exptl.* **14**, 429.
Pressman, D., Eisen, H. N., and Fitzgerald, P. J. (1950). *J. Immunol.* **64**, 281.
Puck, T. T., Wasserman, K., and Fishman, A. P. (1952). *J. Cellular Comp. Physiol.* **40**, 73.
Rammelkamp, C. N., and Bradley, S. E. (1943). *Proc. Soc. Exptl. Biol. Med.* **53**, 30.
Ramsay, J. A. (1949). *J. Exptl. Biol.* **26**, 65.
Ramsay, J. A. (1950). *J. Exptl. Biol.* **27**, 145.
Ramsay, J. A. (1951). *J. Exptl. Biol.* **28**, 62.
Ramsay, J. A. (1952). *J. Exptl. Biol.* **29**, 110.
Ramsay, J. A. (1953a). *J. Exptl. Biol.* **30**, 358.
Ramsay, J. A. (1953b). *J. Exptl. Biol.* **30**, 79.
Ramsay, J. A. (1954). *J. Exptl. Biol.* **31**, 104.
Ramsay, J. A. (1955a). *J. Exptl. Biol.* **32**, 200.
Ramsay, J. A. (1955b). *J. Exptl. Biol.* **32**, 183.
Ramsay, J. A., Brown, R. H. J., and Falloon, S. W. H. W. (1953). *J. Exptl. Biol.* **30**, 1.
Ramsay, J. A., Brown, R. H. J., and Croghan, P. C. (1955). *J. Exptl. Biol.* **32**, 822.
Rappaport, R., Jr. (1955). *J. Exptl. Zool.* **128**, 481.
Rather, L. J. (1952). *Medicine* **31**, 357.
Rector, F. C., Jr., Seldin, D. W., Roberts, A. D., Jr., and Copenhaver, J. H., Jr. (1954). *Am. J. Physiol.* **179**, 353.

Reid, R. T. W. (1956). *Australian J. Exptl. Biol. Med. Sci.* **34**, 142.

Reinhoff, W. F. (1922). *Bull. Johns Hopkins Hosp.* **33**, 392.

Rennick, B. R., Calhoon, D., Gaudia, H., and Moe, G. K. (1954). *J. Pharmacol. Exptl. Therap.* **110**, 309.

Rhodin, J. (1954). "Correlation of Ultrastructural Organization and Function in Normal and Experimentally Changed Proximal Convoluted Tubule Cells of the Mouse Kidney," p. 1. Aktiebolaget Godvil, Stockholm.

Rhodin, J. (1955). *Exptl. Cell Research* **8**, 572.

Rhodin, J. (1956). From a paper read before the Renal Association, Ciba Foundation, London, October 31, 1956.

Richards, A. N. (1938). *Proc. Roy. Soc.* **B126**, 398.

Richards, A. N., and Schmidt, C. F. (1924). *Am. J. Physiol.* **71**, 178.

Richards, A. N., and Walker, A. M. (1927). *Am. J. Physiol.* **79**, 419.

Richterich-van Baerle, R., and Goldstein, L. (1957). *Experientia* **13**, 30.

Richterich-van Baerle, R., Goldstein, L., and Dearborn, E. H. (1956). *Science* **124**, 74.

Richterich-van Baerle, R., Goldstein, L., and Dearborn, E. H. (1957). *Enzymologia* **18**, 190.

Rinehart, J. F., and Farquhar, M. G. (1954). *J. Appl. Phys.* **25**, 1463.

Rinehart, J. F., Farquhar, M., Jung, H. C., and Abul-Haj, S. K. (1953). *Am. J. Pathol.* **29**, 21.

Robertson, J. D. (1953). *J. Exptl. Biol.* **30**, 277.

Robertson, J. D. (1957). *Biol. Revs. Cambridge Phil. Soc.* **32**, 156.

Robinson, J. R. (1950). *Proc. Roy. Soc.* **B137**, 378.

Robinson, J. R. (1953). *Biol. Revs. Cambridge Phil. Soc.* **28**, 158.

Robinson, J. R. (1954a). "Reflections on Renal Function." C. C Thomas, Springfield, Illinois.

Robinson, J. R. (1954b). *Symposia Soc. Exptl. Biol. No.* **8**, 42.

Robinson, J. R. (1954c). *J. Physiol. (London)* **124**, 1.

Rollason, H. D. (1949). *Anat. Record* **104**, 263.

Rollhäuser, H. (1957). *Z. Zellforsch. u. mikroskop. Anat.* **46**, 52.

Rosenberg, T. (1954). *Symposia Soc. Exptl. Biol. No.* **8**, 27.

Rossi, F., Pescetto, G., and Reale, E. (1957). *J. Histochem. and Cytochem.* **5**, 221.

Ruska, H., Moore, D. H., and Weinstock, J. (1957). *J. Biophysic. Biochem. Cytol.* **3**, 249.

Sawyer, W. H. (1957). *Proc. Symposium Colston Research Soc.* **8**, 171.

Sawyer, W. H., and Sawyer, M. K. (1952). *Physiol. Zoöl.* **25**, 84.

Schmidt-Nielsen, B., and Forster, R. P. (1954). *J. Cellular Comp. Physiol.* **44**, 233.

Schmidt-Nielsen, B., Schmidt-Nielsen, K., Houpt, T. R., and Jarnum, S. A. (1957). *Am. J. Physiol.* **188**, 477.

Schmidt-Nielsen, K., and Schmidt-Nielsen, B. (1952). *Physiol. Revs.* **32**, 135.

Schwartz, I. L., and Opie, E. L. (1954). *Federation Proc.* **13**, 132.

Seligman, A. M., Tsou, K.-C., Rutenburg, S. H., and Cohen, R. B. (1954). *J. Histochem. and Cytochem.* **2**, 209.

Sellers, A. L., Roberts, S., Rask, I., Smith, S., Marmorston, J., and Goodman, H. C. (1952). *J. Exptl. Med.* **95**, 465.

Sellers, A. L., Griggs, N., Marmorston, J., and Goodman, H. C. (1954). *J. Exptl. Med.* **100**, 1.

Shannon, J. A. (1933). *J. Cellular Comp. Physiol.* **4**, 211.

Shannon, J. A. (1934). *J. Cellular Comp. Physiol.* **5**, 301.

Shannon, J. A. (1935). *J. Clin. Invest.* **14**, 403.

Shannon, J. A. (1938a). *J. Cellular Comp. Physiol.* **11**, 135.

Shannon, J. A. (1938b). *Proc. Soc. Exptl. Biol. Med.* **38**, 245.

Shannon, J. A. (1938c). *J. Cellular Comp. Physiol.* **11**, 123.

Shannon, J. A. (1938d). *J. Cellular Comp. Physiol.* **11**, 315.

Shannon, J. A. (1940). *J. Cellular Comp. Physiol.* **16**, 285.

Shannon, J. A., and Fisher, S. (1938). *Am. J. Physiol.* **122**, 765.

Shannon, J. A., and Ranges, H. A. (1941). *J. Clin. Invest.* **20**, 169.

Shideman, F. E. (1957). *In* "Metabolic Aspects of Transport Across Cell Membranes" (Q. R. Murphy, ed.), p. 251. Univ. Wisconsin Press, Madison, Wisconsin.

Shideman, F. E., and Rene, R. M. (1951). *Am. J. Physiol.* **166**, 104.

Simer, P. H. (1955). *Anat. Record* **121**, 416.

Sjöstrand, F. S. (1944). *Acta Anat.* **1**, Suppl. **1**, 1.

Sjöstrand, F. S. (1956). *In* "Physical Techniques in Biological Research" (G. Oster and A. W. Pollister, eds.), Vol. 3, p. 241. Academic Press, New York.

Sjöstrand, F. S., and Rhodin, J. (1953). *Exptl. Cell Research* **4**, 426.

Smetana, H. (1947). *Am. J. Pathol.* **23**, 255.

Smith, C., and Freeman, B. L. (1954). *Proc. Soc. Exptl. Biol. Med.* **86**, 775.

Smith, H. W. (1936). *Biol. Revs. Cambridge Phil. Soc.* **11**, 49.

Smith, H. W. (1937). "The Physiology of the Kidney." Oxford Med. Publs., New York.

Smith, H. W. (1943). "Lectures on the Kidney." Univ. Kansas, Lawrence, Kansas.

Smith, H. W. (1951). "The Kidney." Oxford Univ. Press, London and New York.

Smith, H. W. (1952). *Federation Proc.* **11**, 701.

Smith, H. W. (1953). "From Fish to Philosopher," p. 23. Little, Brown, Boston, Massachusetts.

Smith, H. W. (1956). "Principles of Renal Physiology." Oxford Univ. Press, New York.

Smith, H. W. (1957). *Am. J. Med.* **23**, 623.

Smith, H. W., and Clarke, R. W. (1938). *Am. J. Physiol.* **122**, 132.

Smith, H. W., Goldring, W., and Chasis, H. (1938). *J. Clin. Invest.* **17**, 263.

Smith, J. F., and Lee, Y. C. (1957). *J. Exptl. Med.* **105**, 615.

Smith, W. W. (1939). *J. Cellular Comp. Physiol.* **14**, 95.

Solomon, S. (1957). *J. Cellular Comp. Physiol.* **49**, 351.

Sperber, I. (1944). *Zool. Bidrag Uppsala* **22**, 249.

Sperber, I. (1948a). *Ann. Roy. Agr. Coll. Sweden* **16**, 49.

Sperber, I. (1948b). *Nature* **161**, 236.

Sperber, I. (1948c). *Ann. Roy. Agr. Coll. Sweden* **15**, 317.

Sperber, I. (1948d). *Ann. Roy. Agr. Coll. Sweden* **15**, 108.

Sperber, I. (1949). *Ann. Roy. Agr. Coll. Sweden* **16**, 446.

Sperber, I. (1954). *Arch. intern. pharmacodynamie* **97**, 221.

Sternberg, W. H., Farber, E., and Dunlap, C. E. (1956). *J. Histochem. and Cytochem.* **4**, 266.

Straus, W. (1954). *J. Biol. Chem.* **207**, 745.

Sulkin, N. M. (1949). *Anat. Record* **105**, 95.

Suzuki, T. (1912). "Zur Morphologie der Nierensekretion unter physiologischen und pathologischen Bedingungen." Gustav Fischer, Jena.

Swanson, R. E. (1956). *Am. J. Physiol.* **184**, 527.

Swanson, R. E., Hoshiko, T., and Visscher, M. B. (1956). *Am. J. Physiol.* **184**, 535.

Taggart, J. V. (1954). *In* "The Kidney" (A. A. G. Lewis and G. E. W. Wolstenholme, eds.), p. 65. Little, Brown, Boston, Massachusetts.

Taggart, J. V. (1956a). *In* "Enzymes: Units of Biological Structure and Function" (O. H. Gaebler, ed.), Chapt. 16, Pt. IV. Academic Press, New York.

Taggart, J. V. (1956b). *Science* **124**, 401.

Taggart, J. V., and Forster, R. P. (1950). *Am. J. Physiol.* **161**, 167.

Taggart, J. V., and Forster, R. P. (1952). *Methods in Med. Research* **5,** 228.

Taggart, J. V., Forster, R. P., Schachter, D., Kaplan, S. A., and Trayner, E. M. (1953). *Bull. Mt. Desert Isl. Biol. Lab.* **4,** (1), 54.

Thomas, C. A., Jr. (1956). *Science* **123,** 60.

Toussaint, C. L., Walter, R., and Sibille, P. (1953). *Rev. belge pathol. et méd. exptl.* **23,** 83.

Ussing, H. H. (1954a). *Symposia Soc. Exptl. Biol. No.* **8,** 407.

Ussing, H. H. (1954b). *Proc. Symposium Colston Research Soc.* **7,** 33.

Van Slyke, D. D., Rhoads, C. P., Hiller, A., and Alving, A. S. (1934). *Am. J. Physiol.* **109,** 336.

Vernier, R. L., Farquhar, M. G., Brunson, J. G., and Good, R. A. (1956a). *J. Lab. Clin. Med.* **48,** 951.

Vernier, R. L., Farquhar, M. G., Brunson, J. G., and Good, R. A. (1956b). *Bull. Univ. Minn. Hosp.* **28,** 58.

Vernier, R. L., Brunson, J. G., and Good, R. A. (1957). *A. M. A. J. Diseases Children* **93,** 469.

Vimtrup, B. (1949). *Scand. J. Clin. & Lab. Invest.* **1,** 339.

Vimtrup, B., and Schmidt-Nielsen, B. (1952). *Anat. Record* **114,** 515.

von Möllendorff, W. (1927–1928). *Z. Zellforsch. u. mikroskop. Anat.* **6,** 441.

von Möllendorff, W. (1929). *Handb. norm. u. pathol. Physiol.* **4,** 183.

von Möllendorff, W. (1936). *Z. Zellforsch. u. mikroskop. Anat.* **24,** 204.

Wachstein, M. (1955). *J. Histochem. and Cytochem.* **3,** 246.

Wachstein, M., and Meisel, E. (1953). *Stain Technol.* **28,** 135.

Walker, A. M. (1940). *Am. J. Physiol.* **131,** 187.

Walker, A. M., and Hudson, C. L. (1937a). *Am. J. Physiol.* **118,** 153.

Walker, A. M., and Hudson, C. L. (1937b). *Am. J. Physiol.* **118,** 130.

Walker, A. M., and Oliver, J. (1941). *Am. J. Physiol.* **134,** 562.

Walker, A. M., Hudson, C. L., Findley, T., and Richards, A. N. (1937). *Am. J. Physiol.* **118,** 121.

Walker, A. M., Bott, P. A., Oliver, J., and MacDowell, M. C. (1941). *Am. J. Physiol.* **134,** 580.

Wearn, J. T., and Richards, A. N. (1924). *Am. J. Physiol.* **71,** 209.

Webb, D. A. (1940). *Proc. Roy. Soc.* **B129,** 107.

Weiss, J. M. (1955). *J. Exptl. Med.* **101,** 213.

Weiss, L. P., Tsou, K.-C., and Seligman, A. M. (1954). *J. Histochem. and Cytochem.* **2,** 29.

Wells, L. J. (1946). *Anat. Record* **94,** 504.

Welt, L. G. (1956). *Yale J. Biol. and Med.* **29,** 299.

Wesson, L. G., Jr., and Anslow, W. P., Jr. (1952). *Am. J. Physiol.* **170,** 255.

White, H. L. (1929). *Am. J. Physiol.* **90,** 689.

Whittam, R. (1956). *J. Physiol. (London)* **131,** 542.

Whittam, R., and Davies, R. E. (1953). *Biochem. J.* **55,** 880.

Wigglesworth, V. B. (1931). *J. Exptl. Biol.* **8,** 428.

Wilbrandt, W. (1938). *J. Cellular Comp. Physiol.* **11,** 425.

Wilbrandt, W. (1954). *Symposia Soc. Exptl. Biol. No.* **8,** 136.

Williams, J. K. (1956). *Federation Proc.* **15,** 200.

Winton, F. R., ed. (1956). "Modern Views on the Secretion of Urine." Little, Brown, Boston, Massachusetts.

Wirz, H. (1955). *Helv. Physiol. et Pharmacol. Acta* **13,** 42.

Wirz, H. (1956). *Helv. Physiol. et Pharmacol. Acta* **14,** 353.

Wirz, H., and Bott, P. A. (1954). *Proc. Soc. Exptl. Biol. Med.* **87,** 405.

Wirz, H., Hargitay, B., and Kuhn, W. (1951). *Helv. Physiol. et Pharmacol. Acta* **9,** 196.

Yamada, E. (1955a). *J. Biophys. Biochem. Cytol.* **1,** 551.

Yamada, E. (1955b). *J. Histochem. and Cytochem.* **3,** 309.

Zimmerman, K. W. (1929). *Z. mikroskop.-anat. Forsch.* **18,** 520.

Zimmerman, K. W. (1933). *Z. mikroskop.-anat. Forsch.* **32,** 176.

CHAPTER 3

The Blood Cells and
Their Formation[1]

By MARCEL BESSIS[2]

[1] In this chapter we shall study the cells of the blood and their formation in the adult mammal. Detailed descriptions of the cells of the blood in the embryo can be found in the general reviews of Jolly (1923), Sabin et al. (1936), Bloom (1938), Bloom and Bartelmez (1940), Gilmour (1941), Knoll (1927, 1932), and Knoll and Pingel (1949). The formation of the blood cells in animals is discussed in the following papers: Jordan (1938). Winqvist (1954), Wintrobe (1956).

[2] Translated by Eric Ponder, Nassau Hospital, Mineola, New York.

I. The Morphology of the Cells in the Circulating Blood

The cells in the circulating blood are the end results of a series of stages of development of stem cells which are found in the hematopoietic organs. It is necessary always to bear in mind when studying the structure, the movements, and the ultrastructure of the cells of the blood that we are dealing only with the last step of a long series of developmental processes. For simplicity, we shall start by studying the mature cells of the blood; these are classified as erythrocytes, leucocytes (granulocytes, monocytes, and lymphocytes), and platelets.

A. *The Erythrocytes*

The erythrocyte is a non-nucleated cell which is discoidal and biconcave in shape. In man, it measures about 7.5μ in diameter in dried films and about 8.3μ in the fresh state in plasma. Seen with the ordinary microscope, it has a yellow-orange color. With the phase contrast microscope, it appears so dark as to be almost black with a clear center. Strictly speaking, the erythrocyte presents no movement; it is easily deformable, very elastic, and one can see after it has been considerably deformed by circulating through capillaries of a very small diameter that it immediately recovers its discoidal form. The red cell presents a peculiar "scintillation," the cause of which is not entirely understood; this "scintillation," or "flicker," is perhaps the result of a molecular movement of hemoglobin (Ralph, 1947; Pulvertaft, 1949; Blowers *et al.*, 1951; Forkner *et al.*, 1936; Tompkins, 1953).

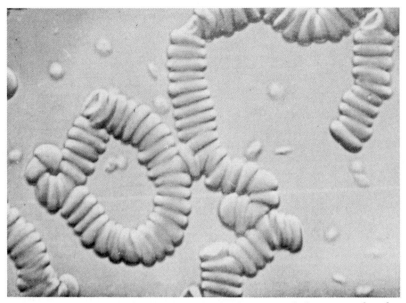

FIG. 1. Appearance of red cells in the fresh state, with formation of rouleaux. Photograph taken with the interference microscope by the method of Nomarski.

1. Rouleaux

Red cells adhere to each other in a characteristic fashion by forming rouleaux (Fig. 1). The formation of rouleaux depends on the quantity and the nature of the proteins of the plasma surrounding the cells. An excess of fibrinogen (in pneumonia, for example) or an excess of globulins (in plasmocytoma, for example) results in a considerable increase in the length of the rouleaux. On the other hand, if the red cells are washed in physiological saline, rouleaux do not form at all. *The rate of sedimentation* of the red cells depends on the formation of rouleaux; it accordingly varies with the quantity of fibrinogen and of globulin in the plasma or serum. The rate of sedimentation is a useful test in clinical medicine because many diseases (infections in particular) are accompanied by an increase in the quantity of the proteins in the plasma (cf. Fahraeus, 1929).

2. Disk-Sphere Transformation

After several hours between slide and coverslip and suspended in their own plasma, or after they have been washed in saline, red cells lose their discoidal form and become spheres covered with tiny spikes; this is called the spherical form or "sea urchin" form. It is due to the absence of a substance which exists in normal plasma (antisphering factor)

TABLE I

DISK-SPHERE TRANSFORMATION

Conditions and substances producing spherical forms or sea urchin forms	Restoration of the normal discoidal form
Washing with isotonic NaCl	Normal plasma: 1/8 to 1/64
Washing with isontonic KI	Albumin: 1/8000
Washing with isotonic KCL	Glucose: 1/40
Traces of lecithin	Gelatin: 1/100 to 1/1000
Traces of the biles salts	Quinine hydrochloride: 1/200
Heating to 50° C.	pH between 3.8 and 4.5
Storage in the icebox from 3 to 15 days	(acid citrate of Wurmser)
Raising the pH between 8 and 9 with bicarbonate	Phenergan: 1/20,000

which has been identified as an albumin (Furchgott and Ponder, 1940; Ponder, 1948). The disk-sphere transformation is reversible in either direction, and Table I (Bessis and Bricka, 1950) shows some of the conditions under which the transformations take place.

The cause of the discoidal form of the red cells, which is so characteristic and which is 20–30% larger than the surface of a sphere of the same volume, is still altogether unknown; it is not due to the structure of the red cell alone, but more probably to physicochemical properties operating at the surface (Ponder, 1948, 1955).

3. Agglutination

Erythrocytes agglutinate when treated in a variety of ways, some of which are specific and others non-specific. Specific agglutination is due to the antibodies of the blood groups, a matter which has been dealt with in extremely important investigations, for, in practice, compatible blood transfusion depends on scientific considerations, particularly those of genetics (cf. Race and Sanger, 1958; Mollison, 1956; Mourant, 1954). Non-specific agglutination takes place in the presence of various metals (Jandl and Simmons, 1957) and in the presence of certain viruses (Melnick, 1956).

When one examines agglutinated red cells with the microscope and tries to separate one cell from others, one realizes that the agglutinated cells are united by elastic filaments (Bessis and Bricka, 1950). When these elastic filaments break, they have all the appearances of "myelin forms"

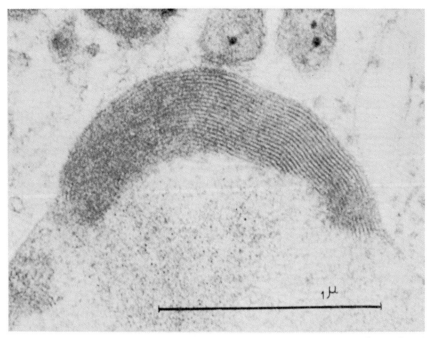

Fig. 2.   Appearance with the electron microscope of a section of a red cell during its digestion in the interior of a macrophage. The successive layers of lipids in the stroma can be seen, these having the appearance of myelin forms.

(Bessis *et al.*, 1951). They accordingly contain a large proportion of lipids.

The electron microscope gives further information about these structures. After they have been treated by gold shadowing followed by moulage, it can be shown that the surface of the erythrocytes is spotted with irregular depressions called "craters." When agglutination occurs, these craters are profoundly modified, and they are the origin of the filaments which are essential to agglutination (Bessis *et al.*, 1951). The craters and the substances with a high lipid content accordingly play an important role in agglutination. Myelin forms have been found in erythrocytes which have been altered during digestion. In this case, the electron microscope shows the myelin forms to have the characteristic alternation of layers of lipoprotein and layers of water (Fig. 2).

4. Hemolysis

When the red cell is changed in a variety of ways, the pigment which it contains diffuses out of it, and the cell is hemolyzed. Hemolysis can be due to a hypotonicity of the surrounding medium (osmotic hemolysis), to physical injury (coagulation or compression), to chemical injury (as in the case of saponin, the bile salts, fatty acids, lecithin, etc., and particularly, hemolysins resulting from antibody formation). The *stroma* which remains after hemolysis appears is seen with the electron microscope as an empty envelope, the surface of which presents folds. In some cases, it can still undergo a reversible disk-sphere transformation (cf. Bessis and Bricka, 1950; Ponder *et al.*, 1952a, b).

The study of hemolysis has resulted in a number of very important and practical researches, because it is one of the most sensitive tests for immune reactions (such as complement fixation). In addition, it has great theoretical interest when the mechanisms which underlie it are studied. Its immunological importance has been dealt with and discussed in the book of Bordet (1939). All that is known about the mechanism of hemolysis can be found in the monographs of Ponder, 1948, 1955.

5. Sectioning of Erythrocytes

Heating to 50° or 60° C. results in red cells taking on a variety of forms; in particular these are elongated forms which end up by dividing into very small globules. The fragments are quite spherical, although they do not hemolyze, at least immediately. By micromanipulation, one can also cut a red cell without hemolyzing the fragments (Comandon and Fonbrune, 1929, Seifriz, 1936) or red cells can be phagocytosed by a leucocyte which soon alters them by cutting them in two (Policard and Bessis, 1953).

6. Structure and Composition of Erythrocytes

Discussions about the structure of red cells are not likely to be soon ended because it is necessary that the hypotheses proposed agree with the experimental results and explain in particular the discoidal form of erythrocytes and the different ways in which they are hemolyzed. Two hypotheses have been formulated: that the structure is a sac containing hemoglobin, or that it is a spongework in which hemoglobin is contained. The structure which agrees best with all the known facts seems to be something between these two, and supposes that the cytoplasm of the cell is condensed at its periphery and is very scanty in the interior. If this is so, the stroma which is observed would be formed. Hemoglobin is to be found in all the fractions of the stroma, although in greater or lesser quantity; in the peripheral regions there is very little of it and it is strongly bound to lipoprotein components, but in the interior it exists as a concentrated solution.

According to this hypothesis, there is no need for the existence of a membrane in the usual sense of the word, but there is a peripheral region of great density and a central region (interior) which is almost liquid. Moreover, a lipid film (or rather a film of mixed lipids, proteins, and glucosides) is to be found at the most peripheral regions of the stroma. This film is very thin. The books of Ponder (1948, 1955) summarize the investigations on the composition of human red cells.

7. Hemoglobin

The hemoglobin in the interior of the red cell exists as an extremely concentrated solution (about 33%); this concentration is so great that if it were slightly increased the hemoglobin would crystallize. Crystallization of hemoglobin in the interior of a red cell is a phenomenon which can occur under conditions which will be discussed later on. Hemoglobin is a chromoprotein formed of a colorless protein matrix (globin), to which four molecules of heme are attached; it gives the characteristic color to the cell.

Heme is the result of the union of protoporphyrin III with an atom of ferrous iron. This iron is bound to four pyrroles of the porphyrin ring. Because there are two residual valences remaining, it can become fixed to the nitrogen of globin and also to oxygen; the latter reversible combination allows of oxygen being transported to the tissues.

8. Different Kinds of Hemoglobin

Hemoglobins may differ between each other because of the structure of their protein component. The content of amino acids in the globin of different species is quite variable.

In man, normal adult hemoglobin is not the only one which may be met with, and at present one knows of fetal hemoglobin, adult hemoglobin, and at least eight abnormal hemoglobins. Even adult hemoglobin is heterogeneous. The best-known and the most important of the abnormal hemoglobins is the hemoglobin of sickle cells which occurs in patients with sickle-cell anemia and which is called hemoglobin S (sickle-cell hemoglobin) or hemoglobin B. As compared with adult hemoglobin, it possesses two important differential characteristics. Its solubility is much less and decreases markedly when it is reduced, the reduced form being more than 100 times less soluble than the oxygenated form. Further, its electrophoretic characteristics are different.

Hemoglobins C and D have been discovered in American Negroes. Hemoglobin C shows an even greater increase in positive charge than hemoglobin B. Its isoelectric point is still higher, and it migrates more rapidly in acid media, whereas in veronal buffer at pH 8.6 and on paper, it scarcely moves from the origin. On the other hand, its solubility is even greater than that of hemoglobin A. Hemoglobin D has the same electrophoretic characteristics as hemoglobin S, from which it cannot be separated by these means. It is characterized by its great solubility in its reduced form, which distinguishes it from the very small solubility of hemoglobin S.

Hemoglobin E has been recently described in inhabitants of Thailand. In electrophoretic patterns, it migrates between hemoglobins S and C. Still other hemoglobins, G, I, J, and K have been recently described, and one may look forward to the description of still others.

The study of the abnormal hemoglobins is not of biochemical interest only. It constitutes the first example of the fruitful idea of "molecular diseases" (Pauling *et al.*, 1949). For further information about the different hemoglobins, the reader should consult the general reviews of Beaven *et al.* (1951), Chernoff (1955), Pauling (1954), Penati *et al.* (1956), Zuelzer *et al.* (1956), Malassenet (1957), and Singer *et al.* (1951).

9. *Intracellular Crystallization of Hemoglobin and Sickle-Cell Formation*

The investigations on sickle-cell anemia and the sickle-cell trait, in which the red cells possess the abnormal hemoglobin S, are very numerous (compare the general reviews referred to above). These are hereditary diseases which occur in a quite marked proportion of certain Negro populations. Sickle-cell anemia is characterized by an anemia together with a variety of symptoms and signs. From a cytological point of view, the characteristic phenomenon observed is called "sickling." The red cells take on bizarre forms as soon as the tension of oxygen is reduced, and they look like sickles or holly leaves. At the same time, they become rigid. The phenomenon is reversible, and when the tension of oxygen is increased, the red cells resume their normal form. The hypothesis has been put forward that the hemoglobin "crystallizes" when it becomes reduced, but recent work has shown that the term "crystallization" is not exact.

Studies carried out by X-ray diffraction (Dervichian *et al.*, 1952) have shown that the "short-order" orientation in the interior of the red cell is smaller in the sickled cell than in the normal erythrocyte. The hypothesis that there is a state of crystallization in the sickled red cell ought accordingly to be abandoned. Ponder (1955), by measuring the specific heat and the heat of compression, has found that hemoglobin S in the reduced state has characteristics very different from that of crystalline hemoglobin. At the same time, it has been known, particularly from the investigations of Ponder (1945), that in some animals and under certain conditions normal adult hemoglobin can crystallize in the interior of the red cells. More recently, it has been shown (Maganelli and Ricciardi, 1956; Ager

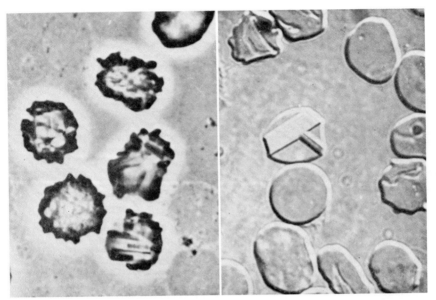

Fig. 3. Appearance with phase contrast (at the left) and with the interference microscope (at the right) of intracellular crystallization of hemoglobin in human cells.

and Lehmann, 1957) that the hemoglobin of normal individuals can crystallize in the interior of the red cells if the latter are treated with a 2% solution of metabisulfite at low temperature. Recent morphological studies have led to the comparison of the phenomenon of intra-erythrocytic crystallization with that of sickle-cell formation: (1) Phase contrast microscopy and the interference microscope show a great difference between sickle cells and red cells in which hemoglobin has crystallized (Figs. 3 and 4). (2) The polarizing microscope allows one to see and to measure the birefringence associated with these two phenomena. In sickle cells, the birefringence could be interpreted as being due to a mass of extremely small crystals or to a birefringence of form. (3) The electron microscope confirms this last interpretation. It shows an absence of crystalline structure. All one sees in sections of sickle cells is a "stratified feltlike structure" and a fibrous structure, which are compatible with the hypothesis of hemoglobin in the state of a gel and which explain the birefringence of form entirely satisfactorily (Bessis et al., 1958).

B. The Granulocytes

The granulocytes, once called polynuclears, are characterized by a nucleus with several lobes and a cytoplasm which contains granules of

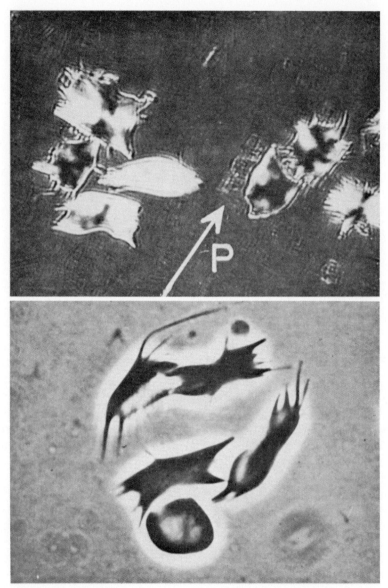

F<small>IG</small>. 4. Above: the appearance of sickled cells with the polarization micro-scope (the arrow indicates the plane of polarization). Below: the appearance of sickled cells with phase contrast.

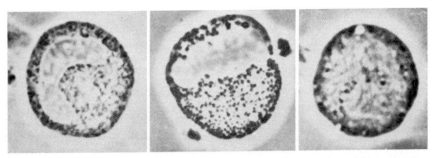

FIG. 5.   Appearance with phase contrast of granulocytes; from left to right, a neutrophile, an eosinophile, and a basophile.

different kinds; these are called (wrongly) neutrophile granules, basophile granules, and acidophile granules (Fig. 5). The number of nuclear lobes in the granulocytes varies in the normal individual from 2 to 6. The percentage of granulocytes, when classified according to the number of the nuclear lobes (the Arneth count, cf. Cooke and Ponder, 1927) varies in many pathological conditions.

The *neutrophile* granules contain ribonucleoproteins and lipids. In addition, they also contain glycogen and many enzymes. Among these, peroxidases are easy to demonstrate by cytochemistry and are often looked for so as to make a diagnosis between immature cells and the cells of the granulocytic series.

The *acidophile* granules are stained a rose-orange color by eosin (eosinophilic granules). The surface of the eosinophile granules stains with Sudan black, which demonstrates the presence of lipids (Vercauteren, 1951). When the eosinophile leucocyte is autolyzed, one often observes crystalline proteins called "Charcot-Leyden crystals" (cf. the recent review of Ayres, 1949, 1950).

The *basophile* granules stain with Giemsa in a metachromatic fashion and with a wine-red color. They are soluble in water, and after being dissolved leave empty vacuoles in the cytoplasm. Basophilic granules do not give Charcot-Leyden crystals, but when treated with distilled water, the granules disappear and spindle-shaped crystals appear; these are specific for basophiles (Bessis and Tabuis, 1955). The basophile granulocytes of the blood are very different from the mastocytes of tissues, with which they have sometimes been confused by older investigators (cf. below).

1. Movements of Granulocytes

a. Movements of pseudopods. Like the other cells of the blood the granulocytes, which are spherical while circulating, are deformed as soon

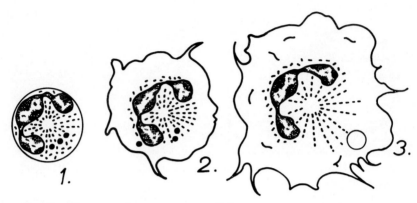

Fig. 6.   Diagrammatic representation of the spontaneous spreading on a plastic surface of a neutrophile granulocyte. Note the aster, the rod-shaped chondrosomes, and a contractile vacuole.

as they touch a solid surface upon which they can creep. The speed of movement of granulocytes varies between 19 and 36 μ per minute (Lewis, 1934; Henderson, 1928; Jolly, 1913; Comandon, 1919, 1929).

b. Spontaneous spreading of granulocytes. Granulocytes possess the property of spreading on a suitable support (Bessis and Bricka, 1949). The cells, initially free in the liquid surrounding them, begin to attach themselves to the support, and veils appear at different points on the periphery so that the cell ends up by being surrounded by a fringe of hyaloplasm with a diameter of the order of 20 μ. In the second stage of this phenomenon, the granules spread progressively into the cytoplasm but do not occupy the periphery of the cell, so that a band of 3–5 μ remains free of granules. This granule-free region, which is made up of pure hyaloplasm, shows active undulation (Fig. 6).

The nucleus has a smooth surface, and in its concavity is found an important mass of granules surrounding the centrosome. Outside this region, the granules lie in a single layer and are more dispersed as one passes outward from the centrosome.

c. Intracytoplasmic movements. Movements of the centrosome. The centrosome appears as a region which is clearer than the surrounding cytoplasm; it contains no granules and has a diameter of from 0.5 to 1.0 μ. The granules which lie near it are arranged radially. The centrosome, described as lying in the concavity of the nucleus, has a to-and-fro movement with a period of the order of 30 seconds and a very variable amplitude (from 5 to 10 μ) (Bessis and Locquin, 1950; Policard and Bessis, 1952). The nucleus is passively affected by the movement of the centrosome, the latter pushing on the nucleus or even deforming it. The granules

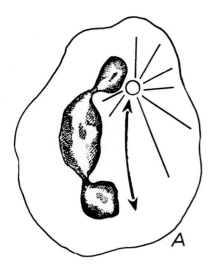
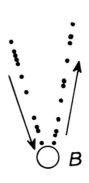

FIG. 7. Diagrammatic representation of the movement of a centrosome in a spread-out granulocyte. A. Diagrammatic representation of the amplitude of movement of the centrosome. B. The trajectory of a neutrophile granule in the zone which is influenced by the centrosome. Each point represents the position at intervals of equal times.

nearest the centrosome seem to be bound to it and move with it, but at greater distances the amplitude of movement of the granules decreases and those at the periphery seem to be scarcely affected by the movement of the centrosome (Fig. 7).

d. Movements of granules. Apart from the displacements which are due to the movement of the centrosome, specific granules (neutrophile and eosinophile) can show a variety of movements, such as Brownian movement, displacements of large amplitude toward the centrosome, etc.

e. Vacuoles. It can be shown that the granulocytes contain four kinds of vacuoles: those which stain with neutral red and which appear and develop *in vitro*; degenerating vacuoles, particularly those near the junction of the nucleus and the cytoplasm (Dustin, 1949); the vacuoles of pinocytosis (Bessis and Bricka, 1952); and contractile vacuoles (Bessis and Locquin, 1950).

Pinocytosis causes vacuoles to appear in a distinctive way. The cell throws out a hyaloplasmic veil and the central region of the veil (the region which becomes a vacuole) becomes clear while the edges become dark. The vacuole in the veil rounds up and reaches the interior of the cell. This phenomenon is very general and is comparable to phagocytosis; it allows the cell to absorb liquids.

C. Lymphocytes

The lymphocytes of the circulating blood measure from 9 to 16 μ in diameter. Their nucleus is generally centrally placed and round, but it may be oval or slightly concave. The cytoplasm is basophilic and shows mitochondria, together with a few azurophile granulations and a centrosome which is particularly clearly defined. There are a very large number of enzymes in the lymphocyte, such as pepsin, nuclease, etc. One frequently sees a small refractile body called a "Gall body," which is a vacuole containing water surrounded by lipids.

D. Monocytes

Many histologists group under the name of histiocytes (or the cells of the histiomonocyte series) a large number of cells such as the reticulum cell, fixed or free, plasmacytes, mastocytes, and monocytes. These cells are generally found in the tissues, but can pass into the blood (particularly monocytes and plasmocytes). Hematologists draw a distinction between all these "histiocytes" and consider them as the end result of specialized cells from which they are derived.

Monocytes are large mononuclear cells, their diameter varying from 15 to 30 μ. Their nucleus is quite large, sometimes round, but more usually irregular. The monocytes differ from lymphocytes in that they can, like the granulocytes, spread on a glass surface. They are also able to phagocytose particles.

Monocytes show the same movements as reticulum cells (the histiocytes of histologists) when in tissue culture; this is not surprising because it is admitted that this is only a matter of their resembling cells to which they are very closely related (Policard, 1925; Thomas, 1938; Chevremont, 1942).

When the monocyte moves, it generally takes on a triangular form, one of the points being directed in the line of motion; the hyaloplasmic veils are produced in the neighborhood of the points. Monocytes differ from granulocytes by these veils appearing even when the cell is not attached to a support.

E. Thrombocytes or Platelets

Thrombocytes are tiny non-nucleated fragments of protoplasm which undergo very rapid changes in shape, etc. In order to obtain a good

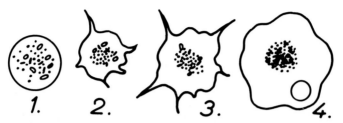

Fig. 8. Diagrammatic representation of the spontaneous spreading of a platelet on glass as seen with phase contrast. Note in 4 a contractile vacuole.

cytologic image, it is necessary to take certain precautions to prevent their agglutination by contact with tissues, glass, and even with air.

On well-prepared films, they appear as little structures measuring from 2 to 5 μ, but there also can be found thrombocytes of large size (macrothrombocytes) which measure as much as 10 μ. They are round or oval when they are examined in the living state in the capillaries. Star-shaped forms, shrunken forms, and agglutinated forms, etc., do not occur except when the platelets have left the vessels, or in pathological conditions. Thrombocytes are made up of a cytoplasm containing reddish-violet granules. These granules are usually found in the center of the cell, but sometimes they are scattered over the whole surface.

1. Spontaneous Spreading of Thrombocytes

Observations made in the living animal lead one to think that the platelets of mammals in the circulation are discoidal in form, but, when they escape from the vessels or into a tube in which they are collected and when they are placed between slide and coverslip to be examined, platelets become star-shaped and are known as dendritic forms. Within a few minutes, the protoplasm of the platelets gives rise to filaments of increasing length which generally reach 3–5 μ. These filaments continually change their form and their position; this phenomenon is too slow to be appreciated by the eye, but it is clearly seen with cinematography accelerated about forty times (Fig. 8). When one of these dendritic filaments touches the glass of the slide or some other surface (the nature of which is of great importance), it adheres to the surface and then other dendritic filaments adhere in a similar way so as finally to give rise to the "spread-out platelet." Such a platelet is completely adherent to the surface, does not show the dendritic filaments, and usually appears round or oval; it appears almost transparent, except in its center where the granulations have collected (Fonio and Schwendener, 1942).

Platelets contain both contractile vacuoles and the vacuoles of pinocytosis (Gonzalez-Guzman and Bessis, 1952; Bessis and Bricka, 1952).

II. GENERAL OBSERVATIONS OF THE FORMATION
OF BLOOD CELLS

The origin of the different lines of development of the blood cells is to be found in the undifferentiated cells of the hematopoietic tissues. These undifferentiated cells form a part of the reticuloendothelial system. Blood cells are continually destroyed and formed again by the cells of a larger system, the study of which accordingly constitutes a fundamental part of blood cell cytology.

Aschoff, in 1913, put into one class all the cells scattered through different tissues which are capable of accumulating in their cytoplasm certain vital stains introduced into the living organism. To these he gave the name of the *reticuloendothelial system*. By doing this, he grouped in one class a large number of cells which had already been separately described by many investigators as macrophages, but Aschoff deserves credit for having concluded that the anatomical and functional aspect of the endothelial system constitutes a unit. In 1914, Kiyono introduced the term *histiocyte*, including under this name cells capable of fixing vital stains. A recent monograph has been devoted to it (edited by Halpern, 1957). The reticuloendothelial or reticulohistiocytic system accordingly includes all the cells capable of fixing vital stains (trypan blue, pyrrole blue, carmine, etc.), and the cells included are the following: (1) the reticulum cells of the hematopoietic organs (lymphatic glands, bone marrow, Peyer's patches, tonsils, etc.); (2) endothelial reticulum cells, such as those which lie in blood sinuses (in the spleen, in the bone marrow, the liver, and cells of Küpffer), as well as in the pulmonary alveoli and the tissues of the conjunctiva, etc.; (3) wandering tissue cells, such as histiocytes, "dust cells," ocrocytes, etc.; (4) the monocyte, which circulates in the blood; and (5) some of the cells of microglia.

The reticuloendothelial system has three principal functions: (1) a metabolic function particularly involving the metabolism of lipids, carbohydrates and glucosides, salts, and water; (2) a function related to immunity (fixation of granules and production of antibodies); (3) a cytopoietic function.

A. *The Hematopoietic Function of the Reticulohistiocytic System*

In the adult, the hematopoietic system is normally localized in the bone marrow, the spleen, and the lymph glands. In these tissues, stem cells of the different developmental lines of the blood cells are found.

Reticulum cells are essentially distinct from the stem cells of each developmental line, and the stem cells *ripen* to give rise to the cells of the
circulating blood; we shall see later the meaning which has to be given
to this statement.

All authors are agreed that it is proved that the mature cells are
derived from stem cells by passing through the stages indicated in
Tables II and III. At the same time, everybody agrees that the stem cells
are formed from reticulum cells, although hematologists still discuss the

TABLE II

BLOOD CELLS

	Lymph glands, etc.	Bone marrow		
Hematopoietic organs	Lymphoblast	Myeloblast	Proerythroblast	Megakaryoblast
	↓	↓	↓	↓
		Myelocyte	Erythroblast	Megakaryocyte
	↓	↓	↓	↓
Blood	Lymphocyte	Granulocyte	Erythrocyte	Thrombocyte

TABLE III

HISTIOCYTIC AND MONOCYTIC CELLS

Reticuloendothelial Tissue	Histioblast	Plasmoblast	Mastoblast
		↓	↓
		Plasmocyte	Mastocyte
	↓		
Blood	Monocyte	The passage of these cells into the blood stream is exceptional	

subject of the existence of totally undifferentiated stem cells which are
intermediate between the reticulum cell and the stem cell as ordinarily
recognized. This has given rise to innumerable schemes and cytogenetic
theories which now are gradually being forgotten. The question turns on
whether one can know if these completely undifferentiated stages really
exist. It is certainly possible that each of the reticulum cells from the

very beginning could develop only into the cells of a single developmental line, but it should be remembered that our present methods of studying morphological detail do not allow us to come to any such conclusion.

B. Differentiation and Maturation

As regards the cause of differentiation and of maturation in the cells of the blood, we know practically nothing, and the mechanism which results in an undifferentiated stem cell becoming sometimes a red cell, sometimes a white cell, or sometimes a platelet escapes us completely. Great hopes were entertained by hematologists when the method of tissue culture of bone marrow was introduced, and it was thought that the results obtained by the addition or the removal of substances easy to obtain *in vitro* would reveal the cause of differentiation. Up to now, what was so promising to begin with has led to disappointment.

It is necessary to separate *differentiation* from the *maturation* of cells (cf. Bessis, 1954). An undifferentiated cell can develop in several ways, whereas a differentiated cell cannot develop except in a single direction; one usually says that it matures. For example, a reticulum cell is an undifferentiated cell, and it can differentiate (according to the environmental conditions) into an erythroblast, a lymphoblast, etc. Once this step is taken, the cell begins its *maturation*.

In losing its potentiality, the cell acquires from this time onward irreversible characteristics, so that its maturation is unidirectional. It may continue to mature and to become functional; on the other hand, if the circumstances are unfavorable, it may not complete its maturation, but it can never return to its early state or change its direction of maturation. *Dedifferentiation* in the sense that a cell becomes the equivalent of an undifferentiated cell does not occur, at least in the case of the cells of the blood.

If we are reduced to pure speculation regarding the causes and the mechanism of differentiation, we possess plenty of information on the maturation of the blood cells. The classical hematological methods, in particular the Romanovski stains, have already furnished a number of important morphological details, but it is principally in the domain of biochemistry that the results obtained during recent years have advanced our knowledge. The reader can find important information in the general reviews of Theorell (1947), of Aschkenasy (1949), and of Errera (1952).

C. Biochemical Phenomena Accompanying Maturation

There are a considerable number of investigations on this subject. On the one hand, they deal with the chemical composition of the blood

cells as determined either by classical biochemical methods or by cyto-chemistry; on the other hand, they deal with the mechanism and the necessary requirements for the synthesis of the various structures found in the blood cells. To illustrate this, we shall give here Table IV, after Errera (1952). This table summarizes much of the cytochemical informa-tion which we have regarding the different stages of development of the blood cells.

The most numerous investigations have been concerned with the quantities of nucleic acid in the nucleus and in the cytoplasm. The essen-tial phenomenon of maturation is a considerable increase of the cyto-plasmic basophilia which precedes the appearance of more specific charac-teristics (the appearance of hemoglobin and of granules), followed by a decrease in basophilia in the course of the further maturation of the cell. Brachet (1952), Theorell (1947, 1950), and others have established that the basophilia is due to ribonucleoproteins. The work of Thorell on isolated cells *in vitro* has led to a quantitative expression of these variations, largely due to cytospectrophotometric studies. There will be found in the general reviews of Aschkenasy (1949, 1952), of Valentine (1951), of Wooley (1949), and of Granick (1952) an account of what we know about the biochemical structure of the blood cells, the requirements for normal hematopoiesis, the role of proteins, of carbohydrates and glucosides, of lipids, of minerals, and of hormones.

D. *Morphological Phenomena Associated with Maturation*

Young cells have a number of distinct characteristics. They are larger than mature cells. Their proportion of nucleus to cytoplasm is large, the chromatin is clear, and its structure is very fine; the young cells also contain one or more nucleoli, and the cytoplasm is basophile and does not possess specific granules.

As the young cells mature, their size becomes smaller, and the nucleo-cytoplasmic ratio becomes less. The nucleus loses its nucleoli, its chro-matin becoming more dense and intensely colored. The cytoplasm changes from being basophilic to being acidophilic, and at the same time it acquires the properties which are specific for one or another line of development. One can judge from the appearance of the nucleus and the cytoplasm whether one is looking at a young cell, a mature cell, or a cell about to die, and it is these characteristics which enable one to determine the different stages of maturation.

The termination *blast* indicates a stem cell. The termination *cyte* indicates a mature cell, and prefixes *pro* and *meta* indicate transitions between cells which are well defined. It should be remembered that all cells do not absolutely correspond to the names which we give them, and very often one hesitates to place a cell as evolving from one stage or beginning to pass into the stage that follows. Morphological modifications do little except to translate the evolution of the functional characteristics belonging to the young cells, which do not reach their perfect state except in cells that are mature.

Recently, following the work of the school of Jacobi, there has been an attempt to establish a nomenclature founded on nuclear volume (Leibetseder, 1952, 1954; Weicker, 1954; Lennert and Remmele, 1958). It is likely that the old, purely morphological classifications will soon be replaced by a classification of this type, based on quantitative measurements.

Law of synchronism in nucleocytoplasmic evolution. During the maturation of the cells of the blood, the development of the cytoplasmic features and those of the nucleus are absolutely parallel. A young cytoplasm corresponds to a young nucleus, and a mature cytoplasm is found when the nucleus is mature. This synchronism varies with each line of cellular development, but it is quite definite in *the normal state* and for a known developmental line (Bessis, 1946). It is, however, disturbed in a variety of pathological states. Of all the kinds of asynchronism which are possible, the most readily recognized is that in which there occurs a maturation of the cytoplasm which is more rapid than that of the nucleus. This occurs every time that there is hyperplasia in a developmental line. A development which jumps over established stages and definite characteristics may appear at a very early stage, earlier than the stage at which the development usually occurs and sometimes even in the stem cells. In certain cases, in particular in the hemosarcomas (sarcomas and leukemias), there is an actual disorderly confusion in the development of the features of the nucleus and the cytoplasm (Bessis, 1954, 1957).

E. Liberation of Mature Cells

The mechanism of the passage of mature cells into the circulation is not yet clearly understood. Most authors think that leucocytes leave the hematopoietic organs as soon as they have arrived at maturity and are capable of their own movements; at the same time, they are thought to have acquired the enzymes necessary for a temporary dissolution of the capillary wall. If this is so, they can enter the circulation by diapedesis. The red cells, which are produced like the other blood cells outside the circulatory system, cannot enter the blood stream in this manner. Doan (1932) and Isaacs (1930) have suggested some hypotheses bearing on this subject.[1]

[1] The entry into the blood stream by diapedesis is actually proved.

FIG. 9. Loss of the nucleus from an erythroblast and movements of the reticulo-cyte. Taken from a film sequence with phase contrast.

III. MATURATION AND ULTRASTRUCTURE OF THE CELLS OF DIFFERENT DEVELOPMENTAL LINES

A. *The Erythrocytic Series*

In the normal adult, the cells of the erythrocyte series are found in the hematopoietic bone marrow. In the embryo and in certain pathological conditions, the liver, the spleen, and other organs may also actively produce the cells of the erythrocyte series. In the embryo, in adults suffering from pernicious anemia or related conditions, a different erythrocytic series is found, i.e., the megalocytic series. This question has been considered in Section II, above.

1. *Maturation*

The erythrocyte series is made up as follows. The stem cell is the proerythroblast which, by maturing, gives rise to basophile erythroblasts; these progressively synthesize more and more hemoglobin, thus passing through the stages which are first called polychromatophilic erythroblasts, and later, acidophilic erythroblasts. By this time, the cell contains almost the same concentration of hemoglobin as the mature red cell. By losing its nucleus, it becomes the erythrocyte which is found in the circulating blood.

The expulsion of the nucleus from erythroblasts has been demonstrated by cinematography (Bessis and Bricka, 1952) (Fig. 9). The successive phases of the phenomenon of the expulsion of the nucleus last about 10 minutes. The cell shows active movements and throws out numerous excrescences, one of which contains the nucleus. After several of these active movements, the nucleus is expelled. The phenomenon resembles that observed after mitosis, when the two newly formed cells try to separate one from the other.

2. Reticulocytes

After the loss of the nucleus, the acidophilic erythroblast becomes a reticulocyte (sometimes called a proerythrocyte). This cell is a young erythrocyte which retains mitochondria together with a small quantity of basophile material which can be shown by staining with vital dyes (neutral red or acridine orange, with fluorescence). This basophile substance is made up of ribonucleoproteins (Dustin, 1941, 1947). It appears as a network in the interior of the erythrocyte which is stained with the vital dyes. As the cell matures, the network disappears, after having first become increasingly thin.

Reticulocytes possess movements of their own, and this may give rise to clover-leaf forms which have three, four, or five segments. After some hours, reticulocytes mature and become erythrocytes which do not have any movements of their own, except for the scintillation or flicker which has already been referred to.

3. Ultrastructure of Erythroblasts. The Iron Cycle as Seen with the Electron Microscope

The ultrastructure of the erythroblast is shown schematically in Fig. 10. A study of these cells at high magnification allows us to describe and to follow the molecule of ferritin, which plays an extremely important role in iron metabolism. The ferritin molecule contains a very high concentration of iron (23%), and since iron is opaque to the rays of the electron microscope, a very strongly contrasted image results. Further, iron is arranged in the ferritin molecule in a special manner; it lies in six masses situated at the corners of a little octahedron of 50 A. on the side. This makes the molecule very easy to identify (Farrant, 1954; Bessis and Breton-Gorius, 1960). This molecule with its characteristic appearance can be found in the interior of cells, and so it is possible to formulate, in this particular case, the cytochemistry of ferritin and allied iron-containing substances on a molecular scale.

a. Digestion of erythrocytes. Appearance of ferritin and hemosiderin. In the bone marrow, the spleen, and the liver, both in man and other mammals, reticulum cells can be seen taking up old red cells. The digestion of the latter lasts about 5–10 minutes, as has been shown by micro-

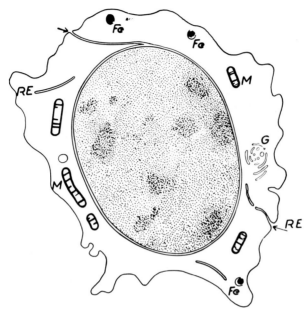

Fig. 10.   Diagrammatic representation of an erythroblast examined in thin section with the electron microscope. RE = endoplasmic reticulum, M = mitochondrion, G = Golgi body, Fe = a mass of iron.

cinematographic studies, for the digested red cell splits into two or more fragments within the cytoplasm (Policard and Bessis, 1953).

The steps in the digestion process are easily followed with the electron microscope. In general, vacuoles which become larger and larger appear in the phagocytosed red cells; sometimes the cytoplasm of the red cell hemolyzes and the remaining stroma can be clearly seen. In both cases, there appear around the fragment of a red cell a considerable number of little iron-containing granules (Figs. 11, A and 12) (Bessis and Breton-Gorius, 1957).

At high magnification and with the electron microscope, these granules show the characteristic appearance of the molecules of ferritin. This allows one to conclude with certainty that the digestion of the red cells by the reticulum cells in the bone marrow gives rise directly to the formation of ferritin molecules (Fig. 11, B).

Often the molecules of ferritin form a mass visible with the ordinary microscope, under which circumstances histologists apply the name hemosiderin to them. In these visible masses, there is a mixture of ferritin molecules and other substances, the nature of which is still undetermined. Sometimes, the masses of hemosiderin are crystalline, under which circum-

MARCEL BESSIS

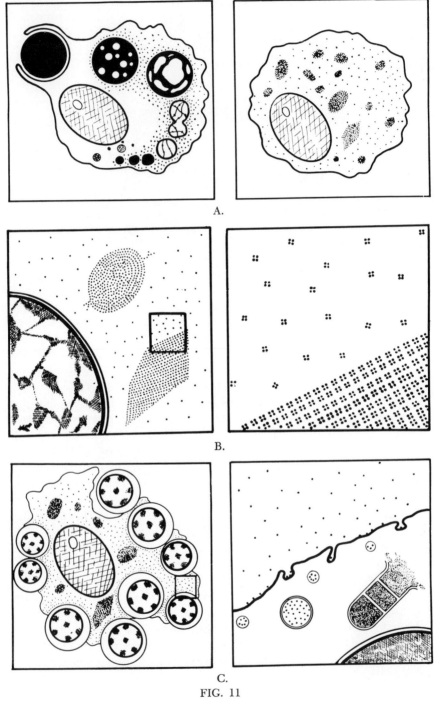

A.

B.

C.

FIG. 11

stances the ferritin is quite pure and comparable to that which one obtains by chemical means (Bessis and Breton-Gorius, 1957) (Fig. 13).

The ferritin and the hemosiderin which one sees in the reticulum cells can come directly from the phagocytosis of red cells or may be the result of simple storage, as ferritin, of the iron carried to the cell by transferrin.

b. The erythroblastic islands. Incorporation of ferritin by the erythro-blasts. The bone marrow contains erythroblastic islands which were described a long time ago by the early cytologists. As a result of what has been seen with the electron microscope, these now appear in a new light (Bessis, 1958). In the middle of these islands, one always finds one or two reticulum cells (Figs. 11, C and 14) filled with ferritin either in its dispersed form or in the form of a mass of hemosiderin. At high magnification, the electron microscope shows that the molecules of ferritin leave the reticulum cell and enter into the erythroblasts by a phenomenon not unlike pinocytosis. Pinocytosis, as we describe it, is a process of incorporation of liquid droplets by a mechanism in which cytoplasmic veils are withdrawn to form vacuoles within the cell. The electron microscope shows that an identical mechanism may occur on a very much smaller scale, the liquid droplets being only a few hundred Ångstroms in diameter and their incorporation in the cytoplasm of the cell resulting from the closing up of very small invaginations.

The incorporation of the molecules of ferritin by erythroblasts differs slightly from this "micropinocytosis" and occurs in the following phases (Fig. 15): (1) The molecules adhere at first to the plasma membrane of the cell, and over a region which is more or less large but which is always submicroscopic. (2) This region is then drawn in, as if aspirated by the cytoplasm, and forms a shallow invagination of the order of about 500 A. (3) By a pinching-in of its edges, this invagination becomes a vacuole, surrounded by a membrane which presents on its interior surface the particles which had adhered to the plasma membrane before the invagination occurred. (4) During the last stage, the membrane of the vacuole disappears, the liquid contained in it is absorbed, and the molecules become included in the matrix of the cell. These molecules are found only in the lower part of the vacuole, while the upper part, formed

---

Fig. 11.   A. Diagram showing the different stages of the digestion of red cells by a reticulum cell and the appearance of ferritin and of hemosiderin. B. Diagrammatic representation of the amorphous or crystalline structure of ferritin and of hemosiderin. These could be the direct result of the digestion of red cells or could be the result of a transfer of iron by the transferrin of the plasma. The ferritin molecule has four iron-containing masses, the diameter of each mass being about 15 A. C. The penetration of ferritin into the erythroblast, its concentration, and its transformation by the mitochondria.

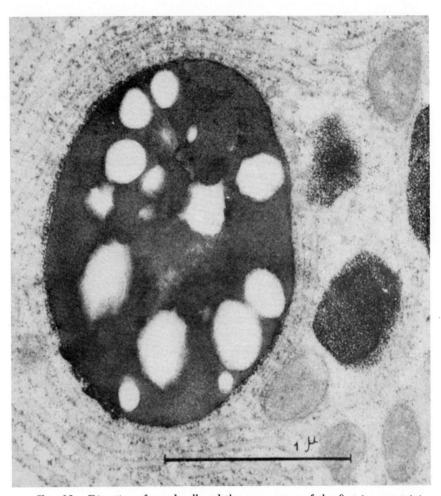

FIG. 12.   Digestion of a red cell and the appearance of the first iron-containing granules. The red cell is very much altered by the enzymes of the reticulum cell and shows large, clear vacuoles. At the periphery an accumulation of black granules which represent the granules of ferritin can be seen. The cytoplasm of the reticulum cell is filled with flat sacs (the endoplasmic reticulum which have their walls lined with the granules of Palade (ribonucleoproteins); these are clearer and larger than the iron-containing granules. Large masses of ferritin granules derived from the digestion of other red cells can also be seen.

by the union of the two borders of the primitive invagination, remains empty. This is a very characteristic appearance. The process differs from pinocytosis, which results in the absorption of a liquid so that the cell appears to "drink" (from the Greek, I drink). To define the process

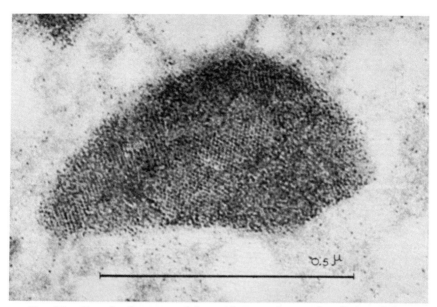

Fig. 13. Growth of a mass of hemosiderin made up of well-aligned ferritin molecules which form true crystals of which the characteristics (spacing of molecules, orientation, etc.) can be identified with those of ferritin crystals isolated by chemical means.

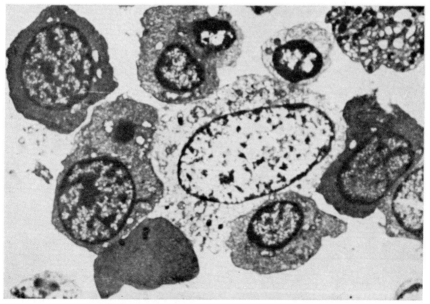

Fig. 14. An erythroblastic island seen with the electron microscope at low magnification.

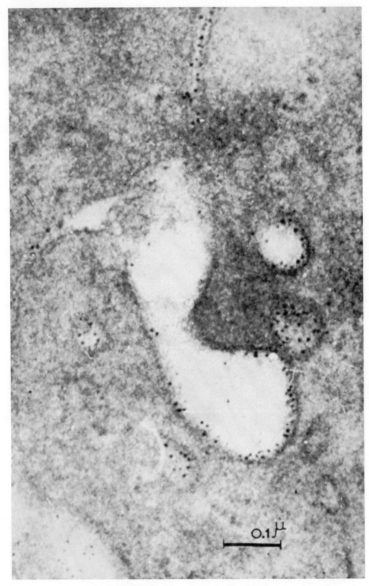

Fig. 15.   Rhopheocytosis. Pinocytosis is the incorporation of liquid in cells by the formation of vacuoles arising at the cell periphery. In these vacuoles iron-containing granules are found (rhopheocytosis).

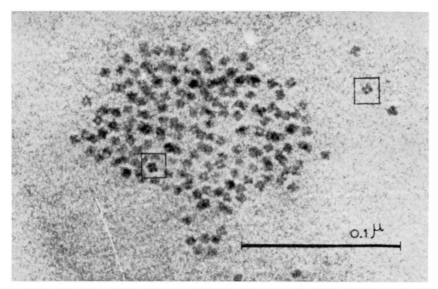

Fig. 16. Mass of ferritin granules. The ultrastructure of the ferritin molecule can be perfectly distinguished by its being made up of four little iron-containing masses. Each one of these has a diameter of 15 A.

described above, Policard and Bessis (1958) have proposed the name "rhopheocytosis" (from the Greek, I aspirate).

The phenomenon of rhopheocytosis is absolutely constant normally and in many diseases. Although its existence is quite certain, we do not know if *all* the iron which is used in the formation of hemoglobin enters the erythroblast by means of this mechanism or if some of it enters the cell in a more direct manner, e.g., by a splitting-off from the transferrin of plasma. It can further be supposed that several mechanisms exist at the same time, in the following way: (1) The iron resulting from the destruction of old red cells goes directly to the erythroblasts which form the erythroblastic island. (2) Transferrin carries iron to the reticulum cells of the bone marrow which transforms the iron into ferritin and distributes it among the erythroblasts of the island. (3) Some of the iron is directly transferred to the erythroblasts by transferrin, several *in vitro* experiments having shown that erythroblasts can actually capture iron from transferrin.

The problem is thus not yet solved; nevertheless, it is certain that molecules of ferritin pass from the reticulum cell to the erythroblast and that one finds in the erythroblast *iron in the form of ferritin* (Fig. 16).

c. Fate of the ferritin in erythroblasts. The probable role of mito-chondria. In the normal erythroblast one finds, shortly after the penetration of the molecules of ferritin, that these often become concentrated to form a larger or smaller mass. The mechanism which produces this mass is not yet known, but in certain cases the mass can be as large as $0.5\,\mu$ in diameter. Rhopheocytosis, by itself, gives rise to little vacuoles not containing more than a dozen iron molecules per section.

These large masses can easily be seen with the ordinary microscope, and the reaction of Perls shows that they contain iron. The cells containing them are the "sideroblasts," but the electron microscope has shown that in reality *all* the erythroblasts are sideroblasts. With the ordinary microscope, only about 30% of the erythroblasts look like sideroblasts; only some of them contain masses larger than $0.2\,\mu$ in diameter.

In the bone marrow of normal individuals, one does not appreciate the meaning of the ferritin granules and these masses. All that one sees is that as the cell matures and as hemoglobin appears, the ferritin molecules disappear. This is a phenomenon which one meets with frequently in certain pathological conditions, and it will be studied later because it seems to have great biochemical importance. It can be shown that ferritin is present in the interior of mitochondria, the importance of which as reservoirs for enzymes is well known; at the same time, Rimington (1957) has recently called attention to the fact that a suspension of mitochondria derived from the liver enables one to obtain certain stages of the biosynthesis of hemoglobin *in vitro*. The fact that the mitochondria contain ferritin is an argument in favor of their role in a chain reaction which leads to the incorporation of ferritin iron into the hemoglobin molecule. Further, ferritin undergoes a remarkable transformation in the mitochondria, a subject which will be referred to later.

As the red cell becomes mature, the mitochondria break up at the stage of the reticulocyte; their contents diffuse out into the medium which surrounds them, at which time they disappear.

d. Anomalies of the incorporation of iron in certain anemias. Anemias exist which do not depend on sideropenia. The most common of these are the thalassemias. In thalassemia major, or Cooley's anemia, the bone marrow contains a markedly increased quantity of iron, while the reticulum cells often contain hemosiderin in a crystalline state, i.e., in a state in which they could act as large depots. The erythroblasts are full of iron, either as dispersed ferritin or as ferritin masses.

Finally, in almost all the polychromatic and acidophilic erythroblasts, one finds molecules of iron in the mitochondria (Figs. 17–19). The way in which these molecules can penetrate and become concentrated in the mitochondria is still unknown. In the mitochondria a transformation of the molecules of ferritin takes place; they lose their characteristic structure and take on a powdery appearance for which we have proposed

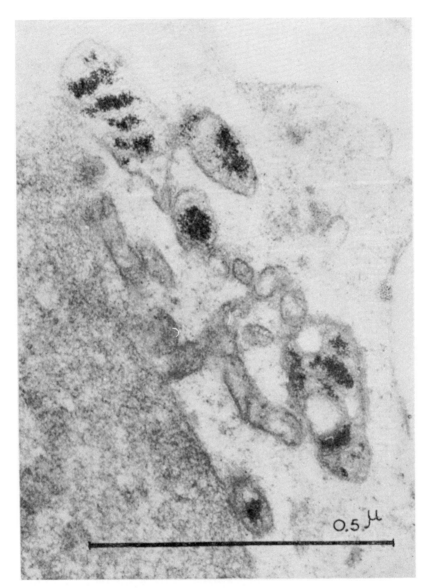

FIG. 17. An erythroblast from the bone marrow of a patient with hypochromic anemia. The mitochrondria are filled with black masses, these corresponding to iron-containing granules. Their continuity is interrupted by the cristae.

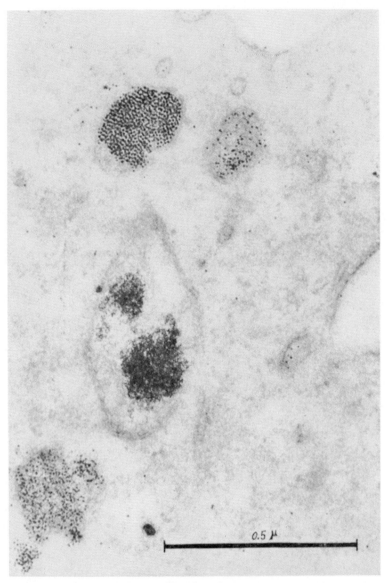

Fɪɢ. 18.   A fragment of an erythroblast showing iron in two forms. Above: a mass of ferritin molecules. In the center: a mitochondrium containing masses of very fine granules.

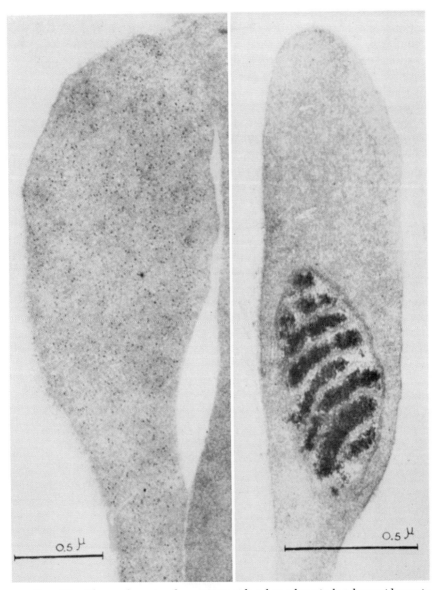

Fig. 19. The erythrocyte of a patient with a hypochromic but hypersiderermic anemia. Note in this cell the presence of many grains of ferritin. The iron captured by this cell has not served for the synthesis of hemoglobin (at the left). At the right: an abnormal erythrocyte from the marrow of the same patient. The thinness of the central zone can be seen very well in this cell, which has been cut transversely. At the periphery there is a mitochondrium in which iron has accumulated between the cristae, giving an appearance like the rungs of a ladder.

the name of "ferriginous micelles." The mitochondria swell, finally break-
ing up and liberating these ferriginous micelles into the surrounding
cytoplasm, but in the reticulocyte and even in the erythrocytes the
quantity of hemoglobin which is formed remains small by comparison
with the normal amount. It appears that there is a blockage of the normal
biosynthesis of iron so that it cannot be incorporated into the molecule
of hemoglobin.

In addition to the thalassemias, there has recently been described
another group of hypochromic anemias in which there is an excess of
hemosiderin. In these cases, the electron microscope reveals a situation
which is very similar to that in thalassemia major. The erythrocytes can
be so poorly hemoglobinized that it becomes difficult to pick out a red
cell in the electron microscope field. On the other hand, the quantity of
molecules of ferritin, either dispersed or in masses, is very large (Fig. 19).

It is important to note that an examination of the bone marrow of
individuals with simple hemochromatosis (without a hypochromic anemia)
presents a very different appearance. There is the usual accumulation of
hemosiderin in the reticulum cells, and the erythroblastic developmental
line is perfectly normal, the ferritin appearing in normal quantities in
the erythroblast and disappearing in the mature erythrocyte. In the simple
hemochromatoses, there is no excess of ferritin in any of the precursors
of the mature red cells.

B. The Granulocytic Series

The stem cells of the granulocytic series are found in the red bone
marrow intermingled with those of the erythrocyte series.

The cells of this developmental line are characterized by the appearance of granules in their cytoplasm; at first these are azurophile (sometimes referred to as undifferentiated), but later they become transformed
into neutrophile, eosinophile, or basophile granules. As regards the function of these granules, our knowledge is still very limited.

The stages of maturation which can be distinguished are those of
the myeloblast, the promyelocyte, the myelocyte, the metamyelocyte, and
the granulocyte. It is at the promyelocyte stage that the granules become
differentiated; this is shown in Fig. 20.

1. The Granules of Leucocytes

These have different appearances according to their nature and their
state of maturity. Neutrophile granules are lined up like grains of rice,
but they are cut in different planes in electron microscope sections and
this gives them a very diversified appearance (Fig. 21).

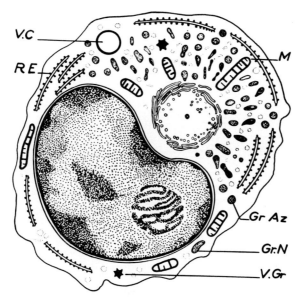

FIG. 20. Section of a promyelocyte as seen with the electron microscope. *VC* = contractile vacuole, *RE* = endoplasmic reticulum, *M* = mitochondrion, *Gr Az* = azurophile granule, *Gr.N* = neutrophile granule, *VGr.* = fatty vacuole.

Basophile granules are usually homogeneous, very deeply staining and contained in vacuoles. Eosinophile granules, when they are mature, show an internal crystalline structure which has an appearance which varies from species to species.

Immature and nonspecific granules (azurophile when stained with Giemsa) are spherical, and it is easy to distinguish them from mature granules and other granules in the white cell.

2. Centrosome and Golgi Body

The region of the centrosome has a special importance in leucocytes. By microdissection, this region appears in the living cell to be viscous and almost rigid. Microcinematography with phase contrast shows several other of its properties, such as its capacity to deform the softer nucleus, the absence of all protoplasmic movements in its interior, the stopping at its periphery of movements of granules and mitochondria, and also regular pendular movements of the entire centrosome, which has a period of oscillation of about a minute (Bessis and Locquin, 1950; Policard and Bessis, 1955).

In sections examined with the electron microscope, the centrosome appears to be a cytoplasmic region which has no mitochondria, no

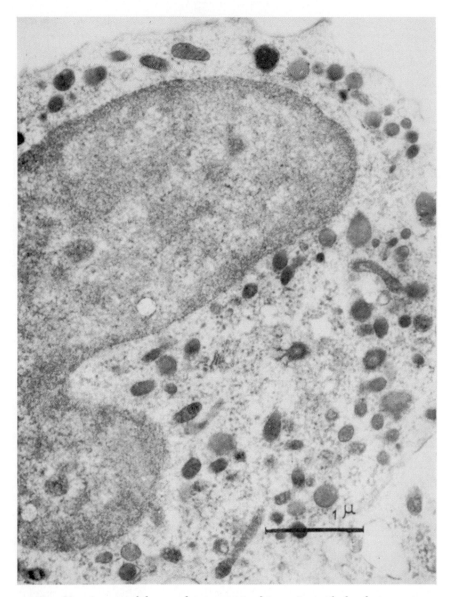

FIG. 21. A neutrophile granulocyte examined in section with the electron microscope. Note the centriole, the canaliculi of Golgi, here poorly developed, the neutrophile granules, and the mitochondria.

granules, and no endoplasmic reticulum. These structures surround it but do not penetrate into it. In the more or less complete circle formed by these structures there can often be seen tiny vacuoles or canaliculi 400–1000 A. thick. In some cases, at the level of these structures there can be seen supporting structures of a dense material some 500–1000 A. thick, and 0.2–0.6 μ long. These have been described by Dalton and Felix (1953) in other cells and represent the Golgi apparatus.

The substance of the centrosome itself is almost entirely homogeneous. In its groundwork, which is dense, there can be observed very fine vacuoles which are indefinitely outlined and which give it a foamy appearance, as well as very fine but denser granules measuring 100–200 A. The centrioles lie at the center of the centrosphere (or according to the nomenclature used, the centrosome).

3. Centrioles

The structure of the centriole has recently been described (Bernhard and De Harven, 1956; Bessis and Breton-Gorius, 1957). Its form is that of a little cylinder about 150 A. long with a wall made up of nine fine tubules. The diameter of these tubules is 200 A. Around the little cylinder are found nine little masses lying at two levels; fusiform fibers appear to arise from these (Bessis and Breton-Gorius, 1958) (Fig. 22). The structure of the centriole is therefore one of extreme complexity. Its study would certainly be very fruitful and would probably explain the peculiar movements of the centrosome. The centriole with nine tubules (or multiples of 9) is to be found in all known mobile structures, such as the tails of spermatozoa, mobile cilia, etc.

4. Ultracentrifugation of Granulocytes

The examination with the electron microscope of centrifuged white cells is very interesting. Ultracentrifugation allows of the displacement of structures in the interior of the cell without killing them, and by doing so leads to a differentiation between artifacts and real structures, for it is evident that the structures which are displaced by the ultracentrifuge exist in the living cell. The method also enables molecules or small structures to be concentrated in the same part of the cell; this results in molecules and fine structures which would not be noticed if they were dispersed being more clearly seen.

Let us recall briefly the results of ultracentrifugation of leucocytes which are afterward examined by the ordinary microscope (Bessis, 1950). The leucocytes take on a pear-shaped form, the wider part being at the pole which receives the greatest centrifugal force. In general, the nucleus occupies a position in the middle of the centrifuged cell, although if the

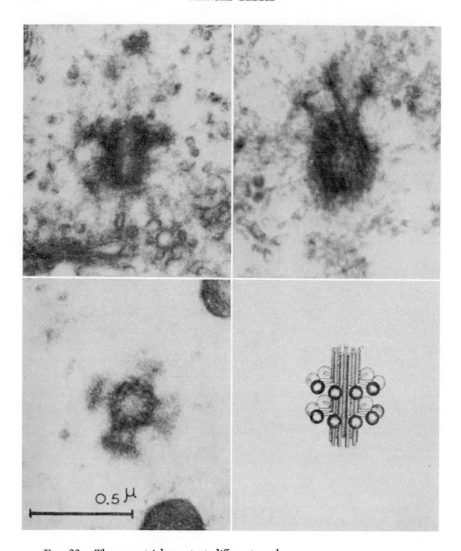

0.5 μ

FIG. 22. Three centrioles cut at different angles.

Upper left: longitudinal section of a centriole of a plasmocyte (normal marrow). At the right: a macelike structure. At the left, two such structures can be distinguished. Note at the foot of the figure the canaliculi of the Golgi body.

Upper right: An oblique section of the centriole of a neutrophile promyelocyte (normal marrow). The radially arranged structure, which is the second crownlike arrangement of the centriole, can be seen clearly, and also three filaments ending in macelike structures can be seen. On the left can be seen very fine canaliculi which appear to be separated from one another.

Lower left: Transverse section of the centriole of a plasmocyte (from a case of plasmocytoma). The tubes are seen here in transverse section and their clear centers centrifugal force is very great, its effects can be seen to affect the chromatin; the upper end of the nucleus is relatively clear, and the lower end contains chromatin masses. The nucleolus is found at the lower part of the nucleus. Granules behave in a manner which differs according to their nature; specific granules are thrown down very rapidly to the lower end of the cell, the mitochondria are found just above them, and lipid granules occupy the upper part of the centrifuged cell. The basophilic material of the cytoplasm is always concentrated in a very limited zone

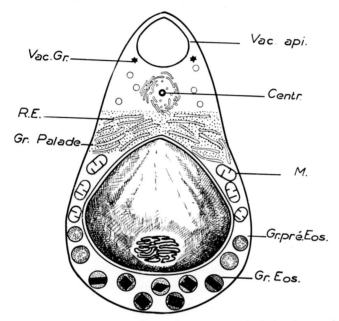

FIG. 23.   Section of an eosinophile promyelocyte which has been subjected to ultracentrifugation. Seen with the electron microscope. *R.E.* = endoplasmic reticulum, *Vac.Gr.* = fatty vacuoles, *Gr. Palade* = granules of Palade, *Gr.pré.Eos.* = pre-eosinophile granule, *Vac. api* = apical vacuole, *Centr.* = centrosome, *M.* = mitochondrion, *Gr. Eos.* = eosinophilic granule.

just above the nucleus when the cell has granules and just below the nucleus when it has none. Sometimes there appears a large vacuole occupying all of the superior pole of the cell. This vacuole often does not stain with Giemsa but sometimes stains a pale rose color; it indicates

---

and their osmic-staining walls can be distinguished, as can the nine tubules arranged in a helix. Around the centriole, properly speaking, there can be seen arranged in the form of a cross a certain number of macelike objects and the filaments which connect them to the cylinder of the centriole.

Lower right: a reconstructed diagrammatic representation of the centriole.

neither degeneration nor death of the cell, for in 10–20 minutes it disappears when the cell is examined in the fresh state, and afterward the cell resumes its apparently normal activity.

The electron microscope has given additional information about several points (Bessis and Thiéry, 1955) (Fig. 23). The fine cytoplasmic structures of leucocytes which have been subjected to ultracentrifugation become arranged in the following manner from below upward: (1) the specific granules, (2) nonspecific granules, (3) mitochondria, (4) the reticular component of the cell, (5) the vacuoles, some of which are derived from the dilation of sacs of reticulum, and (6) fatty granules.

The nucleus is usually situated in the middle of the cell, but it often divides into two parts, the upper one containing the fluid part of the nucleus and the lower one containing dense chromatin. These two parts can separate, one from the other.

The basophilic material in the cytoplasm usually is concentrated above the specific granules. In cells without granules, it lies at the same level as the masses of chromatin. The basophilic material is not always closely associated with the sacs of endoplasmic reticular material.

The zone containing the centrosome and the Golgi apparatus appears always above the nucleus but below the vacuoles at the upper pole. The little vacuoles and the canaliculi which make up the Golgi body are displaced by centrifugation as a single unit.

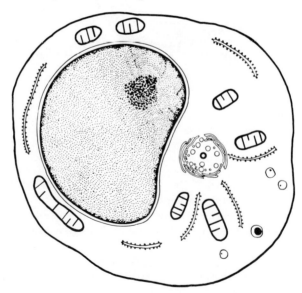

Fig. 24. Representation of a lymphoblast examined in section with the electron microscope.

C. The Lymphocytic Series

The stem cells of the lymphocytic series are found in the lymphoid follicles of lymph glands and in all the lymph nodules which are scattered throughout the body, particularly in the tonsils, the spleen, the thymus, the lymphoid follicles of the intestine, etc. The stages of maturation of this developmental line are the lymphoblast and the large and small lymphocyte. A schematic representation of the ultrastructure of the lymphoblast is given in Fig. 24.

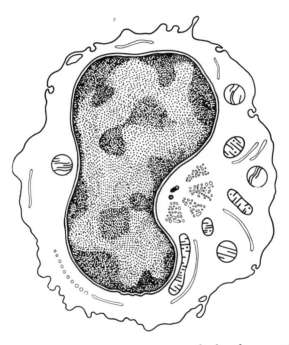

FIG. 25. A monocyte seen in thin section with the electron microscope.

D. The Histiocytic Series

In hematology, the term histiocyte has a different meaning than that which it has in classical histology. There is a distinction between the histiocyte of the tissues which becomes a monocyte (Fig. 25) when in the circulating blood, and two cells which pass into the blood stream only exceptionally but which are very important. These two are the plasmocyte and the mastocyte.

1. The Plasmocytic Series

Plasmocytes are met with particularly where there are reticulohistiocytic cells, i.e., they are very widely scattered. Although their total mass is normally large, it increases greatly in certain diseases and the cells can even pass into the blood stream. The organs which are normally particularly rich in plasmocytes are the lymph glands, the spleen, the skin, the bone marrow, and the intestine.

The two different stages of maturation of this developmental line are the plasmoblast and the proplasmocyte. In mature forms, the oval nucleus lies at one pole of the cell with its axis perpendicular to the axis of the cell as a whole. The cytoplasm is extremely basophilic because it contains an abundance of ribonucleoproteins. It can often be seen to contain canaliculi or irregular vacuoles which may literally riddle the surface of the cell and which correspond to the endoplasmic network. The ergastoplasm of plasmocytes can be clearly seen in the living state by phase contrast microscopy (Fig. 26) (Thiéry, 1955). It appears as a network of sacs and of canaliculi arranged in a parallel fashion around the nucleus and the centrosome. If the cells are compressed or autolyzed, these sacs swell up and become transformed into round vacuoles. Round acidophile bodies (Russell bodies) are sometimes found in the cytoplasm of plasmocytes (Fig. 27). Plasmocytes containing azurophile granules are also frequently found. (See Fig. 28.)

a. Mott cells. In trypanosomiasis, plasmocytes filled with spherical hyalin vesicles have been described in the cerebrospinal fluid. These are mature Mott cells. They are also found in smaller number in other pathological conditions (typhus, malaria, and kala-azar) and also in normal preparations. They also occur in a rare form of myeloma (Bessis and Scebat, 1946). More definite information about the formation of these cells has been provided by electron microscope examination, which has shown that in some cases protein crystals are present in the vacuoles (Thiéry, 1958).

b. Function of the plasmocytes. The presence of numerous plasmocytes in chronic infections and the increase of plasma globulins in plasmocytoma or multiple myeloma have given rise to the idea that these cells play a role in the formation of antibodies and even of normal globulins. Bjorneboe *et al.* (1947; Bjorneboe and Gormsen, 1943) and more particularly Fagraeus in a series of investigations summarized in 1948, have carried out experimental studies of the relation which exists between the production of antibodies and the proliferation of plasmocytes.

Bing and associates (1945) have studied the changes of plasmocytes during immunization both by microspectrographic methods and by microchemical methods. They have found great quantities of ribonucleoproteins accompanying the intense proliferation of these cells during the first few days of immunization; after this, the quantity of ribonucleoproteins

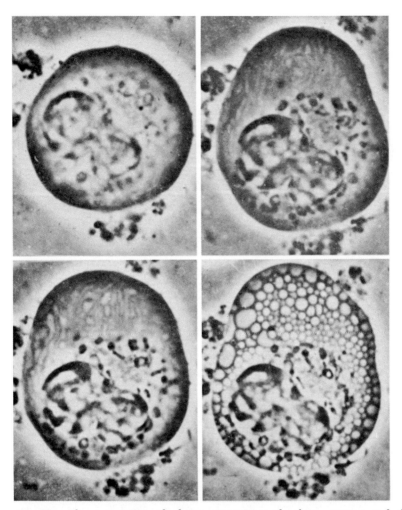

FIG. 26. The appearance of plasmocytes seen with phase contrast and the changes which they undergo between slide and coverslip. Above and to the left: the plasmocyte is in good condition but the successive figures show the vacuolization of the endoplasmic reticulum which even appears vacuolated below and at the right. After Thiéry (1955).

remains constant until the plasmocytes become completely mature. These results appear to show that the formation of antibodies proceeds in a manner identical with the synthesis of proteins in general, i.e., that they use nucleic acids as intermediaries.

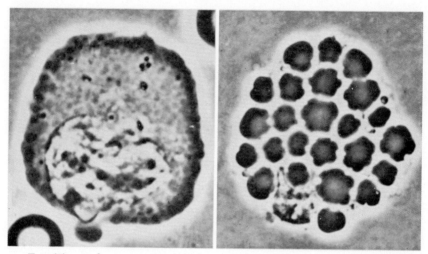

FIG. 27. A plasmocyte showing the development of Russell bodies (seen with phase contrast).

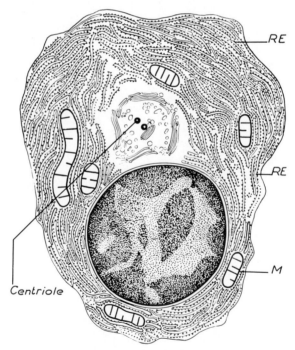

FIG. 28. Section of a plasmocyte seen with the electron microscope. M = mitochondrion, $R.E.$ = *endoplasmic reticulum*.

2. The Mastocytic Series

Mastocytes are another type of cell belonging to the tissues which can pass into the blood. They are found widely throughout the organism, occurring especially in association with connective tissue in the septa and capsule of the liver, in the bone marrow, in lymph glands, and in the spleen. The stages of maturation are the mastoblast and the promastocyte. The cell is characterized by its containing very large metachromatic granules. Recent work of many investigators has shown that they take part in the metabolism of heparin and of histamine (Arvy, 1956; Padawer, 1957).

The granules are partly digested by heparin, after which they show a central mass which is chromatophobic but not metachromatic; its nature is still unknown. The electron microscope (Smith and Lewis, 1957) has shown that they occasionally are laminated.

E. The Thrombocytic Series

Wright was the first to think, in 1906, that platelets originated from large cells which had just been discovered in the bone marrow. These cells are known as megakaryocytes. This hypothesis has been discussed for fifty years and has been corroborated only by indirect observations which could not be verified by the observation of sections or films, these showing nothing but dead cells. In 1956, however, observation on living material by phase contrast (Thiéry and Bessis, 1956a, b) showed that according to the expression of Bard, "it is the giants of the bone marrow which give rise to the dwarfs of the blood."

Megakaryocytes are found in the red bone marrow where they are mixed with the stem cells of erythrocytes and granulocytes. They become hyperplastic as a result of an increase in size and shape of their nuclei without this being accompanied by a division of their cytoplasm. The result is that these giant cells are formed; they can be as large as $300\,\mu$ in diameter with numerous nuclear lobes and numerous centrioles.

The successive stages of the maturation are the megakaryoblast, the basophile megakaryocyte, the granular megakaryocyte (a stage at which fine azurophile granules appear), and the platelet-forming megakaryocyte.

1. Formation of Platelets

In preparations in the living state observed with phase contrast microscopy, the following phenomena can be shown to occur in sequence.

In the cytoplasm of completely mature megakaryocytes small but definite regions can be distinguished, arranged regularly like a ribbon. These are the future platelets. The cell throws out large pseudopods which

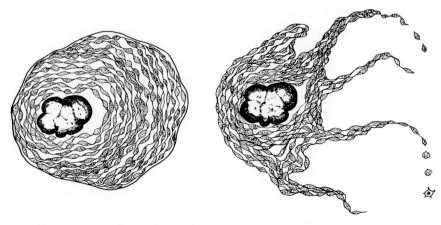

FIG. 29. This figure shows the way in which platelets are liberated from mature megakaryocytes (taken from a film sequence).

become longer and longer, looking like the tentacles of an octopus. The greater part of the cytoplasm of the cell becomes distributed in these cellular extensions which continue to increase in length and to branch until they become filiform. The filaments have little enlargements upon them; these are platelets which are still adherent to one another but which the movements of the blood stream quickly separate (Fig. 29). In this way, each megakaryocyte gives rise to about 2000 platelets.

The time necessary for this development of the platelet depends essentially on the state of maturation of the cell. Some platelet-forming megakaryocytes are transformed into platelets in less than 3 hours, while others, although they appear by phase contrast to be identical with the others, take more than 12 hours to develop into platelets.

It should be noted that the nucleus does not seem to participate directly in the formation of platelets, which are formed from the cytoplasm. The nucleus becomes pycnotic.

The appearance of the octopus with its arms which become finer and finer occurs *in vitro* and in a liquid medium, but seems to be slightly different from what happens in the living animal where the platelet-forming megakaryocyte throws out quite gross extensions, such as are seen in histological preparations of bone marrow and, in particular, in the illustrations of Wright. One can imagine, as Wright did (1906), that as these extensions become longer and longer, they encounter the blood sinuses with which they become entangled, liberating small groups of platelets which are separated and carried away by the blood stream.

2. The Ultrastructure of Megakaryocytes

The basophile megakaryocyte presents nothing particularly interesting. The nuclear membrane is very thick, regular, and sometimes has some speckles of chromatin on its internal side. The region of the centrosome is often very clear; it is a round region surrounded by small vacuoles which are characteristic of the Golgi body. There is not much archiplasm in this basophilic cell, but one can observe some flattened sacs, often parallel to the nuclear membrane; "exoplasm," i.e., a peripheral region of the clear cytoplasm containing neither granules nor mitochondria, is almost always found in the cell at this stage. The ectoplasmic zone often forms vacuoles and appears bulbous. The granules of the cytoplasm are very numerous, very small, very dense to electrons, and smaller than the mitochondria which have their usual structure with their parallel cristae.

In the granular megakaryocyte, these nonspecific granules become very numerous and reach the periphery of the cell. The exoplasm is reduced to a narrow band without granules. The fundamental observation which characterizes this stage is the multiplication and, above all, the elongation of the little vacuoles which appeared in the previous stage. The cytoplasm is completely studded with canaliculi which are often Y-shaped or comma-shaped and which give a characteristic appearance to the entire cytoplasm.

In the platelet-forming megakaryocyte, this system of canaliculi develops still more and becomes arranged so as to circumscribe oval regions which contain three to ten granules. These granules are made up of about 20% of mitochondria and 80% of azurophile granules. In other regions, in which the canaliculi are not present, the granules collect in little groups delimiting little regions which are ultimately isolated by elongated vacuoles (Bessis and Thiéry, 1956).

Some megakaryocytes have been caught at the moment when these little areas, which represent the future platelets, were already liberated. Under these circumstances, the periphery of the cell appears extremely denticulated, having an appearance which recalls the fjords of the North Sea on a map. The platelets themselves form a number of islands in close contact with the edge of the cell. These platelets have exactly the same appearance as the circulating platelets which are found in the blood (Fig. 30). They have the appearance of round or oval masses, with or without dendrites, and made up of a homogeneous cytoplasm which is very finely granular, although one cannot say if this granulation exists in the living state or is due to the action of fixatives. They contain ten to fifty azurophile granules and mitochondria.

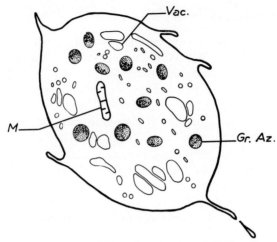

Fig. 30. A platelet examined in thin section with the electron microscope. M =mitochondrion, $Gr. Az.$ =azurophile granule, $Vac.$ =vacuole.

IV. Total Quantities and Renewal of the Blood Cells

In clinical hematology, one customarily relies on cell counts together with the differential count. In hospital services, these constitute almost the only data available to the practitioner, at least in the great majority of cases. However, it is interesting from several points of view to have an understanding not only of the number per cubic millimeter of each kind of cell but of the absolute values which apply to the different kinds of blood cells.

Such quantitative studies allow one to draw up a balance sheet of a given population of cells taken from a given region (for example, from the bone marrow) in a way very like that in which an economist draws up a general balance sheet for a large town; an understanding of the conditions in the town changes with the number of births, the number of deaths, the people who have gone away, or have come to live there. The number of specialized workers, as well as the extent of their education, depends on the needs of the population, and the physical state of the inhabitants depends on the environment (e.g., the available nourishment, the state of the environment, psychological factors, and so on). All the hematological constants which one calls red and white cell counts, differential counts, mitotic indexes, the granulo-erythroblastic indexes, etc., and also the hormonal and other factors which may influence these constants, should be interpreted from this point of view.

The great drawback of these methods of calculation of total values is their lack of precision; in a large proportion of normal individuals, the values which serve as a point of departure vary greatly and cannot be measured except in a very approximate manner (as, for example, the volume of the bone marrow, the number of cells in lymphatic glands, etc.) or must be measured by methods which are too indirect (as, for example, estimates of the time between mitoses, of the time required for differentiation and for maturation, etc.).

The activity with which the different kinds of blood cell are produced is extraordinary. Each day a normal person puts into circulation 250,000 million red cells, 15,000 million white cells, and 500,000 million platelets. The total volume of the cells thus put into circulation each day is of the order of 50 cm.³. Each second, there are liberated 2.5 million red cells, 120,000 white cells, and 5 million platelets.

The functional capacity of the bone marrow is extraordinarily great. While 250,000 million red cells are produced daily by the bone marrow in a healthy man, in response to a variety of stimuli, such as anoxia, a healthy bone marrow is capable of producing seven times this number (Giblett et al., 1956; Berlin et al., 1957). It has been estimated that the reserve of myeloid cells in the marrow of the guinea pig is 100 times the number of granulocytes in its blood (Yoffey, 1956).

The time during which the cells of the blood remain in the blood vessels is far from representing the real duration of their life, for to this must be added the time that these cells take to mature and to die in the hematopoietic system. Further, it has been shown that almost all types of white cells have a long extravascular life. In particular, the lymphocytes pass only about an hour in the blood stream, although the length of their life is about 30 days.

The reader should be warned against placing too much reliance on the figures which are usually given (see Tables V–IX). These figures

TABLE V

WEIGHT OF THE HEMATOPOIETIC ORGANS[a]
Results as a percentage of body weight

1. Weight of the hematopoietic marrow		2. Weight of the lymphopoietic organs	
Subjects	Results as a percentage of body weight	Subjects	Results as a percentage of body weight
Infant	2.6	Rat and rabbit	0.5 to 1.0
Adult	1.7 to 2.9		

[a] Estimate of the volume of the intraosseous spaces, corrected to take account only of the red bone marrow.

TABLE VI

The Number and Total Weight of the Blood Cells of the Hematopoietic Tissues[a]

1. In the bone marrow

Series	Number	Weight (grams)	Series	Number	Weight (grams)
Erythrocytic	6×10^{11}	100	Megakaryocytic	0.1×10^{11}	200
Granulocytic	10×10^{11}	900	Eosinophiles	1×10^{11}	
Lymphocytic	4×10^{11}	100	Basophiles	0.1×10^{11}	
Monocytic	0.5×10^{11}		Hemolyzed cells	3×10^{11}	
Plasmocytic	0.2×10^{11}				

2. Lymphocytic cells in extramedullary tissues[b]

Tissue	Number	Grams
Spleen	1.6×10^{11}	40
Lymph glands	1.6×10^{11}	40
Thymus and tonsils		
Peyer's patches	0.8×10^{11}	20

[a] After Osgood (1954). Patt *et al.* (1957) give the following figures, obtained by calculation, which are different from those of Osgood:

Number of red cells in circulation (man of 70 kg.)	29×10^{12}
Number of erythroblasts in the marrow	230×10^{9}
Number of reticulocytes	350×10^{9}
Total number of granulocytes in the marrow	615×10^{9}

[b] After Osgood (1954).

TABLE VII

Number and Total Volume of the Different Blood Cells[a]

Cells	Number (in a thousand million)	Volume (in cm.³)
Erythrocytes	25,000[b]	2,300
Leucocytes	35	20
Neutrophile granulocytes	24	15
Eosinophile granulocytes	0.7	0.4
Basophile granulocytes	0.01	0.04
Monocytes	7.7	3
Lymphocytes	2.8	3
Platelets	1,500	10

[a] After Osgood (1954).

[b] Or 25×10^{12}. The figure for the total number of cells of an adult man lies between 10^{13} and 10^{14}.

TABLE VIII

DURATION OF THE DIFFERENT STAGES OF MATURATION IN THE
ERYTHROCYTIC SERIES[a]

1. Erythrocytic Series		
Cells	Time of maturation (after Osgood)	Time of maturation (after Patt)
Basophile erythroblasts	48 hours	23 hours
Acidophile erythroblasts	48 hours	23 hours
Reticulocytes	43 hours	35 hours
Erythrocytes	120 days	120 days

2. Granulocytic series	
Stage of maturation	Duration in hours (according to Osgood)
Myelocyte	12 hours
Metamyelocyte	12 hours
Granulocyte stab cells	36 hours
Mature granulocyte	108 hours
Dying and autolyzing	18 hours
Time occupied by mitosis estimated at	16 hours
Total (mitosis deducted)	142 hours

3. Lymphocytic series[b]	
Lymphoblasts and lymphocytes in hematopoietic organs	3 days
Outside the hematopoietic organs	30 days (according to different authors, these cells remain in the blood stream from 30 minutes to 24 hours)

4. Thrombocytic series[c]	
Megakaryocytes	16 hours
Platelets	72 hours

[a] Calculated from observations in the literature.
[b] After Osgood (1956).
[c] After Odell (1956).

TABLE IX

LIFE SPAN OF THE DIFFERENT BLOOD CELLS

Cells	Time	Remarks
Erythrocyte	100–120 days	These results are definite and have been confirmed by different methods
Neutrophile granulocyte	30 minutes (in the blood stream) 4–8 days (total duration of life)	Study still proceeding. The present results are certainly provisional
Eosinophile	8–12 days	
Lymphocyte	30 minutes to 3 hours (in the blood stream) 10–15 days (duration of life)	
Platelets	3 to 8 days	—

represent only orders of magnitude, which can vary at least 100%. The techniques and results pertaining to the material which follows is to be found in the work of Kindred (1942), Osgood (1954), Odell (1956), and Yoffey and Courtice (1956). See also the book edited by Stohlman (1959).

REFERENCES

Ager, J. A. M., and Lehmann, H (1957). *J. Clin. Pathol.* **10,** 336.
Arvy, L. (1956). *Année biol.* **32,** 169.
Aschkenasy, A. (1949). *Exposés ann. biochim. méd.*
Aschkenasy, A. (1952). *Congr. intern. biochim. 2nd Congr. Paris.*
Aschoff, W. A. L. (1913). Cited by Aschoff, W. A. L. (1924). *Ergeb. inn. Med. u. Kinderheilk.* **26,** 1.
Ayres, W. W. (1949). *Blood* **4,** 595.
Ayres, W. W., and Starkey, N. M. (1950). *Blood* **5,** 254.
Beaven, G. H., Hoch, H., and Holiday, E. R. (1951). *Biochem. J.* **49,** 374.
Berlin, N. I., Lawrence, J. H., and Elmlinger, P. J. (1957). *Blood* **12,** 147.
Bernhard, W., and De Harven E. (1956). *Compt. rend.* **242,** 288.
Bessis, M. (1946). *Rev. hématol.* **1,** 45.
Bessis, M. (1950). *Compt. rend. soc. biol.* **144,** 44.
Bessis, M. (1954a). *Rev. hématol.* **9,** 745.
Bessis, M. (1954b). "Traité de cytologie Sanguine." Masson, Paris.
Bessis, M. (1957). *Rev. hématol.* **12,** 142.

Bessis, M. (1958). *Rev. hématol.* **13**, 8.

Bessis, M., and Breton-Gorius, J. (1957a). *Bull. microscop. appl.* **7**, 54.

Bessis, M., and Breton-Gorius, J. (1957b). *Rev. hématol.* **12**, 43.

Bessis, M., and Breton-Gorius, J. (1957c). *Semaine hôp. pathol. et biol.* **33**, 411.

Bessis, M., and Breton-Gorius, J. (1958). *Compt. rend.* **246**, 1289.

Bessis, M., and Breton-Gorius, J. (1950). *Compt. rend.* **250**, 1360.

Bessis, M., and Bricka, M. (1949). *Compt. rend. soc. biol.* **143**, 375.

Bessis, M., and Bricka, M. (1950). *Rev. hématol.* **5**, 396.

Bessis, M., and Bricka, M. (1952). *Rev. hématol.* **7**, 407.

Bessis, M., and Dupuy, A. (1950). *Compt. rend. soc. biol.* **145**, 1509.

Bessis, M., and Gorius, J. (1950). Congrés de Microscopé. *In Rev. Opt.*, p. 650.

Bessis, M., and Locquin, M. (1950). *Compt. rend. soc. biol.* **144**, 483.

Bessis, M., and Scebat, L. (1946). *Rev. hématol.* **1**, 447.

Bessis, M., and Tabuis, J. (1955). *Compt. rend. soc. biol.* **149**, 873.

Bessis, M., and Thiéry, J. P. (1955). *Rev. hématol.* **10**, 583.

Bessis, M., and Thiéry, J. P. (1956). *Sangre* **1**, 123.

Bessis, M., Bricka, M., and Dupuy, A. (1951). *Compt. rend. soc. biol.* **145**, 1509.

Bessis, M., Nomarski, G., Thiéry, J. P., and Breton-Gorius, J. (1958). *Rev. hématol.* **13**, 249.

Bing, J., Fagraeus, A., and Theorell, B. (1945). *Acta Physiol. Scand.* **10**, 282.

Bjorneboe, M., and Gormsen, H. (1943). *Acta Pathol. Microbiol. Scand.* **20**, 649.

Bjorneboe, M., Gormsen, H., and Lundquist, F. (1947). *J. Immunol.* **55**, 121.

Bloom, W. (1938). *In* "Handbook of Haematology (H. Downey, ed.), Vol. 2, p. 865. Hoeber, New York.

Bloom, W., and Bartelmez, G. W. (1940). *Am. J. Anat.* **67**, 21.

Blowers, R., Clarkson, E. M., and Maizels, M. (1951). *J. Physiol. (London)* **113**, 228.

Bordet, J. (1939). "Traité de l'immunité des maladies infectieuses," Vol. 1. Masson, Paris.

Brachet, J. (1952). *Actualités biol.* **No. 16.**

Chernoff, A. I. (1955). *New Engl. J. Med.* **253**, 322.

Chèvremont, M. (1942). Thèse d'agrégation. Liège.

Comandon, J. (1919). *Compt. rend. soc. biol.* **82**, 1305.

Comandon, J., and Fonbrune, P. de (1929). *Arch., Anat., Microscopie,* **25**, 555.

Cooke, W. E., and Ponder, E. (1927). *In* "The Polynuclear Count." Griffin, London.

Dalton, A. J., and Felix, M. D. (1953). *Am. J. Anat.* **92**, 277.

Dervichian, D. G., and Ponder, E. (1950). *Compt. rend.* **224**, 1848.

Dervichian, D. G., Fournet, G., and Guinier, A. (1952). *Biochim. Biophys. Acta* **8**, 145.

Doan, C. A. (1932). *J. Lab. Clin. Med.* **17**, 887.

Dustin, P. (1941). *Acad. roy. Belg.* **27**, 612.

Dustin, P. (1947). *Symposia Exptl. Biol.* **1**, 114.

Dustin, P., Jr. (1949). *Acta Clin. Belg.* **4**, 70.

Errera, M. (1952). *Congr. intern. biochim. 2nd Congr. Paris.*

Fagraeus, A. (1948). *Acta Med. Scand. Suppl.* **204**, 1.

Fahraeus, R. (1929). *Physiol. Revs.* **9**, 241.

Farrant, J. L. (1954). *Biochim. et Biophys. Acta* **13**, 569.

Fonio, A., and Schwendener, J. (1942). "Die Thrombozyten des menschlichen Blutes." Hans Huber, Berne.

Forkner, C. E., Zia, L. S., and Teng, C. (1936). *Chinese Med.* **50**, 1191.

Furchgott, R. F., and Ponder, E. (1940). *J. Exptl. Biol.* **17**, 117.

Giblett, E. R., Coleman, D. H., Pirzio-Biroli, G., Donohue, D. M., Motulski, A. G., and Finch, C. A. (1956). *Blood* **11**, 47.

Gilmour, J. R. (1941). *J. Pathol. Bacteriol.* **52**, 25.

Gonzalez-Guzman, I., and Bessis, M. (1952). *Compt. rend. soc. biol.* **146**, 835.

Granick, S. (1952). *Congr. intern. biochim. 2nd Congr. Paris.*

Halpern, B. N., ed. (1957). "Physiopathology of the Reticulo-endothelial System." Masson, Paris.

Halpern, B. N., and Bessis, M. (1950). *Compt. rend. soc. biol.* **144**, 759.

Henderson, M. (1928). *Anat. Record* **38**, 71.

Isaacs, R. (1930). *Folia Haematol.* **40**, 395.

Jandl, J. H., and Simmons, R. L. (1957). *Brit. J. Haematol.* **3**, 19.

Jolly, J. (1913). *Compt. rend. soc. biol.* **74**, 504.

Jolly, J. (1923). "Traité technique d'hématologie," Vol. 2. Masson, Paris.

Jordan, H. E. (1938). *In* "Handbook of Haematology," Vol. 2, p. 703. Hoeber, New York.

Kindred, J. E. (1942). *Am. J. Anat.* **71**, 207.

Knoll, W. (1927). *Neue Denkschr. Schweiz, naturforsch. Ges.* **64**, 1.

Knoll, W. (1932). *In* "Handbuch der allgemeinen Hematologie," Vol. 1, p. 553. Urban & Schwarzenberg, Berlin.

Knoll, W., and Pingell, E. (1949). *Acta Haematol.* **2**, 369.

Leibetseder, F. (1952). *Acta Haematol.* **8**, 161.

Leibetseder, F. (1954). *Rev. hématol.* **9**, 158.

Lennert, K., Remmele, M. (1958). *Acta Haematol.* **19**, 99.

Lewis, W. H. (1934). *Bull. Johns Hopkins Hosp.* **55**, 273.

Maganelli, G., and Ricciardi, G. (1956). *Progr. med.* **12**, 750.

Malassenet, G. (1957). *Rev. hématol.* **12**, 64.

Melnick, J. L. (1956). *In* "Diagnostic Procedures for Virus and Rickettsial Diseases," 2nd ed. p. 97. Am. Public Health Assoc. New York.

Mollison, L. P. (1956). "Blood Transfusion in Clinical Medicine," 2nd ed. Blackwell, Oxford, England.

Mourant, A. E. (1954). "The Distribution of the Human Blood Groups." Blackwell, Oxford, England.

Odell, T. T., Jr. (1956). *Congr. Intern. Soc. Haemotol., 6th Congr., Boston.*

Osgood, E. E. (1954). *Blood* **9**, 1141.

Padawer, J. (1957). *Trans. N. Y. Acad. Sci.* **19**, 690.

Pauling, L. (1954). *Harvey Lectures* **49**, 216.

Pauling, L., Itano, H. A., Singer, S. J., and Wells, I. C. (1949). *Science* **110**, 543.

Patt, H. M., Maloney, M. A., and Jackson, E. M. (1957). *Am. J. Physiol.* **188**, 585.

Penati, F., Lovisetto, P., Turco, G. L. (1956). "Le emoglobine normali e pathologiche." Edizioni di Haematologica, Pavia, Italy.

Policard, A. (1925). *Compt. rend.* **182**, 168.

Policard, A., and Bessis, M. (1952). *Compt. rend.* **234**, 913.

Policard, A., and Bessis, M. (1953). *Compt. rend. soc. biol.* **147**, 982.

Policard, A., and Bessis, M. (1955). *Exptl. Cell Research* **8**, 583.

Policard, A., and Bessis, M. (1958). *Compt. rend.* **246**, 3194.

Ponder, E. (1945). *J. Gen. Physiol.* **29**, 89.

Ponder, E. (1948). "Hemolysis and Related Phenomena." Grune and Stratton, New York.

Ponder, E. (1955a). *J. Gen. Physiol.* **38**, 575.

Ponder, E. (1955b). *In* "Protoplasmatologia Handbuch der Protoplasmaforschung," Vol. X. Springer, Wien.

Ponder, E., Bessis, M., and Bricka, M. (1952a). *Compt. rend.* **235**, 96.

Ponder, E., Bessis, M., Bricka, M., and Gorius, J. (1952b). *Compt. rend.* **234**, 2645.

Pulvertaft, R. J. V. (1949). *J. Clin. Pathol.* **2**, 281.

Race, R. R., and Sanger, R. (1958). "Blood Groups in Man," 3rd ed. Blackwell, Oxford, England.

Ralph, P. H. (1947). *Anat. Record* **98**, 489.

Rimington, C. (1957). *Rev. hématol.* **12**, 591.

Sabin, F. R., Miller, F. R., Smithburn, K. C., Thomas, R. M., and Hummel, L. E. (1936). *J. Exptl. Med.* **64**, 97.

Seifriz, W. (1936). "Protoplasma." McGraw-Hill, New York.

Singer, K., Chernoff, A. I., and Singer, L. (1951). *Blood* **6**, 413.

Smith, D. E., and Lewis, Y. S. (1957). *J. Biophys. Biochem. Cytol.* **3**, 9.

Stohlman, F., Jr. (1959). "The Kinetics of Cellular Proliferation." Vol. 1. Grune & Stratton, New York.

Thiéry, J. P. (1955). *Rev. hématol.* **10**, 745.

Thiéry, J. P. (1958). *Rev. hématol.* **13**, 61.

Thiéry, J. P., and Bessis, M. (1956a). *Compt. rend.* **242**, 290.

Thiéry, J. P., and Bessis, M. (1956b). *Rev. hématol.* **11**, 162.

Thomas, J. A. (1938). *Ann. sci. nat. Zool.* **11**, 209.

Theorell, B. (1947). "Studies on the Formation of the Cellular Substance During Blood Cell Production." H. Kimpton, London.

Theorell, B. (1950). *Rev. hématol.* **5**, 561.

Tompkins, E. H. (1953). *J. Am. Med. Women's Assoc.* **8**, 301.

Valentine, W. N. (1951). *Blood* **6**, 845.

Vercauteren, R. (1951). *Bull. soc. chim. biol.* **33**, 522.

Weicker, H. (1954). *Schweiz. med. Wochschr.* **84**, 345.

Winqvist, G. (1954). *Acta Anat.* **22**, Suppl. 21.

Wintrobe, M. M. (1956). "Clinical Haematology," 4th ed. Lea & Febiger, Philadelphia, Pennsylvania.

Wooley, D. W. (1949). "The Preservation of the Formed Elements and of the Proteins of the Blood." Harvard Med. School, Boston, Massachusetts.

Wright, J. H. (1906). *Boston Med. Surg. J.* **154**, 643.

Yoffey, J. M. (1956). *J. Histochem. and Cytochem.* **5**, 516.

Yoffey, J. M., and Courtice, F. C. (1956). "Lymphatics, Lymph and Lymphoid Tissue." 2nd ed. Arnold, London.

Zuelzer, W. W., Neel, J. V., and Robinson, A. R. (1956). *In* "Progress in Haematology" (L. M. Tocantins, ed.), p. 91. Grune & Stratton, New York.

CHAPTER 4

Bone and Cartilage

By P. LACROIX

I. INTRODUCTION

Few tissues have recently benefited as much as have bone and cartilage from the combined efforts of investigators using the various types of light microscopes, the electron microscope, qualitative and quantitative radio-isotope procedures, microradiography, X-ray diffraction techniques, histochemistry, *in vitro* cultivation, and transplantation experiments.

Generally speaking, more attention has been paid to the intercellular material than to the cells themselves. Consequently the writing of this essay would have been of no avail without the following assumption: what is meant by "The Cell" should be understood here to be "the cytological level." Moreover, it stands to reason that the cells of bone and cartilage have to be considered together with their intercellular substance, the osteoblasts with the bone being deposited nearby, and the osteoclasts with the bone being destroyed at their contact.

Limitation of space will force us to make choices at every step. The main emphasis will be put on those aspects which are now being developed and which are likely to be of interest to the research worker. All our illustrations and most of our information will be provided by the long bones of higher mammals. First of all, we shall deal with the adult state. We shall then examine how this state is attained and, finally, we shall devote a paragraph to the experimental production of bone and cartilage. We feel confident that this order of presentation will allow us to cover the main cytological subjects and put them in their proper background.

II. Adult Bone

It is usual to separate the study of compact bone from that of cancellous bone because the former lends itself much more easily than the latter to coordinated illustrations. Both types of bone, however, present the same fundamental structure and are only varieties of the same tissue.

A. Compact Bone

The numerous data gathered in recent years on the constitution of compact bone at the cytological level are still far from being clearly linked with each other. For this reason, let us begin with a survey of the subject before going into finer detail.

1. General Features

Without in the least hinting that radioactive isotopes will solve all problems in bone investigation, we must recognize however that they are given the prominent role in present-day research and this chapter must therefore be written accordingly.

To do so, we shall first comment on the histological picture and explain why the corresponding microradiograph should often be preferred to it for reference in the analysis of an autoradiograph.

a. Histology. Transverse sections of the radius of an adult dog are prepared by sawing slices of bone and grinding them under alcohol until they are about $50\,\mu$ thick. They are then treated, without previous decalcification, by the periodic acid-Schiff (PAS) method.

Figure 1 (see Plate I) demonstrates the successive stages of the formation of a Haversian system, or, as it is called now, of an osteon. Of course, a particular region had to be selected in order to present all the interesting aspects within the same field.

In *a* there is a resorption cavity (half of which appears on the illustration) which is a cross section, presenting an irregular and festooned outline, of a tunnel bored longitudinally in the diaphysis.

In *b* we observe the filling of the cavity, operated by the concentric laying down of successive layers of bone. The deposition occurs in two steps: the innermost layers (arrow), just elaborated, are separated by a dark line from the surrounding layers, laid down previously. The PAS method as applied here to undecalcified ground sections makes it easy to demonstrate the succession of two stages which has, until recently, escaped attention. Except for the delicate cementing lines, delineating the osteons from the interstitial lamellae like the cement of a mosaic, the inner ring we speak of is the only structure to contrast in pink against a pale background.

In *c* an osteon is nearing completion, as shown by a very thin PAS-positive inner ring more clearly observed at a higher magnification.

In *d* and *e* two fully deposited osteons seem identical under microscopic examination, although in fact they are not, as will be seen presently.

The observations recorded by Fig. 1 must be supplemented by the study of similar sections, decalcified in ethylenediaminetetraacetic acid (EDTA) at pH 7 and stained by various methods (Vincent, 1954). After toluidine blue staining, the innermost layers are orthochromatic, whereas the outer layers of the same depositing osteon are metachromatic. There are however some variations (Arnold and Jee, 1954; Engfeldt and Hjertquist, 1955) in the results of the toluidine blue method, but, at any rate, it remains true that the tinctorial affinities of the two concentric zones of a building osteon are not the same. Methylene blue extinction does not occur at the same pH for the outer layers (pH 5.6) as for the inner layers (pH 4). Except for the cells, the most strongly basophilic structure is a thin line between the outer and the inner layers. These data are suggestive of a metabolism of mucopolysaccharides in the building up of an osteon.

b. Microradiography. The very section just examined under the microscope had been microradiographed with soft X-rays on a very fine-grain emulsion. The microradiograph (Fig. 2) may be superimposed exactly on the histological picture. The resorption cavity and the Haversian canals appear black on the enlarged negative. Since nearly every cell of Fig. 1

finds its counterpart in a black dot of Fig. 2, correlation is justified at least down to the level of the size of a cell.

With the kind of X-rays that we have used, the microradiograph is a map of the calcium content of the tissue (Amprino and Engström, 1952). A section whose organic matter was removed would give the same picture, and a decalcified section would produce a uniform faint shadow. The whiter a structure appears, the more calcium it contains.

By transferring the indications of the microradiograph to the histological picture, some important implications emerge at once.

The innermost layers of a depositing osteon (*b*, arrow) do not show up on the X-ray plate. They may possibly have begun to calcify, but their calcium content is too low to absorb the soft X-rays used in this case: Bone does not calcify appreciably as soon as it is laid down. It appears first as a "preosseous layer." Intense calcification keeps pace but follows only after a noticeable delay.

PLATE I

FIG. 1. Transverse ground section of the radius diaphysis of an adult dog. PAS method. The successive stages of the deposition of an osteon are indicated by letters. In *a*, a resorption cavity. In *b* the process is in full activity and the arrow points to the most recent layers, which are strongly PAS-positive. In *c*, an osteon nearing completion. In *d* and *e*, two fully deposited osteons. Magnification: ×95.

FIG. 2. Microradiograph of the ground section illustrated by Fig. 1. The picture is a map of the calcium content of the tissue. The innermost layers of a depositing osteon (arrow of Fig. 1) do not show up on the X-ray plate. Osteons optically identical (*d* and *e* of Fig. 1) may differ in calcium content. Magnification: ×95.

FIG. 3. Autoradiograph of compact bone, 10 hours after injection of Ca^{45}. Transverse ground section of the radius diaphysis of an adult dog. The three ring-shaped imprints of localized radioactivity are usually called the "hot spots." Magnification : ×50.

FIG. 4. Microradiograph of the ground section from which the autoradiograph shown in Fig. 3 was obtained. Comparison of both illustrations shows that the hot spots correspond to osteons being deposited. More precisely, Ca^{45} labels the layers calcifying heavily at the time of injection. Magnification: ×50.

FIG. 5. Autoradiograph of compact bone 42 days after injection of Ca^{45}. Transverse ground section of the radius diaphysis of an adult dog. Magnification: ×80.

FIG. 6. Microradiograph corresponding to the autoradiograph shown in Fig. 5. The three ring-shaped imprints of the upper half of Fig. 5 label the peripheral layers of three osteons which are almost completely deposited. Magnification: ×80.

FIG. 7. Transverse ground section of the radius of an adult dog. Gomori technique for alkaline phosphatase. The lacunae and canaliculi are black. The arrow points to a cementing line on either side of which the arrangement of the canaliculi is different. Magnification: ×460.

FIG. 8. Transverse ground section of the radius of an adult dog. Ruppricht's method. On the left, two lacunae belonging to adjacent osteons do not communicate with each other. On the right, a lacuna sends off its canaliculi into both territories. Magnification: ×1300.

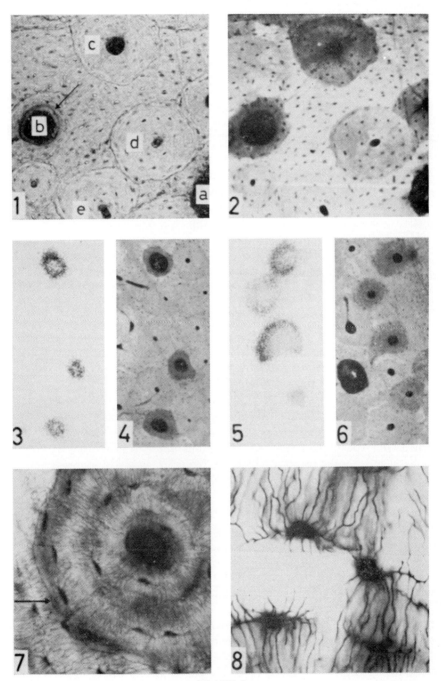

PLATE I

By photometric measurements (Amprino, 1952) it can be estimated that when it begins to calcify heavily (as in the outer zone of *b*) the bone becomes loaded, all of a sudden, to at least three-quarters of its full capacity. The phenomenon is parallel with the change in the staining properties as revealed by histological procedures.

Full calcification is not reached by the time the deposition is completed. Osteon *d* supports this statement. Osteogenesis has certainly ceased in this osteon, which must still calcify further before its calcium content reaches that of osteon *e*.

Recall that osteons *d* and *e* both look alike optically. As far as calcification problems are concerned, the microradiograph is therefore a document indispensable for the correct interpretation of an autoradiograph.

c. Autoradiography. Obviously *radiocalcium* must be given the first place in this study. Its fairly long half-life (152 days) and its low-energy beta radiation are of great advantage for autoradiographic work.

Adult dogs are given an injection of 400 microcuries of Ca^{45} per kilogram of body weight and are sacrificed at various intervals. Transverse ground sections of their radius diaphyses are prepared, microradiographed, and pressed against an emulsion for several days or weeks. The autoradiograph, the microradiograph, and the section itself are studied at the same magnification with the aid of three identical microscopes projecting the three pictures onto the same screen.

The comparison adds some data of dynamic significance to the static configuration of the microradiograph and of the section.

Figure 3 represents the autoradiograph obtained 10 hours after injection. Its meaning is clear when it is superimposed on Fig. 4, which is the corresponding microradiograph. The imprints of radioactivity, the so-called "hot spots," have the shape of rings and are produced, as could be expected, by the osteons which are being deposited. Figures 3 and 4 are similar to those which were the first to be published on the subject (Lacroix, 1953a).

A more accurate localization of the radioactivity at the histological level would be provided by a stripping-film emulsion left on the specimen and examined with it. But close scrutiny of Figs. 3 and 4, together with reference to Figs. 1 and 2, indicate clearly enough that the radioactivity is mainly that of the layer into which the preosseous layer is transformed when it suddenly calcifies and changes its staining reactions.

There are other spots of localized radioactivity which are weaker than those just commented upon: they belong to the osteons which are entirely deposited and which are completing their calcification. For instance, osteon *d* in Fig. 1 would give such a weak spot (Vincent, 1955).

Whether they are strong or weak, the imprints of localized radioactivity are observed on a background of diffuse radioactivity which belongs to the fully calcified bone. For technical printing reasons, this diffuse reaction fails to appear on our illustration. If it did, the localized imprints would be too black and would not show their ring-shaped structure.

The diffuse reaction is considered to be the result of an exchange, that is, of a process in which the withdrawals and additions of ions are statistically equal, so that the weight of the mineral material remains unchanged.

At this point it would be helpful to know exactly the amount of radioactivity contained in the hot spots as compared with the diffuse reaction, but reports on this question (Arnold *et al.*, 1956; Cohen *et al.*, 1957), based on densitometry of autoradiographs, are still conflicting. For the time being we may remark simply that the diffuse reaction covers a very large surface as compared with the small surface of all the hot spots put together and that, consequently, count measurements of whole-bone samples cannot be interpreted as the expression of a single metabolic event.

Let us rather proceed in another direction in following the autoradiograph-microradiograph relationship from week to week. Experiments along this line have been performed in our laboratory by Ponlot (1958) and will be briefly summarized here.

Figure 5 exemplifies the distribution of radiocalcium 42 days after injection. Figure 6 is the corresponding microradiograph.

The three ring-shaped imprints in the upper half of Fig. 5 correspond to the peripheral layers of three osteons which are now almost completely built. Of course the radioactivity has not been displaced at the histological level. The microradiograph shows the layers which have been deposited inside the layers labeled 6 weeks previously. The observation is in agreement with other data on the chronology of the Haversian remodeling which will be dealt with in the subsequent pages.

As already pointed out by Arnold *et al.* (1956) and by Cohen *et al.* (1957) the relative intensities of localized and diffuse reactions does not seem to change markedly in a matter of weeks. The radiocalcium initially incorporated by exchange is supposed to be displaced at the ultrastructural level to nonexchangeable positions. The word "recrystallization" is often used in this connection.

Lastly, in addition to the two kinds of localized reactions already observed in short-term experiments, a third type of imprint is recognized in long-term experiments. Six weeks after injection of radiocalcium, the specific activity of the blood is very low. It is nevertheless sufficient to label faintly the osteons which have just begun to be deposited. Such weak spots are similar to the weak spots belonging to deposited osteons completing their calcification, and the corresponding microradiograph is indispensable for a convincing identification. The faint labeling of new osteons implies that the radiocalcium released by the skeleton is partly re-used. We shall demonstrate in the next paragraph that the phenomenon of delayed redistribution of radiocalcium occurs in a much more obvious manner during growth.

Radiosulfur is almost as good a tool as radiocalcium for the present study (Lacroix, 1954, 1956; Vincent, 1954; Lea and Vaughan, 1957).

Short-term autoradiographs obtained with S^{35} are similar at first sight to those produced by Ca^{45}, but they are not identical. Here, the imprints of radioactivity correspond, not to the layers which begin to calcify, but to the preosseous layer. In adult bone, the metabolism of sulfur precedes that of calcium.

Since the S^{35} autoradiographs are the genuine expression of beginning osteogenesis, they provide us with a reliable method to estimate the chronology of the remodeling. From week to week, we can follow the laying down of new lamellae inside those which are labeled. The investigation reveals that the deposition of an osteon takes about 6 weeks in the adult dog. This figure has recently been confirmed by labeling the calcifying layer with lead detected histochemically (Vincent, 1957a).

Six weeks is a rough estimate because the process of remodeling may stop completely for a while, because osteons vary in size, and because differences presumably exist according to age and species. Further work is needed, but since a small proportion only of the whole skeleton is being destroyed and replaced at a given time, we may already assert with confidence that the histological renewing of the adult skeleton is a very slow process indeed.

Researches with radiosulfur have also taught us that it takes a much longer time for an osteon to reach its full calcification than to be deposited and that the completion of calcification does not proceed at a uniform rate everywhere in the same bone. Autoradiographs obtained 18 weeks after S^{35} administration show that some radioactive osteons are almost fully calcified but that others lag behind. However the process does not require more than six months in the young monkey (Vincent, 1957b).

Many *other isotopes* are stored preferentially in the skeleton. Radiophosphorus has been among the first to be used for autoradiographic study of compact bone (Engfeldt *et al.*, 1952). Strontium, barium, and radium are retained more or less like calcium. The pattern of distribution of several other "bone-seeking" radioelements is not yet known.

Up to now, the study of the unphysiological radioelements has been centered mainly on their role in the hazards of radioactivity. In the future they will undoubtedly contribute particular data on the life of bone, that for technical reasons (half-life, type of radiation, etc.) the "physiological" radioisotopes would be unable to uncover.

2. Cytological Study

We may now deal with our real subject, the study of adult bone tissue at the cytological level.

a. The bone cells. The main data on the bone cells, are, in fact, data on the system of lacunae and canaliculi in which they are contained. This system may be observed in dry ground sections or, better, with the aid of Schmorl's thionine method, but several other methods have been tried (Ponlot, unpublished) in order to check the classical descriptions.

The Gomori technique for alkaline phosphatase is rather interesting (Fig. 7). All the lacunae are black. Each lacuna has the shape of an almond and has a particular orientation with respect to the center of the osteon, that is to the Haversian canal. The short and the intermediate axes of the almond are seen on a transverse section, such as that reproduced by our illustration, the short axis being radial and the intermediate being concentric. The longest axis would show only on a longitudinal section.

Numerous canaliculi leave the lacunae. Nearly all of them assume a radial direction toward the center or toward the periphery of the osteon. Some run straight, some are winding, some branch once or twice. Most of the canaliculi of neighboring lacunae anastomose with each other.

The arrangement of the lacunae and canaliculi in relation to the cementing lines deserves particular attention.

A cementing line (arrow of Fig. 7) separates two pieces of bone tissue of different age. The piece on the right of the illustration is a complete osteon, the more recent of the two, whose outline is unbroken. The piece to the left is older and is that part of the previous bone which was respected when a resorption cavity, now filled with the complete osteon, was extending. On either side of the cementing line, the lacunae and canaliculi are different, testifying to the past history of the region. On the right, the canaliculi leaving the outer aspect of the most peripheral lacunae, instead of reaching the cementing line, loop back toward the center of the osteon. On the left, the canaliculi reach the cementing line, where they end abruptly.

The description holds good for the greater part of the cementing lines. Here and there, however, canaliculi cross the cementing line. To illustrate this point, the method of Ruppricht (*in* Krompecher, 1937) is helpful. Sawn sections of bone are bathed in a saturated solution of basic fuchsin in alcohol, then ground under xylol to the desired thickness; the cavities of the bone tissue are filled with the dye and are easily observed in the section made transparent by the xylol. Figure 8 is centered on a cementing line, as ascertained by the low-magnification picture and by the corresponding microradiograph. On the left, two lacunae do not communicate with each other across the cementing line. On the right, a lacuna sends off its canaliculi into both territories.

Some additional evidence from another material, the human tibia, is not out of place. Phase contrast microscopy provides pictures (Fig. 9, Plate II) in which the innumerable brilliant striae running radially correspond to the canaliculi. Study under high power convinces the observer that interosteonic communications are even more frequent than stained preparations would indicate.

The lacunae and the canaliculi contain the bone cells, or osteocytes, and their processes. Little is known with certainty of their fine structure, owing to the difficulty of applying cytological techniques to such a hard tissue. Each cell has an oval, rather large, hyperchromatic nucleus. Cytoplasmic processes may be followed in the canaliculi. According to Amprino (1950), they are continuous with the processes of adjacent cells except perhaps in very small fragments of osteons. For Lipp (1954), the cytoplasm itself does not extend very far in the canaliculi; but the whole osteocyte with its processes is contained in a limiting membrane which extends farther than the cytoplasm in the canaliculi and which is continuous with that of the other cells. At all events, a system is provided by which the cells not only of an osteon, but of the whole bone are able to "communicate" with each other.

PLATE II

Fig. 9. Transverse ground section of a human tibial shaft decalcified in EDTA and examined in phase contrast. The brilliant striae running radially correspond to the canaliculi of Figs. 7 and 8. Magnification: ×338.

Fig. 10. Same section as in Fig. 9, but photographed in polarized light with nicols crossed. Brilliant, or anisotropic, and dark, or isotropic, lamellae alternate with each other. Magnification: ×338.

Fig. 11. Electron micrograph of a decalcified section of compact bone from an adult human femur. Shadowed with uranium. A large bundle of fibrils courses parallel to the plane of the section. In the lower part of the illustration the fibrils are cut transversely. From Robinson and Watson (1952); courtesy of the authors and of *The Anatomical Record*. Magnification: ×26,000.

Fig. 12. Electron micrograph of an undecalcified section of bone. Thin shaving of the rib cortex of a 40-year-old man. The collagen fibrils are completely masked. The inorganic bone crystals form electron-dense ribbons of bands which run at right angles to the direction of the underlying fibrils. From Robinson and Watson (1952); courtesy of the authors and of *The Anatomical Record*. Magnification: ×26,000.

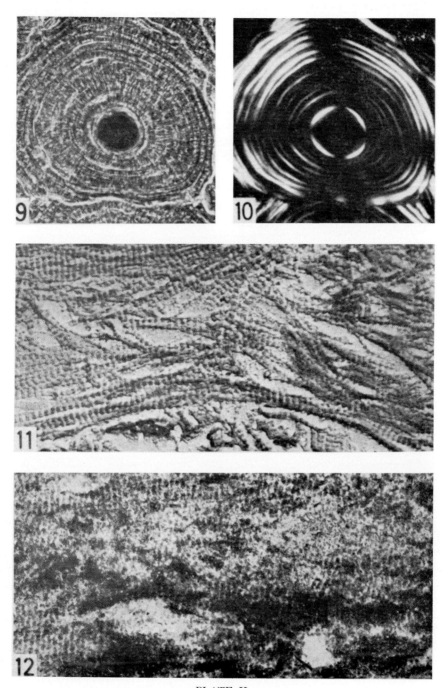

PLATE II

The cell population of adult bone is extremely sparse, but the real intercellular distance is not that between two lacunae, it is that between two canaliculi. The impression conveyed by Figs. 1 and 2 must now be corrected by the information brought by Figs. 7–9.

Empty lacunae are sometimes encountered in normal bone. They are numerous in the vicinity of a recent fracture line. Bone tissue devoid of its cellular components is spoken of as dead bone. Two forms of osteocyte death have been described (Rutishauser and Majno, 1951), a fast one in which the osteocyte just fades out completely in about 6 hours without alteration of the surrounding intercellular substance, and a slower one, lasting about 2 days and accompanied by some digestion of the pericellular structures. In most of the soft tissues, the death of a few cells would pass unnoticed. In bone tissue, the empty lacunae could not possibly be reoccupied by new cells: the whole area of dead bone must be destroyed and rebuilt. The histological replacement being very slow, it follows that the average life span of the osteocytes of adult bone is most likely a good deal of the adult life span of the animal itself.

b. The fibrils. Silver methods demonstrate that collagen fibers are extremely abundant in bone. It is usual to comment upon this fiber structure with the aid of its aspect in polarized light. Figure 10 is a well-known picture representing the very osteon just considered in phase contrast.

Brilliant, or anisotropic, and dark, or isotropic, lamellae alternate with each other. For many years textbooks have stated that the activity or the inactivity on polarized light was simply due to a different orientation of the fibers. But doubts were raised, first in the name of histology (Ruth, 1947) and then from the researches which were conducted with the electron microscope and which we shall summarize now.

Rouiller *et al.* (1952) have studied replicas from bone surfaces which had been polished and etched. They have shown that the stratified structure of an osteon is due to the succession of two kinds of lamellae; one, anisotropic, is packed with collagen fibrils whose general orientation is parallel with the Haversian canal; the other, isotropic, contains fewer fibrils, and these run somewhat radially. Fibrillar continuity is seen everywhere between adjacent lamellae.

Figure 11, from Robinson and Watson (1952, 1955) shows an electron micrograph of a decalcified section of compact bone from an adult human femur. A large bundle of fibrils courses parallel to the plane of the section; in the lower part of the illustration the fibrils are cut transversely. No major difference is noticed between these fibrils and those of other connective tissues. The diameter of the fibrils is fairly uniform and measures about 800 A. There is a large period spacing which ranges from 550 A. to 750 A. and averages about 630 A. Most obvious in the period is a two-band region, or doublet, which is in register across several fibrils. There is even a suggestion of bridging between the doublets of adjacent fibrils.

In the infant, the fibrils are thinner. Five bands may be observed in each period, and it looks as if the doublet banding of the adult is due to the preferential persistence of two of the original five bands.

In very old subjects, the fibrils are thicker but the doublet banding is the same as in the middle-aged adult.

The apparent identity of collagen fibrils in bone and in other tissues has been stressed by Martin (1953) and by Schwartz and Pahlke (1953).

Frank *et al.* (1955), basing their observations on sections of decalcified bone, have published pictures which confirm the general arrangement described by Rouiller *et al.* (1952). They note, however, that variations occur even in the same osteon. Few osteons indeed would present the regular succession of brilliant and dark lamellae that we have selected for our Fig. 10. Differences exist also between species: our illustration is provided by human material, not only in order to correspond with the observations that we are reporting, but also because dog bone would give a more diffuse picture.

c. The bone salts. For more than thirty years, since it was shown that bone salts have a composition similar to that of apatite, various opinions have been held on the real pattern of the mineral phase of bone at the molecular level. For a detailed discussion on the issue, which belongs more to the field of crystallography than to that of cytology, the reader is referred to Dallemagne and Fabry (1956, 1957).

What matters more here is the aspect of the crystals as they are seen in undecalcified sections examined with the electron microscope. In Fig. 12, again from Robinson and Watson (1952, 1955), the fibrils are masked. The inorganic bone crystals form electron-dense ribbons which run across the section at right angles to the direction of the fibrils. These ribbons have a width of 400 A. and are apparently continuous across several underlying fibrils. They are made of elements which seem to be tabular in shape and to have dimensions of about 400 A. by 200–300 A. and by 25–50 A.

The observations of Fernández-Morán and Engström (1956, 1957), based on ultrathin sections obtained with a diamond knife, are different. They speak of rod- or needle-shaped apatite particles approximately 200 A. long and about 40 A. in diameter, and they add that their figures are consistent with previous X-ray diffraction studies. The reports of Ascenzi (1955) are also at variance.

Discrepancies are to be expected at this early stage of a study hampered by many technical difficulties. It is nevertheless fairly certain that the formation of crystals is linked with the periodic structure of collagen and that the total sum of the surfaces of the crystals is enormous, a conservative estimate (Neuman and Neuman, 1953) being 100 square meters per gram of bone salts.

An alternating high and low calcium content from one lamella to the other has been recorded by Engström and Engfeldt (1953). The observation has been confirmed by Vincent (1955).

d. *The ground substance.* From a morphological point of view, the existence of a cementing medium between the collagen and the mineral crystals is far from being evident.

From a chemical point of view, the presence of mucopolysaccharides in bone has been known since the turn of the century and has been recently reaffirmed (Rogers, 1951; Glegg and Eidinger, 1955).

More to the point for us are the researches of Kent *et al.* (1956) because they throw a bridge between biochemistry and autoradiography. The amount of S^{35} retained in bone is very small (0.01 to 0.1% of the injected dose). This small dose is incorporated in bone in two forms: (1) a labile form throughout the bone which is in large part removed by decalcifying agents; (2) a nonlabile form which is not removed by decalcifying agents and which is present in high concentration in the hot spots, i.e., in regions where metachromatic properties toward toluidine blue are linked with calcification processes. The Oxford group has succeeded in locating the nonlabile radioactivity in the carbohydrate fraction as opposed to the collagen and protein fractions. Though the radioactive constituent has not yet been isolated in a pure form, several data suggest strongly that it is chondroitin sulfate or a closely allied substance.

All these facts, on the one hand, and the crystal-collagen relationship observed with the electron microscope, on the other hand, seem to indicate that the successive stages that we have described in the life history of an osteon imply basically a "combination" of some sort between ground substance and collagen as one of the local prerequisites for calcification.

From his studies on *in vitro* calcification of preosseous cartilage, Sobel (1955) concludes "that a complex of chondroitin sulfate with collagen in a critical configuration may be responsible for initiating the calcification." We cannot find better words to convey our feeling when we consider the formation of an osteon. Some reservation remains, however, in our mind, as far as bone is concerned, because in this tissue (quoting the figures of Eastoe and Eastoe, 1954) the amount of mucopolysaccharide-protein complex is so small (0.24%), as compared with the amount of collagen (18%) and of inorganic matter (70%), that it is difficult to visualize the said complex as an all-pervading ground substance.

B. Cancellous Bone

Trabeculae of cancellous bone are made of blocks of lamellar bone united with each other by cementing lines. These blocks present varying calcium contents. As a rule, those which are less mineralized are superficial with respect to the more mineralized.

The remodeling of cancellous bone presents the same stages as in compact bone. New lamellae appear in the form of a preosseous layer. After some lag, they suddenly take about three-quarters of their loading capacity of calcium. Concurrently toluidine blue shows a change in their staining properties identical with that observed in compact bone. Afterward full calcification is reached slowly.

In spite of its different spatial distribution, cancellous bone thus presents a profound similarity with compact bone.

We cannot possibly expand on the subject without the aid of many pictures and, for more information, we prefer to call attention to the researches carried out in our laboratory by Vincent (1955) and to the illustrations of his monograph [see also Engström (1956) for the microradiographic aspect; Trueta and Harrison (1953) for the vascularization].

III. THE GROWTH OF A LONG BONE

Radiocalcium is particularly well suited for an autoradiographic study of bone growth (Comar et al., 1952; Tomlin et al., 1953). In order to improve the resolution of the pictures we (Lacroix and Ponlot, 1958) have found it helpful to embed the bones in methyl methacrylate because the plane of section may then be carefully polished.

The *short-term stage* (Fig. 13, Plate III) is illustrated by the tibia of a 75-day-old puppy which had received 400 microcuries of Ca^{45} per kilogram of body weight and was sacrificed 10 hours later. The radioactivity is distributed in regions of high intensity, appearing dense black on the printed illustration, and in regions of low intensity, which appear grayish.

The significance of the picture does not leave any room for doubt: the strong reaction labels the regions where new bone is being laid down, the weak reaction corresponds to the bone which had been deposited and fully calcified prior to injection. The short-term distribution of radiocalcium obeys the same laws in growing and in adult bone: a fraction of the isotope takes part in the calcification processes, another fraction is exchanged with calcium ions in the old, established bone.

The picture demonstrates the mode of growth of a long bone and saves lengthy descriptions. Here, statements which were previously based on rather indirect proofs become self-evident. For instance, the inner arrow points to an area where new bone is being laid down on the inner

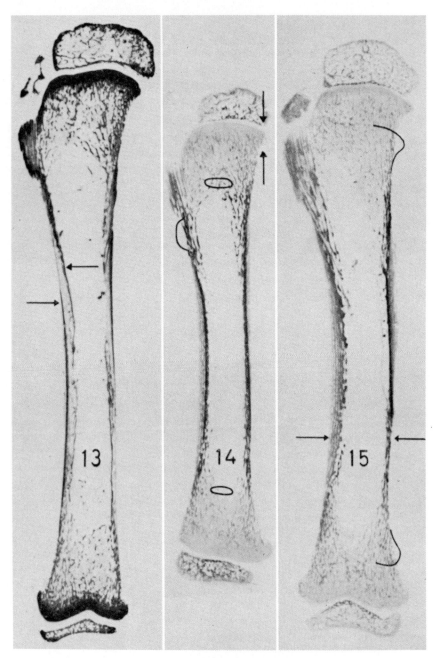

PLATE III

surface of the anterior wall, whereas the outer arrow indicates a region where new bone is deposited on the outer surface. We are now considering a part of the tibia where the middle fifth of the shaft, compared to a cylinder or to a rough prism, is progressively flaring toward the upper metaphysis, more or less like a funnel. The particular autoradiographic reaction of this region has been called the "funnel effect" by Leblond *et al.* (1950). It shows that, strange as it may seem, a long bone reduces some of its diameters at the same time that it enlarges nearly everywhere else.

A more complete understanding of the mode of growth is gained if we now follow the fate of the autoradiographic picture with the passing of time.

Puppies of the same litter received the same dose of radiocalcium at the age of 20 days. The short-term picture is not reproduced here because it is similar to Fig. 13 (but smaller, of course, since the animals were younger).

The *3-week stage* is represented by Fig. 14. On the autoradiograph we have superimposed the outline of the epiphyses and of the tibial tuberosity as they were located at the time of injection.

The combined picture shows that the black imprints of radioactivity correspond to what is left of the original bone tissue. The strongly labeled bone is now almost completely buried in the new bone which has been deposited during the 3 weeks' interval. The new bone is faintly radioactive, and all the more faintly as it is of more recent deposition. The radiocalcium of the new bone comes from the whole skeleton as it was

---

PLATE III

Fig. 13. A 75-day-old puppy was sacrificed 10 hours after injection of radiocalcium. Contact autoradiograph of the tibia. The strong reaction labels the regions where new bone is being laid down. The weak reaction corresponds to the bone which had been deposited and fully calcified prior to injection. Arrows point toward a region commented upon in the text. Magnification: × 2.5.

Fig. 14. Puppy sacrificed 3 weeks after injection of radiocalcium administered when it was 20 days old. On the autoradiograph of its tibia are superimposed the outlines of the epiphyses and of the tibial tuberosity as they were located at the time of injection. What is left of the original bone is strongly radioactive. The new bone is faintly labeled. Arrows indicate a region illustrated by Fig. 17 (Plate IV). Magnification: × 2.7.

Fig. 15. Littermate of the puppy of the preceding illustration injected the same day and sacrificed 4 weeks later. On the autoradiograph are sketched the outlines of the metaphyses as they appear on Fig. 14. In 4 weeks the anterior wall has been almost completely destroyed and replaced, whereas the posterior wall has been less remodeled. Arrows point to a level where the contrast between the two walls is well marked. Magnification: × 2.7.

built at the time of injection and which is being destroyed. Some radio-calcium is eliminated by the kidney or bowels on its way from the destruction sites to the building sites where it is being redistributed.

The histological lability of the growing skeleton contrasts with the histological stability of the adult skeleton, established in the preceding paragraph. Both notions supplement each other and now assume their full meaning.

The epiphyses have been displaced by longitudinal growth at an equal rate at both extremities: The tibial tuberosity has moved apace with the upper epiphysis so that the quadriceps tendon has kept the same relative distances from both extremities. Of course the epiphyses migrate really, whereas the "migration" of the tibial tuberosity is metaphorical and is the result of a resorption on its lower aspect and of a reconstruction on its upper aspect.

The *4-week stage* (Fig. 15) gives us some additional information.

Let us first take advantage of the perfect correspondence of Figs. 14 and 15 to draw on the latter a part of the outline of the former. No comment is needed to understand that growth in length implies a constant peripheral resorption of the metaphysis, that, therefore, the part of the bone between the two vertical arrows in Fig. 14 is destroyed almost as soon as it is formed.

It is clear also that growth in width does not occur at the same rate all around the shaft and that, furthermore, every level differs from the others in this respect.

One level, that indicated by horizontal arrows of Fig. 15 deserves to be specially mentioned. In 4 weeks, the anterior wall has been completely destroyed and replaced by a new one farther forward. On the contrary, the posterior wall has been left undisturbed and contains some structures which, one month before, were part of the metaphysis. These structures had just been produced at that time by the growth cartilage. They are made of endochondral bone, a point which is easily checked and confirmed by histological or microradiographic (Owen *et al.*, 1955) examination. That a limited portion of a growing bone is respected while everything is being destroyed and replaced elsewhere in the same bone is not a mere detail, as we shall see later on.

The peripheral resorption of the metaphysis, the processes expressed by the "funnel effect," the differential growth in width of the shaft, and the relative permanency of some small portions of the bone are some of the mechanisms by which the growing tibia always keeps the same general shape. This notion of constancy of form is evident but alludes nevertheless to the main mystery of bone growth. In the paragraphs to come, dealing with cytological events, we must therefore bear in mind that our analysis concerns phenomena which fit into a well-defined pattern of organization and which somehow are subordinate to it.

At this juncture it is fair to state that, so far, the autoradiographic techniques have not added a single new morphological detail to the body of facts assembled by time-worn methods: markers, vital staining with madder, comparison of standard radiographs, and histological study. The peripheral resorption of the metaphysis is a very old observation. In 1945 (Lacroix, 1945a), we had illustrated the histological basis and the result of the "funnel effect." In the same article, we had shown that a limited part of the diaphyseal tube is made of endochondral bone. In other papers we (Lacroix, 1951) had also described the preferential enlarging of some diaphyseal walls and the apparent migration of the tibial tuberosity. But now, instead of elaborate demonstrations, three autoradiographs in all are sufficient to show convincingly what we meant in our previous writings and to direct us safely to the places where cytological events are bound to show up at their best.

IV. ENDOCHONDRAL OSSIFICATION

The first rudiments of the axial skeleton, of the appendicular skeleton, and of part of the skull are entirely cartilaginous. Their shape is typical at a very early stage. In the human, they begin already to exhibit their proper configuration in the 21-mm. embryo where the humerus, for example, is only 2 mm. long. This means that these miniature skeletal pieces will have to increase their volume at an enormous rate while keeping their general shape and while continuously performing their supporting and metabolic functions.

The mechanisms by which this coordinated growth is achieved are many, and among them endochondral ossification is the major one.

Endochondral ossification is best studied where it is the most active, in the growth cartilages of the long bones, although it occurs in other places such as the epiphyses and in some tuberosities. The data will be divided into several groups, each of them centered on a particular technique.

A. *Optical Microscopy*

To aid in better realization of the functional meaning of a growth cartilage, we shall present the histological counterpart of the autoradiographs of the preceding paragraph. The reader may thus rely in the most concrete manner on the combined illustrations to see how growth results from the cytological events.

Figure 16 (Plate IV) is provided by the upper extremity of the tibia of a 6-week-old dog (the stage of Fig. 14). It represents a longitudinal section in the center of the growth cartilage. Let us analyze it, layer by layer.

In *a* is an irregular plate of bone which belongs to the ossification center of the upper epiphysis. Small remnants of cartilage are still visible. The tissue is being remodeled, as proved by the row of osteoblasts lining the cavity appearing in the field and by the corresponding spotty radioactivity detected by the autoradiograph. On the whole, the plate is a solid structure which distributes evenly to the growth cartilage the stresses of weight bearing supported by the epiphysis and which, on the other hand, transmits to the epiphysis the push of growth developed in the cartilage.

In *b* is the basal zone of the growth cartilage which is strongly united with the epiphyseal bone. Pairs of cells are dispersed within a dark-staining intercellular substance. Many vessels run between the bone and the cartilage.

In *c* is the proliferating zone with its parallel columns of flattened cells piled more or less like stacks of coins. The multiplication of the cells goes apace with the synthesis of a considerable amount of intercellular substance which takes a much lighter stain than in layer *b*.

During the whole period of growth, the cartilage compensates by continuous regeneration for the destruction which it undergoes at its metaphyseal surface. It is therefore surprising that longitudinal sections,

---

PLATE IV

Fig. 16. Longitudinal section of the upper growth cartilage of the tibia of a 6-week-old puppy, the stage of the autoradiograph of Fig. 14. In *a*, the epiphyseal bone. In *b*, the basal zone. In *c*, the proliferating zone. In *d*, progressive hypertrophy. In *e*, the calcified cartilage. In *f*, the endochondral bone. Arrow is referred to in the text. Magnification: ×98.

Fig. 17. Longitudinal section of the upper growth cartilage of the tibia of a 6-week-old puppy. Region isolated by arrows in Fig. 14. Example of the so-called "ossification groove." The deepest part of the groove, the new columns being added at the periphery of the growth cartilage, the "perichondrial ring," and the osteoclasts planing the metaphysis are indicated by arrows. Magnification: ×56.

Fig. 18. Transverse section of the endochondral bone trabeculae which appear in longitudinal section in the two preceding illustrations. The trabeculae are made of a core of calcified cartilage surrounded by a coating of bone, covered in turn by osteoclasts. In the lower left corner an osteoclast is wrapped around the blunt end of a trabecula. Magnification: ×166.

Fig. 19. The central part of the growth cartilage has been removed from the distal extremity of the radius in a newborn rabbit and grafted in the anterior chamber of the eye in an adult rabbit. Result 15 days later. The graft receives its vessels from the iris. It has grown and has differentiated almost as if it were still in its normal place. Magnification: ×18.

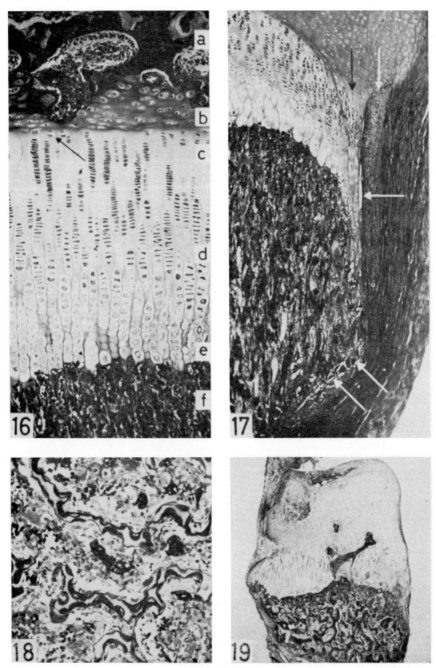

PLATE IV

such as that of Fig. 16, do not show any mitotic figures. In fact, mitoses are numerous in zone c, but they are to be seen only in transverse sections, i.e., in sections perpendicular to the long axis of the bone.

How is it then that the cells arrange themselves in columns whose direction lies perpendicular to that of their mitotic axis? The question leads us to realize that the cells in the columns are not exactly super-imposed and that the classical comparison with a stack of coins is mis-leading. A careful examination of a longitudinal section reveals that each cell has a thick, rounded edge, containing the nucleus on one side and a thin, drawn out edge on the opposite side. The arrangement of the cells in a column is not that of uniform disks but, rather, that of wedges piled on each other, in which the thick part of one alternates with the thin part of the next. This arrangement is clearly seen only in certain groups of cells or in certain parts of a single group, because a suitable longitudinal section must pass through both the thick and the thin parts of adjoining cells. If the section does not pass exactly in the plane mentioned, the cells seem to be of uniform thickness.

It is generally admitted that all the cells in a given column of zone c are the daughters of one pair of cells of the basal zone. The arrow of Fig. 16 allows us to watch one of these pairs, which is just about to pro-liferate.

At a certain stage in their evolution, the cells begin to enlarge. In d we observe this progressive hypertrophy. Most commonly the cells increase in height throughout, but particularly on the side on which they are thinned out, so that they soon lose their cuneiform appearance and ac-quire a uniform height. In this development, the partition which separates two daughter cells and which was at first oblique in relation to the plane of the growth cartilage gradually loses its obliquity and comes to lie in the plane of the cartilage. It sometimes happens that the partition in-creases in obliquity instead of diminishing, and finally comes to lie parallel to the axial bars; this process then ends in a side-by-side arrangement of the two hypertrophied cells in the plane of the cartilage. Lastly, the cells may simply increase regularly in all their dimensions without change in their reciprocal relations: they thus retain the appearance of wedges alternating with one another which was characteristic of their original arrangement.

On the other hand, while the longitudinal dimension of each group is increasing to a great extent, the growth of cell dimensions in the trans-verse direction is slight though recognizable. The bulging lateral walls of the cells indent the longitudinal bars of intercellular substance and thus impart to them that scalloped appearance which they retain even when they have become the axes of endochondral bone trabeculae. The longitudinal bars become thinner as they are drawn out and, consequently, the lateral expansion of the cells does not result in a transverse expansion of the whole cartilage.

The progressive hypertrophy of the cells is such that, when completed, it has multiplied the initial longitudinal dimension of a group of cells by a factor of about four. This process is the main component of the thrust developed in the cartilage, a thrust which is powerful since, in the human, it can break a thick steel staple trying to anchor the epiphysis to the metaphysis.

In e is the zone of calcified cartilage. Silver methods as well as microradiography demonstrate the presence of calcium in the thick longitudinal bars of intercellular substance in a region corresponding to the last two or three hypertrophied cells.

The cells have now reached the end of their evolution. They are destroyed and disappear at the moment when the cavities they occupy are invaded through the thin transverse partitions by buds of vascular and connective tissue. In these buds, the connective cells play the active part, since tissue culture studies of growth cartilages prove that invasion may be accomplished by the cells alone, without capillaries.

In zone f are produced the first trabeculae of endochondral bone. In a fast-growing cartilage, like the one under study, all the longitudinal bars persist to form the core of the bony trabeculae. Osteoblasts arrange themselves like a pseudoepithelium along the festooned contour of the bars and a coating of bone is deposited. Farther down in the metaphysis, some of the trabeculae become thicker and thicker whereas some others are destroyed by osteoclasts. The result of this selection is exemplified by the autoradiographs.

The structure of endochondral bone is more clearly observed in transverse sections (Fig. 18) than in longitudinal sections. The corresponding microradiograph would show that the cartilaginous core is definitely more calcified that its bone coating, a fact which could not have been suspected without this technique.

In addition to the foregoing general description, we feel that we have to insist on some other aspects of the growth cartilage which are still very often misunderstood.

On Fig. 14, two vertical arrows have been added on either side of the posterior border of the growth cartilage. The histological aspect of the region between the arrows is now represented by Fig. 17.

This region is known under the name of Ranvier's "ossification notch" or "ossification groove." The reason is that the very thick periosteum on the right, and the periphery of the growth cartilage on the left, delimit a kind of notch, the deepest part of which is indicated by a white vertical arrow. Since this notch is the section of a circumferential groove, the latter term is equally valid.

In the notch is a very thin lamella of primary bone (white horizontal arrow). It is applied directly to the flank of the growth cartilage and to the most peripheral trabeculae of endochondral bone. It is connected with the periosteum by a padding of young connective cells in which all transitional stages are observed between proliferating periosteal cells and osteoblasts being engulfed in bone. It is growing at its upper extremity. It is being destroyed by osteoclasts at its lower extremity.

From the picture, we may imagine the three-dimensional structure and speak accordingly of the "perichondrial ring of the ossification groove." Growing at its epiphyseal edge while being resorbed at an equal pace at its diaphyseal edge, the ring migrates with the growth cartilage which it encircles—an apparent migration, of course, which expresses a continuous replacement.

The perichondrial ring is given its very simple morphological meaning if we consider its origin in the history of the skeletal piece. At the beginning of ossification, the diaphysis is a tube made up entirely of periosteal bone from end to end. But, as shown by the autoradiographs, the metaphysis has soon to reduce its outer diameters so that the bone may keep its typical shape. The aim of combining our illustrations is to show the process in full activity. As already indicated, superimposition of the outline of Fig. 14 on Fig. 15 proves that the bone tissue projecting like a spur at the posterior aspect of the metaphysis must be eaten away. This localized planing is actually observed in Fig. 17, where two white oblique arrows point to a front of advancing osteoclasts. As a result, the extremity of the periosteal tube is isolated and is separated from the shaft proper by a limited portion made up exclusively of endochondral bone.

Figure 17 may be considered as fairly representative of the perichondrial ring of the ossification groove. In some cases the structure is more distinctly "independent" and calls for a comparison with a ring on a finger. In other cases, particularly in slow-growing extremities, the ring is hardly discernible or is even lacking.

These details had passed unnoticed until we called attention to them in 1945 (Lacroix, 1945a). Since then, illustrations almost identical with our Fig. 17 have been published by Godart (1952) and van Wel (1954). A study of bone growth in the human embryo by Gardner (1956) and preliminary notes by Pratt (1957) also coincide with our description.

Other phenomena are occurring in the ossification groove.

It is clear that transverse enlargement of a skeletal piece keeps in step concomitantly in the diaphysis, in the epiphysis, and in the growth cartilage. Once again, the autoradiographs substantiate this observation, since Figs. 14 and 15 testify to the actual increase in size of the growth cartilages in the anteroposterior direction.

The black vertical arrow of Fig. 17 points to new columns which are being added at the periphery of the cartilage and which are not yet differentiating as completely as the previous columns farther to the left. We believe that this process accounts for the enlargement of the growth cartilage in the transverse directions.

Just one more point about Fig. 17 may be mentioned. A few words of comment on the periosteum which appears in the illustration are appropriate here. Without this fibrous sheath, the growing extremity would be extremely fragile because, as is well known, the junction cartilage-metaphysis is the weak region in a young bone. Fastened to the epiphysis which migrates away from the rest of the bone, the periosteal membrane is continuously under a tension which causes it to elongate at the same rate as the diaphysis. We have shown (Lacroix, 1951), and it has been confirmed by Pinard (1952), that in fact the periosteum is drawn out throughout its length. This mode of growth by stretching being different from that of the subjacent diaphysis, it follows that the periosteal sheath slides slowly on the bone that it encloses. The obliquity of the imbricated bone trabeculae in a very young diaphysis and the arrangement of the nutrient canal are both the consequences of the movement of the periosteum, which, in turn, is determined by the functioning of the growth cartilage.

B. Histochemistry

The growth cartilage is particularly well suited for histochemical studies. In this tissue a given aspect is always observed between the preceding and the following stages and chronological questions are answered with a certainty which is almost unique in histology.

The presence of *glycogen* in the growth cartilage was demonstrated a century ago. The point was investigated carefully by Follis and Berthrong (1949). In young cartilage cells, only a few small granules of glycogen are found. With multiplication of the cells and their increase in size, more numerous and larger granules of glycogen appear, until, in the hypertrophic cells, the cytoplasm is filled with them. But the most mature cells, those closest the invading blood vessels, do not appear to contain glycogen. There is a rough inverse relationship between the presence of glycogen and the presence of lime salts. Further evidence of this inverse relationship has been produced by Follis (1949a) from observations recorded in healing rickets.

Among enzymes, *alkaline phosphatase* has been the most often studied. In the proliferating cells only the nuclei show evidence of phosphatase.

As the cells enlarge, they become progressively richer in enzyme, the increase being apparent first in the nuclei, then in the cytoplasm, and lastly in the matrix. According to Morse and Greep (1951), the zone of calcified hypertrophied cells (zone *e* of Fig. 16) is relatively free of phosphatase; whereas Follis (1949b) considers that this zone is richer than the others. It was maintained originally that phosphatase is associated with the production of phosphate ions which secure the precipitation of calcium in bone salts. Many objections have been raised and present views are so conflicting that a definite short statement is impossible (Bourne, 1956).

A vast amount of work has been done on the *in vitro* calcification of the growth cartilage. The technique has proved extremely useful for investigating a number of enzymatic mechanisms. The papers by Gutman and Yu (1951), Cartier and Picard (1955), Picard and Cartier (1956), Sobel (1955), Sobel *et al.* (1957), and Zambotti (1957) should be consulted in this connection.

On the whole, histochemistry of the growth cartilage has been mainly concerned, up to now, with the biochemistry of calcification. This is, however, only the final stage in the function of the growth apparatus. The preceding stages are at least as important, since they are responsible for the growth itself, but their histochemical study is still in infancy.

C. Autoradiography

The techniques of autoradiography are now coming in the foreground to cover the stages neglected by histochemistry.

The importance of *radiosulfur* for a deeper insight into the metabolism of the growth cartilage has been revealed by Dziewiatkowski. In 1949 (Dziewiatkowski *et al.*, 1949), he had already suggested that the major portion of the labeled sulfur was retained in the cartilaginous tissues of the rat as chondroitin sulfate. In 1951 (Dziewiatkowski, 1951a), he substantiated his view when he isolated from the cartilage chondroitin sulfate with S^{35} incorporated therein. The same year (Dziewiatkowski, 1951b), he demonstrated with autoradiographs that the concentration of S^{35} in the growth cartilage is maximal in the cells undergoing hypertrophy and is markedly lower in the calcified layer. He further showed (Dziewiatkowski, 1952) that the autoradiographic reaction fades in cartilage fixed in a solution of formaldehyde saturated with barium hydroxide, owing most likely to the fact that the barium salt of chondroitin sulfate is soluble in water. Later the same author pursued his work by considering the effect of various factors on the synthesis of S^{35}-labeled compounds in the growth cartilage. More recently (Dziewiatkowski *et al.*, 1957) he again studied the nature of said compounds.

Confirmation and extension of most of these results have been brought by Boström (1953), by Boström and Jorpes (1954), and by Davies and Young (1954).

Focused at first on the reactions of the intercellular substance, attention subsequently shifted to the cells.

Amprino (1955a, 1956) has shown in the chick that there is already a high uptake of S^{35} in the axial region of the limb rudiment, in the precartilage cells which are not yet surrounded by much matrix.

Bélanger (1954, 1956) has studied autoradiographs obtained by coating the sections with a fluid emulsion, a technique allowing a more accurate localization of the radioactivity than contact autoradiography.

Two and six hours after injection of radiosulfate in 4-day-old rats, the radioactivity is found in the cells (except, as stated above, in the fully hypertrophied cells of the calcified layer). At 1 day, and also at 2 days, radioactive sulfur is still detected over the cartilage cells, but an increasing proportion is recognized over the intercellular substance. At 6 and 8 days after the introduction of tracer, the localization of radiosulfate is strictly extracellular. These observations, recorded with sections cut without decalcification, remain the same if the autoradiographs are made with decalcified sections. The sulfate ions are picked by the cells and thereupon incorporated into an organic molecule of the intercellular substance, which would be, in all probability, based on the consensus of previous work, chondroitin sulfate.

If S^{35} is given as labeled methionine, the general pattern of the autoradiograph is the same except that there is a localization of more intense radioactivity in the vicinity of the ossification groove. The observation is of special interest for us since we have shown that increase in size of the transverse diameters of the cartilage occurs in this region. It appears as if S^{35} methionine is able to demonstrate a definite cellular stage which is not visualized with radiosulfate.

The migration of radiosulfate from the cells to the intercellular substance has also been observed by Verne et al. (1956).

Radiocarbon has been used by Greulich (1956) in newborn rats and with the coated method. Close comparison of his results with those of Bélanger is therefore justified.

The distribution of C^{14} is practically identical with that of S^{35}. A comparison of undecalcified and decalcified sections indicates that there is only slight alteration in the distribution and the intensity of the autoradiographic images as a result of exposure of sections to nitric acid. Contrary to what earlier work with C^{14} would have led to expect, the deposition of radioactive carbon occurs primarily in the organic components of the growth cartilage rather than in its mineral portion. Since, moreover, autoradiographs of tissue treated with saliva are similar to those of untreated tissue, glycogen does not contribute appreciably to the autoradiographic picture. In spite of the similarity of the C^{14} reactions in cartilage to those of S^{35}, both at very early and at late intervals, Greulich deems, on the basis of other preliminary observations, that newly synthesized collagen is the material chiefly responsible for the picture obtained with radiocarbon.

D. Electron Microscopy

The first comprehensive papers on the electron microscopy of endochondral ossification are those of Scott and Pease (1956) and of Robinson and Cameron (1956). Considering how interesting they are and how recently they have been published, further developments are certainly to be expected.

The columnar cells, before they begin to enlarge, present an irregular contour and fill completely a lacuna limited by a capsule. The large vesicular nucleus occupies an eccentric area and is in close apposition to the plasma membrane. The cytoplasm consists mainly of undulating and anastomosing lamellae of double membranes very close to each other. The over-all electron density and the compactness of the components indicate a state of low water content. The capsule which surrounds each cell is very thick and does not present any demonstrable structure. More particularly, it does not contain any visible fibrils. On the contrary, the matrix which surrounds the capsule consists of a meshwork of fine fibrils of 100 A. diameter and without periodicity. The absence of obvious periodic banding, except for some exceptional suggestions, is in agreement with the observations of Martin (1954).

At the beginning of hypertrophy, the double membranes are more loosely disposed. The nucleus, now centrally located, is deformed, probably by the internal pressure of the cell. At the termination of hypertrophy, the membranes seemingly rupture, but for a while at least, traces of them persist. The nucleus is shrunken and pycnotic. The cell is obviously in a degenerate condition. Briefly speaking, the decrease in electron density indicates that the hypertrophy is a matter of cytoplasmic hydration.

Calcification starts at the periphery of the cartilaginous longitudinal bars and spreads to its interior. It begins with the appearance of small islands composed of needle-shaped crystals 50 A. wide and 400–500 A. long. The crystals keep their size but, owing to addition of new crystals, the islands coalesce progressively. Eventually the crystals seem uniformly dispersed. The main interest of the process is that the arrangement of the crystals has no apparent relationship to the general direction of the fibrils. To say the least, the relationship which is observed in bone between crystals and fibrils is not demonstrable in calcified cartilage. Consequently we should not take for granted, as most authors do, that the mechanism of calcification are absolutely identical in the two tissues.

Proceeding in their analysis, Scott and Pease (1956) have studied the formation and the destruction of endochondral bone. Let us seize the opportunity here of reproducing their electron micrographs of an osteoblast and of an osteoclast.

Figure 20 (Plate V) represents a part of an osteoblast. The nucleus, with a double membrane, is eccentrically located and most often presents a single prominent nucleolus. Typical mitochondria and a laminated endoplasmic reticulum are observed. Between the cell border and the calcified matrix (appearing in black), there is an interspace of variable width which contains fibrils with the typical 640-A. periodicity of mature collagen. This zone is not yet calcifying and is referred to by Scott and Pease (1956) as the "preosseous zone." Calcification begins with the deposition of aggregates of crystals along the fibrils but, at first, without respect to their periodicity. Later these clusters grow in size and number and eventually line up with collagen bands as described in adults by Robinson and Watson and reported previously in this chapter.

The existence of a preosseous stage in the formation of endochondral bone is most interesting. We have amply commented on such a stage occurring in the internal remodeling (Figs. 1 and 2). We shall demonstrate in the next paragraph (Figs. 24–27, Plate VI) that the deposition of a preosseous layer, calcifying with delay, is also observed in late periosteal ossification.

Osteoclasts are very numerous in endochondral bone (Figs. 17 and 18). They are large cells occupying a pit at the surface of the bone (Howship's lacuna) or wrapped around the blunt end of a trabecula. Sometimes they are divided into two or more lobes connected by a thin filament of cytoplasm, an aspect which becomes intelligible if we take into account the motility of the cell, observed in tissue cultures. They contain an average of between ten and thirty nuclei. Since mitoses are not observed, it is supposed that the osteoclast is formed by the fusion of uninucleated cells.

An intriguing feature of the osteoclast is its striated border confined to the zones where contact between cell and bone is the closest. For some authors, this border is part of the cell itself. For Ham (1953), it consists of exposed bone fibrillae left behind by the isolated departure of the intercellular material. Hancox (1956) is inclined to admit that both explanations have a real basis, and his view is in agreement with the electron microscope image which will be described now.

Figure 21 represents a small part of an osteoclast together with the adjacent resorbing bone. Nuclei do not appear on the picture. The cyto-

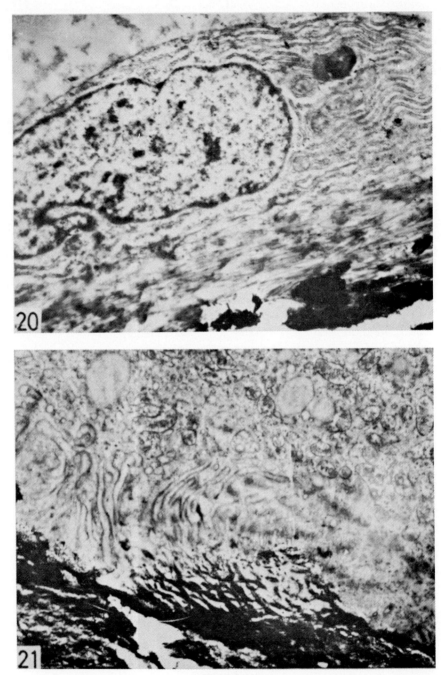

PLATE V

plasm contains vacuoles and numerous mitochondria. There is no conspicuous endoplasmic reticulum. At the surface, the plasma membrane presents an area of intricate infolding. Invaginations of the plasma membrane plunge deeply into the cytoplasm and are associated with vacuoles. Debris of resorption are sometimes seen between the folds of the membrane. The bone being resorbed presents a fringed appearance due to frayed remnants of collagenous fibers.

Such a picture demonstrates that the osteoclast is the agent of bone destruction and suggests that its folded plasma membrane is the specific cellular structure involved in the process. The main problem is now to understand how these cells appear and, above all, how their distribution corresponds so accurately to the pattern of skeletal growth. It seems (Benoit and Clavert, 1947) that the stimulus which determines the formation of the osteoclasts and which attracts them emanates from the fully calcified bone tissue itself.

E. Transplantation Experiments

The last aspect of endochondral ossification which remains to be dealt with is that information provided by transplantation experiments.

We have shown (Lacroix, 1943a) that a growth cartilage grafted in an extraskeletal site proliferates and differentiates almost normally for a while. Here is a typical experiment. The distal growth cartilage of the radius of a newborn rabbit was removed. Its peripheral part was cut off so as to obtain a small cube of pure cartilage completely deprived of any other tissue. It was then grafted in the anterior chamber of the eye of an adult rabbit. Figure 19 illustrates the result 15 days later. The transplant is attached to the iris from which it receives its vessels. On its metaphyseal aspect, it produces endochondral bone trabeculae and slowly pushes then away.

The experiment proves that the growth cartilage is indeed what it looks like, a machinery devised to develop a thrust (see also Ring, 1955).

PLATE V

Fig. 20. Electron micrograph of an osteoblast observed in endochondral bone. The nucleus is eccentrically located. The cytoplasm contains an extensive laminated system. Between the cell border and the calcified matrix in black there is a fibrillar interspace which is not yet calcifying. From Scott and Pease (1956); courtesy of the authors and of *The Anatomical Record*. Magnification: ×16,700.

Fig. 21. Electron micrograph of an osteoclast in contact with resorbing bone. Nuclei do not appear on the picture. At the surface of the cell the plasma membrane is intricately infolded. The bone being resorbed has a fringed appearance due to frayed remnants of collagenous fibers. From Scott and Pease (1956); courtesy of the authors and of *The Anatomical Record*. Magnification: ×20,000.

The statement must be expressed, evident as it is here, because it may be questioned in the lower vertebrates (Policard, 1941).

The same experiment proves also that the mechanisms of endochondral ossification are fundamentally independent from the surrounding tissues and that the growth cartilage is endowed with far-reaching self-determination. Once they have started (as in the pair of cells pointed out by the arrow of Fig. 16), the mechanisms will go on automatically, provided humoral conditions are normal.

In connection with this remark another experiment should be cited. We described in 1943 (Lacroix, 1943b) how a piece of hyaline rib cartilage, transplanted for some time in contact with a growth cartilage acquires the structure of the latter. We suggested then that the fact should be considered as an induction phenomenon.

The idea found an additional support when we observed in 1945 (Lacroix, 1945b) that, in experiments of the kind illustrated by Fig. 19, an ossification groove sometimes regenerates with its typical perichondrial bony ring. The identity of the newly produced ring with the normal structure indicates that induction processes are at play in endochondral ossification.

Further experiments done while trying to isolate a hypothetical organizer, for which the name "osteogenin" was perhaps coined prematurely, were not in agreement with each other (for details, see Lacroix (1951). The situation in 1955 was summed up by McLean and Urist when they said that many of the observed phenomena in the development and growth of bone may be attributed to induction, but that this mechanism is probably not simple enough to be accounted for by diffusion of a specific chemical compound.

In the present state of affairs, we feel that a more simple line of research should be followed. We have reported (Lacroix, 1956) that a piece of growth cartilage identical with that grafted alive (Fig. 19) remains osteogenetic even if it is soaked in alcohol before being grafted. In control experiments, pieces of rib cartilage soaked in alcohol produce merely a few lamellae of bone, in only a minority of cases. These facts preclude, in our opinion, the possibility of considering the irritating action of alcohol as the osteogenetic factor (Heinen *et al.*, 1949). They hint that the contact of young connective cells with the ground substance of the inducing cartilage might transform these cells into osteoblasts (Weiss, 1950). Our pilot experiments are, as a matter of fact, the duplication with growth cartilage of previous experiments (Lacroix, 1953c) performed with callus cartilage (see Figs. 31–33, Plate VII, and their legends). It appears likely, from preliminary notes, that the researches now being conducted by Bridges and Pritchard (1956, 1957) will confirm and extend our findings. It also appears likely that interpretation will be difficult because it has been found that alcohol-fixed muscle tissue is endowed with chondrogenetic properties and, above all, because some results obtained in the rabbit are not reproducible in other species.

V. PERIOSTEAL OSSIFICATION

Growth in width of the long bones is the result of balanced processes involving increase of whole diameters by deposition of new layers of bone on the outer surface of the diaphysis and increase of diameters of the medullary cavity by osteoclastic resorption of the innermost layers of the diaphyseal tube.

Deposition of new layers, i.e., periosteal ossification in the present case, should be considered from its early to its late histological expression.

The *early stages* are illustrated by Fig. 22 (Plate VI), a transverse section at low magnification of the tibia of a newborn puppy.

New trabeculae are being produced under the periosteum. The arrangement of the osteoblasts is that of a one-cell-thick pseudoepithelium carpeting the trabeculae. It is obvious that the trabeculae very soon reach their full thickness, that the size of the osteoblasts diminishes when osteogenesis slackens, and that the osteoblasts are reduced to a few flattened cells when osteogenesis has ceased. The deep intertrabecular spaces contain only a very loose connective tissue with few cells and some capillaries. Osteoclasts are found here and there on the trabeculae bordering the medullary cavity.

The growing trabecula indicated by an arrow on Fig. 22, is reproduced at a higher magnification in Fig. 23. It is a typical example of primary bone with its bundles of fibers woven without any recognizable pattern. It is dark in polarized light with crossed nicols. The osteoblasts originate in the cells proliferating in the deepest layer of the periosteum. They enlarge considerably and come to be aligned on the surface of the trabecula. One of them is being engulfed and will soon be an osteocyte.

Osteoblasts are pear-shaped with their pointed extremity in contact with the trabecula being deposited and the blunt end containing the nucleus. They send off fine cytoplasmic processes anastomosing with those of their neighbors, an observation which is explained by the motility they exhibit in cultures. The nucleus contains three or four nucleoli. Mitotic figures which were observed among periosteal cells are no longer seen at this stage. Adjoining the nucleus, there is a large vacuole which seems to be the negative image of the Golgi apparatus. Mitochondria are extremely numerous and are distributed throughout the cytoplasm, except

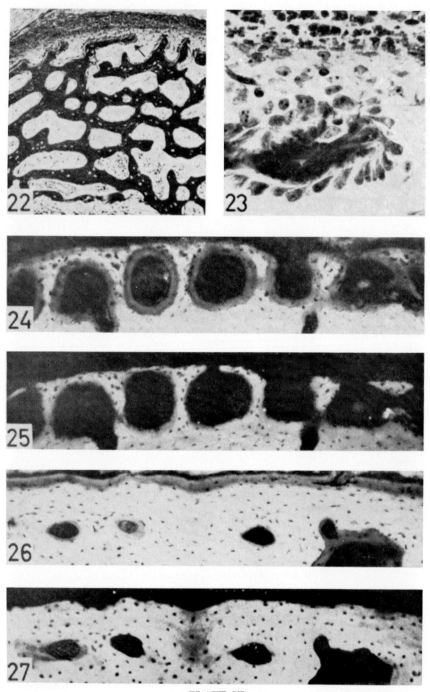

PLATE VI

for the vacuolar region. The cytoplasm is strongly basophilic and contains large quantities of ribonucleic acid (Lison, 1948; Cappellin, 1949; Clavert, 1950; Follis, 1951; Pritchard, 1956). It is now generally considered that the morphological and cytochemical characteristics of the osteoblasts indicate that they participate in the formation of at least some of the material of the trabecula.

Electron microscopy data have been furnished by Fitton Jackson and Randall (1956), who have examined periosteal ossification in the chick embryo. They state that fibrillogenesis occurs in intimate association with the cells. They produce an excellent picture showing a definite relationship between the deposition of small mineral crystals and the periodic banding of the fibrils.

At ages varying from one case to the other, the histological processes resulting in the transverse growth begin to differ from those of the newborn animal.

These *late stages* have been studied in our laboratory by Lea and Ponlot (1958). As shown by Figs. 14 and 15, growth in width may be fast or slow according to the aspect of the shaft which is examined. To the fast and to the slow processes correspond different histological aspects which may often be found opposite to each other in the same cross section. They will be exemplified here with observations provided by the radius of a 5-month-old dog.

The fast process is illustrated by Fig. 24, an undecalcified ground

---

PLATE VI

FIG. 22.   Transverse section of the radius of a newborn puppy. New trabeculae are being produced under the periosteum. The growing trabecula indicated by an arrow provides the following illustration. Magnification: × 84.

FIG. 23.   Region indicated by an arrow in preceding illustration. Osteoblasts originate in the deepest layer of the periosteum and are clustered around the growing trabecula. Magnification: × 630.

FIG. 24.   Transverse ground section of the radius of a 5-month-old dog. PAS method. From right to left are observed the successive stages of the formation of osteons being added at the periphery of the diaphysis. Bone tissue is first laid down in the form of a PAS-positive layer. Magnification: × 133.

FIG. 25.   Microradiograph corresponding to the preceding illustration. The PAS-positive layer appearing on Fig. 24 is not seen on the X-ray plate. Magnification: × 133.

FIG. 26.   Same material and same technique as in Fig. 24. Here no ridges are formed. The thick PAS-positive layer is even and uniform at the surface of the diaphysis. Magnification: × 133.

FIG. 27.   Microradiograph corresponding to the preceding illustration. The PAS-positive layer coating the surface of the diaphysis does not absorb the soft X-rays. Magnification: × 133.

section treated by the PAS method. The sequence of events may be followed from right to left. Parallel ridges begin to appear and become more and more salient, delimiting grooves between them. Each of the ridges assumes the shape of a T whose horizontal bar merges later with those of adjoining ridges. In this way the grooves are transformed into tunnels which are subsequently filled by concentric deposition. The final result is a row of osteons, each giving a typical concentric image when viewed in polarized light. The most outstanding point, which had been overlooked so far, is that osteogenesis begins with the formation of a thick PAS-positive preosseous layer which calcifies suddenly, this being proved by comparison of the section (Fig. 24) with the corresponding microradiograph (Fig. 25).

The slow process is best observed if we superimpose Fig. 26 (obtained like Fig. 24, but from another region of the periphery of the same diaphysis) and Fig. 27, the corresponding microradiograph. Here no ridges have formed. The thick PAS-positive preosseous layer is even and uniform at the surface of the diaphysis. These pictures give us the opportunity of comparing the periosteal preosseous layer with that of a substitution osteon. In the right part of the illustrations, there is a resorption cavity in which deposition has just begun. No difference is noticed between both preosseous layers.

The phenomena illustrated by Figs. 24–27 are extremely important, quantitatively, in the building up of the adult skeletal piece. A large part of the lamellar bone tissue that they leave in their wake will persist in adult life, whereas, nothing will remain of the primary bone elaborated by the initial stages shown in Figs. 22 and 23.

We realize now that there are two types of osteons: those represented by Figs. 1 and 2, which are substituted for pre-existing bone, and those, shown by Figs. 24 and 25, which are added to pre-existing bone.

Before leaving the subject of periosteal ossification, we should like to remind the reader that our presentation follows a particular order aiming at concentrating the most significant data while maintaining unity in the chapter. Periosteal ossification is, however, only one case of intramembranous ossification. Other cases have their own interest. For instance, Bevelander and Johnson (1950) have shown that in the intramembranous ossification of the pig mandibula, phosphatase appears in the spicule before calcification (as in the growth cartilage) and that calcification goes apace with a change in the composition or state of mucopolysaccharides (as, apparently, in the deposition of an osteon and in the late stages of periosteal ossification). Sutural growth also provides excellent opportunities for a cytological study (Pritchard *et al.*, 1956).

VI. Experimental Production of Bone and Cartilage

In the preceding paragraphs, bone and cartilage formation has been studied *in situ*. The time has come now to deal with the same phenomenon as determined in experimental conditions.

A. *Culture of Skeletal Tissues*

A comprehensive review on the subject has been written recently by Fell (1956).

Hanging-drop cultures of small blocks of undifferentiated mesoderm from the limb buds of 3-day chick embryos produce, in about 10 days, nodules of hyaline cartilage completely enveloped by perichondrium.

If entire skeletal primordia are cultivated, they not only differentiate histologically, but also develop their anatomical shape, except for small details. It follows that the general configuration of a bone is basically determined by factors present in the early rudiment and that other factors, such as active supporting function (see Evans, 1957), normal surrounding musculature, vessels, nerves, important as they are later, are not absolutely essential in the first stages of bone development.

For us here, the most significant facts supplied by culture experiments are those concerned with isolated skeletal tissues and with the behavior of their cells.

Periosteum from the chick on the point of hatching gives excellent ossification, provided the delicate osteoblastic layer underlying the thick fibrous coat has not been damaged.

Osteoblasts from the bone tissue itself also ossify *in vitro*. If they are first cultivated, then grafted into the anterior chamber of the eye (Maximow and Bloom, 1957), they produce an ossicle similar to that obtained in grafts of bone marrow or of callus cartilage as described in the next pages.

For the time being, the interest of skeletal tissue cultures lies in the brilliant technical possibilities of investigating the local action on them of various agents (Gaillard, 1953; Fell, 1956; Chèvremont, 1956; Liébecq-Hutter, 1956). But this type of work does not explain why an undifferentiated cell becomes an osteoblast or a chondroblast; in other words, except for unconfirmed reports, bone or cartilage has not been conclusively shown to be produced in cultures of nonskeletal tissue. The grafts of skeletal tissues inside the living organism will now permit us to proceed a little further in the analysis of the factors which initiate osteogenesis and chondrogenesis.

B. Grafts of Skeletal Tissues

The subject is immense indeed and treating all its features would require a monograph. For more data and a detailed bibliography, the reader is referred to our previous publications (Lacroix, 1951, 1953b, c).

1. Grafts of Periosteum

In a 3-month-old rabbit, the periosteum of the medial surface of the tibia is removed without any particle of bone tissue and grafted under the kidney capsule of the same animal. From the fourth day, trabeculae of primary bone are produced (Fig. 28, Plate VII). The osteoblasts come from the periosteum, and, provided early stages are studied, there can be no doubt of their origin. The periosteum contains spindle-shaped connective cells, actively proliferating and showing all stages of transformation in osteoblasts lining the bone trabeculae.

In adult rabbits, bone is also produced. Negative results are explainable by the fact that transplantation sites where the periosteum is wrinkled are unsuitable. The new bone appears about the twelfth day; the final stage is a compact plate of even thickness; its fine structure is intermediate between that of primary and that of lamellar bone. In this case, the properties of the resting osteoblasts have been reawakened.

In very young rabbits, chondrogenesis is sometimes observed together with osteogenesis (Cohen and Lacroix, 1955). Here the properties of the osteoblasts are qualitatively modified from the normal.

PLATE VII

Fig. 28.   Autograft of tibial periosteum under the kidney capsule in a 3-month-old rabbit. Result 4 days later. New bone being laid down by the graft. Magnification: ×80.

Fig. 29.   Autograft of bone marrow under the kidney capsule in a young rabbit. Result 6 days later. An incomplete shell of new bone has been produced by the graft. Magnification: ×30.

Fig. 30.   Detail of the upper part of preceding illustration. Osteoblasts are recruited in a population where grafted cells are mixed with cells streaming from the thickened renal capsule. Magnification: ×190.

Fig. 31.   In a rabbit, the cartilaginous tissue of the callus of a broken rib has been grafted in the knee. Result 21 days later. Buried in the synovial membrane, the graft has produced an ossicle the section of which resembles that of a very young diaphysis. Magnification: ×25.

Fig. 32.   Same experiment as that illustrated by Fig. 31 except that the graft had been put under the kidney capsule. Result 106 days later. Lamellar bone has been substituted for the primary bone of early stages. Magnification: ×80.

Fig. 33.   Same experiment as that illustrated by Fig. 32 except that the cartilage has been killed by alcohol before being put under the kidney capsule. Result 118 days later. Osteogenesis has been induced by the dead tissue. Magnification: ×70.

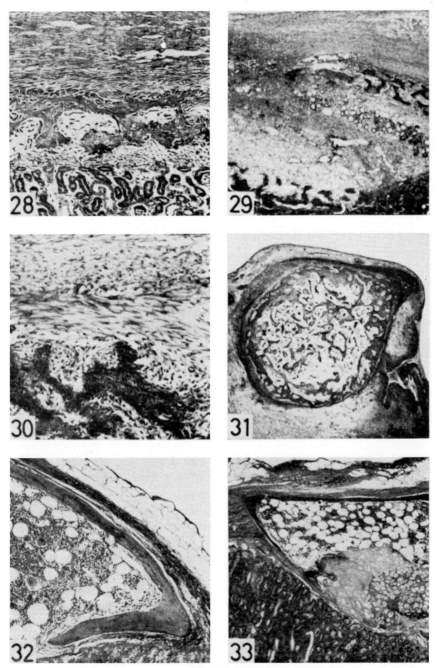

PLATE VII

2. *Grafts of Bone Marrow*

Bone marrow taken out from the radius was grafted under the kidney capsule in a young rabbit. The appearance, after 6 days, is represented by Fig. 29. A bud of granulation tissue was produced by the renal capsule and had invested the marrow which, except for a few doubtful patches, is healthy and abundantly irrigated by large capillaries coming from the capsule. At the periphery of the graft, an incomplete shell of trabecular primary bone, whose wall is of regular thickness, has formed.

Where do the osteoblasts come from? The lower part of the illustration hardly needs a higher magnification to prove that bone has been produced by the grafted marrow. In the upper part, the situation is not the same. Closer examination (Fig. 30) reveals that connective cells produced by the thickened capsule are streaming in between the bone trabeculae. Here the osteoblasts are recruited from a mixed population of grafted cells and cells from the transplantation bed.

The observation leads us to believe that the grafted osteoblasts are able to communicate their bone-forming properties to the nonskeletal connective cells proliferating in the vicinity.

Not all the sites, however, would justify this opinion. In the anterior chamber of the eye, the bone is elaborated only by the grafted marrow without any detectable participation of the surrounding tissues (Danis, 1957). Grafts in muscle (De Bruyn and Kabisch, 1955) do not seem to authorize a clear-cut statement on the issue.

3. *Grafts of the Cartilage of a Fracture Callus*

The consistency of the results obtained with growth cartilage (an example of which is presented by Fig. 19) has led us to investigate a related tissue: the cartilage of a fracture callus.

A rib is broken in an adult rabbit. Eleven days later, the fracture is being repaired by a callus which, at this stage, is made essentially of undifferentiated cartilage produced by the periosteum. It is easy to cut out a block of this cartilage and to graft it under the kidney capsule of the same animal.

The graft elicits a lively ossification as early as the fourth day. On the ninth day, it has produced a mass of endochondral bone. This bone in turn is destroyed and is replaced on or about the twenty-second day by a piece of primary bone, the center of which is hollowed out.

Similar results had been recorded in the anterior chamber of the eye (Urist and McLean, 1952). They have been obtained also in the knee (E. Van Maele and A. Vincent, unpublished), where the graft is buried in the synovial lining and where the ossicle which is found after 3 weeks gives a histological section strikingly identical with that of a young diaphysis (Fig. 31).

Later, the primary trabeculae are slowly replaced by typical lamellar bone similar to that of an adult diaphysis, forming a lacunar shell containing hematopoietic marrow (Fig. 32).

From this short description, it appears that the potentialities of the tissues involved in fracture healing are reminiscent of those of the growing skeletal tissues. In grafts, the cartilage from the fracture callus presents a typical succession of stages which is strongly evocative of the succession observed in bone growth. Repair processes in bone seem to imply a repetition of growth processes.

Yet another interest of the callus cartilage lies in the following observation. When killed by soaking in alcohol for several days, this tissue still is found to be as constantly osteogenic as growth cartilage. New bone appears in the graft being destroyed by connective cells, first primary bone, then lamellar bone. We have noticed aspects (Fig. 33) which deserve to be compared with those produced by living cartilage (Fig. 32) and which offer good prospects for investigating the induction processes.

4. Remarks

The experimental possibilities of grafting skeletal tissues are endless. The grafts may be pretreated by various agents. Associations of tissues may be considered. A variety of sites should be employed. Autografts should be compared systematically with homografts.

This means that, in spite of a bibliography which amounts nowadays to several hundreds of original papers, it is still too early to obtain a coherent view of the biology of skeletal grafts and that any consideration on them should not be accepted incautiously.

With this reservation in mind, the facts that we have presented suggest: (1) that the potentialities of skeletal cells may be quantitatively and qualitatively modified in the grafts; (2) that extraskeletal connective cells may be transformed into osteoblasts under the influence of living skeletal cells; and (3) that this transformation may occur also by the agency of a nonliving factor linked to the constitution of skeletal tissues.

C. Ossification Determined by Nonskeletal Tissues

In cultures or in grafts of skeletal tissues, the production of bone and cartilage is brought about by tissues derived from the skeleton. Another fascinating viewpoint is now provided by experiments where osteogenesis is determined by tissues which, at least apparently, have nothing to do with the skeleton.

The most interesting of these tissues is the urinary bladder epithelium which, grafted in close association with certain fasciae, elicits a constant bone formation. The grafts transform into cysts, and osteogenesis occurs in the connective tissue lying in the immediate vicinity of the proliferating epithelial cells. Some sites respond readily: the sheath of the rectus abdominis muscle, the fascia lata, the subcutaneous tissue or the synovial membrane of the knee. Other sites such as the liver or the spleen do not react at all.

The classical article of Huggins (1931) contains the essentials of the question. It has since been confirmed and extended in a number of papers, one of the latest being that of Johnson and McMinn (1956). None of them though, excellent as some may be, has given any definite explanation of how the urinary epithelium can induce new bone formation. It does seem, however, that there must be some common denominator between this phenomenon and that determined by a grafted skeletal tissue.

VII. Concluding Remarks

Within the space allotted, we have presented a survey of present-day knowledge of the life of bone and cartilage at the cytological level. To conclude our chapter, we should like to show briefly how basic work on skeletal tissues is useful in fostering researches of a more applied character.

A. Problems of Bone Physiology

It has long been recognized that the organism draws from the skeleton the calcium it needs to maintain the calcemia at a constant level. But the physiologists seldom discuss the possibility of locating histologically the available calcium.

When, in 1953 (Lacroix, 1953a), we showed how Ca^{45} is distributed *in vivo* in compact bone, we pointed out that, after all, it was an over-simplification to consider the hot spots as the mere expression of increased mineralization, and that, strictly speaking, they were places where an intense calcium metabolism results in an increased mineralization. We suggested "that the incompletely calcified bone is mainly a metabolic tissue which plays the major part in the calcium exchanges of the skeleton with the body fluids," that it is, in other words, a sort of calcium bank at the disposal of the organism for its everyday deposits and withdrawals. To the histological heterogeneity of bone would correspond a physiological heterogeneity, and the internal remodeling would meet the necessity of maintaining a metabolic skeleton dispersed within the supporting skeleton.

The idea is still for us a hypothesis, but it is now considered as one of the "emerging concepts of the metabolic functions of bone" by Neuman and Neuman (1957). According to these authors, each of the newly deposited crystals is surrounded by a hydration shell which is in equilibrium with the extracellular fluid; isoionic and heteroionic exchanges occur between the crystals and the hydration shells; and, furthermore, certain ions are displaced within the crystals by recrystallization. As the osteon matures, the crystals become larger, less hydrated, and more perfect internally. The rate and extent of diffusion, exchange, and recrystallization, decline. At full mineralization, the osteon has become unavailable to immediate demands.

Whatever may be the fate of our physiological interpretation of the internal remodeling, we should not overlook the fact that this process stores and liberates calcium in a more obvious way since it builds new bone tissue and destroys old bone tissue (Amprino, 1955b).

The trend today (McLean and Urist, 1955; Copp, 1957; Talmage and Elliott, 1957) is to admit that both interpretations should be combined and that calcium homeostasis is controlled by a dual mechanism.

The skeleton is also the reserve for other electrolytes. Sodium is the last, but probably not the least, to come into the limelight (Nichols and Nichols, 1956).

The young skeleton is not very well suited for such studies because the histological expression of its metabolic function is intermingled with that of its growth. This is the reason why, in this chapter, the adult state has been considered so clearly apart from the growing state.

B. Problems of Surgical Pathology

We have indicated how researches now in progress on bone growth and on fracture healing meet on a common ground.

Understanding of other aspects of the response of bone to injury or disease also derive direct benefit from fundamental research. For instance, correlations between Ca^{45} autoradiography, microradiography, and histology prove that, in a broken bone, the remodeling is more widespread than normal (Lacroix and Ponlot, 1957). This fact seems to account for posttraumatic osteoporosis, since it means that more cavities are being bored per volume unit of bone.

C. Problems of Radiation Hazards

Although the organism discriminates between radiostrontium and radiocalcium (Comar and Wasserman, 1958), the evaluation of the dangers of radiostrontium, the most ominous of the fission products, can profit from a better knowledge of the fate of radiocalcium in the skeletal tissues.

On the basis of our illustrations, let us sketch what may be foreseen of the behavior of radiostrontium in function of age.

At an age corresponding approximately to that of the subject of Fig. 14, one month is enough to renew histologically most of the tibia labeled at the time of administration. The radioactivity of the destroyed bone is redistributed in the new bone with some progressive loss. The new bone will in turn be destroyed and its radioactivity again redistributed with concomitant loss. In the growing skeleton, irradiation fades as a whole and becomes more and more diffuse. Moreover, Hevesy (1955) has shown that from birth to adult stage a mouse loses half of its original calcium atoms. Optimism, however, is not justified. Our autoradiographs reveal that the process entails a peculiarity which, trivial in all other respects, is a decisive factor as far as radiation hazards are concerned: a small part of the posterior wall of the tibia (level of arrows in Fig. 15) is less provisional than the rest of the edifice; it escapes destruction, at least for a while, and it is therefore subjected to prolonged and heavy irradiation. Every bone of a growing skeleton has its own particular region endowed with the same relative stability.

We may expect from these considerations that for a given dose of radiostrontium per unit weight, the danger will be highest at a critical period of growth. Before that period, the growth is so fast that the skeleton is renewed entirely from stage to stage. After that period, the initial retention decreases appreciably.

It is indeed most noteworthy that Owen and associates (1957), studying the effect of a single injection of a high dose of Sr^{90} in rabbits, have just stated that one group of animals, aged between 6 and 8 weeks, is far more susceptible to radiation damage than either younger or older animals.

This example has been chosen because it happens to correspond with our illustrations. Other examples might be presented to emphasize that researches on bone, seemingly of purely academic interest, have in fact a direct bearing on the practical problems of atomic age.

Less than ten years ago, the laboratories specially equipped for the study of bone were rather few. Textbooks would desperately convey the feeling that everything had been said and done by the old German masters. The wheel has turned and the "bone field" is now one of the most active areas in biological research. The reasons for this extraordinary revival of interest are obvious. Of paramount importance are the new tools provided and improved every day by the physicists. But there is also the firm belief of many that the subject conceals more implications than those we have just mentioned. We fear and we hope that this essay will already be somewhat obsolete by the time it has taken its printed shape.

Acknowledgment

The author and his co-workers wish to acknowledge with thanks the help they have received from the Institut Interuniversitaire des Sciences Nucléaires of Belgium.

References

Amprino, R. (1950). *Compt. rend. assoc. anat.* **37**, 11.
Amprino, R. (1952). *Z. Zellforsch. u. mikroskop. Anat.* **37**, 144.
Amprino, R. (1955a). *Acta Anat.* **24**, 121.
Amprino, R. (1955b). *Atti soc. ital. patol.* **4**, 9.
Amprino, R. (1956). *In* "Bone Structure and Metabolism" (G. E. W. Wolstenholme, ed.), p. 89. Churchill, London.
Amprino, R., and Engström, A. (1952). *Acta Anat.* **15**, 1.
Arnold, J. S., and Jee, W. S. S. (1954). *Anat. Record* **118**, 373.
Arnold, J. S., Jee, W. S. S., and Johnson, K. (1956). *Am. J. Anat.* **99**, 291.
Ascenzi, A. (1955). *Sci. Med. Ital.* **3**, 696.
Bélanger, L. F. (1954). *Can. J. Biochem. and Physiol.* **32**, 161.
Bélanger, L. F. (1956). *In* "Bone Structure and Metabolism" (G. E. W. Wolstenholme, ed.), p. 75. Churchill, London.
Benoit, J., and Clavert, J. (1947). *Compt. rend. soc. biol.* **141**, 911.
Bevelander, G., and Johnson, P. L. (1950). *Anat. Record* **108**, 1.
Boström, H. (1953). *Arkiv Kemi* **6**, 43.
Boström, H., and Jorpes, E. (1954). *Experientia* **10**, 392.
Bourne, G. H. (1956). *In* "The Biochemistry and Physiology of Bone" (G. H. Bourne, ed.), Chapt. 9. Academic Press, New York.
Bridges, J. B., and Pritchard, J. J. (1956). *J. Anat.* **90**, 563, 593.
Bridges, J. B., and Pritchard, J. J. (1957). *J. Anat.* **91**, 589.
Cappellin, M. (1949). *Chir. org. movimento* **33**, 410.
Cartier, P., and Picard, J. (1955). *Bull. soc. chim. biol.* **37**, 485, 661, 1159, 1169.
Chèvremont, M. (1956). "Notions de Cytologie et Histologie." Desoer, Liège.
Clavert, J. (1950). *Compt. rend. Acad. Sci. (Paris)* **231**, 998.
Cohen, J., and Lacroix, P. (1955). *J. Bone and Joint Surg.* **37A**, 717.
Cohen, J., Maletskos, C. J., Marshall, J. H., and Williams, J. B. (1957). *J. Bone and Joint Surg.* **39A**, 561.
Comar, C. L., and Wasserman, R. H. (1958). *In* "International Conference on Radioisotopes in Scientific Research, 1957." Pergamon Press, New York.
Comar, C. L., Lotz, W. E., and Boyd, G. A. (1952). *Am. J. Anat.* **90**, 113.
Copp, D. H. (1957). *Am. J. Med.* **22**, 275.
Dallemagne, M. J., and Fabry, C. (1956). *In* "Bone Structure and Metabolism" (G. E. W. Wolstenholme, ed.), p. 14. Churchill, London.
Dallemagne, M. J., and Fabry, C. (1957). *Acta Chir. Belg.* **56**, Suppl. 1, 75.
Danis, A. (1957). *Acta Chir. Belg.* **56**, Suppl. 3.
Davies, D. V., and Young, L. (1954). *J. Anat.* **88**, 174.
De Bruyn, P. P. H., and Kabisch, W. T. (1955). *Am. J. Anat.* **96**, 375.
Dziewiatkowski, D. D. (1951a). *J. Biol. Chem.* **189**, 187.
Dziewiatkowski, D. D. (1951b). *J. Exptl. Med.* **93**, 451.
Dziewiatkowski, D. D. (1952). *J. Exptl. Med.* **95**, 489.
Dziewiatkowski, D. D., Benesch, R. E., and Benesch, R. B. (1949). *J. Biol. Chem.* **178**, 931.

Dziewiatkowski, D. D., Di Ferrante, N., Bronner, F., and Okinaka, G. (1957). *J. Exptl. Med.* **106**, 509.

Eastoe, J. E., and Eastoe, B. (1954). *Biochem. J.* **57**, 453.

Engfeldt, B., and Hjertquist, S. O. (1955). *Acta Pathol. Microbiol. Scand.* **36**, 385.

Engfeldt, B., Engström, A., and Zetterström, R. (1952). *Biochim. et Biophys. Acta* **8**, 375.

Engström, A. (1956). *In* "Bone Structure and Metabolism" (G. E. W. Wolstenholme, ed.), p. 3. Churchill, London.

Engström, A., and Engfeldt, B. (1953). *Experientia* **9**, 19.

Evans, F. G. (1957). "Stress and Strain in Bones." C. C Thomas, Springfield, Illinois.

Fell, H. B. (1956). *In* "The Biochemistry and Physiology of Bone" (G. H. Bourne, ed.), Chapt. 14. Academic Press, New York.

Fernández-Morán, H., and Engström, A. (1956). *Nature* **178**, 494.

Fernández-Morán, H., and Engström, A. (1957). *Biochim. et Biophys. Acta* **23**, 260.

Fitton Jackson, S., and Randall, J. T. (1956). *In* "Bone Structure and Metabolism" (G. E. W. Wolstenholme, ed.), p. 47. Churchill, London.

Follis, R. H., Jr. (1949a). *Proc. Soc. Exptl. Biol. Med.* **71**, 441.

Follis, R. H., Jr. (1949b). *Bull. Johns Hopkins Hosp.* **85**, 360.

Follis, R. H., Jr. (1951). *Bull. Johns Hopkins Hosp.* **89**, 9.

Follis, R. H., Jr., and Berthrong, M. (1949). *Bull. Johns Hopkins Hosp.* **85**, 281.

Frank, R., Frank, P., Klein, M., and Fontaine, R. (1955). *Arch. anat. microscop. Morphol. exptl.* **44**, 191.

Gaillard, P. J. (1953). *Intern. Rev. Cytol.* **2**, 331.

Gardner, E. (1956). *In* "The Biochemistry and Physiology of Bone" (G. H. Bourne, ed.), Chapt. 13. Academic Press, New York.

Glegg, R. E., and Eidinger, D. (1955). *Arch. Biochem. Biophys.* **55**, 19.

Godart, H. (1952). *Presse méd.* **60**, 414.

Greulich, R. C. (1956). *J. Bone and Joint Surg.* **38A**, 611.

Gutman, A. B., and Yu, T. F. (1951). *Conf. on Metabolic Interrelations Trans. 3rd Conf.* p. 90.

Ham, A. W. (1953). "Histology," 2nd ed. Lippincott, Philadelphia, Pennsylvania.

Hancox, N. (1956). *In* "The Biochemistry and Physiology of Bone" (G. H. Bourne, ed.), Chapt. 8. Academic Press, New York.

Heinen, J. H., Jr., Dabbs, G. H., and Mason, H. A. (1949). *J. Bone and Joint Surg.* **31A**, 765.

Hevesy, G. (1955). *Kgl. Danske Videnskab. Selskab Biol. Medd.* **22** (9).

Huggins, C. B. (1931). *A.M.A. Arch. Surg.* **22**, 377.

Johnson, F. R., and McMinn, R. M. H. (1956). *J. Anat.* **90**, 106.

Kent, P. W., Jowsey, J., Steddon, L. M., Oliver, R., and Vaughan, J. (1956). *Biochem. J.* **62**, 470.

Krompecher, S. (1937). "Die Knochenbildung." G. Fisher, Jena.

Lacroix, P. (1943a). *Mém. acad. roy. méd. Belg.* [2] **2**, (2).

Lacroix, P. (1943b). *Acta Biol. Belg.* **3**, 93.

Lacroix, P. (1945a). *Arch. biol. (Liége)* **56**, 185.

Lacroix, P. (1945b). *Anat. Record* **92**, 433.

Lacroix, P. (1951). "The Organization of Bones." Churchill, London and Blakiston Div.–McGraw-Hill, New York.

Lacroix, P. (1953a). *Bull. acad. roy. Méd. Belg.* [6] **18**, 489.

Lacroix, P. (1953b). *Acta Chir. Belg.* **52**, 877.

Lacroix, P. (1953c). *Soc. intern. Chir. 15th Congr. Lisbon*, p. 553.

Lacroix, P. (1954). *In* "Radioisotope Conference" (J. E. Johnston, ed.), Vol. 1, p. 134. Academic Press, New York.

Lacroix, P. (1956). *In* "Bone Structure and Metabolism" (G. E. W. Wolstenholme, ed.), pp. 36, 144. Churchill, London.

Lacroix, P., and Ponlot, R. (1957). *Acta Chir. Belg.* **56**, Suppl. 1, 149.

Lacroix, P., and Ponlot, R. (1958). *In* "International Conference on Radioisotopes in Scientific Research, 1957." Vol. 4, p. 125. Pergamon Press, New York.

Lea, L. M., and Ponlot, R. (1958). *Arch. biol. (Liége)* **69**, 455.

Lea, L., and Vaughan, J. (1957). *Quart. J. Microscop. Sci.* **98**, 369.

Leblond, C. P., Wilkinson, G. W., Bélanger, L. F., and Robichon, J. (1950). *Am. J. Anat* **86**, 289.

Liébecq-Hutter, S. (1956). *J. Embryol. exptl. Morphol.* **4**, 279.

Lipp, W. (1954). *Acta Anat.* **20**, 162.

Lison, L. (1948). *Bull. histol. appl. et tech. microscop.* **25**, 23.

McLean, F. C., and Urist, M. R. (1955). "Bone." Univ. Chicago Press, Chicago, Illinois.

Martin, A. V. W. (1953). *Biochim. et Biophys. Acta* **10**, 42.

Martin, A. V. W. (1954). *J. Embryol. exptl. Morphol.* **2**, 38.

Maximow, A. A., and Bloom, W. (1957). "A Textbook of Histology," 7th ed. Saunders, Philadelphia, Pennsylvania.

Morse, A., and Greep, R. O. (1951). *Anat. Record* **111**, 193.

Neuman, W. F., and Neuman, M. W. (1953). *Chem. Revs.* **53**, 1.

Neuman, W. F., and Neuman, M. W. (1957). *Am. J. Med.* **22**, 123.

Nichols, G., Jr., and Nichols, N. (1956). *Metabolism,* **5**, 438.

Owen, M., Jowsey, J., and Vaughan, J. (1955). *J. Bone and Joint Surg.* **37B**, 324.

Owen, M., Sissons, H. A., and Vaughan, J. (1957). *Brit. J. Cancer* **11**, 229.

Picard, J., and Cartier, P. (1956). *Bull. soc. chim. biol.* **38**, 697, 707.

Pinard, A. (1952). *Acta Anat.* **15**, 188.

Policard, A. (1941). "L'appareil de croissance des os longs." Masson, Paris.

Ponlot, R. (1958). *Arch. biol. (Liége)* **69**, 441.

Pratt, C. W. M. (1957). *J. Anat.* **91**, 591, 603.

Pritchard, J. J. (1956). *In* "The Biochemistry and Physiology of Bone" (G. H. Bourne, ed.), Chapt. 7. Academic Press, New York.

Pritchard, J. J., Scott, J. H., and Girgis, F. G. (1956). *J. Anat.* **90**, 73.

Ring, P. A. (1955). *J. Anat.* **89**, 79, 231, 457.

Robinson, R. A., and Cameron, D. A. (1956). *J. Biophys. Biochem. Cytol.* **2**, Suppl., 253.

Robinson, R. A., and Watson, M. L. (1952). *Anat. Record* **114**, 383.

Robinson, R. A., and Watson, M. L. (1955). *Ann. N. Y. Acad. Sci.* **60**, 596.

Rogers, H. J. (1951). *Biochem. J.* **49**, 12.

Rouiller, C., Huber, L., Kellenberger, E., and Rutishauser, E. (1952). *Acta Anat.* **14**, 9.

Ruth, E. B. (1947). *Am. J. Anat.* **80**, 35.

Rutishauser, E., and Majno, G. (1951). *Bull. Hosp. Joint Diseases* **12**, 468.

Schwartz, W., and Pahlke, G. (1953). *Z. Zellforsch. u. mikroskop. Anat.* **38**, 475.

Scott, B. L., and Pease, D. C. (1956). *Anat. Record* **126**, 465.

Sobel, A. E. (1955). *Ann. N. Y. Acad. Sci.* **60**, 713.

Sobel, A. E., Burger, M., Deane, B. C., Albaum, H. G., and Cost, K. (1957). *Proc. Soc. Exptl. Biol. Med.* **96**, 32.

Talmage, R. V., and Elliott, J. R. (1958). *In* "International Conference on Radioisotopes in Scientific Research 1957." Pergamon Press, New York.

Tomlin, D. H., Henry, K. M., and Kon, S. K. (1953). *Brit. J. Nutrition* **7**, 235.

Trueta, J., and Harrison, M. H. M. (1953). *J. Bone and Joint Surg.* **35B**, 442.

Urist, M. R., and McLean, F. C. (1952). *J. Bone and Joint Surg.* **34A**, 443.

van Wel, J. P. (1954). "Bijdrage tot de kennis van de groei der beenderen." Vims, Utrecht.

Verne, J., Bescol-Liversac, J., Droz, B., and Olivier, L. (1956). *Ann. Histochim.* **1**, 191.
Vincent, J. (1954). *Arch. biol. (Liége)* .65, 531.
Vincent, J. (1955). "Recherches sur la constitution de l'os adulte," Editions Arscia, Bruxelles.
Vincent, J. (1957a). *Rev. belge pathol. et méd. exptl.* **26**, 161.
Vincent, J. (1957b). *Arch. biol. (Liége)* **68**, 561.
Weiss, P. (1950). *Quart. Rev. Biol.* **25**, 177.
Zambotti, V. (1957). *Sci. Med. Ital.* **5**, 611.

ADDENDUM

Reports on the progress made since this chapter was written and the corresponding bibliography may be found in:

Ghosez, J. P. (1959). *Arch. biol. (Liége)* **70**, 169.
Lacroix, P. (1959). *Bull. acad. roy. méd. Belg.* [6] **24**, 638.
Leblond, C. P., and Lacroix, P. (1959). *Compt. rend. Acad. Sci. (Paris)* **249**, 934.
Leblond, C. P., Lacroix, P., Ponlot, R., and Dhem, A. (1959). *Bull. acad. roy. méd. Belg.* [6] **24**, 421.
Ponlot, R. (1960). "Le radiocalcium dans l'étude des os," Editions Arscia, Bruxelles, and Masson, Paris.

See also, when they are published, the Proceedings of the Symposium on "Radioisotopes in the Study of Bone" held in Princeton, N.J., in August 1960, and organized by the Council for International Organizations of Medical Sciences.

CHAPTER 5

Skin and Integument and Pigment Cells

By WILLIAM MONTAGNA

I. Introduction

This chapter deals only with the integument of mammals. The structure and function of the various integumentary organs are varied and a brief account of them must be incomplete. Cutaneous organs can produce serous, mucous, lipoidal, or mixed secretions, as well as keratinous products. There is scarcely any morphological or functional resemblance between the sweat glands and hair follicles, or the hair follicles and sebaceous glands. It would be ill advised, however, to emphasize only the differences, because the most important aspect of the cutaneous system would be overlooked. In spite of the apparent chasms of difference between the various appendages, the cells of each of them share all of the basic similarities.

The undifferentiated cells of any of these appendages are structurally similar, and their differentiative potentials are probably identical. Whether an undifferentiated cell differentiates into hair cortex, medulla, inner root sheath, or into a sebaceous cell, or, for that matter, into any cell of the cutaneous system, seems to be determined not so much by the cell itself but by its location and by influences that must come from the outside. If this were not so, cells from hair follicles could not, under certain conditions, transform readily into sebaceous glands (Montagna, 1956a), or the cells of sweat glands transform into epidermal cells (Lobitz and Dobson, 1957). Each undifferentiated epidermal cell must have the equipment to do all of the things that embryonal epidermal cells can do. Some of these potentials must be stronger than others and can be triggered off more readily. For example, any cutaneous cell, given the proper opportunity, undergoes keratinous or sebaceous transformation. The cells in certain localities carry with them a bias that favors transformation in one direction instead of another, but still possess the basic equipotentiality. The most important function of the integument is the formation of the keratinized mantle that surrounds the animal. This involves the transformation of entire cells into tiny particles of fibrous protein. In spite of a wealth of chemical and histological information, it has remained for the electron microscopist to begin to elucidate with conviction the processes of keratinization (Mercer, 1958).

Skin is composed of epidermis and dermis. In the past, for convenience, the two have been dealt with as if they had been parts of different organs. Yet, the epidermis is totally dependent upon the dermis for nutrients and for the evocation and guidance of the different paths of its differentiation. For this reason, and to give it deserved prominence, the dermis is discussed first in this chapter.

II. The Dermis

A. Introduction

The dermis consists of a thin, superficial papillary layer and a thick, deep reticular layer. The papillary layer bears the negative imprint of the underside of the epidermis and forms a continuous sleeve around each cutaneous appendage. Around hair follicles, it forms the connective tissue sheath, and at its base, the dermal papilla. The papillary layer is composed of widely separated, delicate collagenous fibers, very few elastic fibers, reticular fibers, and numerous capillary meshes and loops. The reticular layer is composed of coarse, branching collagenous fiber bundles, mostly directed parallel to the surface. A few perpendicularly directed fibers come from the tela subcutanea, where they branch loosely and are a part of the framework of the fatty layer. Collagenous fibers outline rhomboidal meshes whose orientation indicates the direction of extensibility of the skin. A loose network of reticular fibers is enmeshed between the collagenous fibers.

There are more cells in the papillary layer than in the reticular layer; fibroblasts are the most numerous; a few mast cells may be found, and histiocytes are numerous only under certain conditions. Pigment-bearing cells, or melanophores, and pigment-producing cells, melanocytes, are sparingly distributed in the dermis, although they may be numerous in nevi and moles.

B. Structure and Function

1. The Intercellular Substance

The semifluid, amorphous ground substance that fills the spaces between the fibers contains hyaluronic acid and chondroitin sulfate B (Meyer and Rapport, 1951; Stearns, 1940). It stains with the periodic acid-Schiff technique, suggesting that it contains glycoproteins (Gersh and Catchpole, 1949) and chondromucin (Lillie et al., 1951). The papillary layer contains relatively more ground substance than the reticular layer. The delicate connective tissue beds around sweat glands, sebaceous glands, hair follicles, and particularly around the bulge of the outer root sheath and the bulb of hair follicles usually stain metachromatically (Montagna *et al.*, 1951b), indicating the presence of chondroitin sulfate. The ground substance in the dermal papilla of active hair follicles stains intensely metachromatically.

The specific physiological significance of the ground substance is not fully known. Since material in wounds that stains metachromatically diminishes as the scars mature, the mucopolysaccharides may be tied up with the formation of collagenous fibers (Bunting, 1950; Meyer, 1945; Sylvén, 1941). Hyaluronic acid in connective tissues occurs in greatly hydrated gels and may be involved in water binding. The ground substance contains no free fluid, and even water is bound to hyaluronic acid. Sulfate mucopolysaccharides and their protein complexes could act as cationic exchange resins (Meyer, 1947).

Collagenous fibers are colorless, branching, wavy bands 1–15 μ in width. Each fiber is a bundle of parallel, unbranching fibrils, embedded in and held together by a cementing substance that forms a membrane around each fiber. Weak acids and alkalis swell the fibers and make them transparent; strong acids and alkalis destroy or macerate them; they are digested by pepsin but are unaffected by trypsin. Tannic acid and the salts of heavy metals combine with collagen and form leather, a tough, insoluble substance. When microincinerated, collagenous fibers leave a considerable amount of blue ash residue, indicative of sodium and potassium.

Seen under the electron microscope, collagenous fibers are composed of fibrils surrounded by various amounts of an amorphous substance, the quantity of which is greater in the skin of the young than in that of adults. This substance probably contains acid polysaccharides and protein, and filaments measuring less than 50 A. (Gross, 1950). The widths of collagen fibrils range from 700 to 1400 A., the majority being around 1000 A. (Gross and Schmitt, 1948); they tend to be finer in infants than in adults. The fibrils are characteristically cross-striated by evenly spaced bands, the distance between which averages 640 A. Between these bands are found smaller cross striations. Bundles of adhering fibrils show that the cross bands are in almost perfect correspondence. The fibrils are composed of finer filaments aligned parallel to the axis of the fibril.

So called because they branch and form a network, reticular fibers are found throughout the connective tissue. A dense bed of them is found in the upper part of the papillary layer underneath the epidermis. The greatest preponderance of reticular fibers is in the papillary layer and in its extensions around the cutaneous appendages. The fibers are of different thicknesses, and at least the larger ones seem to consist of bundles of finer fibrils.

Reticular fibers are argyrophile; they are reactive to the periodic acid-Schiff technique, and they stain poorly or not at all with most connective tissue stains. Paper chromatographic analyses of "reticulin," a substance which reticular fibers yields on boiling, show large concentrations of galactose, glucose, and mannose, and smaller amounts of fucose (Glegg et al., 1953). The four sugars mentioned above are also present in collagen, but in much smaller amounts.

The electron microscope shows that reticular fibrils are one-third or one-half the width of collagenous fibrils; otherwise they are similar to collagenous fibrils (Gross, 1950; Gross and Schmitt, 1948). Since the physical and chemical properties of reticulin and collagen are similar, reticular fibers are probably "precollagenous" elements. Collagenous fibers may develop by the coming together of very small component units formed through a lateral association of "protofibrils" and a subsequent progressive deposition of molecular collagen on their surfaces (Porter and Vanamee, 1949).

Elastic fibers in the dermis are coarse, branching, cylindrical, or flat ribbons scattered between the collagenous fibers. The diameter of the fibers ranges in size from barely resolvable with the light microscope to several microns in width. Elastic fibers are relatively chromophobic but can be stained selectively with orcein or resorcin fuchsin. They have little affinity for basic or acid dyes and stain weakly with acid dyes buffered to pH 2.0. When skin is boiled in $0.1 N$ sodium hydroxide, the residue that is left is a mixture of albuminoids and scleroproteins, elastin, which comes from elastic fibers (Lowry et al., 1941). Under the electron microscope elastic fibers seem not to have a fibrillar structure.

2. The Cells

The fibroblast, or a cell that structurally is indistinguishable from a fibroblast, seems to be the parent cell from which all other cells in connective tissue arise. Fibroblasts are ubiquitous in the dermis; they are larger in the reticular layer and resemble mesenchymal cells. Their large oval nucleus is stippled with very delicate chromatin particles and contains one or more large nucleoli. Under ordinary light the cytoplasm of the living fibroblast is amorphous, but under the phase contrast of dark-field illumination it is seen to contain mitochondria, granules, and vacuoles of varying sizes. Mitochondria, in the form of rodlets and filaments, are concentrated around the nucleus, and only a few of them are seen in the cytoplasmic processes. The cytoplasm of fibroblasts stains a very pale color with basic dyes; that of the fibroblasts in the papillary layer is more basophile. Although fibroblasts must be the cells responsible for the deposition of the fiber elements of the dermis, the specific way in which this is done is not yet known.

Although certain cytological features have been ascribed to histiocytes, they are applicable only to those cells which are in the process of phagocytosis. Normal healthy skin seems not to contain histiocytes, but if particulate matter has been injected into the skin, histiocytes can be seen laden with particles. Two general types, large and small histiocytes, can be recognized in the dermis. The large cells phagocytose large and small particles, but the small cells, only the small particles. When the particles are too large to be engulfed, several histiocytes gather around them and become fused into multinucleated foreign-body giant cells.

A wide variety of cells can be transformed to histiocytes. They arise from fibroblasts, lymphocytes, and Schwann cells. In tissue culture, skeletal muscle fibers and subcutaneous connective tissue are readily transformed into histiocytes. For these reasons, it is unlikely that histiocytes comprise a clear-cut cellular system; they are cells characterized only by a particular functional state that Chèvremont calls "l'état histiocytaire" (Chèvremont, 1948).

Mast cells in tissues can be recognized by the presence in their cytoplasm of granules that stain metachromatically. According to this criterion cells indistinguishable from fibroblasts in shape and size, but containing few such granules, must also be considered to be mast cells. It is not always possible, however, to be sure if these are mast cells or histiocytes that have phagocytosed mast granules. "Typical" mast cells are large and rounded, usually with one small nucleus, and the cytoplasm is filled with coarse granules.

Mast cells, most numerous in the papillary layer, may be found anywhere in the dermis. They are arranged in concentric circles around the walls of small blood vessels, as if they arose from fibroblast-like perivascular cells (Fawcett, 1954).

Mast cells are not identical in all of the tissues and differ even in the same animal. For example, the mast cells in the connective tissue around the abdominal viscera of man contain lipid granules (Rheingold and Wislocki, 1948), but those in the dermis do not. The mast cells of the dog contain demonstrable phospholipid granules (Montagna and Parks, 1948); those of the rat do not. All mast cells contain alkaline and acid phosphatases, and abundant esterase (Montagna, 1956a); they also contain aminopeptidases (Braun-Falco and Salfeld, 1959). They are reactive to the benzidine peroxidase technique and sometimes to the M-Nadi reagent (Compton, 1952; Montagna and Noback, 1948). The granules always contain disulfide groups (Montagna *et al.*, 1954) and readily accumulate radioactive sulfur after the injection of S^{35} (Asboe-Hansen, 1953).

Various physiologic, pathologic, and pharmacologic factors have profound effects on the morphology of mast cells. Mast cells are large and laden with granules in thyroid deficiency, but they become small with sparse granulation after the administration of thyroxine (Asboe-Hansen, 1950). Their number increases after the injection of thyrotropic hormone (Asboe-Hansen and Iversen, 1951). Treatment with corticotropin and cortisone causes degranulation; the cells become vacuolated and attain bizarre shapes (Asboe-Hansen, 1952; Videbaek *et al.*, 1950). Multiple mast cell tumors rapidly regress and disappear by treatment with cortisone (Bloom, 1952). Intradermal injections of hyaluronidase cause a partial degranulation of mast cells, but a number of unrelated irritants do the same thing (Sylvén and Larsson, 1948). Exposure of skin to X-irradiation is followed by a degranulation of mast cells and subsequently by an increase in their number (Sylvén, 1940).

Mast cells secrete heparin (Mergenthaler and Paff, 1956). The cells with strongly metachromatic granules that also stain well with the periodic acid-Schiff method are believed to be the more mature types that contain the highly esterified form of heparin, and those with weakly staining granules, the more immature types that contain heparin monosulfate (Drennan, 1951).

Mast cells also contain and release histamine. They increase in number in several itching skin diseases (Asboe-Hansen, 1951), and skin from sites of urticaria pigmentosa, which abounds in mast cells, contains two and a half times more histamine than the skin immediately adjacent to the lesion. Mast cell tumors from the dog and from man also yield large quantities of histamine (Riley and West, 1952, 1953). In dogs it is possible to predict mast cell tumors and metastases by the demonstration of high blood histamine and heparin values (Cass *et al.*, 1954). In rats injected intravenously with a fluorescent histamine liberator, the fluorescence can be seen localized first in the mast cells, and especially in those within the loose areolar tissue around the peritoneum. When later the mast cells break up, the histamine content of the tissue drops (Riley, 1954, 1959).

III. The Epidermis

A. *Microscopic and Submicroscopic Anatomy*

The epidermis is a stratified squamous epithelium composed of a living stratum Malpighii, which rests upon the dermis, and a dead, horny, superficial stratum corneum. In the stratum Malpighii the stratum basale,

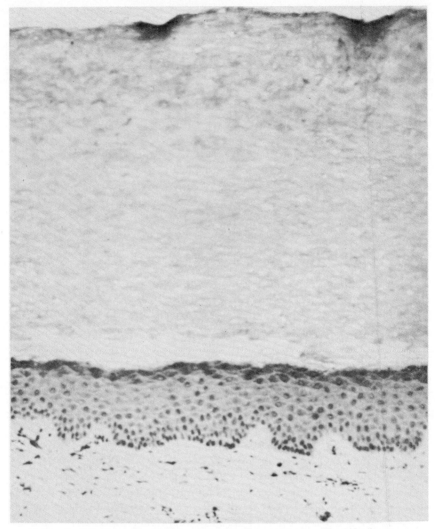

Fig. 1. Entire epidermis from the sole of man showing all of the layers.

or germinativum, is one cell in thickness, and the stratum spinosum is of variable thickness. Between the spinous layer and the corneal layer is the stratum granulosum composed of cells that contain granules stainable with basic dyes (Figs. 1 and 2). A stratum lucidum, which is seldom colored with histological stains, is located above the stratum granulosum of the epidermis of the palms and soles (Fig. 2). The stratum corneum is composed of flattened, keratinized cells. It is very thick in the palms and

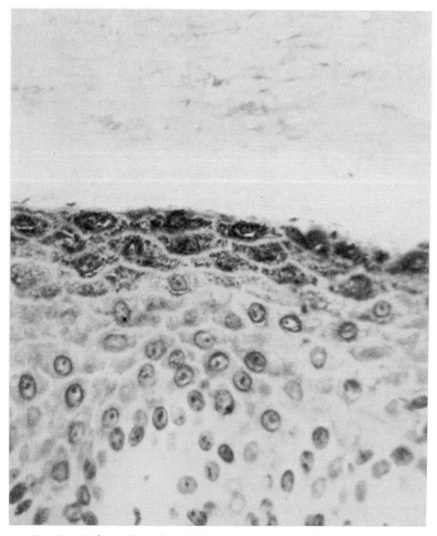

FIG. 2. Epidermis from the sole of man showing clearly the stratum granulosum and the stratum lucidum.

soles, and its cells, unlike those in the stratum corneum elsewhere in the body, are large and firmly cemented together (Figs. 1–3).

The cuboidal, columnar, or fusiform cells in the basal layer may have cytoplasmic processes that grow into the papillary layer of the dermis. The cells in the second layer often have a cytoplasmic process squeezed between the basal cells. The cells in the stratum spinosum are increasingly flattened horizontally as they rise to the surface.

In histological sections the dermoepidermal junction of human skin appears as an undulating line. Epidermal cones and ridges of different sizes project into the dermis, enclosing between them dermal papillae. The epidermal cones and ridges are really systems of branching ridges and papillae of different sizes. The underside of epidermis that has been separated from the underlying dermis by maceration with acetic acid or trypsin has on the underside blisterlike protuberances and is pitted by craters (Horstmann, 1957). The form of the epidermal ridges and papillae are related to the arrangement of the hair and sweat glands, and there are characteristic regional differences in the architecture of the underside. The skin of most laboratory mammals, such as the mouse, rat, and rabbit, has no epidermal ridges and the dermoepidermal junction is flat. The skin of animals whose epidermis is smooth on the underside usually contains numerous, delicate hair follicles which may take the place of the ridges and papillae in anchoring the epidermis to the dermis.

The epidermis is separated from the dermis by a matting of argyrophile and PAS-reactive reticular fibers, the meshes of which contain the cytoplasmic processes of the basal epidermal cells, when these are present. There is no homogeneous membrane that corresponds to a true basement membrane. The electron microscope reveals no typical basement membrane in mammalian epidermis such as that found under the skin of amphibians (Weiss and Ferris, 1954; Porter, 1954; Palade, 1953, 1955). A submicroscopic membrane about 350 A. thick follows the lower contours of the basal cells and is separated from them by a space about 300 A. To distinguish this from the basement membrane in other organs, Selby (1955) calls it dermal membrane. The microscopically visible basement membrane is composed of fibers that have the same electron microscope properties as collagenous fibers and are considered to be reticulin, or young collagen fibers.

Epidermal cells are not completely attached to adjacent cells. Contact with other cells is made at small areas of adhesions that give rise to characteristic cytoplasmic processes called intercellular bridges. These processes are pulled in when living cells are dissociated and the plasma membrane remains smooth. Intercellular bridges are best defined in the stratum spinosum, and they gradually become less clear in the stratum granulosum. At the place of contact between two intercellular bridges is a spindle-shaped swelling called the node of Bizzozero, or desmosome.

In the cytoplasm of epidermal cells are seen delicate striations, the tonofibrils, which are oriented toward the intercellular bridges. These are anisotropic and stainable with Heidenhain's hematoxylin.

The electron microscope shows that intercellular bridges are small areas of contact shared by adjacent cells (Selby, 1955, 1956; Odland, 1958; Palade, 1953, 1955; Porter, 1954; Weiss and Ferris, 1954) where the plasma membrane of both cells is thickened, forming the nodes of Bizzozero. There is no continuity of the cytoplasm across the bridges. At the place of contact, the two surfaces each have a row of dense, elongated granules. Skeins of cytoplasmic fibroprotein filaments, approximately 100 A. wide, sweep toward the intercellular bridges and are attached to the granules. These filaments are also attached to dense rodlets arranged in a row in the plasma membrane of the basal cells that face the dermal membrane. In stained histological preparations, a clumping of these filaments gives rise to the tonofibrils visible under the light microscope.

The basal cells of the thick epidermis from the palms and soles shows odd, undulating filaments of Herxheimer, often contained in the basal cytoplasmic extensions into the dermis. Under the electron microscope these appear as thick bundles of filaments similar in composition to the tonofibrils (Selby, 1955).

Mitochondria are exceedingly difficult to demonstrate in the cells of the epidermis. They can be shown only in tissues postchromed a very long time (Favre, 1950). In the rare good preparations, numerous, small mitochondria are distributed around the nucleus. Mitochondria appear to become fragmented and disappear in the cells of the stratum granulosum, although good mitochondria are found there with the electron microscope. Under the electron microscope (Porter, 1954; Selby, 1955) mitochondria in the epidermal cells are similar to those found in other tissues; they have ridges that protrude from the inner surface toward the interior, forming the cristae. Each mitochondrion is surrounded by numerous small protein granules, which probably represent the earliest precursors of keratin.

Little is known about the Golgi apparatus in the epidermis. In osmic acid preparations a juxtanuclear network, or a group of rodlets, is seen in the cells of the basal layer. This polarity is gradually lost in the cells of the spinous layer, where the Golgi apparatus is dispersed irregularly, and it is no longer demonstrable in cells of the stratum granulosum.

The electron microscope has failed to show a typical Golgi apparatus in epidermal cells (Selby, 1955), although a complex submicroscopic structure that corresponds to it has been demonstrated in sebaceous cells (Palay, 1957). Skin is difficult to fix well, and poor fixation could fail to preserve such delicate structures. Since the Golgi apparatus contains some lipids, it can be demonstrated with sudan black (Montagna, 1950).

The cells of the Malpighian layer are intensely basophile. When stained with basic dyes, the basophile substance is concentrated in the basal layers and fades gradually upward. This substance is destroyed by ribonuclease and represents largely ribonucleic acid (Montagna, 1956a). This is confirmed by observations with the electron microscope, which shows the cytoplasm of epidermal cells rich in dense, fine granules of ribonucleoproteins (Mercer, 1958). Neither in the basal layer nor at levels farther up are these particles associated with the system of membranes forming an endoplasmic reticulum. Abundant ribonucleoprotein particles, or strong basophilia in stained preparations, and the absence of an endoplasmic reticulum are features of well-differentiated cells with great capacity for protein synthesis, but having no way of getting rid of it.

The cells in the stratum granulosum contain granules of keratohyalin (Fig. 2). The granules, which may be stained with most basic dyes, are surrounded by a film of ribonucleoprotein, and they are rich in calcium (Opdyke, 1952). Long considering keratohyalin to be a precursor of keratin, the microscopist has failed to produce convincing evidence for it. Unlike keratin, which begins to be synthesized as filaments and then fibrils in the lower layers of the epidermis, keratohyalin is synthesized farther up. Under the electron microscope, keratohyalin granules seem to be fibrous and similar to the trichohyalin granules in the cells of the inner root sheath of hair follicles (Mercer, 1958). The keratin of the epidermis, then, must be composed of two distinct elements, fibrous keratin, similar to hair cortex keratin, and fibrous keratohyalin.

B. Histochemical Properties

A large number of biologically active substances can be demonstrated in the epidermis with histochemical methods. Strong succinic dehydrogenase activity is found in the basal layers; the activity diminishes gradually in the upper layers and becomes extinct in the stratum granulosum. Monoamine oxidase is distributed in the same way as succinic dehydrogenase. Cytochrome oxidase is abundant only in thick epidermis. Very large quantities of acid phosphatase are present throughout the epidermis, up to and including the stratum granulosum. Small amounts of alkaline phosphatase are also present; human epidermis has virtually none. The epidermis abounds in lipases and nonspecific esterases. Whereas lipase is evenly distributed from the basal layer to the stratum granulosum, the esterases are mostly concentrated in a band of cells immediately above the stratum granulosom (Fig. 3); the reaction stops suddenly above this band (Montagna, 1956a). These enzymes may serve to hydrolyze lipids during keratinization and split off the free fatty acids known to be abundant in the surface lipids. Small amounts of cholinesterases are also localized in the epidermis, particularly in that of the mouse and the rat. The epidermis abounds in β-glucuronidase; the reaction is concentrated in a band identical to that described above. Phosphorylase

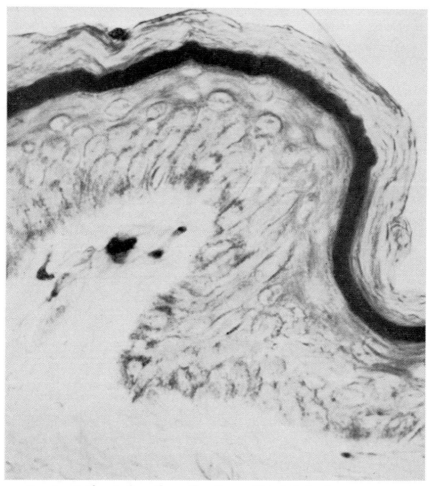

Fig. 3. Epidermis from the axilla of man. The dark band represents a concentration of esterases just above the stratum granulosum. The stratum, in contrast to that of the sole, is thin and flaky.

activity, profuse in the epidermis of primates, is largely absent in that of other animals studied. Thiol groups are evenly distributed from the basal layer to the stratum granulosum, but the keratohyalin granules are always unreactive. A concentration of thiol groups is found in a narrow band of cells above the stratum granulosum, which corresponds to the cells rich in acid phosphatase, esterases, and β-glucuronidase. Variable small amounts of thiol groups may be found in the stratum corneum, particularly in that of the palms and soles. The stratum corneum is rich in disulfide groups.

Mammalian epidermis is relatively free of histochemically demonstrable glycogen, although that of human skin usually shows some. Glycogen may be seen in the cells around the pilosebaceous orifices and the openings of sweat glands; it is always found in the epidermis of skin crevices that enclose blocked keratinized material. Any injury or irritation that disturbs the normal rate of keratinization and/or mitotic activity causes an immediate accumulation of glycogen in the epidermis. When glycogen is stored in the basal cells, mitotic activity always stops; when mitotic activity is restored, glycogen disappears (Lobitz and Holyoke, 1954). The epidermis contains traces of mucopolysaccharides that stain metachromatically. It also contains, in the intercellular spaces and in the intercellular bridges, substances other than glycogen, which are stainable with the periodic acid-Schiff technique (Montagna, 1956a).

All of the cells of the Malpighian layer contain perinuclear lipid granules. The stratum corneum contains abundant lipids released by the cells in the final process of keratinization. Cholesterol can always be demonstrated in the stratum corneum.

Microincinerated epidermis leaves different amounts and qualities of mineral ash. In general, all of the layers of the epidermis contain the white ashes of calcium and magnesium. The basal layer contains mostly the bluish ashes of sodium and potassium.

C. The Replacement of Cells

Mitotic activity in the epidermis takes place largely, if not entirely, in the basal layer. This, therefore, is the true stratum germinativum of the epidermis. The mitotic cells that are found in the spinous layer are really the basal cells in other levels. When human epidermis is grown *in vitro*, only the basal layer survives and grows (Pinkus, 1938). All other cells, therefore, are postmitotic.

The cells in the epidermis of the footpads of rats undergo 5.24% mitosis during 24 hours. A number of mitoses equal to 100% of the cells would require about 19 days (Storey and Leblond, 1951), which is the approximate time required for the renewal of the entire stratum germinativum and stratum granulosum. This also corresponds to the time required for each cell to travel from the stratum germinativum to the lower level of the stratum corneum. Heat speeds up epidermal mitotic activity, and in rats kept in cages with the floors heated to 25 or 30° C., renewal time of the epidermis is about 7 days.

Mitotic activity in the epidermis takes place in diurnal cycles; it is maximal during sleep and rest and minimal during the periods of bodily activity. In mice and rats, which are nocturnal animals, the period of greatest mitotic frequency occurs at midday and the period of minimal frequency at night. In human epidermis mitosis is higher in the night hours than in the morning hours (Cooper, 1939; Cooper and Schiff, 1938).

Mitosis is understood best in the epidermis of the mouse (Bullough, 1946, 1948a, b, 1949a, b, c, d, e, 1950a, b, 1952a, b, 1953, 1954a, b; Bullough and Ebling, 1952; Bullough and Green, 1949; Bullough and Van Oordt, 1950). Mitotic frequency extends from 10 A.M. to 4 P.M. with a peak at approximately 1 P.M. Excessive muscular exercise, starvation, shock, or extreme cold are accompanied by abnormal depressions of the mitotic rate, probably caused by a drain of the sugar reserves of the body. Both adrenaline and cortisone have a powerful antimitotic action *in vivo* and *in vitro,* perhaps by interfering with carbohydrate metabolism. Thyroxine decreases, testosterone increases the rate of mitotic activity, and together they induce only a moderate increase (Eartly *et al.,* 1951). Androgenic and estrogenic hormones have a profound effect upon mitosis. The epidermal mitotic rate of female mice can be stimulated by injections of estrone (Argyris, 1952; Astbury and Woods, 1930; Bullough, 1947). Peaks of activity are obtained normally in females in the third day of diestrus and in early estrus. Both glycogen and estrone are mitogenic, but the maximal stimulation obtained by an increased glycogen concentration is small compared with that obtained with estrone.

The duration of each mitotic division in male and female mice is about 2½ hours. Glycogen and androgen both increase mitotic rate, but neither has an effect on the duration of each division. Estrogen, however, not only increases the number of divisions, but also reduces the duration of each to less than 1 hour (Bullough, 1953). When the skin of mice is placed in a phosphate-buffered saline medium in an oxygen gas phase, with glucose as substrate, the addition of estrone doubles the mitotic rate; L-lactate or pyruvate also double the rate, and the addition of estrone provides no further stimulus. Estrone, then, seems to act on pyruvate, involving hexokinase, and facilitates the reaction of glucose to glucose-6-phosphate.

Most of the factors that influence mitosis *in vivo* can be eliminated by studying pieces of skin in culture media that contain no unknown substances (Bullough, 1954a; Bullough and Green, 1949). Diurnal mitotic cycles are eliminated in this way and similar results are obtained from the epidermis taken from mice at any time. The nutritional state of the animal, and stress and shock, all of which depress mitosis in animals, do not affect the rate of the epidermal mitosis which develops subsequently *in vitro.* Estrogen, however, has a continued effect and the skin of female mice reflects, *in vitro,* the phase of the estrous cycle of the animal. With glucose as a substrate in the culture media, insulin increases mitotic activity while growth hormone inhibits it. When both hormones are used together they counteract each other. The point of action of these two hormones seems to be the glucokinase system.

In human epidermis local bursts of mitotic activity seem to be provoked by the removal of all or a part of the horny layer. Single layers of horny cells can be removed from the skin with Scotch tape; by repeating this process one may remove as many of the layers as he wishes (Pinkus, 1951, 1952). The removal of only four layers leads to appreciable local mitotic activity, suggesting that the replacement of cells is probably guided by the rate of their loss. Mitosis is a complex phenomenon, and when one considers the numerous factors that influence it, studies that do not take them into account are confusing.

IV. The Pilary System

A. Microscopic Anatomy

Hair follicles may be considered to be long, tubular holocrine glands whose secretion, the hairs, are cylinders of compactly cemented keratinized cells. The base of hair follicles is swollen into an onion-shaped bulb; this is hollowed out into a domed cavity that contains loose connective tissue, the dermal papilla (Fig. 4). Hair follicles grow at a slant, and bundles of smooth-muscle fibers, the arrectores pilorum muscles, extend at an acute angle from the surface of the dermis to the bulge on the side of the follicle, below the level of the sebaceous glands.

Hair follicles consist of concentric layers of tissue around the hair. An inner root sheath surrounds the hair, and an outer root sheath, continuous with the surface epidermis, is on the outside. The inner root sheath is composed of Henle's layer, one cell in thickness, on the outside, Huxley's layer, one or more cells in thickness, in the middle, and the cuticle of the inner sheath. This is a single layer of flattened, imbricated cells, whose free borders are directed downward and are interlocked with the cuticle cells of the hair, which are directed upward. A connective tissue sheath around the follicles is composed of an inner circular and an outer longitudinal layer. A hyalin, noncellular glassy, or vitreous, membrane separates the connective tissue sheath from the outer root sheath. The connective tissue sheath is continuous above with the papillary layer of the dermis, and with the dermal papilla at the base of the follicle.

Being cyclic organs, hair follicles have periods of activity and rest. In the human scalp, for example, the growing period is very long and the period of rest is short; in the other parts of the body, however, the periods of rest are usually longer than those of growth.

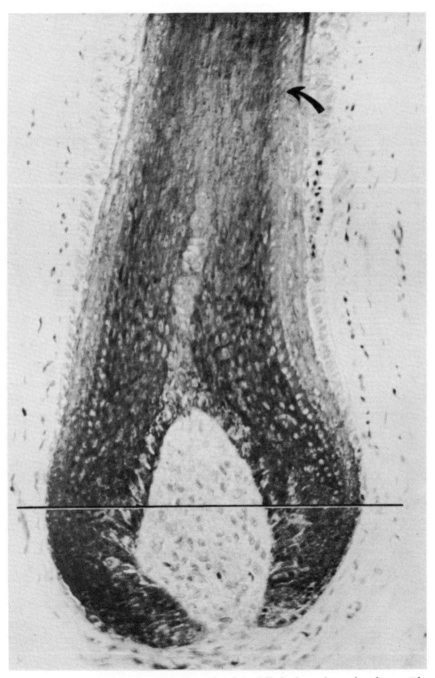

FIG. 4. Longitudinal section through a hair follicle from the scalp of man. The line separates the matrix from the upper bulb.

Resting follicles are one-half to one-third the length of active ones, and they have no bulb. The hairs in quiescent follicles have a clavate base surrounded by an epidermal sac; a cord of small, undifferentiated cells, the hair germ, projects downward from the middle of the epidermal sac and maintains contact with the dermal papilla. Delicate keratinized rootlets, radiating from the enlarged end of the club hair, anchor it to the epithelial sac. A capsule of partially keratinized cells, corresponding to the inner sheath of the active follicles, separates the epithelial sac from the club. The epithelial sac and the hair germ correspond to the outer root sheath of active follicles.

During its life cycle a hair follicle may undergo drastic changes. The small follicles in the face of young human males, or in the axillary and pubic regions of males and females, for example, are transformed at the time of puberty into large follicles that produce coarse hairs. Conversely, in the balding scalp the large follicles may regress to very tiny follicles that produce barely visible hairs.

Hairs may be straight, wavy, or crimped. Deflections in the hair bulb (Mercer, 1953), the eccentric position of the hair in the follicle, and asymmetric keratinization of the hair fiber may bring about crimping, or waving, in hair fibers. The exact factors responsible for the curling of hairs are not known.

1. The Bulb

The bulb has a lower region of undifferentiated cells and an upper region where the cells differentiate into the inner root sheath and the hair (Fig. 4). A line drawn across the widest diameter of the papilla would separate the two regions at the critical level (Auber, 1952). Below this line is the matrix, composed of mitotically active cells. In pigmented hairs the pigment stops at this line and the matrix is rarely pigmented. Cells move up in single rows from the matrix (Fig. 5); they increase in volume and become elongated vertically. Above the constriction of the bulb, the cells are conspicuously larger and longer. Numerous delicate, acidophile fibrils are formed in the cytoplasm and are piled up against the cell membranes. Farther up, the fibrils are coarser and more numerous and stain with basic dyes. In the keratogenous zone (Giroud and Bulliard, 1930), the cells become hyalinized, and distinct fibrils can be seen only with certain techniques or under polarized light. Depending upon the length of the follicle, the upper limit of the keratogenous zone is at approximately one-third of the way between the tip of the dermal papilla and the surface of the skin. Above the keratogenous zone is the fully keratinized hair.

Melanin is distributed very precisely in the follicle. Only the medulla

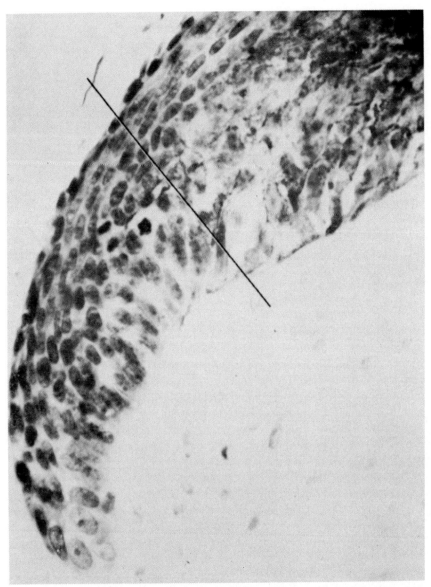

FIG. 5. Longitudinal section through part of the matrix of a hair follicle from the scalp. The line separates the matrix from the upper bulb.

and the cortex of the hair contain pigment. Most of the columnar cells that line the dome of the dermal papilla are melanin-producing, dendritic melanocytes; their dendrites are insinuated between the cells in the upper bulb that are differentiating into the cortex and medulla of the hair. The matrix is free of it (Fig. 5), as are also the cuticle cells of the cortex and the inner and outer root sheath. Traces of pigment may occasionally be seen in the outer root sheath, and even very small dendritic cells may be found there. This condition is common in the hair follicles of some mammals.

Cells move up from the matrix and increase in volume many-fold. Large numbers of them are funneled up through the relatively narrow neck of the bulb. The rate of growth of the hair is maintained with greater ease than if the matrix and upper bulb were of the same diameter as the hair. These cells are probably under pressure as they pass through the neck of the bulb; Henle's layer of the inner root sheath forms the solid outer surface of the funnel by becoming keratinized as soon as its cells rise from the matrix. The outer sheath, the vitreous membrane, and the connective tissue sheath on the outside, all give sturdiness to the funnel. As the cells move up from the matrix they outline patterns of shearing lines (Fig. 5).

2. The Hair

Hairs have a cuticle on the outside, a cortex, and usually a medulla, in the center. The cuticle is a single layer of flattened, imbricated, trans-lucent cells with the free margins directed toward the tip of the hair. The bulk of most hairs consists of cortex. The medulla, in the center, is dis-continuous or absent in very fine hairs. The cells of the medulla are large, loosely connected, and partially keratinized. Large intra- and intercellular air spaces in the medulla give sheen and color tones to the hair by modi-fying the reflection of light. In most hairs the medulla is small, but in rodents it makes up the greater volume of the hair.

The cells of the cuticle of the hair move up from the matrix in a single row. In contrast to the cells of the cortex they contain no melanin. In the upper part of the bulb the cuboidal cells become columnar, with the long axis oriented radially. A short distance above this, their outer edges are tipped upward and, since they are broader than they are high, when their outer edges are tipped apically, the cells are automatically imbricated. The cells become flattened as their orientation shifts from horizontal to vertical. In the upper half of the follicle the cuticle cells are fully keratinized and adhere to the outside of the cortex.

3. The Root Sheaths

The inner root sheath arises from the peripheral and central parts of the matrix. When these cells move in the upper bulb they become larger and acquire basophile-staining trichohyalin granules. The cells of Henle's layer begin to differentiate as soon as they pass into the upper bulb from the matrix. In progressively higher levels the trichohyalin granules coalesce, form homogeneous globules and parallel rods, and finally completely fill the cells. In Huxley's layer, trichohyalin granules first appear in the cells in the upper part of the bulb, at a level where the cells of Henle's layer are already keratinized. Some of the cells in Huxley's layer remain unkeratinized and send lateral cytoplasmic processes across the keratinized Henle's layer and make contact with the cells in the axial border of the outer root sheath.

The cells of the cuticle of the inner root sheath are keratinized very slowly and begin to form trichohyalin granules about halfway up in the follicle. They become flattened and reoriented vertically together with the cuticle cells of the cortex. Above the middle of the follicle, the three layers of the inner root sheath become fused into one single, hyalin layer.

The inner root sheath is eliminated in the pilosebaceous canal. It is not known if this is accomplished by physical or chemical agents. Since it is interlocked with the hair, the inner root sheath must grow at approximately the same rate as the hair. Henle's layer on the outside slides against the axial border of the outer sheath which is stationary. The interphase of these two layers is smooth and partially keratinized, facilitating the upward movement of the inner sheath.

The outer root sheath covers the entire follicle all the way to the tip of the bulb. Around the bulb it is very thin and its cells are elongated. Above the bulb the cells are larger and the sheath is pluristratified. At the level of the middle third of the follicle nearly all of its cells are riddled with vacuoles. The cells of the outer root sheath are in contact with each other through the intercellular bridges. The more axially located cells have well-developed tonofibrils. Minute cytoplasmic processes project laterally from the peripheral cells through the thick vitreous membrane above the bulb.

4. The Vitreous Membrane

The vitreous or glassy membrane is thick around the middle part of the follicle and thickest around the upper bulb; it is thin around the lower bulb and nearly invisible in the papilla cavity. Around the lower third or half of the follicle the membrane consists of two layers; around the upper part it is single-layered. The outer layer envelopes the entire follicle and is continuous with the basement membrane of the epidermis. It consists mostly of delicate longitudinally arranged collagenous fibers.

The inner layer around the lower half of the follicle is a tangled skein of fibrils wound horizontally between the cytoplasmic processes of the basal cells of the outer sheath (Montagna, 1956a). During the transition from active to inactive follicles, the skein of fibrils becomes very thick, but the outer hyalin membrane remains unchanged. At the end of the transition, when the follicle is quiescent, the hypertrophied glassy membrane becomes fragmented and is resorbed. Resting hair follicles are surrounded by a thin, single-layered membrane.

5. The Dermal Papilla

The dermal papilla is the loose connective tissue that is enclosed in the bulb of active follicles. In the larger follicles the papilla is vascularized by variable tufts of capillaries; in very small ones there are no capillaries. In the resting follicle the papilla rests free at the base of the hair germ as a cluster of cells with dense, round nuclei and barely discernible cytoplasm. In a growing follicle the papilla is larger, contains more ground substance, and its cells are spaced farther apart. The cells have large, ovoid nuclei and pale-staining, vacuolated cytoplasm. The number of papilla cells seems to remain constant, and what changes do appear during the cycles of growth are probably due to changes in the size of its cells, the withdrawal of the capillary plexus during the periods of quiescence, and the changes in the amounts of intercellular substances. During periods of growth the ground substance in the dermal papilla stains metachromatically; it is periodic acid-Schiff reactive, and contains alkaline phosphatase. During periods of quiescence these properties are lacking, but alkaline phosphatase activity is still demonstrable.

B. Electron Microscopy

Under the electron microscope the indifferent cells of the matrix, or the presumptive cells of the hair and the inner root sheath show typical mitochondria with cristae mitochondriales, agranular vesicles, and many small dense particles of ribonucleic acid; the cells contain no keratin, pigment granules, or endoplasmic reticulum (Birbeck and Mercer, 1957a, b, c).

In the upper limits of the matrix in the mid-bulb region, adjacent cells in small localized areas of the presumptive cortex show some adhesions, but there are still gaps between the cell surfaces, and these gaps are penetrated by the processes of the melanocytes. The cells are said to phagocytose the pigment-bearing tips of the melanocytic processes and become pigment-bearing cells. Fibrous keratin is first seen in these cells as loose, parallel strands of fine filaments about 60 A. in diameter. The filaments, the mitochondria, and the nucleus become oriented parallel to the long axis of the follicle. In the region immediately above the constriction of the bulb, a dense, amorphous intracellular substance forms between the fine filaments and cements them into bundles of fibrils. At progressive higher levels in the follicle, more filaments and interfilamentous cement are formed until the entire cell is packed with fibrils. The final step in keratinization is the addition of an interfilamentous cement substance and the consolidation of the fibrils. Keratin, then, is a complex of fine filaments (alpha filaments) embedded in an amorphous substance (gamma keratin). The amorphous substance has a higher content of cystine than the fibrils.

As soon as they begin to differentiate just above the level of the matrix, the presumptive cells of the cuticle of the cortex show greater density than the cells of the cortex; their cell membrane becomes very smooth and a layer of cement firmly attaches them to one another. There are no gaps between the cells, and the processes of the melanocytes cannot penetrate between them; the pigment-bearing tips of melanocytes cannot be phagocytosed, and the cells of the cuticle remain unpigmented. These particular changes take place precociously, but keratinization takes place late in the cuticle cells. The keratin formed is amorphous, like the gamma fraction found between the fibrils in the cells of the cortex. In the final stage of keratinization the cells have parallel laminae surrounded by an outer keratinized layer. The inner part of the fully keratinized cell is insoluble in keratinolytic solvents.

The three concentric cylinders of cells of the inner root sheath arise from the peripheral portion of the matrix. These cells, like those of the hair cuticle, are held together by an adhesive cement that first appears in the middle and upper parts of the bulb. The characteristic intracellular product in all three layers is trichohyalin. This, unlike the cortical hair keratin, which originates as filaments, appears first in the form of amorphous droplets. Trichohyalin appears first in the cells of Henle's layer at the level of the neck of the follicle in the upper part of the bulb. Synthesis and transformation of trichohyalin in the cells of Huxley's layer, and in those of the cuticle of the inner sheath, lag behind those in the cells of Henle's layer and begin at higher levels above constriction of the bulb. The amorphous trichohyalin droplets are transformed to fibrous plates farther up in the follicle. In cross sections the fibrous plates appear as corrugated sheaths about 100 A. in width. At the same time that the trichohyalin is being transformed, a dark cement substance similar to that found between the cells of the hair cuticle appears in the spaces between the cell membranes.

From these observations it is clear that several keratins with strikingly different properties are formed in hair follicles.

C. Histochemical Properties

The active hair follicles of all animals are rich in glycogen, largely stored in the cells of the outer root sheath, and particularly in the cells immediately above the upper bulb. These cells also contain abundant mucopolysaccharide that stains metachromatically. The outer root sheath is rich in succinic dehydrogenase, monoamine oxidase, β-glucuronidase, esterases, phosphorylase, and acid phosphatase. All of the cells of the outer root sheath contain variable numbers of sudanophile lipid granules, particularly numerous in the region above the constriction of the bulb. The region of the differentiating cortex of the hair above the bulb, in the keratogenous zone, contains large amounts of demonstrable protein-bound, water and alcohol-insoluble sulfhydryl groups (Montagna, 1956a). Above the keratogenous zone the reaction stops abruptly. The fully keratinized hair is rich in disulfide groups. The quantities of all these substances are reduced in quiescent hair follicles.

Three substances, always present in the dermal papilla of active hair follicles, are absent in that of quiescent ones. The papilla of growing follicles is not only large, but it contains mucopolysaccharides that stain metachromatically; it contains substances reactive to the periodic acid-Schiff methods and much alkaline phosphatase. The first two of these properties are lost during the periods of rest. A residual amount of alkaline phosphatase may still be found in the resting papilla of some animals but not in that of others.

D. Growth and Differentiation

Each hair follicle is a biologic system that repeats nearly its entire life cycle with each cycle of growth. During the periods of growth, it grows at top speed and during rest it is torpid. During the transition from a growing to a resting condition, the matrix and the differentiating regions of each follicle degenerate and the follicle undergoes a retrograde differentiation and becomes a simple primordium. At the completion of growth the entire lower part of the follicle is broken down. When growth is initiated in a quiescent folicle, the first step is the building of a bulb from indifferent cells, requiring several days to accomplish; the hair can be elaborated only after the bulb has been established. This metamorphosis is similar to the process that takes place when the follicles are first formed in the embryo, and with each hair cycle the follicles recapitulate the embryonal events. Few, if any, of the cells of the matrix are salvaged at the end of the period of growth. The cells of the so-called hair germ in a quiescent follicle seem to be derived from the cells of the outer root sheath. These cells must revert to indifferent types that are left behind as the seed for the next hair generation. The dermal papilla must be the inducer of differentiation, but too little is known for certain about its role.

Once the matrix is established its cells proliferate constantly, move up, and form the hair and inner root sheath. The patterns of keratinization and the types of keratin formed in the different layers of the hair and in the inner root sheath are quite different even though the cells that make these structures all come from the same pool of indifferent cells in the matrix.

In the upper part of the bulb the cells begin to undergo keratinous differentiation, obtain pigment, and grow many times their original volume. The final phases of keratin differentiation in the hair takes place in the keratogenous zone, above the constriction of the bulb. The cells of the outer root sheath around this zone contain more esterases, β-glucuronidase, glycogen, and phosphorylase than those at other levels. The follicle is surrounded at this level by a very thick glassy membrane. These features suggest that the outer root sheath at this level is a very active tissue. It has been shown in the follicles of the sheep that a strong accumulation of radioactive particles can be recovered first around the general region of the keratogenous zone only 6 minutes after the injection of cystine labeled with S^{35}. This is also the region of the follicle surrounded by the densest capillary networks. These points emphasize the fact that the region of the upper part of the bulb is an important site of exchange.

Exchange must also take place through the dermal papilla. The amount of vascular tissue in the dermal papilla is dependent on its size; the wider a papilla is, the larger are the capillary tufts it contains. The dermal papillae of active human follicles are very wide and contain numerous capillary loops. During periods of quiescence either the tissue of the papilla disengages itself from the capillary tufts by flowing away from them, or the capillaries atrophy. The changes in the size of the papilla are largely due to the increase and decrease in the amount of capillary tissue and to the increase and decrease in the size of the papilla cells.

V. The Sebaceous Glands

A. Microscopic and Submicroscopic Anatomy

Sebaceous glands are more numerous in the skin of some mammals than in that of others, and the skin of whales, dolphins, and porpoises has none. Sebaceous glands grow from the upper part of the outer root sheath of hair follicles and pour their secretion into the pilary canal. In the eyelids, the nipples, the preputium, the labia minora, the lips, and the buccal mucosa, sebaceous glands may be free from hair follicles. They may grow in such abnormal sites as the parotid and submandibular glands

(Andrew, 1952), in the cervix uteri (Dougherty, 1948), and in the mammary glands.

Sebaceous glands are composed of clusters of acini whose ducts converge toward a central collecting duct. When removed *in toto* from the skin by enzymatic digestion (Hambrick and Blank, 1954) and colored with Sudan colorants, sebaceous glands look roughly like bunches of grapes. Their gross shape is determined by their relative abundance in the area and by the nature of the dermis in which they grow.

The ducts of sebaceous glands are composed of stratified squamous epithelium continuous with the outer root sheath of hair follicles or with the surface epidermis. The surface of the stratified squamous epithelium of the duct always shows a small amount of keratinization. At the periphery of the sebaceous acini the cells are small, flattened, and resemble epidermal cells. The cells are progressively larger and misshapen toward the center of the acini, and in the center they are very large and laden with lipids; their cytoplasm is gradually reduced to a flimsy framework between and around the fat droplets. When sebaceous cells have manufactured the secretion product, they die, become fragmented, and all of their remains are incorporated in the secretion or sebum. This process is known as holocrine secretion.

In the peripheral cells the nucleus is strongly basophile and Feulgen positive. During sebaceous differentiation it becomes larger and progressively less basophile and weakly Feulgen positive, and at the completion of sebaceous maturation the nucleus crumbles and is no longer stainable.

The cytoplasm of the undifferentiated peripheral cells is full of basophile granules; these diminish and seem to disappear during sebaceous transformation. Many mitochondria are found in the peripheral cells crowded in the perinuclear cytoplasm, where lipid droplets develop near them (Montagna, 1955). When the volume of the cells increases, the population of mitochondria remains the same, and consequently their relative numbers decrease. In mature sebaceous cells mitochondria are stranded between the lipid globules, and in dying cells they crumble and disappear (Fig. 6).

Under the electron microscope (Palay, 1957), fine nucleoprotein granules, whose distribution corresponds to that of the cytoplasmic basophilia, are seen dispersed over the ergastoplasmic membranes of the undifferentiated peripheral cells. Small clusters of agranular reticulum consisting of small packets of parallel tubular elements and vesicles surround the nucleus. These cells also contain a few lipid droplets and granules. In the cells that have begun to differentiate, the agranular reticulum increases and eventually permeates the entire cytoplasm, as

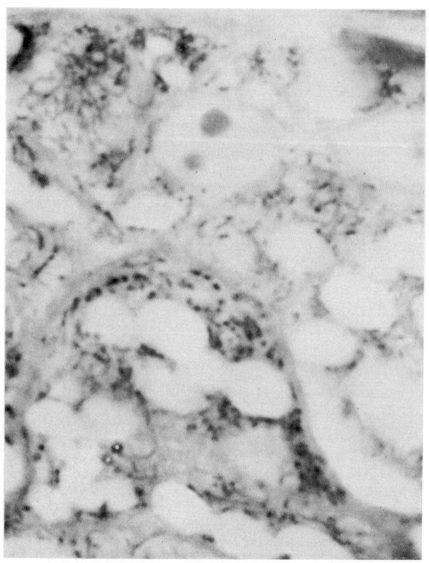

FIG. 6. Mitochondria in sebacious cells, from a gland in the axilla of man. These bear no relation to the large lipid vacuoles.

numerous lipid granules become scattered through the cell. These granules consist of loosely and irregularly arranged networks of strands; they are apparently suspended in a clear circular or elliptical space separated from the rest of the cytoplasm by unusual membranes; each layer of these membranes consists of a bidimensional grid composed of tubules laid roughly at right angles to one another in the plane of the layer (Fig. 7).

The "secretory" droplets are probably formed in the cytoplasmic spaces that are enclosed by the agranular reticulum. As lipid is accumulated the cytoplasm becomes compressed into the interstices between the "secretory" droplets. In fully differentiated cells only thin plates of cytoplasm containing mitochondria, fine granules, and tubules separate the droplets. The irregular strands within the droplets disappear and the husks of the compressed membranous grids become thinner.

The husk of the multiple agranular membranes that surrounds each developing lipid droplet corresponds to the osmiophile bodies described by others and must represent the Golgi apparatus (Fig. 7). Sebaceous transformation, then, seems to occur within the Golgi apparatus, perhaps by a chemical alteration of the membrane substance, by a segregation of materials produced in the cytoplasm and concentrated within the membranes. Mitochondria have little topographical relation to the lipid droplets and remain intact even in the mature sebaceous cells, as seen under the light microscope. The basophile elements of the cytoplasm decrease only in the over-all concentration, as the cell becomes distended by the lipid droplets. The cytoplasm between the droplets seems to be just as densely granular in mature cells as in the undifferentiated peripheral cells.

B. Histochemical Properties

Sebaceous glands contain a number of enzymes whose concentrations are different in different species. Cytochrome oxidase is very intense in the peripheral cells, the reaction fading in the differentiating cells; similarly, succinic dehydrogenase activity is abundant in the indifferent peripheral cells and diminishes as the cells undergo sebaceous transformation (Braun-Falco and Rathjens, 1954; Formisano and Montagna, 1954; Montagna and Formisano, 1955). Benzidine peroxidase activity is found in the sebum and in the differentiated sebaceous cells, the indifferent peripheral cells showing practically no reaction (Montagna, 1956a). Carbonic anhydrase activity is seen in the peripheral, undifferentiated cells, but not in the differentiated, lipid-laden ones (Braun-Falco and Rathjens, 1955). Phosphorylase is found in the glands of man and

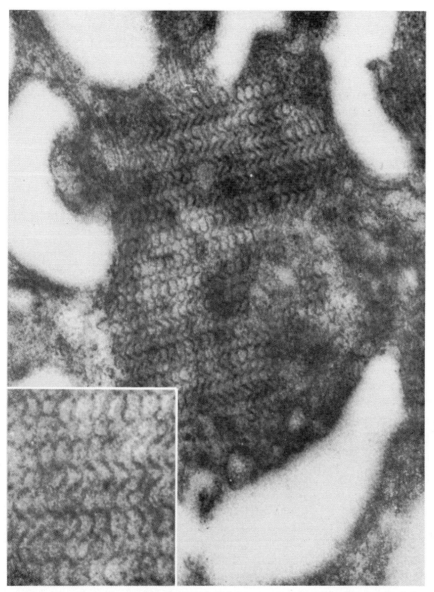

FIG. 7. Electron micrograph of the Golgi apparatus in a sebaceous cell of the rat. In the lower left is a larger detail of the bidimensional grid. Courtesy of Dr. S. L. Palay (1957).

other primates. Alkaline phosphatase activity is nearly absent from the glands of man, mouse, hamster, rabbit, and guinea pig, but abundant in the peripheral cells of those of the cat, dog, seal, rat, etc. In most species the glands have large amounts of acid phosphatase (Moretti and Mescon, 1956). The glands contain lipases, nonspecific esterases, and esterolytic enzymes (Montagna, 1956a; Braun-Falco, 1956a); those of some species even contain cholinesterases. The glands have large amounts of β-glucuronidase (Braun-Falco, 1956b; Montagna, 1957), and abundant monoamine oxidase (Montagna and Ellis, 1959).

The sebaceous glands of man contain glycogen (Montagna *et al.*, 1952a), but those of all other mammals studied do not. The distribution of glycogen is characteristically opposite to that of the lipids, suggesting that in man the first lipid differentiation takes place through a conversion from glycogen. This is also found in the glands of other primates.

The differentiating sebaceous cells contain variable small amounts of zinc (Braun-Falco and Rathjens, 1956).

The principal role of sebaceous glands is to synthesize, store, and secrete lipids. Sebum is composed of unusual substances not found anywhere else in the body (Lederer and Mercier, 1948). It differs from species to species, and cholesterol and free fatty acids are perhaps the only substances that are found in nearly all species (Wheatley, 1952, 1953).

The cells of the ducts and the undifferentiated sebaceous cells all have very small perinuclear lipid granules, which become increasingly numerous and larger. The lipid droplets in any mature cell are of nearly equal size, but they are characteristically different in different animals; they are large in the glands of man and the seal but small in those of the mouse and the hamster.

Sebaceous lipids contain triglycerides, phospholipids, and esters of cholesterol. Changes take place in the sebaceous lipids when the mature cells break down and after the sebum becomes aged; these changes are probably brought about by the lipases and esterases.

Viewed under polarized light, the variable amounts of anisotropy largely correspond to the distribution of the cholesterol esters. In the sebaceous glands of man only the sebum consistently contains birefringent lipids; in those of the rat, dog, and cat birefringent lipids are also present in the degenerating and mature sebaceous cells.

Under near-ultraviolet light (3600 A.) sebaceous glands emit a yellow-to-orange light. The fluorescence corresponds largely to the distribution of the anisotropic cholesterol esters. The sebum in plugged glands sometimes emits a reddish fluorescent light due to porphyrin.

In summary, sebum contains histologically demonstrable cholesterol esters, some phospholipids, fatty acids, and triglycerides. The sebum in the excretory ducts is demonstrably different from the sebum just formed in the centers of the glands.

C. Growth and Differentiation

Since they are holocrine glands, sebaceous glands must replace their cells as rapidly as they lose them in secretion. Mitotic activity occurs in the undifferentiated cells in cyclic rhythms similar to those in the epidermis (Bullough, 1946).

Sebaceous glands are in a state of change. Epithelial buds grow from the walls of the excretory ducts, undergo differentiation, and form new sebaceous units (Montagna, 1956a). These expand, encroach upon nearby ones, and fuse with them. Sebaceous kernels may develop anywhere in the gland, outside or in, where there are undifferentiated cells.

Any cell of epidermal origin can undergo sebaceous transformation. Isolated fragments of hair follicles in the dermis can change into sebaceous cysts, which later may become cornified. Sebaceous metaplasia is encountered very often in the cells of induced papillomas and carcinomas in the mouse. Single cells or clusters of cells in the surface epidermis occasionally undergo sebaceous transformation when the skin is abraded or irritated. During the repair of experimental shallow wounds in human skin, in which the distal part of the pilary canal is removed, the undifferentiated peripheral cells of sebaceous acini flow to the surface of the wound and form epidermal cells; some of them may later differentiate into sebaceous cells on the surface of the wound (Eisen *et al.*, 1955).

The rates of growth and differentiation of the glands are regulated by hormones. The glands are well developed, but relatively small, in infancy and become large at puberty. The glands of eunuchs and eunuchoids are small and underdeveloped (Hamilton, 1941). The administration of testosterone or progesterone to rats produces pronounced enlargements of the sebaceous glands. The removal of the pituitary and of the adrenal glands from ovariectomized rats brings about an atrophy of sebaceous glands, which is not counteracted after the administration of progesterone, and which is counteracted only slightly by testosterone (Lasher *et al.*, 1954).

In the skin of the mouse, sebaceous glands are destroyed when polycyclic hydrocarbons are applied topically (Bock and Mund, 1956). The glands regrow very rapidly from the cells of the outer root sheath of hair follicles when the hair follicles are growing. When resting, the glands do not regrow until after the follicle becomes active again (Montagna and Chase, 1950). There is, then, a coordination of growth and function between the sebaceous glands and the associated hair follicles.

VI. The Sweat Glands

A. *Introduction*

Sweat glands are numerous in the skin of man, the pig, cat, seal, and horse, but they are absent from the skin of the rabbit. Many common mammals have very few sweat glands and some have none. The friction areas of the skin of plantigrade and digitigrade mammals have numerous sweat glands, and in rodents, this is the only place where they are found.

According to their structure and function, sweat glands are separable into apocrine and eccrine types (Schiefferdecker, 1922). In spite of some general structural similarities, these glands are profoundly different organs.

B. *Apocrine Glands*

1. *Microscopic Anatomy*

These occur in the skin of most mammals, and they are numerous in the horse, swine, cat, seal, etc. In the human body apocrine glands are found in the axilla, mons pubis, external auditory meatus, and circumanal area. Occasionally, they may be found anywhere on the body. In the skin of primates apocrine glands seem to have been lost progressively as eccrine glands have developed in their place (Schiefferdecker, 1922). The frequency of distribution of apocrine sweat glands in the human body is subject to individual variations, but they are believed to be more numerous in Negroes than in Caucasians and more numerous in the female than in the male of either race (Homma, 1926).

Apocrine glands are simple, tubular glands (Horn, 1935; Sperling, 1935) with the compactly coiled and dilated secretory portion extending to the lower part of the dermis and the subcutaneous fat (Fig. 8). The ducts lie close and parallel to hair follicles and open into the pilary canal close to the duct of the sebaceous glands. The ducts occasionally open directly onto the surface. The epithelium of the secretory tubule is simple columnar, but that of the duct is composed of two layers of cuboidal cells.

The terminal portion of some of the cells in the secretory coil projects into the lumen, but that of others has a cuticular border. Some segments of individual tubules, or entire tubules, may be dilated and lined with low, simple cuboidal or squamous epithelium. Sandwiched between the secretory epithelium and the thick, hyalin basement membrane is a layer of myoepithelial cells.

The spindle-shaped myoepithelial cells have many of the character-istics of smooth-muscle fibers. Their axis is oriented roughly parallel to

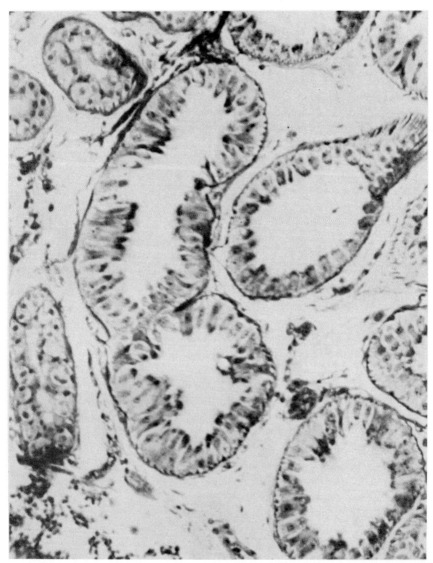

FIG. 8. Apocrine sweat glands in the skin of the axilla of man. The small tubules on the left margin of the figure are eccrine glands.

that of the tubule; they are best developed in those glands that are lined with the tallest epithelium. In the dilated tubules with flat epithelial cells, myoepithelial cells are not visible. The cytoplasm of the myoepithelial cells contains delicate anisotropic, longitudinal fibrils. Myoepithelial cells contract in peristaltic waves like muscle cells (Hurley and Shelley, 1954).

The large and spherical nucleus of apocrine cells is located at the base of the cells and stains well with most basic dyes and with the Feulgen reaction. One or two large nucleoli are flanked by two strongly basophile, Feulgen-positive granules. Mitotic activity is not common.

Numerous and large mitochondria are aggregated in the basal part of the cytoplasm of the secretory cells (Minamitani, 1941a; Ota, 1950). They are sparse or absent in a clear area above the nucleus and absent in the terminal, luminal part of each cell (Fig. 9). The tall columnar cells are full of cytoplasmic granules above the nucleus and contain fewer mitochondria. There seems to be a close relationship between mito-chondria and the secretory activity in each cell, but mitochondria do not change into granules. The number and the shapes of mitochondria in each cell change according to the size of the cell and according to the number and size of the granules in the cell.

The Golgi apparatus, consisting of twisted osmiophile rodlets and granules between the nucleus and the free border, extends over an area about the size of the nucleus (Melczer, 1935; Minamitani, 1941a; Ota, 1950). This is the area of the negative image of the Golgi apparatus (Fig. 9). The Golgi element is interposed between the ripened secretion granules distally and the nucleus proximally. Secretion granules are formed below the Golgi apparatus, perhaps at the expense of the mitochondria, and become ripened when they come in contact with the Golgi element (Minamitani, 1941b). During the ripening process some granules acquire pigment.

The production of a yellow or brown pigment is one of the most characteristic features of apocrine glands. The glands of older animals contain more pigment than those of young ones, and there are also individual differences. The source, the nature, and the function of these pigments are not known. The ceruminous glands of man contain and secrete large amounts of pigment (Montagna et al., 1948).

The human axillary apocrine glands usually contain large amounts of yellow and brown pigment granules, mostly insoluble in lipid solvents. The larger brown ones are basophile; the small yellow granules are acidophile. The brown pigment in the apocrine glands of man (1) is insoluble in water and in organic solvents; (2) is not easily depigmented with oxidizing agents; (3) is fluorescent, relatively sudanophile, acid-fast, and moderately reactive to the plasmal reaction; (4) is only mildly argyro-

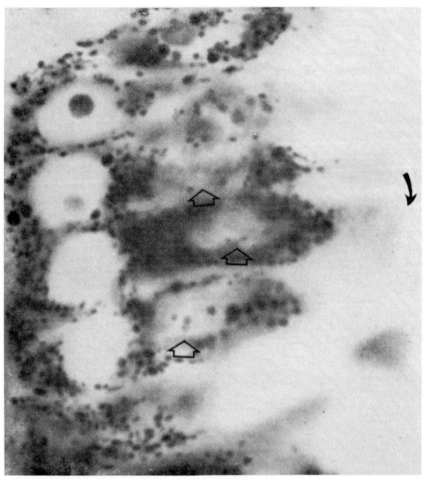

FIG. 9. Mitochondria in the cells of axillary apocrine sweat glands. The hollow arrows indicate the negative image of the Golgi apparatus; the solid arrow indicates the unstained distal border of one cell.

phile; (5) is periodic acid-Schiff reactive, basophile when stained with basic dyes, and slightly acidophile; and (6) does not contain iron. The yellow pigment is also insoluble, but it has no lipid and is not fluorescent; it is not periodic acid-Schiff reactive and is acidophile. In addition, it often contains iron.

The distal cytoplasmic border of apocrine cells is free of pigment or other particulate cellular materials. Since the secretion product within normal tubules rarely shows visible pigment, or strongly autofluorescent and sudanophile material, if the pigment is secreted within the lumen of the glands it must undergo some changes.

2. Histochemical Properties

The presence of intraepithelial ionic iron is a distinctive feature of human apocrine glands. Abundant in some specimens, iron is scarce or absent in others. Adjacent tubules, and even different coils of the same gland, may have different amounts of it. In the same glands some cells may be rich in iron and morphologically similar cells may possess none. Granules of iron are found alongside and above the nucleus, clustered around the negative image of the Golgi apparatus. Regardless of the amount of iron in the epithelium, the clear or flocculent content of the lumen seems always to be free of it (Montagna et al., 1953b). Iron is present in the small, yellow pigment granules; the large, brown ones may contain a thin film of it at their periphery, or they contain none. The amount of yellow or brown pigment does not indicate the presence of iron, but iron is abundant only in the small yellow granules and only if the large brown ones are also present in the same cells.

The cytoplasm at the base of each cell is stippled with many fine, basophile granules that contain ribonucleic acids; the cytoplasm lateral to and above the nucleus is weakly basophile. The negative image of the Golgi region is chromophobic and the terminal cytoplasm is nearly so. The concentration of basophile substances in the cytoplasm of secretory cells is inversely proportional to the number of mitochondria present.

Apocrine glands contain a moderate amount of histochemically demonstrable succinic dehydrogenase (Montagna and Formisano, 1955), frequently localized around the nucleus or above it; the basal part of the cells, the terminal cytoplasmic border, and the myoepithelial cells contain none. The cells of the excretory duct show a strong enzyme reaction. Apocrine glands abound in β-glucuronidase; no other cutaneous organ is as rich in this enzyme (Montagna, 1957).

The excretory duct, from its orifice in the pilary canal to its junction with the secretory coil is rich in phosphorylase activity. In contrast, the secretory coil has no enzyme activity (Ellis and Montagna, 1958). In some of the primitive primates, however, the apocrine glands have moderate to intense phosphorylase activity.

Apocrine glands have moderate concentrations of alkaline and acid phosphatases, distributed diffusely in the cytoplasm of the secretory cells and occasionally in the terminal cytoplasm of the very tall epithelial cells (Bunting et al., 1948). The base of each cell and the myoepithelial cells contain alkaline but no acid phosphatase. All apocrine glands characteristically contain a great deal of esterase and lipase activity. The glands of man and the horse have strong concentrations of amine oxidase (Hellmann, 1955; Shelley et al., 1955).

The secretory cells rarely contain glycogen (Montagna et al., 1951a; Montagna, 1956b). In the glands of man the myoepithelial cells of excessively dilated tubules may contain some glycogen. Apocrine glands contain small amounts of sulfhydryl and disulfide groups distributed diffusely in the cytoplasm of the secretory cells. Both substances are concentrated in the myoepithelial cells.

Although it is generally agreed that the apocrine glands of man and other mammals are innervated by adrenergic nerves (Rothman, 1954), there is no proof that they have any nerves around them at all. We have, however, found nerves containing cholinesterase around the axillary apocrine glands of Negroes but not around those of Caucasians.

3. Patterns of Secretion

The free border of most tall cells terminates in a brush border composed of cytoplasmic processes so small that they are barely visible under the light microscope. The electron microscope shows that these are, indeed, microvilli. Secretion may occur by an exudation of liquids through the brush border.

In spite of the general belief that the activity of the axillary organs of women corresponds to the cyclic menstrual changes, no changes have been found in biopsy specimens removed from the same women at weekly intervals during the menstrual cycle (Klaar, 1926; Montagna, 1956b). Even during pregnancy the glands exhibit a normal individual range of variation similar to that found in nonpregnant women. The axillary apocrine glands of mentally deficient women with a reasonably normal menstrual cycle, are very small (Shelley and Butterworth, 1955), indicating that ovarian hormones in sufficient amounts to maintain the menstrual cycle are not sufficient for the development and maintenance of the axillary glands. Experimental treatment with a number of hormones including testosterone, progesterone, estradiol, and pituitary growth hormones is without effect on the axillary apocrine glands (Shelley and Cahn, 1955b). Castration and menopause bring on changes quickly to some glands, but others in the same field remain normal. Ovarian hormones may play a role in initiating their maturation and maintaining in part their functional state, but other hormones, perhaps from the pituitary and the adrenal cortex, may be more closely implicated in guiding the function of apocrine sweat glands. This receives support from observations in the axillary glands of aging men, which show fewer regressive changes than those of women.

C. Eccrine Glands

1. Microscopic Anatomy

Most mammals have eccrine sweat glands only in the soles and in the pads of digits, and some, such as the rabbit, have none. Eccrine glands are more numerous in the skin of the higher primates, and they are best developed and most numerous in the skin of man.

Eccrine sweat glands are simple tubes that extend from the epidermis to the lower part of the dermis, where they are coiled. The duct is twisted in the epidermis; it is straight in the dermis and coiled deep in the dermis. The basal coil of each gland is composed of nearly equal parts of duct and secretory portions. The outer diameter of the duct is usually smaller than that of the secretory segment, its lumen is larger, and its bore does not vary. The duct is lined with an epithelium composed of two layers of cuboidal cells of about equal size; the secretory epithelium is a single layer of pyriform cells of different sizes. Large cells are crowded toward the base of the tubule and the small ones are pushed to the surface. All cells, however, reach the surface and rest on a layer of myoepithelial cells.

The nucleus of the larger cells is located at the base, and that of the smaller cells is near the distal end of the cytoplasm. The smaller cells are full of basophile cytoplasmic granules and have been named dark cells (Montagna et al., 1953a). The larger cells contain sparse, mostly acidophile, granules and have been called clear cells (Fig. 10). The luminal edge of all epithelial cells has a more or less distinct hyalin cuticular border.

Loosely dovetailed myoepithelial cells are interposed between the basement membrane and the secretory cells and are aligned parallel to the axis of the tubule. The myoepithelial cells in eccrine glands are similar to those in apocrine glands.

The thick basement membrane around the secretory segment is yellowish, hyalin, and refractile in unstained preparations. It is bi-refringent under polarized light when the tubules are sectioned transversely and stains like collagenous fibers. It is composed of a thin inner, periodic acid-Schiff-reactive layer and a thicker unreactive outer layer. The inner layer is more strongly argyrophile than the outer layer. Around the excretory duct the basement membrane is very thin.

The transition between the secretory segment of the tubule and the duct is indicated by a reduction in the thickness of the basement membrane and the appearance of two layers of cuboidal cells. The basal cells are cuboidal; the surface cells are also cuboidal, but they are slightly flattened, and the side which faces the lumen is differentiated into a

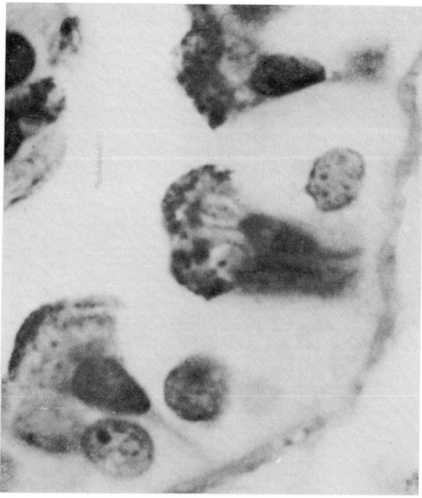

FIG. 10.   Dark and clear cells in an eccrine gland from the axilla of man.

hyalin, cuticular border. This cuticle is a modification of the cytoplasm at the luminal edge of each cell and shows many minor structural variations in the sweat glands of normal skin (Holyoke and Lobitz, 1952).

On the surface of the human palms and soles, the openings of the ducts are very regularly spaced in the center of the epidermal ridges (Hambrick and Blank, 1954). As soon as they enter the epithelial ridges the ducts become coiled. The coils are small at the base and become progressively larger as the terminal part of the spiral approaches the surface of the epidermis.

The cells of the duct are not pigmented. The nuclei of the secretory cells are round or oval and very compact; they stain well with the Feulgen technique. Most of the nuclei are stippled with granules and contain two or more small nucleoli crowded together. The presence of more than one nucleus in one cell has given rise to the belief that the cells divide by amitosis (Ito and Iwashige, 1951), but in spite of this belief recognizable amitosis has never been found.

Mitosis does not take place often in normal eccrine sweat glands (Montagna et al., 1953a), but when secretory coils are injured their cells do divide mitotically (Lobitz and Dobson, 1957). The cells of the ducts are relatively mitotically inert, but the basal cells proliferate readily after injury (Lobitz et al., 1954). Injury gives great impetus to the proliferative ability of the cells near the injury.

Relatively coarse mitochondria are aggregated around the nucleus of secretory cells and of the cells of the duct (Fig. 11). In clear cells they tend to concentrate in the basal part. In the supranuclear cytoplasm, long filaments often radiate toward the luminal border. Mitochondria are numerous in some cells but not in others. They are often looped around chromophobe secretion droplets, suggesting that they may mediate or aid in the formation of secretion material.

A delicate, osmiophile network, the Golgi apparatus, can be demonstrated over the nucleus of the dark cells (Ito, 1943). Sometimes the network descends to the sides of the nucleus, and rarely it may be found under the nucleus. Since its shape changes after local injections of pilocarpine, the Golgi element may be implicated in secretion (Melczer, 1935).

2. Histochemical Properties

The cytoplasm of the cells of eccrine glands in human skin is stippled with delicate granules and a few coarse granules that contain lipid. Slight variations occur in the amount of lipid in different individuals and in individual glands. The glands of children have few lipid granules; those of adults have more and those of old subjects may abound in them (Kano, 1951). Lipids seem to be stable, additive components of eccrine cells; they are not secreted, even when the glands are stimulated to sweat (Shelley and Mescon, 1952).

Most eccrine glands contain a diffuse yellow pigment in their cytoplasm, probably a carotenoid (Bunting et al., 1948). They also have coarse pigment granules that resist extraction, contain no iron, are not reactive to the periodic acid-Schiff test, and are firmly bound to a lipid. Under near-ultraviolet light the pigment fluoresces with a yellow or orange light.

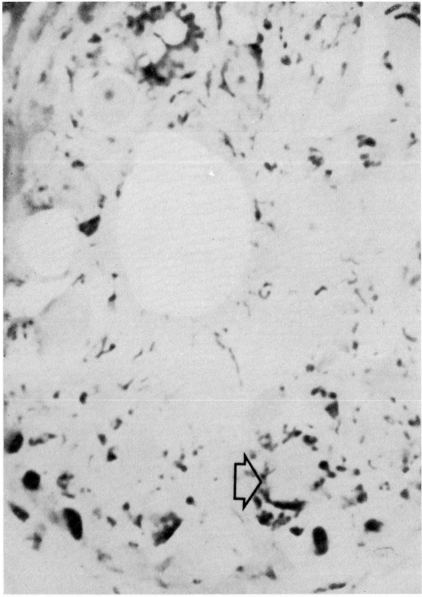

FIG. 11. Mitochondria in the cells of an eccrine gland from the axilla of man. The arrow indicates mitochondria clustered around a nucleus.

Eccrine sweat glands contain glycogen (Tsukagoshi, 1951; Montagna *et al.*, 1952a), more of it in the clear cells than in the dark cells. There is a disparity in the distribution of glycogen in the cells in any one tubule. Some cells are laden with it and others may contain none. In the clear cells glycogen is piled up on the walls of intercellular and intracellular canaliculi. Injections of cholinergic substances into the skin, or profuse thermal sweating, abolish temporarily the glycogen from the secretory cells (Shelley and Mescon, 1952). Glycogen is restored after variable recovery periods which depend on the severity and the duration of the stimulus.

The epithelium of the duct of the sweat glands routinely contains glycogen (Montagna, 1956a; Lobitz *et al.*, 1955), mostly in the basal cells. The distribution of glycogen in the surface cells is unique (Lobitz *et al.*, 1955). A fine line of glycogen marks the separation between the cytoplasm and the cuticle.

All of the cells of the sweat glands, including the duct and the secretory segment, contain abundant phosphorylase activity (Braun-Falco, 1955; Ellis and Montagna, 1958). No other cutaneous appendage is so rich in this enzyme. Granules in the apexes of the dark cells are reactive to the periodic acid-Schiff test, but they are resistant to digestion with saliva and stain metachromatically; therefore, they may contain mucopolysaccharides (Montagna *et al.*, 1951b). These granules are numerous in most specimens, but in some they may be practically absent. In some cells these granules almost fill the cytoplasm; in others they may be aggregated only in the perinuclear area or in the apex of the cytoplasm. The lumen of the secretory coil is generally free of this substance, but a thin layer of it coats the cuticle of the duct all the way to the orifice (Formisano and Lobitz, 1957). The lumen of the ducts of some glands is often filled with material similar to that of the granules just described. A weak to moderate alkaline phosphatase reaction is concentrated in the myoepithelial cells and in the clear cells. Acid phosphatase reaction is localized mostly in the apical border of the secretory cells. The entire gland also shows a moderately strong concentration of β-glucuronidase (Montagna, 1957).

Esterase activity is moderate to strong in the dark cells and weak in the clear cells. Strong esterase activity is obtained when the indoxyl acetate and the AS acetate techniques are used (Braun-Falco, 1956a). The clear cells selectively show more AS esterase than the dark cells do. The nerve fibers that surround the secretory tubules contain specific cholinesterase (Hurley *et al.*, 1953). This has been observed in the glands of man and in those of other animals (Hellmann, 1952, 1955). On denervation, all enzyme activity disappears (Hellmann, 1952). The cholinesterase-reactive nonmyelinated fibers form a tightly wound skein around the secretory coil; no nerves are found around the duct (Montagna and Ellis, 1959).

The duct and secretory coil abound in monoamine oxidase, cytochrome oxidase, and aminopeptidase (Montagna and Ellis, 1959).

The entire length of eccrine sweat glands, whether in man or other animals, has a larger concentration of succinic dehydrogenase than any other epidermal appendage. The reaction is most intense in the secretory coil (Braun-Falco and Rathjens, 1954; Montagna and Formisano, 1955).

The cytoplasm of the secretory cells is delicately stippled with granules of sulfhydryl and disulfide groups more intensely concentrated in the basal portion of the cells. The myoepithelial cells show a somewhat stronger reaction than the other cells. The cells in the duct are weakly reactive, but the cuticular border of the luminal cells is fairly distinct.

3. Patterns of Secretion

Individual glands do not secrete equal amounts of sweat, and some glands have different levels of responsiveness to stimuli. In response to heat, the human body sweats profusely on the forehead, neck, dorsal and ventral part of the trunk, and the back of the hands, and much less on the cheeks, the sides of the trunk, and the extremities. The palms and soles show the least response to thermal stimuli, but they are the first to respond to psychic and sensory stimulation (Rothman, 1954).

The formation of cytoplasmic granules which are presumably secretory, seems to take place in the cytoplasm above the nucleus; this is seen particularly well in the dark cells, where the granules are coarse and numerous. Granules are found near the mitochondria, but these elements exist in rather confined quarters and even unrelated things could be crowded together. A direct transformation from mitochondria to granules has not been ascertained.

Between the clear cells are complex systems of intercellular canaliculi. Canaliculi contain alkaline phosphatase, and glycogen is piled up against their walls. The electron microscope shows many microvilli protruding within the walls of the canaliculi.

The histology and histochemistry of the cells of the excretory duct of eccrine glands suggests that the duct has an active function, probably in the resorption of water (Lobitz and Mason, 1948). The duct is longer than the entire secretory segment, and would seem to be wasteful if it had no function (Montagna, 1956a).

Dyes or iron, forced through the ducts by electrophoresis, eventually escape to the surrounding dermis through the wall of the duct (Araki and Ando, 1953). Water and water-soluble substances are probably reabsorbed by the duct. Urea secreted by the secretory segment becomes more concentrated in the duct, presumably by the reabsorption of water from the solution (Nitta, 1953; Schwartz *et al.*, 1953). The distribution of glycogen in the duct under different physiological states seems to reflect its activity (Lobitz *et al.*, 1955). Severe sweating removes all glycogen from the cells of the duct, whereas moderate sweating depletes only the glycogen in the luminal cells. During sweating the delicate band of glycogen in the luminal cells disappears. When, after sweating, the glycogen in the secretory segment is reduced, that in the coiled segment of the duct is considerably reduced, but that in the basal cells of the straight segment is only slightly affected. In continued profuse sweating, glycogen disappears from the secretory segment and from all of the cells of the straight segment of the duct.

VII. The Nails

A. *Gross and Microscopic Anatomy*

Nails are dense, translucent plates of cornified cells on the dorsal, distal phalanges of the fingers and toes. They are deeply convex from side to side and gently convex from front to back. The distal edge grows free beyond the digit; the rest is tucked into a depression of the skin, the nail fold. This grows over the proximal border of the nail, forming the nail wall, whose free end over the nail lamina is the eponychium. The proximal end of the nail, under the eponychium, has a crescent-shaped whitish area called the lunula. The epidermis of the nail fold continues under the nail, forming a nail bed. This is continuous anteriorly with the epidermis of the volar side of the digits. A pad of keratinized cells, the hyponychium, grows under the angle of the free edge of the nail.

The surface of the nail plate is grooved by delicate, longitudinal parallel lines that correspond to deep corrugations on the underside. These ribs fit into grooves in the epithelium of the nail bed, which is the negative image of the underside of the nail (Horstmann, 1955). These ridges can be seen particularly well in transverse sections or in completely excised nails.

The nail continues underneath the nail fold as the nail root. Its proximal edge is a thin wedge surrounded above and below by a thick epidermal bed, the matrix, or germinative organ. The matrix corresponds to the combined stratum germinativum and stratum spinosum of the epidermis or to the entire bulb of hair follicles. Numerous mitotic figures can be seen only in the cells in the lower part of the matrix; the new cells move up in streams to the root, outlining curved paths. The matrix extends above and below the proximal border of the nail, and for this reason it is divided into a dorsal matrix, at the base of the nail, continuous with the epidermis of the eponychium, and a thick basal matrix on the ventral side. The nail bed has been designated as the sterile matrix, but it is not a part of the germinative matrix.

The basal matrix is much thicker than the nail bed and is sharply demarcated from it. The nail plate is stratified; the main body of the nail is the dorsal nail; there is an intermediate nail of varying thickness and a ventral nail that is seen clearly only in the free portion of the nail. In the attached part, the ventral nail cannot be differentiated from the nail bed. This stratification of the nail lamina can be demonstrated with appropriate staining techniques, with protein silver methods and with polarized light.

In the matrix, a gradual transformation occurs from undifferentiated cells at the periphery to fully keratinized ones in the nail proper. As the cells transform they attain visible, stainable fibrils, in a way similar to that which occurs in the keratogenous zone of hair follicles. The keratinization of the nail is different from that which occurs in other cutaneous organs. The cells, for example, all retain their nuclei which can be demonstrated as isotropic elements under polarized light.

The nail bed under the attached nail rarely shows differentiation. Although opinions vary, it has been assumed that it is static and that the nail glides over it. However, the two structures are so firmly bound that if the nail is pulled out, the nail bed remains attached to it. Therefore, the nail bed must grow and move forward in the same way that the inner root sheath of growing hair follicles does. Even when parts of the nail plate are removed, leaving the nail bed exposed, it moves forward (Pillsbury et al., 1956). Usually, the epithelium of the nail bed does not become keratinized under the attached nail, but in certain abnormal conditions, or after injury, it forms keratohyalin granules and produces a horny layer under the nail plate. Abnormally thick and opaque nails are often the result of keratinization of the nail bed. Since the nail bed has no proliferative capacity of its own, it must be produced by the basal matrix. Complete keratinization is delayed until just before the nail bed reaches the free border, when it forms the ventral nail and part of the callosity of the hyponychium.

B. Histochemical Properties

The matrix of the nail contains ribonucleoprotein; the greatest concentration of it is in the cells of the germinative part, and the differentiating cells contain progressively less. The nail bed also contains some ribonucleoprotein, but much less than the matrix does. The matrix and the nail bed stain a mild metachromatic color with toluidine blue, indicating the presence of some acid mucopolysaccharide; there is no meta-chromatic staining in the nail plate. The nuclei of the matrix are strongly Feulgen reactive; their reactivity gradually diminishes as the cells become differentiated; the nuclei of the cells in the nail plate are unreactive, except a few in the intermediate and ventral nails.

The matrix shows an intense reaction for succinic dehydrogenase. The pattern of distribution of the enzyme is similar to that in the epidermis, being strongest in the germinative portion and gradually becoming extinct in the differentiating cells. There are no traces of alkaline phosphatase in the nail matrix or in any of the epithelial struc-tures associated with the nail. Strong acid phosphatase reaction, how-ever, is found in the differentiating region of the matrix and in the entire epithelium of the nail bed. The enzyme is concentrated at the junction of the nail with the epithelial elements. Esterase activity can be demon-strated in every cell in the matrix, but the reaction is never strong. The keratinizing area of the matrix, however, has a relatively strong reaction. Although the ventral nail is strongly active along its entire length, the nail bed epithelium to which it is attached contains no demonstrable esterase. A single row of strongly reactive cells neatly marks off the boundary of the ventral nail and the other nail elements which are com-pletely unreactive. The distribution of β-glucuronidase in the matrix and the nail bed is like that of the esterases.

The distribution of all these enzymes in the nail-forming organ is like that found in the epidermis of the skin. Acid phosphatase, esterase, and β-glucuronidase are concentrated just above the stratum granulosum in the epidermis where the final stages of keratinization take place.

The nail-forming organ contains very little glycogen. There is none in the matrix, but the epithelium of the nail bed usually contains some; a few glycogen-rich cells in the upper border mark a straight line under-neath the nail lamina. The epithelium of the nail bed near the junction with the epidermis is always rich in glycogen. The ventral nail contains a small amount of glycogen.

In frozen sections colored with Sudan black, only the cells in the nail bed and those of the ventral nail contain some demonstrable lipids.

The entire proximal end of the nail root, including the upper part of the matrix and the keratinizing nail plate, is intensely reactive for sulfhydryl groups; this, which corresponds to the keratogenous zone of Giroud et al. (1934), is composed of strongly reactive fibrils. The nail bed has a relatively weak reaction and the dorsal nail shows no reaction. The ventral nail has strong but diffuse reaction. All of the parts of the nail lamina give an intense reaction for -S-S- groups.

C. General Considerations

The nail plate is composed of different keratins. In some ways the generation, formation, and arrangements of the keratin layers is similar to that found in the hair and hair follicle. The ventral nail, which is formed by the keratinization of the nail bed before the nail plate emerges free at the end of the phalanx, resembles the inner root sheath. The nail bed is a slow-keratinizing layer of the nail and is formed by the proliferation of the ventral matrix, at about the same rate as the dorsal and intermediate nails. The dorsal nail is invested by a closely adhering thin film of keratin formed by the eponychial fold.

The difficulties in obtaining material and the difficulties in handling it are responsible for the lack of knowledge of the growth and differentiation of the nail and its cytology and histochemistry. Only Lewis (1954), Pillsbury *et al.* (1956), and Horstmann (1955) have studied the nail with new methods. It is likely that when we know more about it, the nail-forming organ will prove to be the most exciting of the cutaneous organs.

It is misleading to compare the nail lamina with the stratum lucidum or the stratum corneum of the epidermis. The structure of the nail keratins is not like that found elsewhere in the cutaneous system (Horstmann, 1957), and the apparatus that forms the nail, like that which forms hairs or the surface horny layers, is a special organ; the complex end products it forms have characteristic architecture and chemical composition, quite different from the others. The formation of keratinized structures is a delicately controlled system of events that is different in different structures. The fact that the nail matrix can form the nail lamina, composed of several layers of different keratins, attest to the precision of guidance of the nail matrix as well as to the total lability that is inherent within it.

VIII. SUMMATION

Each cutaneous appendage has distinctive structural features of its own. These features, however, are borne only by the fully differentiated cells; undifferentiated cells from all of them are structurally and potentially similar. The formation of relatively dissimilar substances by them is brought about by agents that specifically evoke one potentiality while suppressing others.

The same type of cell can form substances as different as sebum and keratin; these are the two functions most readily performed by epidermal cells. Cells from sebaceous glands normally show this bimodality. The innermost layer of cells in the ducts is always keratinized, whereas other undifferentiated cells from the same source can undergo sebaceous differentiation. Various disturbances can cause a transformation of nearly all the undifferentiated sebaceous cells into keratinized cells, and when normal conditions are restored, keratinization becomes limited again to the cells lining the excretory duct. The cells of the pilary units also share the ability to undergo sebaceous or keratinous transformation. Fragments of hair follicles stranded in the dermis of the skin of hairless mice give rise to sebaceous cysts and these in turn become keratin cysts.

The cells of the hair follicles and sebaceous glands, and those of the sweat glands readily form epidermis in wound healing (Eisen *et al.*, 1955; Gianni, 1951). When the epidermis is destroyed, fragments of injured hair follicles at the periphery of the wound contribute cells for the restoration of a covering epithelium. Whole hair follicles at the edge of a wound unfold and the cells flow over the wound (Argyris, 1953). The cells of injured secretory coils of eccrine sweat glands can also transform into typical epidermal cells with intercellular bridges (Lobitz and Dobson, 1957). Keratinous or "squamous" changes can take place even in the secretory coil of apocrine glands (Shelley and Cahn, 1955a). These phenomena have long been recognized, but they have been dismissed as metaplasia, which explains nothing. These examples, and others, demonstrate that cells derived from the epidermis are labïle in the adult. Even those cells that have undergone specific transformation, such as the cells of sweat glands or those of the outer root sheath of hair follicles, may regress to indifferent types and then transform again into a different type. This is particularly evident in wound healing, in which the cells of pilosebaceous units and the sweat glands form a surface layer of cells that is indistinguishable from epidermal cells. The inescapable conclusion is that in the adult, regardless of the bias that the cells of the epidermal family may have acquired toward a specific differentiation, they remain basically equipotential. This means that agents that induce and guide differentiation probably reside in the dermis. The best example of this is the relation of the dermal papilla to hair follicles. The follicle remains an organized unit only as long as it remains in contact with the dermal papilla; it perishes when this contact is lost (Montagna and Chase, 1956).

The possible role of the dermis in guiding the differentiation of epidermal appendages is rarely emphasized. In the embryo the mesenchyme always participates in the morphogenesis of organs, and in some organs, mesenchymal differentiation even precedes that of the parenchyma. Parenchymal anlagen require a well-prepared stromal field upon which they may grow. The inductors may require for their action a specific organization of the responding system. Perhaps they act by repatterning the system rather than by directly transforming its cells (Borghese, 1950).

Even *in vitro,* pure cultures of epidermal cells become organized into epidermis only when they are mixed with mesoderm (Fazzari, 1951).

In the dermis, fibroblasts, like epidermis cells, are probably multi-potential cells, the particular modulations of which may be evoked by the specific physicochemical conditions of the environment. The fibroblast need not be considered to be exactly the same type of cell at all times. Under certain conditions, fibroblasts may become undifferentiated, and differentiate again into the different types of connective tissue cells. The fibroblast is a cell delicately sensitive to the demands of the organ; to remain attuned with these demands it must undergo countless modulations. The two principal component elements of skin, the dermis and epidermis, then, both exhibit lability, modulation, and unity.

IX. THE PIGMENT CELLS

A. *General Introduction*

Melanin is a yellow-to-black pigment related to the metabolism of tyrosine. It is resistant to nearly all chemical agents and is not modified even by concentrated acids. Melanin is argentaffin, reducing ammoniacal silver nitrate without the intervention of a reducing agent.

In normal human epidermis, melanin is present in varying amounts in the basal layers of the Malpighian layer; in heavily pigmented skin, melanin granules are present throughout the stratum spinosum and the stratum corneum. When abundant, melanin granules are evenly and diffusely distributed in the cytoplasm of epidermal cells, especially in those of the basal layer; when less abundant, the granules form supra-nuclear caps. Melanin is not produced by the epidermal cells but by dendritic cells, the melanocytes. These cells have arisen from the neural crest and migrate to the epidermis, where they occur couched between the basal cells; in hair follicles they reside in the upper part of the bulb, and are neatly distributed over the dome of the papilla cavity. The dermis normally contains few melanocytes, with the exception of nevi which abound in them.

The number of melanocytes differs in corresponding areas of different human beings, and even in the same individual they vary from area to area. They are most numerous in the skin of the head and the neck, where about 2000 to 4000 may be found per square millimeter of surface epidermis. They are scantiest in the skin of the thigh and arm, where they average about 1000 per square millimeter (Szabo, 1954). Melanocytes are concentrated in the epidermal ridges or rete pegs; few of them are found between the ridges. Melanocytes are usually dendritic with branches radiating in all directions between the surrounding epidermal cells. They may also be smaller with only two terminal processes, resembling fibroblasts. This dimorphism is particularly striking in the skin of the cheek and forehead, where the smaller fibroblast-like melanocytes are more numerous than the larger, dendritic ones (Szabo, 1954).

B. Microscopic and Histochemical Properties

Melanocytes can be demonstrated selectively with the dopa (dihydroxyphenylalanine) reaction of Bloch (1917). This reaction stains the cytoplasm of the epidermal dendritic cells gray, grayish brown, or black, whereas the epidermal cells remain uncolored (Laidlaw and Blackberg, 1932; Radaeli, 1951). Dopa-negative cells that contain melanin are the melanophores. There appears to be a sequence of centripetal maturation of melanin in the dendritic melanocytes, the most distally located granules in the dendrite being the most mature (Masson, 1948).

Some dendritic cells in the epidermis have a cell body smaller than normal melanocytes and the nucleus less well defined. These cells of Langerhans have a clear, not basophilic cytoplasm, usually collapsed around the nucleus. In the upper levels of the epidermis, cells of Langerhans are always unpigmented, regardless of the degree of pigmentation of the cells around them. Mitotic divisions have never been found in their nuclei. Structurally, Langerhans cells resemble melanocytes (Billingham and Medawar, 1948a, b, 1953) and seem to be "effete" melanocytes which, having discharged or lost their pigment, participate in the outward movement of epidermal cells and are eventually sloughed off with the keratin layer.

Melanocytes are present even in albino skin, but the granules they produce are not pigmented.

C. Electron Microscopy and Melanin Formation

The numerous pigment granules in melanocytes mask nearly all cytological details. Little of value, therefore, has been reported about the internal structure of these cells. The electron microscope has furnished the only real clue to their cytology. Melanocytes can be recognized under the electron microscope because they lack the bundles of cytoplasmic filaments and fibrils found in epidermal cells, and they have numerous dense melanin granules in the cytoplasm and in the dendritic processes. Fragments of cytoplasmic processes lie in the intercellular spaces between epidermal cells, or cortical cells in the hair roots. The melanocytes in either locality are relatively alike. Active melanin formation takes place in a specialized region of the cytoplasm, usually distal to the nucleus, which contains numerous mitochondria and small vacuoles (Birbeck et al., 1956). The center of this zone is free of melanin granules, but at its periphery small vacuoles about 0.5 μ in diameter have delicate shells near the surface or they contain convoluted strands and lamellae of dense material (Fig. 12). These appear to be the early stages of formation of melanin granules. The deposition of more dense material and the thickening of the shells results in the formation of granules that are about twice the size of the original vacuoles. When their internal structure can be seen in optimal preparations, mature melanin granules are formed of folded membranes or lamellae, between the folds of which are deposited very fine, dense particles. The amelanotic melanocytes in albino skin also have aggregates of ovoid or rod-shaped bodies about 0.5 μ in length that contain strands of lamellae comparable to those found in colored melanocytes, but these show no deposition of dense material; this is probably due to the virtual absence of tyrosinase. The folded membranes and shells, then, are the protein skeletons of the pigment granules. Melanin granules are formed by the deposition of indole-5,6-quinone upon a protein skeleton, to which it becomes firmly attached.

The mitochondria in melanocytes are numerous and distinct, particularly around the zone of melanogenesis. They undoubtedly play some role in the formation of melanin granules, but the role seems to be indirect. The Golgi apparatus seems to be more directly concerned with the formation of the granules. This is composed of parallel double membranes associated with cisternae. The cisternae of the Golgi material tend to form a series of vacuoles; some of these contain melanin granules. The wall of vacuoles which contain immature melanin granules are often drawn out as if they had been pinched off from the Golgi apparatus. Melanin granules formed by epidermal melanocytes are smaller than those formed by hair follicle melanocytes. The pigment granules in the melanocytes in the skin of Negroes are larger and more numerous than those in the skin of Caucasians.

D. The Spread of Pigment

The problem of transferring melanin granules from the dendrites of melanocytes to the epidermal cells or the cells of the cortex is not fully understood. Melanocytes have been compared to glandular cells whose cytoplasmic processes inject the melanin into the epidermal cells by a cytocrine process (Masson, 1948). The observations of the electron microscopist on the melanocytes of hair follicles seem to indicate that the recipient cells actually phagocytose the pigment-bearing tips of the dendrites of melanocytes. This is the most plausible explanation given of this phenomenon. Whether or not pigment is transferred only in this way, must be confirmed.

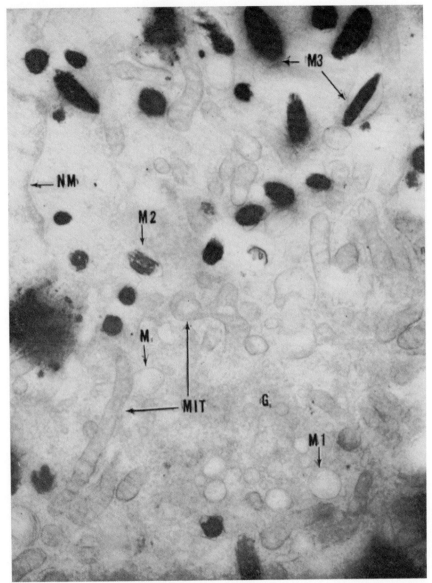

FIG. 12.   Electron micrograph of a melanocyte from a human hair follicle. Only a small part of the cell is shown. *NM*, nuclear membrane; *G*, Golgi zone, with numerous small vesicles; *MIT*, mitochondria; *M1, M2, M3*, melanin granules in various stages of increasing melanization. Courtesy of Drs. N. A. Barnicot and M. S. C. Birbeck.

REFERENCES

Andrew, W. (1952). *J. Gerontol.* **7**, 178.
Araki, Y., and Ando, S. (1953). *Japan. J. Physiol.* **3**, 11.
Argyris, T. S. (1952). *J. Natl. Cancer Inst.* **12**, 1159.
Argyris, T. S. (1953). Thesis, Brown University, Providence, Rhode Island.
Asboe-Hansen, G. (1950). *J. Invest. Dermatol.* **15**, 25.
Asboe-Hansen, G. (1951). Thesis, Rosenkilde and Bagger, Copenhagen.
Asboe-Hansen, G. (1952). *Proc. Soc. Exptl. Biol. Med.* **80**, 677.
Asboe-Hansen, G. (1953). *Cancer Research* **13**, 587.
Asboe-Hansen, G., and Iversen, K. (1951). *Acta Endocrinol.* **8**, 90.
Astbury, W. T., and Woods, T. J. (1930). *Nature* **126**, 913.
Auber, L. (1952). *Trans. Roy. Soc. Edinburgh* **62**, 191.
Billingham, R. E., and Medawar, P. B. (1948a). *Heredity* **2**, 29.
Billingham, R. E., and Medawar, P. B. (1948b). *Brit. J. Cancer* **2**, 126.
Billingham, R. E., and Medawar, P. B. (1953). *Phil. Trans. Roy. Soc. London* **B237**, 151.
Birbeck, M. S. C., and Mercer, E. H. (1957a). *J. Biophys. Biochem. Cytol.* **3**, 203.
Birbeck, M. S. C., and Mercer, E. H. (1957b). *J. Biophys. Biochem. Cytol.* **3**, 215.
Birbeck, M. S. C., and Mercer, E. H. (1957c). *J. Biophys. Biochem. Cytol.* **3**, 223.
Birbeck, M. S. C., Mercer, E. H., and Barnicot, N. A. (1956). *Exptl. Cell Research* **10**, 505.
Bloch, N. (1917). *Arch. Dermatol. u. Syphilis* **124**, 129.
Bloom, F. (1952). *Proc. Soc. Exptl. Biol. Med.* **79**, 651.
Bock, F. G., and Mund, R. (1956). *J. Invest. Dermatol.* **26**, 479.
Borghese, E. (1950). *J. Anat.* **84**, 303.
Braun-Falco, O. (1955). *Arch. klin. u. exptl. Dermatol.* **202**, 163.
Braun-Falco, O. (1956a). *Arch. klin. u. exptl. Dermatol.* **202**, 153.
Braun-Falco, O. (1956b). *Arch. klin. u. exptl. Dermatol.* **203**, 61.
Braun-Falco, O., and Rathjens, B. (1954). *Dermatol. Wochschr.* **130**, 1271.
Braun-Falco, O., and Rathjens, B. (1955). *Arch. klin. u. exptl. Dermatol* **201**, 73.
Braun-Falco, O., and Rathjens, B. (1956). *Arch. klin. u. exptl. Dermatol,* **203**, 130.
Braun-Falco, O., and Salfeld, K. (1959). *Nature* **183**, 51.
Bullough, H. F. (1947). *Nature* **159**, 101.
Bullough, W. S. (1946). *Phil. Trans. Roy. Soc. London* **B231**, 453.
Bullough, W. S. (1948a). *Proc. Roy. Soc.* **B135**, 212.
Bullough, W. S. (1948b). *Proc. Roy. Soc.* **B135**, 233.
Bullough, W. S. (1949a). *J. Exptl. Biol.* **26**, 83.
Bullough, W. S. (1949b). *Nature* **163**, 680.
Bullough, W. S. (1949c). *J. Exptl. Biol.* **26**, 261.
Bullough, W. S. (1949d). *J. Exptl. Biol.* **26**, 287.
Bullough, W. S. (1949e). *Brit. J. Cancer* **3**, 275.
Bullough, W. S. (1950a). *J. Endocrinol.* **6**, 340.
Bullough, W. S. (1950b). *Acta Physiol. et Pharmacol. Neerl.* **1**, 357.
Bullough, W. S. (1952a). *Biol. Revs. Cambridge Phil. Soc.* **27**, 133.
Bullough, W. S. (1952b). *J. Endocrinol.* **8**, 265.
Bullough, W. S. (1953). *Ciba Foundation Colloq. Endocrinol.* **6**, 278.
Bullough, W. S. (1954a). *Exptl. Cell Research* **17**, 176.
Bullough, W. S. (1954b). *Exptl. Cell Research* **17**, 186.
Bullough, W. S., and Ebling, F. J. (1952). *J. Anat.* **86**, 29.

Bullough, W. S., and Green, H. N. (1949). *Nature* **164**, 795.

Bullough, W. S., and Van Oordt, G. J. (1950). *Acta Endocrinol.* **4**, 291.

Bunting, H. (1950). *Ann. N.Y. Acad. Sci.* **52**, 977.

Bunting, H., Wislocki, G. B., and Dempsey, E. W. (1948). *Anat. Record* **100**, 61.

Cass, R., Riley, J. F., West, G. B., Head, K. W., and Stroud, S. W. (1954). *Nature* **174**, 318.

Chèvremont, M. (1948). *Biol. Revs.* **23**, 267.

Compton, A. S. (1952). *Am. J. Anat.* **91**, 301.

Cooper, Z. K. (1939). *J. Invest. Dermatol.* **2**, 289.

Cooper, Z. K., and Schiff, A. (1938). *Proc. Soc. Exptl. Biol. Med.* **39**, 323.

Dougherty, C. M. (1948). *J. Pathol. Bacteriol.* **60**, 511.

Drennan, J. M. (1951). *J. Pathol. Bacteriol.* **63**, 513.

Eartly, H., Grad, B., and Leblond, C. P. (1951). *Endocrinology* **49**, 677.

Eisen, A. Z., Holyoke, J. B., and Lobitz, W. C., Jr. (1955). *J. Invest. Dermatol.* **25**, 145.

Ellis, R. A., and Montagna, W. (1958). *J. Histochem. and Cytochem.* **6**, 201.

Fawcett, D. W. (1954). *Anat. Record* **188**, 297.

Fazzari, I. (1951). *Rass. clin. sci.* **27**, 355.

Favre, M. (1950). *Ann. dermatol. syphilig.* **10**, 241.

Formisano, V. R., and Lobitz, W. C., Jr. (1957). *A. M. A. Arch. Dermatol.* **75**, 202.

Formisano, V. R., and Montagna, W. (1954). *Anat. Record* **120**, 893.

Gersh, I., and Catchpole, H. R. (1949). *Am. J. Anat.* **85**, 457.

Gianni, A. (1951). *Boll. ist. sieroterap. milan.* **30**, 151.

Giroud, A., and Bulliard, H. (1930). *Arch. morphol.* **29**, 1.

Giroud, A., Bulliard, H., and Leblond, C. P. (1934). *Bull. histol. appl. physiol. et pathol. et tech. microscop.* **11**, 129.

Glegg, R. E., Eidinger, E., and Leblond, C. P. (1953). *Science* **118**, 614.

Gross, J. (1950). *Am. J. Pathol.* **26**, 708.

Gross, J., and Schmitt, F. O. (1948). *J. Exptl. Med.* **88**, 555.

Hambrick, G. W., Jr., and Blank, H. (1954). *J. Invest. Dermatol.* **23**, 437.

Hamilton, J. B. (1941). *J. Clin. Endocrinol.* **1**, 570.

Hellmann, K. (1952). *Nature* **169**, 113.

Hellmann, K. (1955). *J. Physiol. (London)* **129**, 454.

Holyoke, J. B., and Lobitz, W. C., Jr. (1952). *J. Invest. Dermatol.* **18**, 147.

Homma, H. (1926). *Bull. Johns Hopkins Hosp.* **38**, 365.

Horn, G. (1935). *Z. mikroskop. anat. Forsch.* **38**, 318.

Horstmann, E. (1955). *Z. Zellforsch. u. mikroskop. Anat.* **41**, 532.

Horstmann, E. (1957). "Die Haut." *Handb. mikroskop. Anat. Menschen.* **III/3**, 1.

Hurley, H. J., Jr., and Shelley, W. B. (1954). *J. Invest. Dermatol.* **22**, 143.

Hurley, H. J., Jr., Shelley, W. B., and Koelle, G. B. (1953). *J. Invest. Dermatol.* **21**, 139.

Ito, T. (1943). *Okajimas folia anat. japon.* **23**, 273.

Ito, T., and Iwashige, K. (1951). *Okajimas folia anat. japon.* **23**, 147.

Kano, K. (1951). *Arch. histol. japon.* **3**, 91.

Klaar, J. (1926). *Wien. klin. Wochshr.* **39**, 127.

Laidlaw, G. F., and Blackberg, S. N. (1932). *Am. J. Pathol.* **8**, 491.

Lasher, N., Lorincz, A. L., and Rothman, S. (1954). *J. Invest. Dermatol.* **22**, 25.

Lederer, E., and Mercier, D. (1948). *Biochim. et Biophys. Acta* **2**, 91.

Lewis, B. L. (1954). *A. M. A. Arch. Dermatol.* **70**, 732.

Lillie, R. D., Emmart, E. W., and Laskey, A. M. (1951). *A. M. A. Arch. Pathol.* **52**, 363.

Lobitz, W. C., Jr., and Dobson, R. L. (1957). *J. Invest. Dermatol.* **28**, 105.

Lobitz, W. C., Jr., and Holyoke, J. B. (1954). *J. Invest. Dermatol.* **22**, 189.

Lobitz, W. C., Jr., and Mason, H. L. (1948). *A. M. A. Arch. Dermatol.* **57**, 907.

Lobitz, W. C., Jr., Holyoke, J. B., and Montagna, W. (1954). *J. Invest. Dermatol.* **22**, 157.

Lobitz, W. C., Holyoke, J. B., and Brophy, D. (1955). *A.M.A. Arch. Dermatol.* **72**, 229.

Lowry, O. H., Gilligan, D. R., and Katersky, E. M. (1941). *J. Biol. Chem.* **139**, 795.

Masson, P. (1948). *N. Y. Acad. Sci., Spec. Publ.* **4**, 15.

Melczer, N. (1935). *Dermatol. Wochschr.* **100**, 337.

Mercer, E. H. (1953). *Textile Research J.* **23**, 388.

Mercer, E. H. (1958). *In* "Biology of Hair Growth" (W. Montagna and R. A. Ellis, eds.), p. 91. Academic Press, New York.

Mergenthaler, D. D., and Paff, G. H. (1956). *Anat. Record* **126**, 165.

Meyer, K. (1945). *Advances in Protein Chem.* **2**, 249.

Meyer, K. (1947). *Physiol. Revs.* **27**, 335.

Meyer, K., and Rapport, M. M. (1951). *Science* **113**, 596.

Minamitani, K. (1941a). *Okajimas folia anat. japon.* **21**, 61.

Minamitani, K. (1941b). *Okajimas folia anat. japon.* **20**, 563.

Montagna, W. (1950). *Quart. J. Microscop. Sci.* **91**, 205.

Montagna, W. (1955). *J. Invest. Dermatol.* **25**, 117.

Montagna, W. (1956a). "Structure and Function of Skin." Academic Press, New York.

Montagna, W. (1956b). *Ciba Foundation Colloquia on Ageing.* **2.** *Ageing in Transient Tissues,* p. 188.

Montagna, W. (1957). *J. Biophys. Biochem. Cytol.* **3**, 343.

Montagna, W., and Chase, H. B. (1950). *Anat. Record* **107**, 83.

Montagna, W., and Chase, H. B. (1956). *Am. J. Anat.* **99**, 415.

Montagna, W., and Ellis, R. A. (1959). *Minerva Dermatol.* **34**, 475.

Montagna, W., and Formisano, V. R. (1955). *Anat. Record* **122**, 65.

Montagna, W., and Noback, C. R. (1948). *Anat. Record* **100**, 535.

Montagna, W., and Parks, H. F. (1948). *Anat. Record* **100**, 297.

Montagna, W., Noback, C. R., and Zak, F. G. (1948). *Am. J. Anat.* **83**, 409.

Montagna, W., Chase, H. B., and Hamilton, J. B. (1951a). *J. Invest. Dermatol.* **17**, 147.

Montagna, W., Chase, H. B., and Melaragno, H. P. (1951b). *J. Natl. Cancer Inst.* **12**, 591.

Montagna, W., Chase, H. B., and Lobitz, W. C., Jr. (1952a). *Anat. Record* **114**, 231.

Montagna, W., Chase, H. B., and Melaragno, H. P. (1952b). *J. Invest. Dermatol.* **19**, 83.

Montagna, W., Chase, H. B., and Lobitz, W. C., Jr. (1953a). *J. Invest. Dermatol.* **20**, 415.

Montagna, W., Chase, H. B., and Lobitz, W. C., Jr. (1953b). *Am. J. Anat.* **92**, 451.

Montagna, W., Eisen, A. Z., and Goldman, A. S. (1954). *Quart. J. Microscop. Sci.* **95**, 1.

Moretti, G., and Mescon, H. (1956). *J. Invest. Dermatol.* **26**, 347.

Nitta, H. (1953). *Nagoya Med. J.* **1**, 59.

Odland, G. F. (1958). *J. Biophys. Biochem. Cytol.* **4**, 529.

Opdyke, D. L. (1952). Thesis, Washington University, St. Louis, Missouri.

Ota, R. (1950). *Arch. histol. japon.* **1**, 285.

Palade, G. E. (1953). *J. Histochem. and Cytochem.* **1**, 188.

Palade, G. E. (1955). *J. Biophys. Biochem. Cytol.* **1**, 59.

Palay, S. L. (1957). *In* "Frontiers in Cytology" (S. L. Palay, ed.), p. 305, Yale Univ. Press, New Haven, Connecticut.

Pillsbury, D., Shelley, W. B., and Kligman, A. M. (1956). "Dermatology." W. B. Saunders Co., Philadelphia, Pennsylvania.

Pinkus, H. (1938). *Arch. exptl. Zellforsch. Gervebezücht.* **22**, 47.

Pinkus, H. (1951). *J. Invest. Dermatol.* **16**, 383.

Pinkus, H. (1952). *J. Invest. Dermatol.* **19**, 431.

Porter, K. (1954). *Anat. Record* **118**, 433.

Porter, K. R., and Vanamee, P. (1949). *Proc. Soc. Exptl. Biol. Med.* **71**, 513.

Radaeli, G. (1951). *Boll. soc. ital. biol. sper.* **27**, 713.

Rheingold, J. J., and Wislocki, G. B. (1948). *Blood* **3**, 641.

Riley, J. F. (1954). *Lancet* **I**, 841.

Riley, J. F. (1959). "The Mast Cells." E. S. Livingstone Ltd., Edinburgh and London.

Riley, J. F., and West, G. B. (1952). *J. Physiol. (London)* **117**, 72P.

Riley, J. F., and West, G. B. (1953). *J. Physiol. (London)* **119**, 44P.

Rothman, S. (1954). "Physiology and Biochemistry of the Skin." Univ. Chicago Press, Chicago, Illinois.

Schiefferdecker, P. (1922). *Biol. Zentr.* **42**, 200.

Schwartz, I. L., Thaysen, J. H., and Dale, V. P. (1953). *J. Exptl. Med.* **97**, 429.

Selby, C. C. (1955). *J. Biophys. Biochem. Cytol.* **1**, 429.

Selby, C. C. (1956). *J. Soc. Cosmetic Chemists* **7**, 584.

Shelley, W. B., and Butterworth, T. (1955). *J. Invest. Dermatol.* **25**, 165.

Shelley, W. B., and Cahn, M. M. (1955a). *Cancer Research* **15**, 671.

Shelley, W. B., and Cahn, M. M. (1955b). *J. Invest. Dermatol.* **25**, 127.

Shelley, W. B., and Mescon, H. (1952). *J. Invest. Dermatol.* **18**, 289.

Sperling, G. (1935). *Z. mikroskop. anat. Forsch.* **38**, 241.

Stearns, M. L. (1940). *Am. J. Anat.* **66**, 133.

Storey, W. F., and Leblond, C. P. (1951). *Ann. N.Y. Acad. Sci.* **53**, 537.

Sylvén, B. (1940). *Acta Radiol.* **21**, 206.

Sylvén, B. (1941). *Acta Chir. Scand.* **86**, 1.

Sylvén, B., and Larsson, L. G. (1948). *Cancer Research* **8**, 449.

Szabo, G. (1954). *Brit. Med. J.* **I**, 1016.

Tsukagoshi, N. (1951). *Arch. histol. japon.* **2**, 481.

Videbaek, A., Asboe-Hansen, G., Astrup, P., Faber, V., Hamburger, C., Schmith, K., Sprechler, M., and Brøchner-Mortensen, K. (1950). *Acta Endocrinol.* **4**, 245.

Weiss, P., and Ferris, W. (1954). *Exptl. Cell. Research* **6**, 546.

Wheatley, V. R. (1952). *Soc. belge de Dermatol. et de Syphilig.* p. 90.

Wheatley, V. R. (1953). *St Bartholomew's Hosp. J.* **57**, 5.

CHAPTER 6

Antibody Formation

by PHILIP D. McMASTER

I. THE SCOPE OF THE CHAPTER

The human and animal body invaded by foreign materials or infectious organisms defends itself in a variety of ways. Of these probably the most interesting is the production of newly formed substances, antibodies, which react against the invading substances, termed antigens.

The formation of antibody will be this chapter's theme, and its purpose will be to present a brief survey of the findings which form the background for contemporary thought about the sites of antibody formation. The organs and especially the cells which are either known to produce antibodies or strongly suspected of forming them will be considered, and an effort will be made to state impartially what we believe the authors themselves wished to convey. No attempt will be made to cite all papers written on this subject since excellent reviews are already at hand by Doerr (1950a, b, c) and by Ehrich (1956) and others. Emphasis will be placed upon recent work and new techniques, although something must be said of important older papers even though the conclusions drawn by the authors may be tenable no longer.

It is to be stressed that the cells involved in the processes of immunity differ from those engaged in antibody formation. The former may act by removing the infectious agents, by digesting them after phagocytosis, or by walling them off from the rest of the body; these activities will not be discussed.

In mentioning the cells suspected of forming antibody in the reviews that follow, confusion can best be avoided by employing the nomenclature used by the original authors. When desirable this can in turn be compared with the nomenclature employed in Table I, which has been reproduced from the recent review by Taliaferro (1949), on the "Cellular Basis of Immunity."

A. The Accumulation of Evidence for the Existence of Antibodies

Although it was recognized in ancient times that recovery from certain diseases conferred immunity to future attacks, the notion of experi-

TABLE I

CONNECTIVE TISSUE CELLS INVOLVED IN IMMUNE REACTIONS[a, b]

I. Predominantly fixed cells

  A. Fibroblasts and endothelial cells

  B. Macrophages —   RES,‡ restricted sense (Aschoff)

    1. °Reticular cells of reticular organs

    2. °Littoral cells of sinuses of reticular organs and of sinusoids of the liver, adrenal, and hypophysis

    3. °Adventitial cells (Maximow's undifferentiated pericytes)

    4. †Histiocytes of ordinary connective tissue and of the lamina propria and interstitial connective tissue of various organs, e.g., macrophages of skin, stroma cells of intestine, septal cells of lung, and glial phagocytes of brain   RES,‡ broad sense (Aschoff)

II. Free cells

  A. †Inflammatory macrophages

  B. †Intermediate polyblasts (cells transitional between nongranular leucocytes and inflammatory macrophages)

  C. Nongranular leucocytes

    1. †Monocytes (Ehrlich's transitional cells and large mononuclears.

    2. °Lymphocytes (including hemocytoblasts of myeloid tissue)

    3. Plasma cells

  D. Granular leucocytes

    1. Heterophiles (granules characteristic of the species, e.g., neutrophiles, pseudoeosinophiles)   Microphages (Metchnikoff)

    2. Eosinophiles

    3. Basophiles

Right-side vertical bracket labels: Macrophage system (Metchnikoff) — Lymphoid-macrophage system (Taliaferro and Mulligan)

[a] From the *Annual Review of Microbiology* (1949) 3, 159, with the kind permission of Dr. William H. Taliaferro and the publishers.

[b] Symbols: ° = Marked mesenchymal potencies demonstrated; † = restricted mesenchymal potencies demonstrated; ‡ = reticuloendothelial system.

mentally or clinically induced immunity arose through the work of Jenner (1789) on vaccination to cowpox. Only a century later did the concept of antibodies, as we conceive of them, take form, from the pioneer studies of such men as Pasteur (1880) on chicken cholera, von Behring and Kitasato (1890) and Ehrlich (1910) on passive immunization, Bordet (1909) on agglutination, and Kraus (1897) on the precipitation reaction. At about the same time Pfeiffer and Proskauer (1896) showed that antibodies had to do with the globulin content of serum, and Ehrlich (1910) considered them as having specific chemical groups, a concept which led to the modern work of such men as Landsteiner (1945), Heidelberger (1939), Tiselius (1937), and many others.

II. THE RETICULOENDOTHELIAL SYSTEM AND ANTIBODY FORMATION

The demonstrations of the existence of antibodies led almost at once to curiosity about the possible sites of their origin. From the work of Metchnikoff (1887, 1888, 1901), Wyssokowitsch (1886), Pfeiffer and Marx (1898a, b), the two Wassermanns (1898, 1899), and many others, as well discussed by Jaffe (1931) and by Sherwood (1951), it became common knowledge that large numbers of phagocytic cells in the spleen, liver, lymph nodes, bone marrow, in the endothelium of blood vessels and scattered throughout the body—and now known as the reticuloendothelial system (RES)—were capable of seizing upon bacteria, particulate matter, or certain antigenic substances. Accordingly, the early workers assumed that these cells must form antibody, and this belief was encouraged by the finding that "blockade" with inert particles cut down antibody formation. A voluminous literature arose on the subject (see Bloom, 1938a; Mann and Higgins, 1938; Cannon *et al.*, 1929, and especially Jaffe, 1931).

A significant study appeared by Sabin (1939), who injected rabbits with an alum precipitate of a colored antigenic azoprotein, prepared by Heidelberger *et al.* (1933) and Heidelberger and Kendall (1934). It entered cytoplasmic vacuoles in the macrophages of the liver and omentum, became decolorized, and, when these cells shed their cytoplasm, antibodies appeared in the blood. Sabin suggested that the cells formed globulin from the material taken up and then liberated the antibody by shedding the cytoplasm. Polymorphonuclear leucocytes also took up the azoprotein and were then engulfed by macrophages, so that the latter obtained even more antigen after the polymorphonuclear cells had acted upon it. She suggested further that the phenomena might have represented the preparation of antigen for other cells rather than the direct formation of antibodies and that the shedding of cytoplasm by the macrophages and the appearance of antibodies in the blood might have been related only in time. This work has been well discussed by Doan (1939) and Wright and Doan (1953). Clearly the observations were correct—however, others have suggested that the findings fit the latter interpretation, that some sort of preparation was taking place. Indeed, prior to this work others had doubted the antibody-forming capacities of the phagocytic cells. For example, Ehrich and Voigt (1934) showed that injections of small amounts of *Staphylococcus* vaccine yielded much formed antibody but only a slight RES response. By contrast large doses brought about much proliferation of RES cells but only a little antibody.

Much more will be said below about the role of the RES and of macrophages in considering various sites of antibody formation and the types of cells suspected of taking part in it.

III. Evidence Suggesting the Formation of Antibodies within Whole Organs

It would seem wise, before undertaking a discussion of the types of cells now believed to form antibodies, to present first some findings which clearly pointed to certain organs as the sites at which antibodies appeared. It was, of course, by this means that some indication was obtained about the cell types involved in the process.

A. The Criteria Necessary for the Proof of Antibody Formation within Certain Organs

The presence of antibody in a tissue is no proof of its formation at that site. The fact is well known that blood vessels of inflamed tissues, even those that have merely been injected, are much more permeable than normal. Consequently, they allow materials in the blood, which would otherwise not escape from the vessels, to localize and accumulate in the injured regions. Under these circumstances antibody, found in high titer in injured tissues, may readily have been formed elsewhere in the body and have seeped into the tissues from the blood. Further, as will become clear below, if a single type of cell or a single tissue is claimed as the site of antibody formation, one must inquire into the portal of entry of the antigen and the amount of antigen given. A little antigen injected into the skin or blood will drain to the nearest lymph node or the spleen, respectively, which will then show more antibody than the blood or other organs. Larger amounts of antigen will pass through these organs and stimulate many sites of antibody formation throughout the body, leading to a higher titer of antibody in the blood than in lymph nodes or spleen.

B. Antibody Formation within the Spleen

1. Antibody in Splenic Extracts

A large volume of excellent work—much of it done nearly half a century ago and now so well known that no review of it is necessary—has shown that antibodies form in the spleen. In the early work of Pfeiffer and Marx (1898a, b) Wassermann (1898), and Deutsch (1899), high titers of antibody were found in spleen extracts following intravenous injections of various antigens, but, for reasons just given, such findings did not show that antibody was actually formed in the organ. It remained for others to prove the point (for discussion of this work see Perla and Marmorston, 1941; Taliaferro, 1949; Topley and Wilson, 1955). Recently T. N. Harris *et al.* (1948) found high titers of antibody in splenic extracts from rabbits injected intravenously with various antigens and low titers after subcutaneous injections although the serum titers were high, showing that extrasplenic regions of the body had formed antibody.

2. *Effects of Splenectomy*

Removal of the spleen, following antigenic stimulation, leads to a decrease in antibody formation. This was shown by Tizzoni and Cattani (1892), Deutsch (1899), Roth (1899), Hektoen (1909, 1920), Luckhardt and Becht (1911), and others. Motohashi (1922) brought out the fact that intravenous injections of small amounts of antigen elicit clear antibody formation in the spleen, but large amounts mask it. Wolfe *et al.* (1950), too, cut down the formation of precipitins in chickens by splenectomy and found the relative reduction greatest if only a little antigen was given.

Human beings deprived of the spleen formed less hemolysin and hemagglutinin (Rowley, 1950a) than normal persons, especially if small injections of the antigens were given, and almost no antibody appeared in splenectomized rats when only small amounts of antigen were given intravenously (Rowley, 1950b). When the antigen was given intraportally, intraperitoneally, or intradermally to the splenectomized rats, they formed antibody quite as well as normal animals. Clearly the site at which antibody formation occurs depends upon the route taken by the invading antigen. Similar findings have been reported by Wissler *et al.* (1953).

The excellent studies of Taliaferro and Taliaferro (1950, 1951b, 1952) and Taliaferro (1956, 1957a, b) throw much light upon splenic antibody formation by contrasting the output of hemolysin by splenic and nonsplenic sources in rabbits given sheep red cells. Splenectomy within 4 days after immunization cuts down the output to about 20%; performed after the sixth day it is ineffective, indicating that most of the antibody forms during a short period. After single injections of antigen the spleen puts out large amounts of antibody for several days and then seems to stop forming antibody abruptly when the circulating antibody reaches its peak titer. Thereafter a decline of antibody titer seems to be the result of a lack of antigen for stimulation, because a second splenic response, like the first, can be brought about by a second injection of antigen. The authors suggest that antigen, or perhaps an antigen-induced product is exhausted during antibody formation. Later other nonsplenic sources continue to form it for a long period. Unexpectedly, splenectomized rabbits, repeatedly stimulated with antigen, put out more antibody than intact animals, apparently from nonsplenic sources. By and large the spleen seems to be responsible for the initial antibody formation, following intravenous injections of antigen. Thereafter nonsplenic sources carry on the process. Much more will be said about the spleen below.

C. Lymph Nodes, Lymphoid Tissues, and Antibody Formation

What can be said about true lymphatic tissue and antibody formation?

1. Early work

Nearly seventy years ago Oertel (1890) and Welch and Flexner (1891) described pathological changes appearing in the lymph nodes of humans and animals, respectively, that had been infected with diphtheria organisms. Councilman (1906–1907) noticed that lymphocytes collect about areas of inflammation. Flemming (1865), Matko (1917–1918), Hellman (1921), Hellman and White (1930), Ehrich (1929a, b, 1946), and others described the cellular reactions in nodes during the progress of infections; changes to be discussed later.

2. Inhibition of Antibody Formation by Destruction of Lymphoid Tissue

Early in the present century it became obvious that X-rays reduce lymphoid tissue and decrease the antibody response (Helber and Linser, 1905; Linser and Helber, 1905; Heineke, 1905; Warthin, 1906; Benjamin and Sluka, 1908; Hektoen, 1915, 1918). The spleen, the lymphoid tissue, and the bone marrow seemed to be the sites of antibody formation. Murphy and associates (Murphy, 1914, 1926; Murphy and Ellis, 1914; Murphy and Taylor, 1918; Taylor et al., 1919; Murphy and Sturm, 1925, 1947) studied the suppression of antibody formation by X-ray and concluded that the spleen and lymphoid tissue was of most importance. Further, Murphy and Sturm (1919) and Nakahara (1919) stimulated lymphoid tissue by dry heat and found an increase in antibody formation. They also (see Murphy, 1926) stressed the importance of large collections of lymphocytes about transplanted tumors as one of the mechanisms of tumor regression. These findings were confirmed by much work, references to which appear in the review by Taliaferro and Taliaferro (1951a). Bunting (1925) considered the lymphocyte as an antibody-forming cell, the macrophage merely as a phagocyte capable of eliminating bacteria. A later section will devote itself to further particulars about irradiation effects and antibody formation.

3. Positive Evidence of Antibody Formation within Lymph Nodes

A few years later, lymph nodes nearest to the sites of intradermal injections of antigen were shown to be capable of forming antibodies (McMaster and Hudack, 1935; McMaster, 1941–1942, 1946) in high concentration at the same time that only faint traces of antibody were present in the blood and there was none in extracts from other organs, including lymph nodes elsewhere in the body. The possibility that the antibodies found in the regional, draining nodes had resulted from the seepage of antibody formed elsewhere in the body was ruled out by injecting two different antigens, one in each ear of mice. The injections rendered the blood vessels in both ears and in the regional nodes of both sides more permeable than normal, so that these regions on both sides had equal opportunity to allow circulating antibodies to seep out of the blood and localize in the nodes or in the ear tissue. Agglutinins to the antigen injected in the right ear appeared first, in high concentration, in the lymph nodes draining that ear and in the blood in traces, but not in the lymph nodes draining the other (left) ear nor in other organs. In the cervical nodes of the left side antibodies appeared first to the antigen injected in the left ear, and not to that injected on the right side. In later experiments (McMaster and Kidd, 1937) similar findings were obtained using vaccine virus as antigen. The findings were confirmed by Burnet and Lush (1938), deGara and Angevine (1943), T. N. Harris et al. (1945), Burnet and Fenner (1949), T. N. Harris and S. Harris (1949, 1950), S. Harris and T. N. Harris (1949, 1950), Amano et al. (1951), Dougherty et al. (1945a, b).

Osterlind (1938) reported that the chief reaction to injections of diphtheria toxin appeared in the local node. Oakley et al. (1949) found the draining lymph nodes in rabbits, guinea pigs, and horses producing antitoxin following secondary stimulation by subcutaneous injections of alum-precipitated diphtheria toxoid. Since small injections of antigen occasionally result in antibody formation in distant nodes, Sjövall (1936) suggested the possibility that lymphocytes circulate from one node to another.

Studying the secondary response to diphtheria toxoid injected into the foot pads of rabbits, Stavitsky (1954) removed the draining nodes before the second injection, thereby reducing the secondary response. Removal of the spleen before a second intravenous or intrasplenic injection also reduced the antibody response. When antigen was injected in the foot pad both the popliteal lymph node and the spleen responded; removal of both decreased the antibody response, but after intravenous injections removal of the node had little effect.

IV. The Types of Cells Involved in Antibody Formation

A. *Lymphocytes*

1. *Evidence for Antibody in Lymphocytes*

When it became clear that antibodies form within the spleen and lymph nodes, suspicion naturally fell first upon the lymphocyte. Rich *et al.* (1939) described a large mononuclear cell with basophilic cytoplasm, termed by them the "splenic tumor cell." It appeared in the red pulp of the enlarged spleens and in the medullary regions of lymph nodes of rabbits given multiple injections of sterile foreign proteins. They considered it a lymphoblast probably involved in the mechanisms of immunity. These cells have since been given many designations by others.

An extensive study of the relationship of lymphoid cells to antibody formation was then undertaken by Ehrich and Harris and their co-workers (Ehrich and Harris, 1942, 1945; Ehrich, 1946; Ehrich *et al.*, 1946; T. N. Harris *et al.*, 1945; T. N. Harris and Ehrich, 1946). When antigen was injected directly into the peritoneal cavity, the granulocytes and macrophages recovered in the resulting exudates contained practically no antibody although the peritoneal fluids obtained held much. Nonspecific irritation of the peritoneum following intravenous injections of antigen yielded similar findings. Clearly antibody, formed elsewhere in the body, had seeped into the abdominal cavity. Ehrich *et al.* (1946) injected two different antigens into the foot pads of rabbits and found much antibody in the popliteal lymph nodes, but only little at the injection sites, although macrophages and granulocytes were present. The antibody content of lymph node extracts, of the lymph and its cells, and of the serum were compared at various intervals with the cellular changes in the nodes (Ehrich and Harris, 1942, 1945). A lymphocytic hyperplasia preceded an increase in the cells in the efferent lymph and an increase in its antibody titer. More antibody was present in the cells than in the lymph itself (T. N. Harris *et al.*, 1945). When the cells of the efferent lymph from one side, injected with a given antigen, were incubated with cell-free lymph from the other side, injected with a second antigen, antibody escaped from the cells to the lymph, but there was no passage of antibody from the lymph to the cells. Following the injection of killed *Shigella* organisms into the foot pads of rabbits, soluble antigenic material appeared in the draining nodes. They suggested (T. N. Harris and Ehrich, 1946; Ehrich and Harris, 1945; Ehrich, 1956) that RE cells, probably the macrophages, split up the aggregated antigens into active antigenic material which stimulated antibody formation by other cells. Ehrich (1956, p. 177) concluded that the RES prepares antigen but the system itself has nothing to do with antibody synthesis. Accordingly, the depression of antibody formation afforded by blockade of the RES arises from interference with the capture of antigen and perhaps with its preparation for antibody-forming cells.

After injecting the foot pads of rabbits with various antigens, T. N. Harris and S. Harris (1949) compared, at various intervals, the change in the ribonucleic and deoxyribonucleic acid content of the draining nodes with the cellular changes appearing in sections of the same nodes when stained with methyl green and pyronine for evidences of protein synthesis. The staining reaction for ribonucleic acid appeared in the cortical parts of the nodes and increased from the second to the fifth day, reaching its maximum just before the peak of antibody formation and appearing first in reticulum cells, then in transitional cells, next in young, and finally in mature lymphocytes. Plasma cells were present, but, since the cellular reaction consisted chiefly of a diffuse lymphocytic hyperplasia, T. N. Harris and S. Harris (1949) considered the lymphocyte important in antibody formation.

The activity of the lymphocyte during the regression of transplanted tumors, described previously by Murphy (1926) has been restudied by Kidd and his colleagues (Kidd and Toolan, 1949, 1950; Ellis et al., 1950; Toolan and Kidd, 1949). The transplantation of mammary carcinoma or lymphosarcoma cells of mice to other, different but resistant, mice produces changes in the germinal centers of the draining lymph nodes, and the proliferation of cells, apparently plasma cells. The transplanted cells become enveloped and destroyed by lymphoid cells.

Wesslen (1952a) brought lymphocytes from the thoracic ducts of tuberculous animals into contact with tuberculin. They were lysed, although similar cells from normal animals were unharmed. He also (Wesslen, 1952b) collected lymphocytes from thoracic duct lymph of immunized rabbits. These were 5–6% large lymphocytes, the remainder small lymphocytes. No plasma cells were seen. He found that lymphocytes in the form in which they reach the blood stream apparently have no demonstrable antibody, although it may be present in the plasma. Nevertheless, they formed antibody in tissue cultures, and, when transferred to guinea pigs, rendered the animals anaphylactically sensitive. Accordingly, in these experiments the lymphocytes seemed to be capable of antibody formation.

Prior to this Miller and Favour (1951) had removed leucocytes from human patients with tuberculosis and also from sensitized guinea pigs and had separated them into suspensions, one rich in lymphocytes, the other rich in neutrophiles. The lymphocytes, when washed and suspended in the plasma of normal subjects, seemed to develop a cytolytic factor apparently originating from the cells.

B. Depression of Antibody Formation by Lymphocytotoxic Substances

1. The Effects of Nitrogen Mustard

Spurr (1947) found nitrogen mustard toxic for lymphoid tissue and capable of suppressing antibody formation, but Phillips *et al.* (1947) reported that the substance, though exhibiting its damaging effect, failed to delay the antibody response to ricin or to reduce the level of antibody in the serum. More recently, however, both Bukantz *et al.* (1949) and Schwab *et al.* (1950) reported not only the suppression of antibody formation by nitrogen mustard, but the reduction of hypersensitivity reactions as well.

2. Some Effects of Pituitary and Adrenocortical Hormones on Lymphocytes, Lymphoid Tissue, and Circulating Antibody

Dougherty and White (1943, 1944) described an acute lymphopenia appearing in mice after single injections of pituitary-adrenotropic hormone or adrenal cortical hormone. Within a few hours the lymphocytes showed karyorrhexis, pyknosis, and shedding of cytoplasm—phenomena described by Dougherty and White (1945) as "dissolution of the lymphocytes." Immature plasma cells (Dougherty and White, 1946a) increased within the nodes and small lymphocytes developed basophilia, indicating an increase in nucleoproteins or at least protein synthesis. The β- and γ-globulins of the blood increased at the same time, suggesting that these proteins had been released through the action of the hormones. The presence of γ-globulin within lymphocytes was also reported by Kass (1945). Prior to this work similar changes in lymphoid tissue had been described by Selye (1941–1942, 1950; Selye *et al.*, 1936) following "stress" with its hormonal effects. Ehrich (1956, p. 206) in discussing these matters suggests that the lymphocytolysis and plasmacytolysis in stress may readily set free antibodies or other globulins but that it has not yet been adequately proved.

Dougherty *et al.* (1944, 1945a), who had reported the presence of antibodies concentrated in lymphoid cells obtained from lymph nodes of mice immunized with sheep red cells, injected (Dougherty *et al.*, 1945b) a filtrate of *Staphylococcus aureus* (*Micrococcus pyogenes* var. aureus) culture into mice repeatedly and transplanted pieces of lymphosarcoma into them while the antigen administrations continued. Extracts of the transplanted tumors, which consisted chiefly of lymphocytes resembling reticular cells, contained antibody globulin. When the administration of antigen was stopped after transplanting the tumor, the growing tumor cells again showed antibody, which, the authors feel, must have been derived from the host's lymphocytes.

Since lymphocytes from immune animals seemed to contain antibody, Dougherty et al. (1945a) and Chase et al. (1946) injected cortical hormones into immunized animals and reported an increase of antibody in the blood at the same time that the dissolution of the lymphocytes was at its height. In further work, Dougherty et al. (1945a) reported anamnestic responses occurring after injections of these hormones into previously immunized rabbits showing at the time no circulating antibody. This work, repeated by Hammond and ·Novak (1950) gave similar results. Dougherty and White (1946b) also obtained anamnestic reactions by means of sufficient X-irradiation to bring about dissolution of lymphocytes and lymphopenia. From this work, summarized in several reviews, White and Dougherty (1946) and White (1947–1948, 1948) postulated a hormonal control over the level of blood lymphocytes and the liberation of antibody from them. It is to be stressed that these authors stated that the lymphocytes contained antibody, but not that they formed it, and they emphasized the fact that the primary sites of antibody formation were still a matter of conjecture (White and Dougherty, 1946).

3. Opposing Views concerning Hormonal Control of Antibody Release

Murphy and Sturm (1947) suggested that cortical hormone is not essential for the release of antibody since, in adrenalectomized rabbits— in which neither cortical hormones nor the pituitary adrenotropic hormones could have any effect—a hypertrophy of the lymphoid system appeared and the circulating lymphocytes increased. Injections of antigens increased them further. Rabbits, injected with horse serum and deprived of their adrenals after 10 days, showed unusually strong precipitin reactions.

Eisen et al. (1947), measuring circulating antibody quantitatively in adrenalectomized rats repeatedly injected with cortical hormones during immunization, reported the antibody like that of controls, which got no hormone, indicating that the release of antibody is not dependent upon adrenal cortical influence. Fischel et al. (1949) failed to find, with quantitative methods, an anamnestic rise in circulating antibody following treatments with either X-ray or ACTH, although, like most workers, they found the number of circulating lymphocytes decreased, presumably by dissolution which should have liberated antibody. Herbert and de Vries (1949) found this effect in human beings, but no consequent rise in circulating antibody, and Spurr (1947) also found no anamnestic response in previously immunized animals given nitrogen mustard. Valentine et al.

(1948) reported the rate of the production of lymphocytes in the thoracic duct to be uninfluenced by cortical hormone in either normal or adrenalectomized cats.

4. Cortisone and Antibody Formation

Soon after the first active preparations of adrenal cortical hormones or cortisone reached the laboratories, conflicting reports began to appear concerning their effects upon antibody formation. These controversial papers will not be discussed since there are excellent reviews of the work by Fischel (1953), Kass and Finland (1953), and Fischel et al. (1951). It is now generally agreed that antibody formation in animals is suppressed if large doses of cortisone are given in proper time relationship to the antigenic stimuli and in such ways that its action is continuous—as shown by the works of Germuth and Ottinger (1950), Germuth et al. (1951), Hayes and Dougherty (1952), Dougherty (1953), Stavitsky (1953), Moll and Hawn (1953), Dews and Code (1953), Janeway (1953), Newsom and Darrach (1954), Berglund (1956a, b), Taliaferro (1957a), and others now to be mentioned. Fischel et al. (1951) report that the reduction of circulating antibody is not due to its rapid destruction by cortisone. In large doses cortisone decreases the lymphoid tissue greatly (Antopol, 1950; Björneboe et al., 1951; McMaster and Edwards, 1957a), yet it does not seem to suppress antibody formation by destruction of protein antigens or by deranging the uptake or storage of these by the RE cells (McMaster and Edwards, 1957b).

Berglund and Fagraeus (1956) demonstrated an inhibition of hemolysin formation by cortisone. Cells from spleens or thymus glands given 2 or 5 hours after giving the antigen protected antibody formation; given one or several days later, the cells had no effect. Apparently they exerted a function or produced some factor necessary for some early cortisone-sensitive phase of antibody formation.

In tissue culture Trowell (1952a) observed a deleterious effect of cortisone upon antibody formation by lymph node cells, and Mountain (1955a) also found cortisone inhibitory. This hormone interferes with the growth and migration of splenic cells in cultures (Holden et al., 1953) and exerts a detrimental effect on lymphocytes (Trowell, 1953). Clearly, its action in depressing antibody formation is not yet explained.

5. Negative Findings concerning the Role of Lymphocytes and Antibody Formation

Rats deprived of 90% of their lymphoid tissue by Andreasen et al. (1948) showed no fall in plasma proteins, but in such experiments the functional capacity of the remaining tissue could not be determined.

Craddock *et al.* (1949) gave typhoid vaccine to cats, but, following the use of cortical hormone or X-ray, found no rise of antibody in the thoracic duct lymph and no evidence of antibody in the lymphocytes within the duct. However, the lymphocytes of the thoracic duct might have given up antibody before they reached the vessel. Habel *et al.* (1949) injected the foot pads of rabbits with two different antigens, one on each side, but after studying the titer of serum, lymph, cells of the lymph, and washed cells of the nodes, found no evidence of antibody formation in the lymph nodes and none of antibody transport by lymphocytes. These authors considered the high antibody titer of the lymph to be the result of seepage from the blood. Very large amounts of antigen were used in these experiments and, as discussed earlier, it might have flooded the body and stimulated antibody formation from many sites other than the lymph nodes draining the sites of injection of antigen. Erslev (1951) extracted the leucocytes of hyperimmunized rabbits and found no significant amounts of circulating antibody although there was much in the plasma. Apparently circulating lymphocytes, after coming from the nodes and appearing in the blood, have no antibody.

T. N. Harris and S. Harris (1950) have brought out technical reasons for these negative findings. They reported a loss of antibody from washed lymphocytes, and noted, to, that lymph nodes draining injected areas showed relatively more antibody than other organs when small doses of antigen were used, which did not flood the body to stimulate many sites of antibody formation. Further, S. Harris and T. N. Harris (1949), when they injected different strains of influenza virus into opposite foot pads of rabbits, found that antibodies to the strain injected on the right side appeared in the right draining node in far higher concentration than in the left node, and vice versa. If one assumes that the low antibody titer in the left node to antigen injected on the right side was due to seepage from the blood, which is debatable, then it is obvious that the high titer of antibody to antigen injected on the right side could not have been so derived.

C. Plasma Cells

1. The Characteristics of Plasma Cells

Waldeyer (1875) seems to have been the first to speak of plasma cells, but they were probably not the cells that were described in 1890 by Ramon y Cajal and termed plasma cells by him (Ramon y Cajal, 1896, 1906), by Unna (1891), and by (Marschalko (1895). Since, even today their development is not well understood, and the subject has been well reviewed by Ehrich (1953a, b; 1956, see especially pp. 153–170), only a few remarks will be made, where appropriate, in the various reviews below. It should simply be pointed out here that, in the spleen, bone marrow, and lymph nodes, as stressed by Marshall and White (1950) and by Taliaferro (1949), large basophilic cells appear which have been given many names by different authors. Taliaferro and Mulligan (1937), Bloom (1938a, b), and Conway (1939) term them large lymphocytes; they are called plasmablasts by Marshall and White (1950), lymphoblastic plasma cells by Huebschmann (1913), developing plasma cells by Koluch *et al.* (1947), immature plasma cells by Fagraeus (1948a), myeloblasts by Hewer (1930), and the acute splenic tumor cell by Rich *et al.* (1939).

Indeed, from their experimental and clinical studies, Good and his co-workers (Good, 1947, 1950; Good and Good, 1949; Good and Campbell, 1905; Campbell and Good, 1949, 1950) have suggested the development of plasma cells from several multipotent connective tissue cells, invasive lymphocytes, reticulum cells of the pia arachnoid—in states of allergic encephalitis (Good, 1947, 1950)—from macrophages or even from Kupffer cells of the liver. However, as will appear below, the cells designated as plasma cells in the more recent articles are, at least in their mature and not too early form, well recognizable by their staining reactions and their appearance under the phase microscope.

Pappenheim (1901, 1902) seems to have been one of the first to suggest that the basophilic material seen in these cells signifies a rich supply of nucleoprotein, now generally recognized as ribonucleic acid. Since then numerous authors, by spectrophotometric and chemical methods (Caspersson, 1936; Brachet, 1940; Caspersson and Theorell, 1941; Mirsky, 1951–1952) and by phase microscopy (Moeschlin, 1940–1941, 1941, 1947, 1949; Moeschlin and Demiral, 1952; Moeschlin *et al.*, 1951), have established the cytoplasmic characteristics of protein synthesis taking place in various cells. This phenomenon appears to be going forward in plasma cells (Theorell, 1944; Bing *et al.*, 1945) rather than in lymphocytes, which are poor in basophilic cytoplasm. Brausteiner *et al.* (1953) showed by the electronmicroscope that plasma cells, like hepatic cells and other protein formers, have bundles of filaments which are lacking in lymphocytes. Moeschlin *et al.* (1951) showed by phase microscopy that certain drop-like granules appear in plasma cells of the spleens of rabbits, but not in lymphocytes, during the period in which antibody is forming. These structures are characteristic of other protein-forming cells.

2. Clinical Studies

Clinicians were the first to become interested in the relationship of plasma cells to immune processes and antibody formation. Renn (1912) and Huebschmann (1913) observed collections of plasma cells in the tonsils and spleens of patients during the course of infectious conditions. They considered the plasma cells to be related to antibody formation. Since that time many clinicians or pathologists, such as Perlzweig *et al.* (1928), Müller (1932), Osgood and Hunter (1934), Markoff (1937), Bing and Plum (1937), Bing (1940a, b), Andersen and Bing (1944), Kagan (1943), Lowenhaupt (1945), and others, have related plasmacytosis with high blood globulin levels and antibody formation. The list of clinical conditions in which they appear can be found in the monograph of Ehrich (1956, pp. 183 *et seq.*). Bing (1940a, b) suggested that plasma cells of reticuloendothelial origin accounted for blood globulin, and Anderson *et al.* (1948) reported changes in the serum γ-globulin which reflected antibody formation against group A streptococci. Good and Campbell (1950) suggested that the plasmacytosis of the bone marrow, which is accompanied by an increase in blood globulin in patients passing through acute stages of rheumatic fever and streptococcal pharyngitis, is to be correlated with antibody formation. Lowenhaupt (1945) suggested a relationship between plasma cells and abnormal globulin production in a discussion of proliferative lesions in multiple myeloma, and, in the course of a study of the origin of abnormal plasma proteins in patients with multiple myeloma, Miller *et al.* (1952) showed by electrophoresis that the proteins of the myeloma plasma cells were the same as those of the serum proteins which seemed to account for the hyperglobulinemia.

3. The Relationship of Plasma Cells to Protein Synthesis and Antibody Formation As Shown by Experimental Studies

In the laboratory, Downey (1911), Doan *et al.* (1930), and also Miller (1931) reported the presence of many plasma cells in various animal infections, and Rich *et al.* (1939) described "splenic tumor cells" in the spleens of animals repeatedly injected with foreign proteins. These cells, now considered by many to be early plasma cells, were related by them to the processes of immunity.

Koluch (1938) and Koluch *et al.* (1947) sensitized rabbits and later elicited anaphylactic shock. They described the appearance of bone marrow plasma cells from reticular plasma cells and believed that this transformation held a direct relationship to antibody formation. Shortly thereafter Björneboe and Gormsen (1941, 1943) and Björneboe *et al.* (1947) performed their classical experiments in which great collections of plasma cells appeared in the spleens, livers, medullary cords of lymph nodes, and even in the fatty tissues of the renal pelves of rabbits given multiple injections of antigenic material from a number of types of the pneumococcus. They consider the plasma cell as the source of antibody. The resulting great increase in blood globulin in their experiments was attributed by them wholly to antibodies. At about the same time Heinlein (1943) hyperimmunized rabbits but did not find plasma cells increased, nor did he consider them important in antibody formation. Bing *et al.* (1945) confirmed Björneboe's and Gormsen's work and further, by microspectrographic and microchemical means, they obtained evidence of protein formation in developing plasma cells.

Fagraeus (1946, 1948a, b) observed a development of reticulum cells to transitional cells, to immature and mature plasma cells in the red pulp of the spleens of rabbits which had been injected with various antigens 2 or 3 days previously. Although findings obtained with tissue cultures will be discussed later, it may best be said here that Fagraeus (1948a) cultured small pieces of the red pulp from these spleens, containing chiefly plasma cells, and also pieces of the white pulp, made up principally of follicular tissue containing chiefly lymphocytes. She found much antibody formed by the red pulp in proportion to the number of plasma cells. The pieces made up chiefly of lymphocytes yielded only a little antibody. On the fourth day after secondary antigenic stimuli the difference was greatest especially in the pieces with immature plasma cells. She concluded that antibodies are formed by reticuloendothelial elements which develop into a type of cell with the morphological characteristics of plasma cells. The work of others with tissue culture techniques will be described below.

Ehrich *et al.* (1949) injected the foot pads of rabbits with various antigens, and compared, by techniques like those used independently by T. N. Harris and S. Harris (1949), the content of the draining lymph nodes for ribonucleic and deoxyribonucleic acids and the cellular changes observed after staining sections with methyl green and pyronine. The increase in ribonucleic acid appeared on the fourth day, not as early as the second day, as found by T. N. Harris and S. Harris (1949). It reached a maximum on the sixth day, at the same time that the antibody also reached its peak. The cellular reaction consisted chiefly of plasmablasts, then of mature plasma cells, which contained ribonucleic acid. The medullary portion of the node showed most of the reactions rather than the cortex, as found by Harris and Harris. After the peak of antibody formation had been reached, the lymphocytes showed their greatest activity, indicating that the plasma cells had been responsible for the antibody formation, not the lymphocytes.

Ehrich has commented (Ehrich *et al.*, 1949; Ehrich, 1956) on his previous reports in which the lymphocytes in the efferent lymph flowing from lymph nodes were reported to contain more antibody than the lymph itself (Ehrich and Harris, 1942, 1945). He suggests that plasma cell antibody was taken for lymphocyte antibody since plasma cells must have been present with the lymphocytes. Ehrich looks upon the mature plasma cell as the source of antibody, not—as believed by Fagraeus—the immature one. He places antibody synthesis in the cytoplasm of the plasma cells, occurring, like the synthesis of hemoglobin in red cells, during the process of maturation but not arriving at its peak of concentration until maturity is reached. A brief paper (Ehrich, 1953a), gives an excellent summary of his views favoring the plasma cells as the important former of antibodies.

Moeschlin et al. (1951) observed plasma cells in the spleens of actively immunized rabbits and confirmed the studies of Fagraeus (1948a). Early in the process of immunization these cells—observed under the phase microscope—showed dark protoplasmic granules which disappeared later in the possess when antibodies were found in the blood. Since these granules were like those already seen by Moeschlin (1949), in myeloma cells during the appearance of the special proteins found in the blood in this condition, the authors suggest that the granules may indicate antibody formation. The results of lymph node and sternal punctures from patients with rubeola led Moeschlin (1940–1941, 1947) to conclude that plasma cells arise from plasmablasts that are distinct from lymphoblasts. He raises the question (Moeschlin et al., 1951) whether certain reticulum cells of the spleen, bone marrow, or lymph nodes—mentioned by Ehrich et al. (1949) as the sources from which plasma cells develop—may process the ability to take up catabolic products of antigens and then, in a secreting phase, to become plasma cells; or, whether it is more probable that certain reticuloendothelial cells may not first become macrophages while other reticular cells become plasma cells.

Keuning and van der Slikke (1950) and Thorbecke and Keuning (1953) studied the formation of antibodies in cultures of tissue from the spleens, bone marrow, lymph nodes, and thymus glands of rabbits following immunization. In this work, which will be discussed in a later section, they found, as did Fagraeus, that the white pulp of the spleen formed little antibody while the red pulp yielded a greater amount. In cultures that formed antibody there was an increase in γ-globulin. Thorbecke (1954) and Thorbecke and Keuning (1956) found aggregates of plasma cells in the spleens, bone marrow, lymph nodes, and livers of rabbits during immunization. These cells did not seem to arise from reticulum cells as believed by Fagraeus (1948a) and by Marshall and White (1950). Instead "plasmablasts" with basophilic cytoplasm appeared lying together in groups in the red pulp, 2–3 days after injection of antigen. In the following days they became smaller, appearing like immature plasma

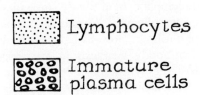

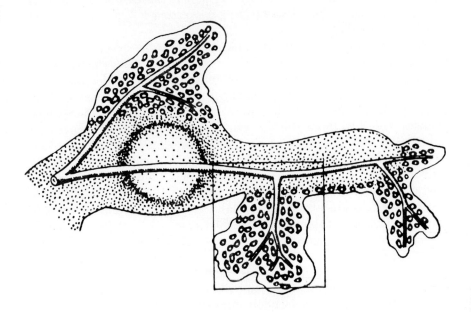

Fig. 1. A diagrammatic sketch, reconstructed from serial sections from the spleen of a rabbit during immunization, to show the localization of young plasma cells bordering the white pulp and surrounding the smallest arterioles emerging from it. Reproduced from the paper by Thorbecke and Keuning (1956) with the kind permission of the authors and the publisher.

cells, finally assuming the appearance of Marschalko, mature plasma cells. They bordered on the white pulp in the spleen and lay about the little arterioles and arterial capillaries which emerge from white pulp, as shown in Figs. 1 and 2, both of which are reproduced here from Thorbecke and Keuning (1956). Other cells like plasmablasts appeared scattered in the splenic white pulp, but true plasma cells were not seen in these situa-

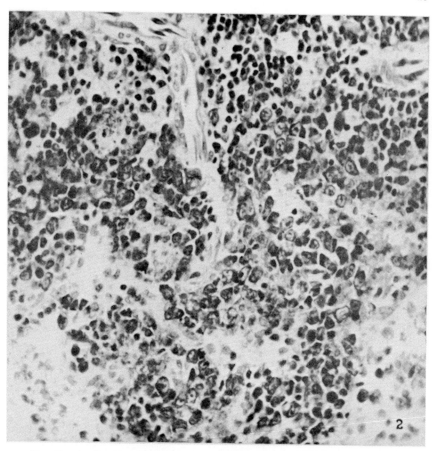

Fig. 2. A photograph of the area of spleen boxed in the rectangle drawn in Fig. 1. Plasmablasts and immature plasma cells surround the arteriole that appears in the upper central part of the picture. Reproduced from the paper by Thorbecke and Keuning (1956) with the kind permission of the authors and the publisher.

tions. Typical reaction centers appeared in the Malphigian corpuscles of the spleen, but only after the highest rate of antibody formation had occurred. They feel that it is highly unlikely that antibody forms in the white pulp since the reaction does not coincide with the production of antibody, as observed also by Ehrich and Harris (1942), Ringertz and Adamson (1950), and Marshall and White (1950).

In their earlier work Thorbecke and Keuning (1953) found antibody formed in cultures of bone marrow taken from immunized animals although the tissues showed no lymph follicles or reaction centers. There was no antibody formation in the liver. However, later work (Thorbecke

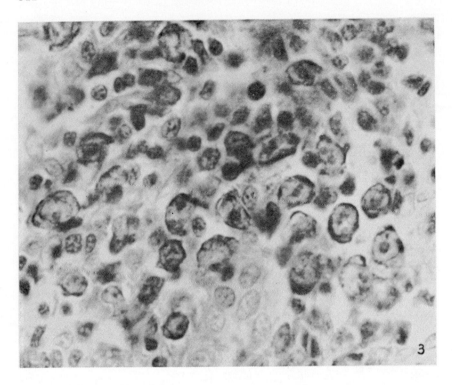

FIG. 3.   The plasma cell reaction with plasmablasts in a periportal space of the liver of a splenectomized and subsequently immunized rabbit. Reproduced from the paper by Thorbecke and Keuning (1956) with the kind permission of the authors and the publisher.

and Keuning, 1956) showed that splenectomized rabbits, given paratyphoid B bacterin or horse serum formed small amounts of antibody in the organ. Clumps of mature and immature plasma cells together with lymphocytes appeared in the periportal connective tissue of the livers from which material was taken for the cultures. Figure 3, also reproduced from their paper, shows well the appearance of these collections of cells. Presumably the plasma cells were the source of the antibody.

Matsumura (1949a, b) and Matsumura et al. (1952) found plasma cells making up 8% of all the cells in lymph-draining nodes during antibody formation. Other Japanese workers (Hanaoka et al., 1950; Shimura et al., 1950) have also found antibody formation associated with plasma cell proliferation,

Askonas and White (1956), after injecting antigen into the foot pads of rabbits, report more antibody in the contralateral lymph nodes than in the draining popliteal nodes. These authors feel that macrophages do not form antibody since the granulomas that formed at the site of injection were composed almost entirely of macrophages which held little or none of it.

The studies of Coons et al. (1955), Leduc et al. (1955), White (1954), White et al. (1955a, b), and many others, indicating antibody formation by plasma cells, will be discussed below.

4. Direct Observational Evidence for the Presence of Antibodies in Plasma Cells

Reiss et al. (1950) suspended washed cells from the popliteal lymph nodes of rabbits previously injected in the foot pads with typhoid or brucella antigen. The organisms added to these suspensions became agglutinated around certain lymphoid cells which were identified as plasmablasts and immature plasma cells. Mature cells showed little or no reaction, and immature and mature lymphocytes, like polymorphonuclears and macrophages, showed none. By contrast, Hayes et al. (1951) spread, on glass slides, dried films of tissues or organ imprints from mice immunized with *Salmonella typhimurium*, and added some of the bacterial antigen to the preparations which were then sealed with vaseline and incubated. Bacterial aggregates appeared around lymphocytes.

Moeschlin and Demiral (1952) washed suspensions of splenic cells taken from normal rabbits and from rabbits 5—7 days, or several weeks, after reinjections of typhoid-paratyphoid bacterin. The cell suspensions, taken when circulating antibody was high, that is 5–7 days after reinjection of the antigen, mixed with killed bacilli agglutinated them in the fluid of the suspension within a few minutes. Granule-containing plasma cells like those considered by Moeschlin et al. (1951) to contain antibody, as already mentioned, often showed agglutination of the organisms on their surfaces. Nevertheless, the authors did not consider the phenomenon as specific as the clumping of the bacteria between the cells, since the former phenomenon might be due to surface-tension effects.

Plasma cells, like lymphocytes, are generally considered to be nonphagocytic. However, and it seems very important for hypotheses of antibody formation, Moeschlin (1951), Moeschlin et al. (1951), and Dubois-Ferrière (1951) have shown phagocytic activity by "plasmocellular reticular cells."

5. Agammaglobulinemia, Lack of Plasma Cells, and the Absence of Antibody Formation

Recently Bruton (1952), Bruton et al. (1952), and Janeway et al. (1953) discovered certain patients lacking plasma and tissue γ-globulin. The latter authors report that in this condition, agammaglobulinemia, the patients form no antibody in their blood or tissues when stimulated with antigen. They succumb readily to bacterial infections, but they resist viral infections without forming globulin antibodies (Good et al., 1957). Some other defense mechanism must account for the resistance. Although agammaglobulinemia patients do not form antibodies, they show some of the cellular changes that appear in the immune response. Following intracutaneous injections of antigenic substances, lymphocytes in the draining nodes increase and shed their cytoplasm, the reticulum of the nodes increases, and the organs enlarge, but no plasma cells appear. Good (1954), Good and Varco (1955a, b), and Good et al. (1957) report that homotransplants of skin continue to grow well in agammaglobulinemic patients instead of regressing as they do in normal subjects. Good et al. (1957) suggest that the lack of plasma cells correlated with these phenomena links them most closely with antibody formation. By contrast, a state of hypersensitivity can be induced in patients with agammaglobulinemia and even transferred to the skin of normal subjects by the leucocytes of the patient. More will be said below of the work of these authors in considering the transfer of lymphoid tissues from donors to recipients.

D. The Eosinophile and Antibody Formation

Eosinophiles have long been associated with conditions of hypersensitivity, allergy, stress, parasitism, antibody formation, and other pathological states. The subject, however, is far too controversial to discuss in a few words. Instead the reader is referred to recent reviews by Speirs (1955, 1957, 1958) for references to older work, or to papers by Ringoen (1921), Biggart (1932), Redd and Vaughan (1955), or Campbell (1943). Only some of the most recent work of Speirs will be mentioned here.

Speirs and Wenck (1955) find that intraperitoneal injections of tetanus or diphtheria toxoid into immunized mice tempts eosinophiles to accumulate in the peritoneal fluid. This phenomenon does not occur in non-immunized mice; it apparently accompanies some early stage of antibody formation in the secondary response. The eosinophiles (Speirs, 1955; Speirs and Wenck, 1955) are attracted to the sites of antigen fixation in sensitized animals, rather than to antigen in a primary response. In sensitized animals these cells collect within lymph nodes draining injected areas of skin, or in the spleen following intravenous injections. The cell accumulations precede or coincide with the appearance of antibody. It is of further interest, as noted by Speirs and Wenck (1955), that the techniques of Coons et al. (1955) and of White et al. (1955a), for staining antibody in plasma cells with fluorescein-coupled proteins, also stain eosinophiles.

Ragweed pollens given to pollen-sensitized mice by Speirs and Dreisbach (1956) and Speirs et al. (1956) caused an increase in the circulating eosinophiles greater than that of other blood leucocytes, and a great increase in the eosinophiles of peritoneal fluid. This increase was found in animals sensitized to albumin as well as to pollen, and it occurred, too, in adrenalectomized mice, showing that it was not a "stress" reaction. Speirs et al. (1956) also correlated the formation of active antibody with an increased number of eosinophiles. Speirs (1956, 1958) reported further that X-irradiation, given before an injection of antigen, reduced the accumulation of eosinophiles much as it has been found to reduce antibody formation, as will be described in a later section of this chapter. That is to say, X-ray reduced the numbers of eosinophiles and mononuclears in the peritoneal fluid and prevented an increase in their numbers in response to injections of antigen, both in immunized and nonimmunized mice. The radiosensitive phase of antitoxin formation, which will be described below, corresponded to the period during which the accumulations of eosinophiles where inhibited by X-ray. Speirs believes therefore that eosinophiles are related to some step in antibody formation in the secondary response, since they accumlate before, or at, the time of peak antibody response, and fail to accumulate in passively immunized animals or under conditions which suppress antibody formation. The administration of cortisone, which, as was discussed above, inhibits antibody formation, also inhibits the eosinophile reaction just described (Speirs and Wenck, 1957; Wenck and Speirs, 1957). Speirs reports further that the eosinophiles, in animals that have been repeatedly injected with antigen, after coming in contact with it, are, taken up by macrophages and other phagocytes. Thereafter antibody appears in the serum. Since the accumulation of eosinophiles precedes the formation of antibody, and since factors that inhibit or augment antibody formation have a similar effect upon the accumulation of eosinophiles at the sites of fixed antigen, Speirs (1957, 1958) believes that eosinophiles play an important part in antibody formation. He suggests that the eosinophiles react with antigens to produce a product which, after the cells are phagocytosed, can be passed on to antibody-forming cells.

V. Some Cellular Changes Associated with Antibody Formation in the Spleens and Lymph Nodes of Immunized Animals

The cellular changes in the spleens and lymph nodes of animals during the processes of immunization have been so recently and fully discussed in the excellent monograph by Ehrich (1956), with a full bibliography, that only brief mention of a few recent papers needs to be made. Hellman (1921) and Hellman and White (1930) put forward the idea that the secondary nodules, described by Flemming (1885) as areas for the development of normal lymphocytes, were actually reaction centers responding to antigenic or toxic stimuli. Hellman and White (1930) suggested that the secondary nodules of both lymph nodes and spleens constituted sites of antibody formation. A year earlier than this Ehrich (1929a, b, c, d) described the lymphocytic hyperplasia in rabbits that followed subcutaneous injections of staphylococci, and rejected the original conception that the site of formation of lymphocytes lies in the secondary nodules of Flemming, since, when the lymphocytosis in the blood was at its highest, the nodules were wanting.

Sjövall and Sjövall (1930) found germinal centers appearing 5–7 days after injections of antigen and subsiding in 10–20 days, and Grégoire (1932) also described increases in the number of these centers after injections of horse serum into guinea pigs. A few years later Glimstedt (1936) raised guinea pigs under sterile conditions and found no germinal centers. In the past year Thorbecke et al. (1957) have found in "germ-free" chicks some plasma cells, some γ-globulin, and a few secondary nodules in lymphoid tissue, but fewer than appear in chicks raised in the ordinary way. The secondary nodules of the germ-free chicks might have been formed as responses to nonbacterial antigens of the food. Osterlind (1938) found, in rabbits injected with diphtheria toxoid, the greatest number of germinal centers present when antibody was highest in the blood. On the other hand, Trowell (1952a, b) reported that the lumbar and sacral lymph nodes of rats, showing no germinal centers, furnished lymphocytes which formed antibody in tissue culture. Other work on this theme by Ehrich and Harris (1942), Rich et al. (1939), Dougherty and White (1945, 1946a, b), and others has already been discussed or will be mentioned below.

Briefly, it is generally agreed that the usual response to antigenic stimuli involves, as stressed by Ringertz and Adamson (1950), several cell systems which are interrelated, that is to say reticulum cells, lymphocytes, and plasma cells. The first response in a node is an influx of polymorphonuclear cells. Ehrich et al. (1945) noted the phenomenon, and Smith and Wood (1949a, b), after injecting pneumococci in the footpads of animals, described the capture of the bacteria by the macrophages of the sinuses within 10 minutes, and the appearance of the polymorpho-nuclears in a few hours. Besides the changes in the nodules, mentioned above, there follows a procession of cellular events variously described by different authors. A diffuse lymphoid hyperplasia appears (Ringertz and Adamson, 1950) which is soon overshadowed by an increase in plasma cells. These authors suggest, in relation to this plasma cell response, the possibility that immature plasma cells develop from immature lympho-cytes. They suggest further that, if this is true, the immature lymphocyte is the precursor of the plasma cell and the lymphoid nodular centers may take part directly in antibody formation by furnishing immature lympho-cytes which—under the influence of antigen—are becoming plasma cells, instead of continuing to develop into mature lymphocytes. Their studies, two years earlier (Ringertz and Adamson, 1948), upon human lymph node material taken at necropsy after various infectious conditions led them to believe that plasma cells may be derived from reticulum cells.

Twelve to thirty-six hours after injecting rabbits with *Bacterium monocytogenes*, Conway (1939) observed the disappearance of small lym-phocytes from spleens and lymph nodes leaving the RE cells so in-ordinately apparent that the false appearance of a hyperplasia resulted. The development of macrophages from lymphocytes was reported. She depicted (Conway, 1937, 1938, 1939) lymphatic tissue as a stroma of reticular cells, fibers, and free cells which may appear to be nodular, dense, or loose. Each type may turn into the others, the differences merely reflecting changes in the proportions of the fixed and free cells. She described cells termed monocytoid lymphocytes, which represent stages in the transformation of lymphocytes to monocytes.

Taliaferro and Mulligan (1937) have also shown that macrophages in the spleen may arise from lymphocytes during malarial infections. Similar possibilities were discussed earlier by Maximow (1902).

Parsons (1943), studying the lymph nodes of mice in a variety of conditions, stresses the cellular changes in the reticulum which are coincident with the disappearance of lymphocytes. She finds no evidence of the development of plasma cells from lymphocytes; instead proliferat-ing reticulum cells seem to be the precursors of plasma cells. Although Maximow and Bloom (1957) derive the plasma cell from the hemocyto-blast which may become the lymphoblast, Parsons sees no evidence for it. The fixed reticulum cells she speaks of seem to be identical with Maximow's fixed undifferentiated mesenchymal cells which can develop into various blood or tissue cells.

Marshall and White (1950), after using single or multiple intravenous injections of various antigens, recognize two types of reactions, the first being a stimulation of primitive reticular cells which leads to the development of plasma cells in the spleen, lungs, liver, and bone marrow. Within 3 or 4 days "activated reticulum cells" with basophilic cytoplasm appear in small foci in the splenic pulp and in the sinuses near the arteries. Later they have the appearance of the cells classified by these authors as plasmablasts. They are probably the same cells that have been termed "splenic tumor cells" by Rich *et al.* (1939), myeloblasts by Hewer (1930), histiocytes by Goldzieher (1927), and transitional cells by Fagraeus (1948a). These cells change later to immature, and finally to mature, plasma cells. In the second type of reaction, "activated reticulum cells" appear in the Malphigian centers which later become composed of lymphocytes. Marshall and White (1950) suggest that antibody formation is not the sole function of either of these types of cells, at least in their mature forms. Since, as they point out, the earliest antibodies can be found 3–4 days after injecting antigen—before the mature forms have appeared and while the cellular reaction shows chiefly only activated reticulum cells or plasmablasts—they propose that one should not speak of plasma cells or lymphocytes as forming antibody, but only of the whole plasma cell or lymphocyte reaction as resulting in its formation.

Most authors have distinguished between the changes occurring in lymphoid tissue after single, as opposed to multiple, injections of antigen. For details, see the papers just reviewed. McNeil (1948), studying the cellular reactions to single injections of antigen, stressed the appearance of pseudoeosinophiles on the first day, then of lymphoblasts, then plasma cells and lymphocytes. The lymphoblasts, appearing from the fifth to seventh days, were characterized as basophilic lymphocytes, which he states later (McNeil, 1950) were indistinguishable from plasma cells. Multiple injections of antigen (McNeil, 1950) gave reactions with marked hyperplasia of reticulum. He concludes that antibody formation is a complicated pluricellular reaction which may require several systems of cell types.

When Marshall and White (1950) gave multiple injections of antigen, the "activated reticulum cells" disappeared from the centers. The lungs showed collections of plasmablasts, plasma cells, small lymphocytes, and reticuloendothelial cells, and the liver and bone marrow contained plasma cells. Collections of plasma cells like these could explain most extrasplenic or extralymphoid antibody formation. For example, Taliaferro and Taliaferro (1950, 1951b) showed that single injections of red cells into rabbits led to antibody formation almost wholly in the spleen, whereas multiple injections led to extrasplenic activity.

Although the observations of Fagraeus (1948a, b) and Thorbecke (1954) have already been discussed, it should be recalled that Thorbecke found big, round cells with strongly basophilic cytoplasm, plasmablasts, in groups within the splenic red pulp and in the white pulp around arterioles. Plasma cells did not seem to arise there. Reaction centers did not appear until the sixth day and their occurrence did not parallel antibody formation.

Recently, too, Björneboe et al. (1951) have described cellular changes in the spleens of mice. Plate 1 of their paper illustrates the phenomenon better than words.

VI. Antibody Formation in Tissue Cultures, Tissue Transplants, and Transferred Cells

A. In Tissue Cultures

Many workers have tried to find antibodies in cultures of normal tissues, to which antigens have been added, ever since Carrel and Ingebrigsten (1912), among the earliest in this field, cultured bone marrow and lymph node tissue of normal guinea pigs with red blood cells of the goat and reported the appearance of hemolysins. Most, however, have been unable to entice cultures to form antibodies under these circumstances. This generally negative finding has important implications, indicating that antibody formation is not the simple result of an antigenic stimulus and the cell's response. By contrast, it was well known even to the earlier workers that tissues taken from animals previously injected with antigens will form antibodies in culture (Ludke, 1912; Pryzgode, 1913; Reiter, 1913; Meyer and Loewenthal, 1927; Poleff, 1928; Kimura, 1932). A few examples of later work, gathered from a wealth of observations, will illustrate these points.

Splenic tissue, taken by Parker (1937, 1950) from rabbits 2 or 3 days after injecting them with foreign erythrocytes, formed antibody, but failed to make any when taken after only 1 day. Reactions within the immunized host seemed to be necessary before antibody could be formed. Rous and Beard (1934) obtained pure cultures of Kupffer cells by an ingenious method. The livers of rabbits injected with magnetic iron and later with antigen were perfused and massaged, while the perfusate passed through glass tubing over a strong magnet. The iron-containing Kupffer cells, collected by the magnet and grown in culture, produced no antibody (Beard and Rous, 1938), possibly because they suffered injury from the iron particles.

Fastier (1948) made fluid cultures of granulocytes, macrophages, bone marrow cells, and splenic cells of rats and allowed them to phagocytose killed *Salmonella schottmülleri* (*S. paratyphi B*) organisms. No antibody was found in the fluid surrounding the phagocytic cells, nor did it appear when lymphocytes from the thymus glands were added. Considering the negative results of other authors with cells of the thymus, this is not surprising.

Doan *et al.* (1950), with others, added to tissue cultures of normal splenic cells the azoprotein antigen of Heidelberger *et al.* (1933) and Heidelberger and Kendall (1934), previously employed by Sabin (1939), as described earlier. It was taken up only by macrophages which developed cytoplasmic vacuoles and began to shed their cytoplasm from the eighth to the fourteenth days. In the interval precipitins appeared in the culture fluid. The cells then returned to a resting stage. Doan *et al.* (1950) and Wright and Doan (1953), prompted by their own observations of the large numbers of young and mature reticuloendothelial phagocytes which exert a macrophagic function in the red pulp of the spleen by the phagocytosis of red cells, granulocytes, thromboblasts, bacteria, and other potential antigens, suggest that the hypothesis of Sabin (1939), already discussed, remains the most tenable explanation of antibody formation up to the time of their publication.

Roberts *et al.* (1948, 1949) obtained much antibody from incubated mesenteric lymph node tissues of rats if the antigen had previously been injected intraperitoneally into the animals. More was obtained from splenic tissue if the antigen was given intravenously. These authors suggest that the route of entry of the antigen influences the relative importance of the splenic and nodal tissue in the formation of antibody.

Ranney and London (1951) cultured not only splenic tissue, but hepatic tissue as well, from rabbits immunized with killed pneumococci. Radioactive antibody appeared in both after adding C^{14} carboxy-labeled glycine to the cultures.

Wesslen (1952b) cultured thoracic duct lymphocytes from rabbits immunized with typhoid organisms and found antibody formation. If the cells were lysed none was formed.

Trowell (1952a, b) cultured lymphocytes from lymphoid tissues of rats, on a medium containing cotton wool, and obtained excellent antibody formation.

The tissue culture findings of Fagraeus (1946, 1948a, b) have already been mentioned in discussing plasma cells and antibody formation. Keuning and van der Slikke (1950) extended her experiments. Tissue bits of red pulp of spleens, taken 4 days after the last of several injections of paratyphoid bacterin, had, at the time of transfer, twice as much antibody as bits of white pulp. After tissue culture the ratio was 10 to 3. Some antibody was also formed by the white pulp. Splenic cells were separated by differential centrifugation into one fraction containing mostly large cells and another containing small lymphocytes. Upon culture more antibody was formed by the large cells which were lymphoblasts, reticuloendothelial cells, and plasma cells. The authors trace antibody formation to lymphoblasts which mature either into plasma cells with much antibody or into lymphocytes with little. More recently Thorbecke and Keuning (1953) have cultured bone marrow, splenic, and lymphatic tissues from rabbits injected with antigen. More antibody was formed by the splenic tissue than by the others and more by the splenic red pulp than by the white pulp. There was a correlation between the number of young plasma cells and the titer of antibody. It is of particular interest that these workers correlated the amount of antibody formed by bone marrow cultures with the number of plasma cells present. Further, antibody was formed although lymph follicles and reaction centers were absent. In the lymph nodes the antibody titer was higher, after subcutaneous injections, on the injected side than on the uninjected one. Little antibody was found after intravenous injection. More will be said below about the findings with liver and thymus.

Mountain (1955a) obtained excellent antibody formation in cultures of minced splenic tissue from rabbits given typhoid bacterin intravenously. Its formation was suppressed by sodium cyanide, by salts of heavy metals and by a variety of other substances including cortisone, as mentioned elsewhere. A positive correlation was found (Mountain, 1955b) between the weight of the spleen and the antibody-forming capacity of a given mass of the spleen.

Stavitsky (1955b) and Wolf and Stavitsky (1956) took fragments of lymph nodes from rabbits 3 days after a "booster" dose of diphtheria toxoid and incubated them in a special culture medium. After 20 hours, five to a hundred times more antitoxin appeared in the medium than could be obtained from the fragments. As found earlier by Parker (1937, 1950), it was necessary to wait 3 days after injecting the antigen before removing the tissues from the donors, in order to get antitoxin formation in the cultures. More will be said below about the work of Wolf and Stavitsky.

Steiner and Anker (1956) cultured splenic cells from rabbits showing a secondary response to bovine serum albumin, as antigen. When grown upon a cellophane membrane, below which the nutrient medium was agitated, the highest titers were obtained from cells taken from their hosts 3 days after the secondary challenge with antigen. No antibody was formed by cells taken after only 1 day or by cells from normal animals if antigen was added to the cultures. These findings confirm those of Parker (1937, 1950), already described.

Antibody formation in sliced granulomatous tissue (Askonas and Humphrey, 1955) will be discussed below.

It is of much interest in relation to the preceding tissue culture work that a lipopolysaccharide obtainable from gram-negative bacteria (Johnson *et al.*, 1956) shows a capacity, when injected into rabbits, to increase antibody formation. Following this lead, Stevens and McKenna (1957a, b) and McKenna and Stevens (1957a, b) found antibody formed by spleen fragments after both primary and secondary injections of protein antigens. Even the primary responses occurred very rapidly, apparently intracellularly, without any latent period. Spleen, removed only 2 hours after a secondary injection, formed antibody in culture. Further, 10 μg. of a lipopolysaccharide endotoxin from typhoid bacilli, given to the rabbits 24 hours before taking the tissue, enhanced antibody formation eightfold. McKenna and Stevens (1957a, b) found splenic explants, removed from normal rabbits that received only endotoxin, capable of forming antibody after 1 hour of incubation with bovine γ-globulin.

B. By Tissue Fragments

Topley (1930) injected normal rabbits with minced tissue from the spleens of other rabbits injected 24 hours previously with paratyphoid bacilli. The recipients formed antibodies more rapidly than they would have if antigen had been given instead. Apparently the antibody was produced by cells of the transferred tissue, not by any antigen contained in it or by the recipients' cells, since a subsequent injection of antigen yielded no secondary response in the recipients. Dougherty *et al.* (1945b) (see also White and Dougherty, 1946) transferred lymphosarcoma tissue, and Fagraeus and Grabar (1953) transplanted splenic fragments, both from immunized rabbits, and obtained antibody formation in normal recipients. Live cells were necessary because the transfer of lysed cells gave no antibody.

Oakley *et al.* (1954) injected rabbits with diphtheria or tetanus toxoid and reinjected them a month later into the interscapular fat or into the foot pads. The fat and the popliteal nodes, transferred into the omenta of normal rabbits, formed antitoxins. Oakley and his co-workers, Warrack and Batty, felt that the result with the fat could be attributed to an influx of lymphoid cells into the fatty tissue prior to its transfer. After transfer these cells formed the antibody.

Bits of splenic and lymph node tissue, Peyer's patches, and thymus glands of mice, given two injections of tetanus toxoid, when transferred to the anterior chamber of the eyes of X-irradiated mice, were found capable by Hale and Stoner (1953, 1954) of forming antitoxin in small amounts. Injections of toxoid given the recipients 10 days later stimu-lated much more antibody formation by the transplants.

Studies on transplantation immunity dealt with elsewhere in this book, but not in this chapter, have shown that lymph node cells from one strain of mice transferred to a different strain, rendered tolerant to, and bearing skin grafts from, the first strain, will survive and form antibodies to the tissue antigens of the graft and destroy it.

Good (1955) and Good and Varco (1955c) found no antibody responses in a patient with acquired agammaglobulinemia following injections of typhoid-paratyphoid vaccine. By contrast, after the implantation of normal lymph nodes, injections of the vaccine gave rise to much antibody for 2 months. Thereafter further antigenic stimulation yielded no response. Martin *et al.* (1957) obtained similar evidence of antibody formation in an agammaglobulinemic patient for an even longer period, 100–110 days, by the transfer of lymph nodes from a normal, but tuberculin-positive donor. The recipient of the tissue, who had been tuberculin negative before receiving the lymphoid material, became tuberculin positive within 2 days.

C. By Cells Transferred from Donors to Recipients

Deutsch (1899) seems to have been one of the first to transfer cells to study antibody formation. He used splenic cells from animals infected with typhoid organisms and found agglutinins in the sera of the recipients. Presumably, he transferred antigen as well as splenic cells.

Landsteiner and Chase (1942) transferred hypersensitivity to simple chemical compounds from sensitized guinea pigs to normals by means of cells of the peritoneal exudates obtained from the sensitized animals. Chase next found (1943, 1945, 1946, 1947, 1948, 1950a, b, 1951, 1953) cells of inflammatory peritoneal exudates, buffy coats, spleens, and lymph nodes of guinea pigs, sensitized to tuberculin or to picryl chloride, able to transfer sensitivity of both the immediate and the delayed types. Heat or freezing destroyed this capability, and extracts of the cells showed no antibody. When more cells were transferred from guinea pigs sensitized to picryl chloride than were required to yield skin reactions in challenged recipients, the uteri of the animals showed anaphylactic sensitivity (Schultz-Dale technique) as early as the second or third day after the transfer.

The transfer of tuberculin sensitivity to normal animals has been con-firmed by Cummings *et al.* (1947), Stavitsky (1948), Kirchheimer and Weiser (1947), Metaxas and Metaxas-Bühler (1948), and, since then, by many others. Lawrence (1949, 1952, 1954, 1955, 1957) produced local cutaneous hypersensitivity to tuberculin in normal subjects by the intradermal injection of whole leucocytes taken from tuberculin-positive patients, and further, he transferred delayed skin sensitivity to strep-tococcal M substance and to tuberculin with disrupted white cells. Again, the transfer of leucocytes from atopic skin-sensitive patients to normal subjects by Waltzer and Glazer (1950) has led to the immediate whealing type of sensitivity upon challenge.

Chase (1951, 1953), Wager and Chase (1952), and Chase and Wager (1957) transferred splenic and lymph node cells from rabbits injected with diphtheria toxoid to normal animals. In some of these experiments the recipients showed two bursts of antibody formation, the first being the antibody formed by the transferred cells, the second having the charac-teristics of an active sensitization. Chase (1953) obtained further evidence that the cells, transferred from immunized guinea pigs to normal guinea pigs, actually formed antibody, since the avidity of the antibody of the recipients duplicated the quality of the donors' antitoxin. Chase and Wager (1957) conclude therefore, that the antibody found following the transfer of cells was not elicited by antigen contained in the cells.

Wesslen (1952b) collected the cells (5–6% large lymphocytes, the remainder small ones but no plasma cells) from thoracic ducts of im-munized rabbits. These cells, transferred to guinea pigs, rendered the animals anaphylactically sensitive.

Stavitsky (1952, 1954, 1956) transferred, to normal rabbits, cells from the popliteal lymph nodes and spleens of other rabbits, 3 days after the latter had been injected in the foot pads, for the second time, with alum-precipitated diphtheria toxoid. Both types of cells yielded antibody in the recipients. However, following intravenous injections of the antigen, only splenic cells served. The speed of the production of antibody in the hosts indicated its formation by the transferred cells. In a later paper, Stavitsky (1957) has given an excellent discussion of these findings and of similar observations by others. He has also (Stavitsky, 1954, 1956) called attention to the much larger amount of antibody formed in the recipients of cells than could have been present in the cells themselves, as stressed also by Roberts and Dixon (1955) in a paper to be reviewed below.

S. Harris and T. N. Harris (1951) and T. N. Harris et al. (1954b, S. Harris et al., 1954) transferred cells of the popliteal lymph nodes of rabbits injected in the foot pads with dysentery bacilli. Agglutinins appeared in the recipients earlier than is usual after injections of the antigen alone. Transferred lymph node cells of the donors, injected with antigen in the foot pads, gave much higher antibody titers than splenic cells, but the latter, from donors injected intravenously, gave better anti-body formation in the recipients than cells from lymph nodes. To rule out the possibility of antibody formation by the recipients, S. Harris and T. N. Harris (1954) and T. N. Harris *et al.* (1954a) transferred similar cells to X-irradiated recipients unable to form antibody. They found approximately a 4-day interval elapsing between injection of the antigen and full maturation of the antibody, but it made no difference whether the cells to be transferred remained in the donor or were given earlier to the recipient. Indeed, cells transferred only 10 minutes after injecting antigen into the donor yielded antibody in the recipients. Further, lymph node cells from normal donors (S. Harris and T. N. Harris, 1955), when simply incubated with dysentery bacilli and washed, or merely exposed to a soluble form of the antigen and washed (T. N. Harris *et al.*, 1955), formed antibody when transferred to normal recipients. Control experiments, which varied the number of cells transferred and the time of exposure to the soluble antigen, suggested that the antibody formation in the recipients was not brought about by the transfer of the antigen, but by the exposed cells. This work has been well reviewed in short papers by T. N. Harris and S. Harris (1955, 1956).

Next, normal lymph node cells, 90–99% of them cells of the lymphocytic series, when incubated with filtrates from dysentery organisms subjected to tryptic digestion, were found capable, by S. Harris *et al.* (1956), of forming agglutinins in X-rayed recipients. Excess antiserum added to the cells or the filtrates, before incubation, cut down the antibody formation, but it had no effect when added after only 5 minutes of incubation. Apparently, some reaction took place between the antigen and the cells, since little or no free antigen could have been transferred with them. Circulating leucocytes, 63–93.5% lymphocytes, 4.5–39% neutrophiles, and 1–4.5% monocytes, similarly treated and transferred (T. N. Harris *et al.*, 1956) yielded some antibody, but less than after the transfer of cells from lymph nodes. Cells of peritoneal exudates, taken 2 days after injecting oil and consisting chiefly of neutrophiles, gave no antibody when transferred, but cells obtained after 9 days, chiefly monocytes, gave a little in 46% of the recipients and as much, in 30% of them, as with the transfer of lymph node cells.

In later work T. N. Harris *et al.* (1957) injected normal rabbits with leucocytes of the blood or from peritoneal exudates of other, donor rabbits. Four to eleven days later—and 1 day after X-raying these recipients—they transferred, from the same donors to the same recipients, lymph node cells which had been incubated with soluble antigenic material in the usual way. Under these circumstances little or no antibody appeared in the recipients. Evidently the preinjection of the donors' leucocytes had altered, in the recipients, the environment of the subsequently transferred lymph node cells so that antibody formation was inhibited. The finding showed that, in the preceding experiments (S. Harris *et al.*, 1956; T. N. Harris *et al.*, 1956) the transferred lymphoid cells had apparently been responsible for the antibody that appeared.

Roberts and Dixon (1955) used the cell transfer technique to learn more about the immune response to soluble foreign proteins. They transferred lymph node cells—82–91% lymphocytes, 2–5% plasma cells, and 6–14% reticuloendothelial cells—to X-rayed recipients from rabbits previously immunized with radioactive (I^{131}) bovine γ-globulin or serum albumin. Within an hour antigen was also injected, and, as a result, the transferred cells formed antibody as a secondary response in their new hosts. Similar cells from normal rabbits, under the same circumstances, formed none. Cells taken from either normal or immunized animals, when incubated with the soluble protein antigens and transferred, also formed none, apparently behaving differently from those treated with bacillary antigens, which did form antibody in the studies of T. N. Harris *et al.* (1954b; S. Harris *et al.*, 1954). Since approximately 90% of the transferred cells were lymphocytes, the authors were inclined to attribute the antibody formation to these cells or to forms that might be derived from them following their injection into the recipients. Dixon *et al.* (1957a) found peritoneal cells (71% macrophages) and lymph node cells (90% lymphocytes) giving, in similar transfer experiments, similar secondary responses. Further, the transfer of twice as many lymph node lymphocytes as were present in the peritoneal exudates—or as many macrophages as were present in the suspensions of lymph node cells—failed to form comparable antibody responses. The work indicated clearly that both peritoneal macrophages and lymphocytes were capable of transferring the ability to make secondary antibody responses after transfer.

An important observation, which goes far to explain these findings and those of much of the other work cited above, was reported by Roberts *et al.* (1957) and Dixon *et al.* (1957b). The sites to which both macrophages and lymphocytes had been transferred contained, in the experiments like those just described that gave good antibody yields, collections of pyroninophilic cells which appeared after the fourth day. Plasma cells appeared at the sixth day. The authors state that the transferred, peritoneal macrophages and lymph node lymphocytes probably developed into the observed plasma cells.

Very recently, while this chapter was in proof, Neil and Dixon (1959), in a similar study with transferred lymph node cells, have obtained definitive proof that the transferred cells formed antibody as they developed, at the transfer sites, into preplasma cells and later into plasma cells. By applying the Coons fluorescence technique, they found the transferred cells showing the specific fluorescence for antibody on the third day, while exhibiting the morphological characteristics of transitional or preplasma cells. Maximum antibody was found on the fifth and sixth days when there was a predominance of preplasma and plasma cells.

D. By the Transfer of Nuclear Fractions

Sterzl (1955) reports that, in his hands, young rabbits 5–7 days old do not form antibody to doses of 10^7 to 10^9 S. *paratyphi* organisms. However, splenic tissue, containing no antibody but taken from adult donor rabbits 48–72 hours after giving them similar doses of the organisms, will produce antibodies in the young animals when transferred to them intraperitoneally. He feels that a precursor is transferred and that the tissues of the recipients help in some way to complete antibody formation. Sterzl and Hrubešová (1956) report the transfer of the antibody-forming capacity to the young rabbits by nucleoprotein fractions of splenic cells prepared, from adult donor rabbits injected 48 hours previously with the paratyphoid organisms, at a time when the donors' tissues show no antibody. Both ribo- and deoxyribonuclear proteins yield positive results unless they are obtained within 24 hours after injecting the donors. Mitochondrial fractions give more antibody than the nuclear fractions. The 5–7-day-old rabbits do not respond to the antigen alone. More will be said of these findings below when discussing the effects of X-irradiation upon antibody formation.

VII. Local Antibody Formation at the Sites of Introduction of Antigenic Substances

The presence of antibody, in the tissues into which antigens are first introduced, frequently occurs under certain conditions which render it unlikely that its localization is due to the seepage of antibody formed elsewhere in the body. However, it is believed by many that local antibody formation is brought about by the entry into the sites of antigenic stimulation of the types of cells that have already been considered, that is to say, by plasma cells, lymphocytes, macrophages, or eosinophiles. The most convincing work, for reasons already discussed, seems to be that which has employed two or more antigens. Readers interested in local antibody formation should consult the important papers of Freund (1927, 1929a, b, 1947) and of Westwater (1940a, b) to become familiar with the changes in the blood to tissue ratio of antibodies under various conditions.

A. In the Eye

After injecting various antigens into the eye, several authors, Römer (1901), Hektoen (1911), Seegal and Seegal (1931, 1934, 1935), Seegal *et al.*

(1933), Seegal and Wilcox (1940), usually found higher titers in the blood than in the organ. Scamburow (1932), after injecting one eye with antigen, occasionally found more antibody in the uninjected eye than in the blood. Thompson *et al.* (1936, 1937) and Thompson and Olsen (1950) reported the presence of precipitins in corneal tissue both earlier and in higher concentration than in the blood, bone marrow, or spleen. Later Thompson and Harrison (1954) injected two different antigens, one in each eye. X-ray or strontium radiation given to the head had no effect upon antibody formation, but radiation to the whole body, with the head shielded, prevented all antibody formation in the cornea. More recently Thompson *et al.* (1957) have again demonstrated apparent local antibody formation in the corneal tissue of rabbits and have given an excellent discussion of the difficulties to be met with in attempts to study local antibody formation. They compared the results of injecting small and large amounts of either soluble or alum-precipitated protein antigens. Small amounts (0.1–0.05 mg.) of alum-precipitated egg white injected into the cornea yielded higher titers of antibody in this tissue than in the serum. Larger amounts, especially of soluble antigen, gave opposite results. The authors stress the importance of the site of antigen injection, its amount, and the type of antigen employed in determining the results. After injecting two different antigens into the corneal tissues of the opposite eyes, the antibody titers against the specific antigen injected in either eye were higher than those against the antigen injected in the other eye. Control experiments showed that the findings were not attributable to the concentration of antibody produced elsewhere in the body.

B. In the Skin

Skin, injected with antigen and containing numbers of macrophages mobilized by previous injections of the same antigen, was found by Cannon and Pacheco (1930), Cannon and Sullivan (1932), and by Walsh *et al.* (1932) to yield higher titers of antibody than the blood serum, liver, or spleen. Hartley (1940) induced the mobilization of large numbers of macrophages and only a few polymorphonuclears or lymphocytes in the skin of animals by injections of aluminum hydroxide gel. The subsequent introduction of vaccine virus, adsorbed on the gel, into the regions of skin containing the mobilized macrophages led to the appearance of neutralizing antibodies, apparently formed locally, before they were present in the blood. He concluded that the macrophages formed the antibody. Local formation of viral neutralizing substances in skin was shown by Ørskov and Andersen (1938a) to appear earlier and in greater concentration than in the serum. Following intratesticular injection of vaccinia virus into rabbits, Ørskov and Andersen (1938b) also demonstrated the appearance of antibodies sooner and in higher titers in the testicles than in the serum, and they concluded that the primary formation of virus-neutralizing material takes place in the injected tissue.

Freund *et al.* (1952) examined the fluid in areas of inflammation produced by repeated injections of diphtheria toxoid and Freund's adjuvant into the skin of a horse. This fluid, which contained mononuclears, neutrophiles, lymphocytes, red blood cells, and many eosinophiles, contained also two to seven times as much antitoxin as the blood, whereas other areas injected with broth and adjuvant contained less than the blood. They concluded that the work supported the concept that antibodies can form at sites of injection of antigen and adjuvant, and that the latter evokes an accumulation of cells which can form antibody.

Oakley *et al.* (1949) injected alum-precipitated diphtheria or tetanus toxoids (APT) as a secondary stimulus, subcutaneously into horses, rabbits, and guinea pigs. Antitoxin formed locally in the skin if the areas had been sufficiently stimulated to produce granulomata. To rule out seepage they used two antigens—as had been done by McMaster and Hudack (1935), by Ehrich and Harris (in their many papers cited above), and by Walsh and Cannon (1938). Oakley and his co-workers injected diphtheria APT in one leg and tetanus APT in the other. In the skin of the side injected with diphtheria APT, the ratio of diphtheria to tetanus antibody was higher than in the blood and lower than in the skin of the side injected with the tetanus APT. Local antibody formation in slices of granulomata produced with Freund's adjuvant has also been reported by Askonas and Humphrey (1955).

C. In the Nasal Mucosa

Walsh and Cannon (1938), after injecting into rabbits one antigen intranasally and another intraperitoneally, or reversing the process, found the splenic titer of the antigen introduced intraperitoneally higher than its intranasal titer, and vice versa for the antigen introduced into the nose, although the titers of both antigens were often equal in the blood.

D. In the Gastrointestinal Tract

Davies (1922) found fecal agglutinins in patients with dysentery. Inoculation of the stomach of guinea pigs with cholera vaccine by Burrows *et al.* (1947a, b) and Burrows and Havens (1948) led to the appearance in the feces of antibodies which arose more rapidly and reached a peak sooner than those of the serum. They believe that fecal antibody is derived from cells in the intestinal mucosa. Howie *et al.* (1953) and Duncan *et al.* (1954) found *Clostridium welchii* (*C. perfringens*) in the stomach contents of patients after partial gastrectomy and a neutralizing substance for the organisms' alpha toxin in the stools, but none in the blood. When Digby and Enticknap (1954) gave rabbits a preparatory "vaccination" against *C. septicum* through an appendicular pouch, rich with lymphoid cells, the animals survived subsequent infection; given through a cecal pouch there was very little effect. These authors do not claim that local antibody formation occurred.

E. In Certain Other Organs

Oakley *et al.* (1951) have provided evidence that antibody may be formed in fat or voluntary muscle of rabbits and guinea pigs after a secondary injection of APT. It is apparently not formed by the fat or muscle cells but by cells of the granulomata that develop about the APT. Batty and Warrack (1955) injected both diphtheria and tetanus APT subcutaneously into rabbits, and one month later introduced one or the other of these antigens directly into various organs. The resulting tetanus to diphtheria antitoxin ratios obtained, under these conditions, gave no signs of antibody formation in the liver, but indicated its formation in the mammary glands, the spleen, the uterus, vagina, and appendix. Batty and Warrack feel that local antibody formation may occur within the cavities of organs lined with mucosa after antigen has been introduced therein.

The several papers just cited tend to confirm earlier work of Pierce (1947, 1949) and of Kerr and Robertson (1947, 1953), who found antibody in the walls of the uterus and vagina of cattle after introducing *Trichomonas foetus* and *Brucella abortus* into the organs; and also the findings of Naylor and Caldwell (1953), who compared the titer of flagellar *Salmonella* antibodies in the urine and serum of patients receiving typhoid "vaccine" and of subjects who were carriers of the disease.

F. In the Udders of Cattle

More than forty years ago Giltner *et al.* (1916) injected *Brucella abortus*, as antigen, into the "milk cistern" of one quarter of a cow's udder. The following day agglutinins appeared in the milk from that quarter, and subsequently from the other quarters. Years later Campbell *et al.* (1950) demonstrated an exceedingly heavy plasmacytosis of the bovine udder during colostrum secretion. Petersen and Campbell (1955) infused, into cows' udders, organisms foreign to the species and found antibodies in the milk in 24 hours. Similar findings were obtained by Mitchell *et al.* (1953) with viruses. More recently Campbell *et al.* (1957) instilled live cultures of *Salmonella pullorum* into the mouths of calves. When such a calf nursed from two quarters of the udder of a lactating cow, antibodies appeared in the milk in a few hours.

G. In the Liver

As early as 1925, Jones injected killed hog cholera bacilli into the mesenteric veins of rabbits and reported that antibody was apparently formed in the liver. Braude and Spink (1951) found an increased activity of nonparenchymatous cells of the livers of mice following splenectomy and injection with *Brucella* organisms. Lowenhaupt (1945) had observed a like phenomenon in patients with multiple myeloma. Acting upon the findings of Braude and Spink, Thorbecke and Keuning (1956) splenectomized rabbits, gave them paratyphoid bacterin, and found aggregates of immature and mature plasma cells and some lymphocytes in the periportal spaces of the liver, as illustrated in Fig. 3. Tissues from such livers formed antibody in cultures whereas, in their earlier work (Thorbecke and Keuning, 1953), liver tissue from intact animals formed none. They believe that the antibody demonstrated in the later experiments was formed by the aggregates of plasma cells. The work of Miller and Bale (1954) and of Miller *et al.* (1951, 1954), which will be discussed below, goes far to show that, under ordinary circumstances, antibody is not formed in the liver. On the other hand antibodies may be formed within the liver by invading cells, or perhaps by Kupffer cells. The true hepatic parenchyma cell does not seem capable of forming antibodies. Cheng (1949) found that damage to the liver of rats had no effect upon the synthesis of plasma proteins.

H. In Bone Marrow

So much has been said already about myeloma cells and marrow plasma cells that only a little will be added here concerning recent work. Koluch (1938) and his co-workers (Koluch *et al.*, 1947) noticed changes in plasma cells of the bone marrow of sensitized animals, and related the changes to antibody formation. Sédallion *et al.* (1939a, b) gave rabbits secondary injections of diphtheria or tetanus antitoxin. When the livers or kidneys of these rabbits were connected to the circulation of similarly immunized animals, no rise in antibody level occurred, showing that the organs formed no antibody. If isolated hind legs were perfused in the same way, large increases in antibody concentration occurred, which, these authors believed, were produced in the bone marrow.

Taliaferro and Taliaferro (1957) transferred minced bone marrow or spleen to normal rabbits from rabbits undergoing a secondary response to bovine serum albumin. Gram for gram both tissues passed on to the recipients the ability to form equal amounts of antibody. Since the rabbit's bone marrow far outweighs its spleen, the former may be of greater importance for antibody formation than the latter. Askonas and White

(1956), using guinea pigs, also found bone marrow capable of forming more antibody than the spleen.

It is to be recalled that Thorbecke and Keuning (1953) showed that cultures of bone marrow, taken from rabbits given killed paratyphoid organisms, formed antibody, presumably because of collections of plasma cells. They pointed out that no lymph follicles and no reaction centers were found in this tissue, which formed antibody without them. Jeschal (1953) obtained plasma cells of bone marrow from patients with scarlatina, diphtheria, and pharyngitis, and agreed with the findings of Moeschlin et al. (1951), already cited, that the characteristic granules to be seen in such plasma cells under the phase microscope indicate the formation of antibodies.

I. In the Thymus Gland

Most workers are agreed that little or no antibody is formed in the thymus gland. Even thymus glands of hyperimmunized rabbits were found by Björneboe et al. (1947) to hold no more antibody than muscle, and T. N. Harris et al. (1948) also reported the tissue inactive in the processes of antibody formation. However, intraocular transplants of thymic tissue from mice immunized with tetanus toxoid were found, by Stoner and Hale (1955), capable of forming tetanus antitoxin in cobalt 60 γ-irradiated recipients when the latter were injected intravenously with toxoid. Dixon et al. (1957a) in work already cited, transferred cells of thymus glands of previously sensitized rabbits to X-rayed recipients. When the latter were injected with antigen, these cells gave a relatively very poor response.

J. In the Central Nervous System

High titers of antibody in the central nervous system have been reported by a number of workers, and good evidence has been presented to show that, under certain circumstances, the findings cannot be attributed to seepage of antibody into the tissues with consequent local accumulation (Olitsky et al., 1943; Morgan et al., 1942; Morgan and Olitsky, 1941; Schlesinger et al., 1944; Schlesinger, 1949a, b). However, whether antibody appears in the central nervous system because of the infiltration of antibody-forming cells, such as plasma cells, or through direct, local formation by reticuloendothelial elements, or by microglia cells or astrocytes, or merely by seepage, has been debated for years. For references and a brief discussion of the subject, the reader is referred to a recent review by the present author (McMaster, 1953, pp. 34–36).

VIII. Evidence Concerning the Cells Engaged in Antibody
Formation as Obtained by a Variety of Techniques

A. *Irradiation Experiments*

1. *The Suppression of Antibody Formation by X-Rays and Other*
   Forms of Irradiation

Some of the early work on the suppression of antibody formation by
X-rays has already been mentioned. Nevertheless a few of the more recent
experiments should now be discussed to indicate something about the
types of cells that are injured or destroyed when antibody formation is
suppressed or decreased by X-rays. Craddock and Lawrence (1948),
irradiating rabbits 8 hours prior to the injection of antigen, noticed a
temporary depression of antibody formation. When X-rays were applied
after antibody formation had started, damage to lymphoid tissue failed
to lower the peak antibody titer. Accordingly they suggested that the
reduction of antibody caused by irradiation was not due wholly to
destruction of lymphoid tissue.

The time relationships between the giving of antigen and of irradiation as well as the type of antigen given are extremely important in pre-
dicting the effect that may occur upon antibody formation, but the matter
is too complex to discuss fully. The reader should consult the papers by
Kohn (1951), Taliaferro and Taliaferro (1951a, 1954), Taliaferro *et al.*
(1952), Taliaferro (1957a, b), Dixon *et al.* (1952, 1956), Wissler *et al.* (1953),
and Talmage (1955). By and large it can be said that radiation of about
400–500 r., given from 12 to 48 hours before an injection of antigen,
markedly decreases antibody formation. X-ray, given at longer periods
before injections of antigen, leads to a progressively decreasing inhibition.
X-ray, only 5 hours before antigen, reduces antibody formation (Dixon
et al., 1952), but given with the antigen, or from 6 hours to 6 days afterward, it has little effect. That is to say, the antibody response is resistant
if X-ray is given after antigen or just a few hours before it (Taliaferro
and Taliaferro, 1951a, 1954; Taliaferro *et al.*, 1952; Kohn, 1951; Dixon
et al., 1952; Wissler *et al.*, 1953; Hale and Stoner, 1954; Williams *et al.*,
1956; Fitch *et al.*, 1953, 1956). An excellent discussion of the different
effects upon antibody formation that can be obtained by varying the time
of irradiation in relation to the administration of antigen has been recently
given by Taliaferro (1957a) and Fitch *et al.* (1956).

Even when 800 r. of total body X-irradiation was given to rabbits 3
days after antigenic stimulation (Maurer *et al.*, 1953), antibody formation
was little affected in spite of extensive damage to the lymphoid tissue,
spleen, and bone marrow. Indeed antibody was formed in normal

amounts, and it was indistinguishable from that of nonirradiated rabbits. In this radio-resistant phase, the small lymphocytes were almost all destroyed, but plasma cells could be found surrounding splenic arterioles.

In spite of severe destruction of lymphoid tissue, Parsons (1943) reported massive proliferations of plasma cells in lymph nodes of X-rayed mice, a phenomenon seen previously in man by Ross (1932) following radium treatments, by Liebow *et al.* (1949) after atomic radiation, and by Thomas and Brüner (1933) in rats subjected to chronic radium poisoning. Marshall and White (1950) found proliferation of plasma cells in the mesenteric and popliteal lymph nodes of rabbits although there was an almost complete destruction of lymphocytes produced by 400 r. of X-irradiation. The plasma cells showed no appearance of damage. As already mentioned, similar effects were obtained with nitrogen mustard.

Fitch *et al.* (1953, 1956) showed a correlation existing between antibody formation in X-rayed rats and a hyperplasia of pyronine-staining cells originating from large immature cells of the red pulp. The cellular increase appears 48 hours after injection of the antigen and reaches a maximum on the fourth day. Thereafter most of the cells disappear, but some become mature plasma cells after the peak of antibody titer has been reached. The effect of X-ray seems to depend upon the time in this cycle of proliferation at which the radiation has its effect. X-ray seems to have no effect upon the fixation of antigen.

An excellent discussion of some of the dynamics of the effects of radiation upon antibody formation, but too complex to summarize in this place, has been given by Schweet and Owen (1957).

Williams *et al.* (1956) correlated the cellular changes in the spleens and lymph nodes with changes in the production of tetanus antitoxin in mice subjected to cobalt γ-irradiation before and after secondary antigenic stimuli with tetanus toxoid. They saw an early destruction of both lymphocytes and plasma cells, but the cells most affected, during the period in which antibody would presumably form, appeared to be the undifferentiated cells of the phagocytic or reticuloendothelial system. They suggested that the findings showed these cells—the "multipotent" units which may differentiate into transitional cells, lymphocytes, plasma cells, or phagocytic cells under various stimuli—to be capable of forming antibodies in the absence of lymphocytes and plasma cells. Hale and Stoner (1954) showed that 700 rep. cobalt 60 γ-radiation did not increase the susceptibility of mice to infection with PR8-A influenza virus, but dosages of 475–875 rep. inhibited antitoxin formation in mice given a secondary antigenic stimulus of alum-precipitated tetanus toxoid.

Differences seen between the response of rats and rabbits to X-rays and spleen shielding, or following variations in the route of the injection of antigen, have been discussed in the papers just mentioned, as also in others to be reviewed below.

2. Some Effects of Lead Shielding of Various Organs During X-Irradiation

Jacobson et al. (1949, 1950a, b) X-rayed rabbits and shielded the spleens or appendicular regions with lead. The shielded animals formed hemolysins to sheep red cells although lymphoid tissue elsewhere in the body was temporarily destroyed. Later, Jacobson and Robson (1952) irradiated rabbits with shielded spleens, removed the organs 24 hours later, and injected antigen after another 24 hours. Antibody was formed, apparently through the release of some humoral or cellular products, which must affect the remaining antibody-forming tissues in the body. Wissler et al. (1953) obtained similar effects of spleen shielding in rats. Taliaferro and Taliaferro (1956), however, using careful quantitative methods confirmed in principle the findings of Jacobson and Robson (1952), but found much less protection. When the shielded spleen, protected from 500 r. irradiation given to the rest of the animal, was left in the body 3–5 days, the procedure failed to increase the peak titer of antibody. A full and excellent discussion of these matters is given by Taliaferro and Taliaferro (1956).

Süssdorf and Draper (1956) found, as had many others before them, that a splenectomy performed prior to an injection of antigen depressed antibody formation. They reported, further, that an appendectomy had little effect. Nevertheless, shielding the appendix against X-rays gave more protection to the antibody-forming capacity than shielding the spleen. Presumably the protection of the antibody-forming tissues must have been indirect, brought about by the protection of mobile cells that might repopulate the antibody-forming areas, or by the shielding of some humoral material. Shielding the livers also protected the antibody-forming capacity. The authors do not infer from this finding that the livers formed antibody, but rather that the lead shielding exerted the indirect effect just mentioned. There seemed to be no correlation between the protection obtained by shielding a tissue and its antibody-forming ability. It is of further interest that direct radiation of 1400 r. given to the spleens, or livers, or appendicular regions did not depress antibody formation.

3. Enhancement of Antibody Formation by Local Irradiation

X-rays have long been used clinically to treat local infections, and local irradiation, unlike total body irradiation, seems to enhance antibody formation. Taliaferro and Taliaferro (1954) noticed that 10–25 r. of X-rays increased the mean peak titer of antibody. Graham *et al.* (1956) and Graham and Leskowitz (1956) injected antigens into rabbits in one thigh and X-rayed it 30 minutes later. More antibody appeared locally than in nonirradiated controls. There was no increase in antibody when the antigen was injected into one thigh and the other was X-rayed. The authors feel that the antibody was formed locally. Very large doses of X-ray, 2000–10,000 r. given only to the spleens of rabbits by Taliaferro and Taliaferro (1956) resulted in peak titers of antibody higher than those obtained in nonirradiated, control animals (see also Taliaferro, 1957a).

4. Restoration of Antibody-forming Ability following Irradiation

As indicated earlier, total body X-irradiation exerted a strong suppression of antibody formation apparently by injury to some very early event in the process. It seemed possible that the suppression might result from the prevention of the formation of substances which initiate antibody formation. Light was thrown upon the matter by Taliaferro and Taliaferro (1956), and Jaroslow and Taliaferro (1956), who gave 400 r. of X-irradiation to rabbits, and 1 day later—when the animals' antibody-forming capacity was maximally damaged—injected them intravenously with mixtures of antigen and splenic cells taken from either normal rabbits or mice. The procedure restored antibody formation in the X-rayed rabbits. Further, the restorative material seemed to be nonspecific, and living cells were not required, since extracts of normal mouse spleens, suspensions or extracts of HeLa cells, or even yeast autolyzates were effective. Taliaferro (1957a) has indicated that the restorative principle may be a product of the action of specific nuclease on ribonucleic or deoxyribonucleic acid. Taliaferro (1957b) has further suggested that the work of Jaroslow and Taliaferro (1956) may throw some light on the mechanisms accountable for the formation of antibody by the X-rayed rabbits which received mixtures of lymph node or splenic cells and antigenic material from dysentery bacilli, in the work of S. Harris and T. N. Harris (1955) and T. N. Harris *et al.* (1955, 1956; S. Harris *et al.*, 1956) as mentioned earlier, and of the 5-day-old rabbits in the experiments of Sterzl and Hrubešová (1956), which, as also described above, received nuclear and microsomal cell fractions from immunized donor rabbits, but without antigenic materials.

Berglund (1956a, b), as already mentioned, stressed the fact that cortisone, like X-ray, seems to affect the early phase of antibody formation, and Berglund and Fagraeus (1956) gave splenic cells or cells from the thymus gland to cortisone-treated animals soon after injecting them with antigen. The cells prevented the reduction of antibody formation by the hormone.

According to Smith and Ruth (1956) the recovery of the ability of irradiated mice to produce antibody to sheep red cells is unchanged by injecting the animals with homologous bone marrow and spleen cells soon after X-ray, although the cells apaprently exert a favourable effect upon survival to lethal doses of X-ray.

5. The Effect of X-Rays upon the Mortality Rate of Immunized and Subsequently Challenged Animals

Silverman and Chin (1955) studied the mortality rate instead of the antibody titer of immunized and subsequently X-rayed mice given 10 MLD (minimal lethal dose) of tetanus toxin. The larger the dose of X-ray the greater was the inhibition of the immune response, and irradiation was equally effective when given either before or after injecting antigen. These authors discuss at some length the reasons for the apparent differences between their findings and those of others who studied the antibody-forming response. Shechmeister et al. (1955) found a protective effect by spleen shielding upon the death rate of mice infected with Escherichia coli.

6. The Lack of Antibody Formation in X-Rayed Animals Compared with That in the Newborn

Very recently a most interesting paper by Dixon and Weigle (1957) has thrown light upon the factors that prevent antibody formation in X-rayed, adult animals and in the newborn. From donor rabbits, stimulated with antigen, these workers took lymph node cells capable of forming either primary or secondary responses, and transferred them to adult, normal or X-rayed recipients, the latter incapable of forming antibodies. They got good antibody formation, and the cells elicited, at the injection sites, collections of plasma cells which seemed to arise in large part from the transferred lymph node cells. By contrast, the lymph node cells yielded no antibody when they were transferred to newborn recipients within 2 hours or 1 day after antigenic stimulation of the donors. When transferred after 3 days they formed antibody. In the newborn recipients, plasma cells failed to appear in significant numbers.

From this evidence, as suggested by Dixon and Weigle (1957), the inability of newborn and X-irradiated animals to form antibody can be attributed to different mechanisms. The X-irradiated animals apparently lack cells which can form antibodies, and if such cells are supplied antibodies form. The newborn rabbits even if supplied with such cells do not provide a proper environment for antibody formation.

B. The Enhancement of Antibody Formation by the Adjuvants of Freund and Others

Procedures for augmenting antibody formation by the addition of substances to antigens have been of great value to immunology, but as yet they have not thrown enough light upon the cytological mechanisms of antibody formation to warrant more than a few remarks in this chapter. The subject is well reviewed by Freund (1947, 1956). It is of interest to cytologists that soluble antigens yield higher antibody titers if precipitated or rendered particulate, as in the studies of Glenny *et al.* (1926, 1931), who used potassium aluminum sulfate and other precipitants mixed with diphtheria toxoid, as antigen. They demonstrated a slow liberation of the antigen by removing the regions injected with the alum precipitates and by producing secondary reactions in guinea pigs injected with the material, a finding which was confirmed in various ways by Harrison (1935) and by Faragó (1935). The work of Cannon and Pacheco (1930) and of Hartley (1940), already cited, showed that injections of aluminum hydroxide gel into the skin of animals induced large accumulations of macrophages, and that subsequent injections of vaccine virus, adsorbed to the gel, into such sites produced high, local titers of neutralizing substances.

Earlier Lewis and Loomis (1924, 1925) found the "allergic irritability" of animals increased by tuberculosis. Guinea pigs given tubercle bacilli reacted to unrelated antigens more actively than those given only the antigens.

A few years later Freund *et al.* (1937) and Freund and McDermott (1942) emulsified aqueous solutions of various antigens in paraffin oil containing killed tubercle bacilli. This adjuvant mixture, added to the antigens, enhanced antibody formation and increased sensitization without inducing tuberculosis. Freund (1947, 1951) and Freund *et al.* (1948) showed that bacteria, other than the tubercle bacillus, such as *Mycobacterium butyricum* and *M. phlei,* served well in the adjuvant mixtures. At first it was believed by him and many others, among them Landsteiner (1945), Halbert *et al.* (1946), Casals and Freund (1939), and Ehrich *et al.* (1945), that the water and oil phases of the adjuvant seemed to delay the absorption and elimination of the antigen, or that prolonged storage of the antigen occurred in macrophages. Later, however, Freund (1951) and Freund and Lipton (1955) found that the adjuvant effect was not confined to the site of injection; instead there was obvious dissemination of the antigen through the lymphatics to the lymph nodes, the spleen, and even to the lungs. Freund (1956) attributes much of the adjuvant action not only to these factors, but also to the collection of cells about the adjuvant, mostly monocytes, large mononuclears, epithelioid cells, giant cells, young and mature lymphocytes, and plasma cells.

White *et al.* (1955b), using the fluorescent antibody technique of Coons *et al.* (1955), investigated the active adjuvant effects of the tubercle bacillary wax of Anderson (1929), which had effects like that described by Raffel *et al.* (1949). They found, at the sites of injection, as had Freund, who used the complete adjuvant, collections of macrophages and other cells which did not show the presence of antibody. Instead the increased antibody, demonstrable after using the wax, seemed to be formed by collections of plasma cell elements that appeared in the lymph nodes, spleens, and livers of the experimental animals. The wax, like the complete adjuvant, exerted a pronounced stimulatory effect on the reticuloendothelial system, locally and in remote organs. Clearly the macrophages had much to do with the reaction, but they did not have antibody.

Fink *et al.* (1955) have used Freund's adjuvant to induce antibodies in mice to homologous tumors.

A paper by Johnson *et al.* (1956) should be consulted for further data on adjuvants, especially since lipopolysaccharide endotoxins obtained from several gram-negative organisms are described, one of which, given to rabbits, enhances two- to fortyfold the antibody formation to a protein antigen. The authors believe the endotoxins to be responsible for a general activation of protein synthesis, part of which may be antibody. This opinion seems to stand in agreement with those of Freund (1956) and White *et al.* (1955b). A lipopolysaccharide product of Johnson *et al.* (1956) was added by Stevens and McKenna (1957a, b) and McKenna and Stevens (1957a, b), together with antigen, to cultures of tissues taken from normal animals. As a result antibody formed in the cultures.

C. *Further Data Contrasting the Behavior of Plasma Cells with That of Other Cells in the Immune Response*

1. *With Lymphocytes*

In spite of the overwhelming evidence for the formation of antibodies by plasma cells, and more evidence will appear below, it should not be held, as yet, that only plasma cells have this capacity. The cases for the lymphocyte and the macrophage are far from closed. One only needs to recall the differences of opinion cited above between the reports of Ehrich *et al.* (1949) and T. N. Harris and S. Harris (1949), who performed similar experiments with similar techniques, one group stressing the role of plasma cells, the other that of lymphocytes. The work with the transfer of cells from lymph nodes by the Harrises and with thoracic lymphocytes by Wesslen, both already cited, is to be recalled together with the other findings already outlined, which implicate the lymphocyte. Not mentioned as yet is a paper by Dougherty (1948), who injected mice once only with sheep red cells or *Staphyloccoccus* toxin. The animals showed no more plasma cells than nonimmune mice, but the lymphocytes contained antibody 7 days after injection of the antigen. By contrast, when two injections were given many plasma cells appeared within 6 hours. He suggests that a morphologic alteration, following upon an antigen-antibody reaction within the cytoplasm of the lymphocyte, may lead to the formation of mature plasma cells as a type of allergic reaction. Clearly the lymphocyte must be considered not only as involved in some way in antibody formation, but, possibly, as a former of antibody.

2. With Macrophages

Accumulations of cells, chiefly macrophages, at the sites of antigen injection, whether in granulomata or not, have been repeatedly mentioned above as a probable source of the local formation of antibody (Cannon and Pacheco, 1930; Hartley, 1940; Oakley *et al.*, 1949; Freund, 1956). DeGara and Angevine (1943) extracted spleens, bone marrow, liver, and skin sites following injections of antigen. Antibody was found in all before it appeared in the blood. The lymph nodes showed little. They believed that antibody was formed locally at the sites of inflammation within the various organs probably by macrophages. The negative findings of Ehrich and Harris (1945) and T. N. Harris and Ehrich (1946) with macrophages from the foot pads of rabbits at the sites of antigen injection, and of White *et al.* (1955a, b) using the fluorescein technique, and of Ehrich *et al.* (1946) with macrophages of peritoneal exudates, have already been mentioned. Roberts (1955) found that antigen-containing macrophages from peritoneal exudates formed no antibody in 6 days of tissue culture, while macrophages from rabbits immunized by intraperitoneal injections of bacteria definitely contained antibody, but only a little. Exudate cells of normal rabbits, allowed to ingest bacteria (10^9) and then injected intraperitoneally into other normal rabbits, produced weak agglutinins in their sera after 10 days, but none at 4 days. Other rabbits given the same number of bacteria formed much more antibody. The behavior of transferred macrophages, described by Roberts and Dixon (1955) and Roberts *et al.* (1957), goes far to clear up their relationship to cells of the lymphoblast and plasmablast type.

As far back as 1888 Metchnikoff reported that lymphocytes of the blood of animals infected with tuberculosis migrated into the lesions and formed large mononuclear cells that later became macrophages. Indeed the origin of macrophages from lymphocytes and monocytes has been repeatedly described by many authors, among them Maximow (1902,

1923) and Bloom (1928). In a review of the functions of white blood cells, Rebuck (1947) gives scores of references to the subject. Further, he describes, in this and in a later paper (Rebuck and Crowley, 1955), an ingenious technique by which lymphocytes of man can be seen, in warm-stage preparations, to hypertrophy into macrophages. He agrees that other sources of macrophages are to be found in reticulum cells, histio-cytes, clasmatocytes, and monocytes. In the lungs too, after infection with pneumococci (Loosli, 1942), macrophages seem to stem from lympho-cytes and monocytes.

Lurie (1942) obtained macrophages from normal rabbits and from others immunized to tubercle bacilli. When these cells had engulfed tubercle bacilli, either in the presence of immune serum or of normal serum, they were transplanted to the anterior chambers of the eyes of normal rabbits. There, in the absence of immune body fluids, the cells continued to inhibit the multiplication of the bacilli, without actual antibody formation.

Whatever may be said of macrophages eventually, it is clear that, whether or not they form antibody, they are undoubtedly active in many conditions in which resistance to infection appears. Taliaferro and Mulligan (1937), Taliaferro and Taliaferro (1955), and Taliaferro (1956), discussing the acquired immunity to malaria, point out that the concept of the reticuloendothelial system, described by Aschoff (1924), is very similar to the macrophage system described by others, as shown in Table I. It includes macrophages of connective tissue, primitive cells that become macrophages, reticular cells of lymph nodes, bone marrow, and the splenic red pulp, littoral cells of venous sinuses, and Kupffer cells of the liver. These cells retain an ability to develop into other types of cells, and lymphoid cells may undergo a heteroplastic development into plasma cells or macrophages (Taliaferro, 1956), as first described by Maximow (1902) and by Bloom (1938a). Since Taliaferro and Mulligan (1937), adopt the view of Maximow (1902, 1907) that lymphoid cells that enter into an area of inflammation may become macrophages, they con-sider it difficult to separate the lymphoid system from the macrophage system. Taliaferro (1956) also stresses the importance of a particular cell of the free mesenchymal lymphoid reserves—a very large lymphocyte or hemocytoblast. After various immunization procedures it has been termed by different authors, a myeloblast, the acute splenic tumor cell, a lymphoblastic plasma cell, a plasmablast, or an immature plasma cell. This cell, because of its virtuosity, may readily, as a plasmablast, be especially involved in antibody formation (Taliaferro, 1949).

Wissler *et al.* (1957) and La Via *et al.* (1956) stress the interrelation-ship of several types of cells in the processes of antibody formation: the macrophage as a phagocyte and the primitive reticular cell as the precursor of antibody-forming cells. They also call attention to the fact that the intense pyroninophilia observed in plasma cells—although related, of course, as shown by Brachet (1940, 1952) and Caspersson (1936) to ribonucleic acid and protein synthesis—does not necessarily constitute evidence of antibody formation.

Following single intravenous injections of typhoid bacterin into rats, Wissler et al. (1957) described cells which appeared in the red pulp of the spleen after a latent period of 48 hours, cells with large vesicular nuclei and little pyroninophilia, and apparently derived by mitosis from primitive fixed reticular cells. Shortly these cells developed an intensely pyronine-staining cytoplasm, but they did not resemble mature plasma cells. They appeared before antibody was found in the blood. Six or seven days after the antigen injection these cells disappeared without evidence of cell death, and many small cells with scanty cytoplasm were present, as though they had developed from the former by the shedding of cytoplasm. The authors found a correlation between the rise and fall of circulating antibody and the increase and decrease of the pyroninophilic cells, and the latter seemed to be comparable to those termed "transitional cells" by Fagraeus (1948a) and the large immature cells described by Leduc et al. (1955). Wissler et al. suggest that the principal antibody-forming cell is this large pyroninophilic one which seems to arise by mitosis from primitive fixed reticular cells which are not phagocytic.

IX. Some of the Newer Views of the Processes of Antibody Formation

The dynamic processes of antibody formation are, of course, closely associated with other forms of protein synthesis, but no complete hypothesis of antibody production can be made until more is known about the cellular sites from which antigens exert their first stimuli and until it is definitely determined whether or not the processes can continue in the body after antigen has been destroyed. The clarification of these points awaits experimental evidence, and until it comes to hand the biochemical hypotheses of antibody formation remain so speculative and controversial—and further they have been so frequently described—that a restatement of them in this chapter, which deals with the evidence for the formation of antibody by certain cells or organs, seems to be superfluous. It needs only to be said that the well-known "template" hypothesis, as put forward from the laboratories of Pauling (1940), Mudd (1932), Breinl and Haurowitz (1930), and Haurowitz (1952), assumes that antigens must persist in the cells that form antibody. On the other hand, much evidence to be discussed below indicates a rapid destruction of antigen and has led to the equally well-known "adaptive enzyme" hypothesis (Alexander, 1931–1932; Burnet and Fenner, 1949; Burnet, 1956) to explain the formation of antibody in the absence of antigen.

A. Tracer Antigens and Their Fate in the Body

To attack this problem workers have coupled traceable materials to antigens and injected them into animals. As will be seen below some of these antigens seem to presist much longer than others.

1. Metals as Tracers

More than twenty years ago Haurowitz and Breinl (1932) and Haurowitz and Kraus (1936) coupled metals to antigenic proteins and found the tracer materials largely in the RE cells throughout the body, especially in the Kupffer cells of the liver and in the spleen and lymph nodes, but also in other cells. McClintock and Friedman (1945) coupled pneumococcal antibody to uranium, or polonium, or to the dye, malachite green, and demonstrated, by X-ray or inspection, that the coupled substances became localized, after injection into animals, in the sites at which a corresponding antigen had been injected.

2. Azoprotein Antigens

Colored azoprotein antigens, which can be seen within cells, were first used by Sabin (1939), as already described, and also by Smetana and Johnson (1942), Smetana (1947), Gitlin (1950), Kruse and McMaster (1949), and others. After injection all appear largely in RE cells though they may be present in other cells such as those of the kidney tubules. Figures 4a and 4b and, at increasingly higher powers, Figs. 5–7 illustrate the distribution of a blue azoprotein (Kruse and McMaster, 1949) in a mesenteric lymph node of the mouse after intravenous injection (see legends for explanation). The dye-protein lies in the cytoplasm of macrophages and of endothelial cells of the sinuses and trabeculae. The lymphocytes contain none, but the cells holding the antigen surround the lymph nodules and lie closely associated with the plasma cells that develop later in this region. In the spleen the antigen-containing cells lie in the red pulp close to the lymph follicles, and in the liver the Kupffer cells are crammed with the blue protein, but none appears in the parenchymal cells. In all the cells in which it appears it lies in the cytoplasm, often close to the nuclei, but apparently, as far as observations go at present, not in them. This blue azoprotein, injected into mice, which form antibody poorly, seems to persist for weeks, because liver tissue containing

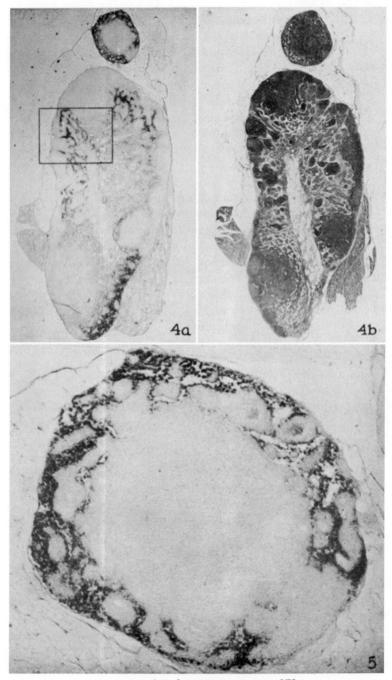

Figs. 4 and 5 description on page 378.

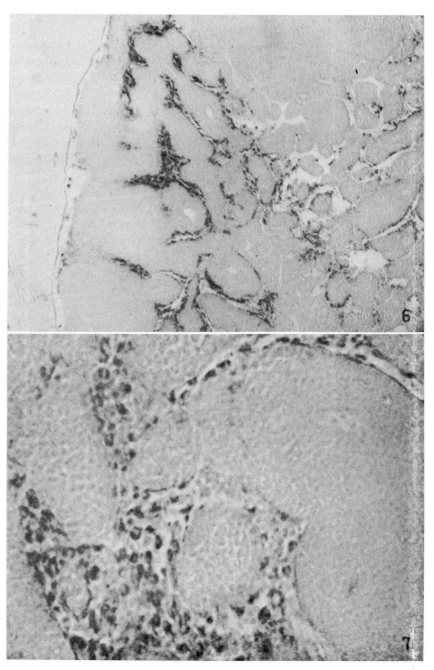

FIGS. 6 and 7 description on page 378.

it, when transferred to normal mice, renders the recipients anaphylactically sensitive to injections of an appropriate antiserum (McMaster and Kruse, 1951). By contrast, a foreign-protein antigen, injected into rabbits, a species which forms antibodies well (McMaster and Edwards, 1957b), does not persist beyond the second week in a form sufficiently intact to stimulate the formation of precipitating antibody when liver tissue, presumably containing the antigen, is taken from such rabbits and transferred to others prepared to give a secondary response. Nevertheless, liver tissue of similar donor rabbits, taken after 6 weeks and transferred to mice, renders the recipients anaphylactically sensitive to the original antigen given to the donors. (McMaster *et al.*, 1955).

3. *The Detection of the Sites of Localized Antigen by Fluorescent Antibody*

Coons and his associates (Coons *et al.*, 1942, 1951, 1955; Leduc *et al.*, 1955; White *et al.*, 1955a, b; Coons and Kaplan, 1950; Coons, 1954) have devised a sensitive tracer technique to identify antigens localized in cells. Frozen sections of tissues, suspected of containing antigen, are thawed and the appropriate antibody coupled with fluorescein is poured over the sections. The resulting antigen-antibody reaction, fixing the fluorescent antibody, demonstrates unlabeled antigens not only in the cells that take up labeled antigens, but in many others, particularly in lymphocytes and apparently within the nuclei as well as the cytoplasm. More will be said of this technique below.

Fig. 4. Unstained (a) and stained (b) transverse sections of the mesenteric lymph node of a mouse 48 hours after an intravenous injection of the azoprotein. Magnification: × 12.

Fig. 5. The round nodule at the top of Fig. 4, a as it appears at higher magnification and unstained in a different section. The black regions represent the blue azoprotein. The lymph follicles and lymphocytes in the medullary portion of the tissue contained no color, although they were intimately surrounded by cells containing the colored antigen. Magnification: × 75.

Fig. 6. The photograph shows the region that is included in the small rectangle drawn in Fig. 4, a. The blue antigen, black or dark gray in the figure, stands out in the cells of the sinuses and endothelium against the light gray of the lymphocytes. Magnification: × 90.

Fig. 7. A photograph of another section from the same lymph node. This section, too, was unstained. At this magnification, a black-and-white print does not show the sharp contrast seen in the actual section because the lymphocytes appear gray. Neverthless the darker appearance of endothelia and cells in the sinuses indicates the presence of the blue antigen. It seems to lie within the cytoplasm of the cells. Magnification: × 350.

Figs. 1–7. Reprinted with the permission of the *Journal of Experimental Medicine* (Kruse and McMaster, 1949).

Gitlin *et al.* (1953) used fluorescein-labeled antibodies against human plasma proteins and detected the homologous proteins in tissue sections.

Cohen *et al.* (1955), using fluorescent antibody and tissue cultures, demonstrated the antigenic material of measles in the nuclei and the cytoplasm of kidney cells from infected monkeys. Buckley *et al.* (1955) grew liver tissues from mouse embryos in culture, which, after inoculation with psittacosis virus, showed by fluorescence the intracellular presence of antigen in the cytoplasm in most of the outgrowing cells.

4. The Use of Fluorescent Antigens

Following the work of Coons *et al.* (1942, 1951, 1955) and others to be discussed below, Schiller *et al.* (1953) conjugated bovine serum albumin with fluorescein to form a fluorescent antigen instead of fluorescent antibody. This material, injected into rats, assumed a distribution like that described above for the azoprotein antigens. The fluorescent antigen appeared in the cytoplasm only, not in the nuclei, nor did it show itself in liver parenchyma cells, or lymphocytes. Nakano (1955a, b, c; 1956) injected fluorescent horse serum protein into mice and saw the strongest fluorescence in the Kupffer cells of the liver. There fluorescence persisted for more than 3 months. Nakano also found traces of fluorescence in the spleen for equal periods of time. In the past year Mayersbach (1957) injected fluorescein-coupled egg albumin intraperitoneally into rabbits and saw fluorescence, 30 minutes later, in the cytoplasm but not in the nuclei of Kupffer cells. By the second hour it was present in the nuclei. It remained for 26 days. The author feels that this finding is to be attributed to slow absorption of the antigen from the peritoneum.

5. The Use of Antigens Labeled with Radioisotopes

Space permits only some brief allusions to the extensive studies by immunologists using isotopically labeled antigens. For details the reader should seek the reviews by Haurowitz (1953), Haurowitz and Crampton (1952a, b), Taliaferro (1949), Coons (1954), and also the many excellent papers by Dixon and his co-workers (Dixon, 1954, 1957; Dixon and Talmage, 1951; Dixon *et al.*, 1951a, b, 1953, 1956; Dixon and Maurer, 1955).

Libby and Madison (1947) demonstrated the presence of radioactive tobacco mosaic virus, after injection into mice, chiefly in the liver and spleen, presumably in RE cells. They estimated that practically all was broken down by the sixteenth day. Fields and Libby (1952) suggest that antigens of large molecular size show greater association with the mitochondrial fraction of cells than antigens of smaller size. Using the electron microscope, Erickson *et al.* (1953) demonstrated radioactive tobacco mosaic virus in liver extracts for 9 days and rod-shaped viruslike materials for 15 days following a single injection of the antigen. Radioactivity was found chiefly in the microsomal fractions (Erickson *et al.*, 1957)—as shown earlier by Haurowitz (1952, 1953), and Crampton and Haurowitz (1950). Some appeared in the nuclear fractions, which they consider to be an artifact since the so-called nuclear fractions contained many whole cells, mitochondria, and microsomes as well. However, the mitochondrial association with the antigen is considered as actual, since, as the mitochondria in the preparations dried out and burst, the rods of the virus became visible. They suggest that antigenic particles smaller than TMV rods might enter into nuclei.

Gavosto and Ficq (1953, 1954a, b) injected mice intravenously with TMV labeled with C^{14}. Radioautographs, to localize the antigen, and sections, stained with methyl green and pyronine or Giemsa, to indicate protein synthesis, showed tracks emanating from bone marrow cells, identified as reticular cells, histiocytes, or young plasma cells. Most of the activity was in the splenic red pulp and a few cells in the lymph follicles. In the liver the Kupffer cells, endothelial cells, and the cytoplasm of some parenchymal cells showed it. After 15 days there was much radioactivity, but longer periods were not studied. Rats given typhoid bacterin and later C^{14}-labeled glycine showed five times as much radioactivity in the medullary portions of the lymph nodes as in the cortical portions. It apparently came from reticular basophilic cells, cells of the walls of the sinuses, young basophilic elements of the follicular centers, and plasmacytes. Lyphocytes showed little. These authors feel that the new cells, RE cells with basophilic cytoplasm (protein synthesis), were the ones that took up the C^{14}-glycine and the tobacco mosaic virus, and that they may form antibodies.

Dixon *et al.* (1951a, b, 1953), Dixon and Talmage (1951), Latta (1951), and Coons *et al.* (1942, 1951, 1955), using proteins labeled with radioisotopes or marked by other means, found no storage of protein antigens in the body.

By contrast, Haurowitz and Crampton (1952b), Crampton and Haurowitz (1952), and Crampton *et al.* (1952, 1953), after injecting various radioactively labeled antigens, found radioactivity in the liver for weeks and in the spleen for somewhat shorter periods. It was present, after the first few minutes following injection, chiefly in the mitochondrial fractions. Ingraham (1951a, b) injected mice with sulfanilic acid-azo-bovine γ-globulin labeled with S^{35}. He found traces of the label in their livers for more than 200 days after injection, but active antigens in extracts of the liver for only 17 days. Thereafter the tissues showed the presence of degraded, radioactive material. Friedberg *et al.* (1955) and others have injected animals with protein antigens labeled with radioactive materials, and judge from the retained radioactivity that antigenic matter may persist longer than is usually believed. However, only the labels serve as tracers; fully antigenic material has not been demonstrated. A discussion of these and other relevant findings can be found in papers by McMaster and Edwards (1957b) and by Dixon (1957).

Garvey and Campbell (1954, 1956) studied the immunological nature of "antigen material" remaining attached to the tracer label following repeated injections of S^{35}-labeled hemocyanin p-azophenylsulfonate into rabbits. An initial rapid loss of antigen became slower and finally constant after 4 weeks. The third day after the last of several injections of antigen, both precipitating and nonprecipitating antigen was present as shown by inhibition tests. After 5–6 weeks only nonprecipitating antigens remained. Although the physical properties of the antigen persisting in the liver changed, it was still capable of reacting with antibody after 42 days. The authors concluded that persisting antigenic fragments may play a more important role than the original antigen in the formation of antibody. To learn more about the nature of the changes in persisting antigenic material, Garvey and Campbell (1957) also labeled bovine serum albumin with S^{35} and injected it into rabbits. Not only did the label remain in the liver tissue for 140 days, associated with material immunologically related to the original antigen, but this material, when extracted from the liver after 3 weeks, was capable of anaphylactically sensitizing guinea pigs. In this respect the material behaved like the antigenic material found by McMaster and Kruse (1951) in the livers of mice weeks after the injection of an azo-bovine globulin. Garvey and Campbell (1957) suggested the possible retention of antigenic material for years.

6. The Persistence of the Pneumococcal Polysaccharide Antigen and "Immunological Paralysis"

Felton (1949) injected mice with large amounts of pneumococcal polysaccharide and later found the animals incapable of subsequent immunization when reinjected with the antigen. He termed this state "immunological paralysis" and believed it to be brought about by some effect of the persisting antigen upon the antibody-forming cells. That large amounts of antigen, not metabolized by the body, might combine with antibody as it was formed, was also considered by him as well as by others, and shown to be the case by Dixon et al. (1955). Kaplan et al. (1950) using the fluorescence technique could demonstrate the pneumococcal polysaccharide persisting, after injection into animals, for several months, and Hill et al. (1950) found the polysaccharide of the Friedlander bacillus persisting in the mouse for 33 days. Ørskov (1956) has reported an immunological paralysis of rabbits induced by a strain of *Klebsiella* organisms which have very heavy capsules.

Dixon and Maurer (1955) injected rabbits repeatedly with large amounts of foreign proteins and demonstrated an immunological paralysis which endured while the antigens remained detectable, and did not inhibit immunological responses of the animals to other antigens. The unresponsiveness seemed to be related to the persistence of the antigens in the hosts, but did not seem to be due to the neutralization of the antibody as it formed.

B. The Detection of Intracellular Antibody by the Coons Technique

Coons *et al.* (1955) have ingeniously modified their technique for antigen detection so as to detect antibody. Sections of tissues from immunized animals are overlayered with solutions of the appropriate antigen, which becomes held through a precipitin reaction with any antibody that may be present. The bound antigen can then be detected by overlayering the section with fluorescent antibody.

In rabbits, hyperimmunized by intravenous injections of foreign protein, antibody appeared in aggregates of plasma cells in the red pulp of the spleen, in the medullary cords of the lymph nodes, in connective tissue of the liver, in the portal areas, and in submucosal regions of the ileum. A little was occasionally seen in lyphoid follicles, so that a slight contribution by lymphocytes to antibody formation could not be ruled out. In lymph nodes of rabbits Leduc *et al.* (1955) demonstrated antibody appearing during the differentiation of a cell family, the mature member of which is the plasma cell. It appeared first within the cytoplasm and often in the large nuclei of immature, typically hematogenous stem cells, with basophilic cytoplasm, lying in the medullary areas of the draining nodes. White *et al.* (1955a, b) found, as already discussed, antibody-containing plasma cells in local granulomata developing 14 days after subcutaneous injections of alum-precipitated antigen. Antibody-containing plasma cells appeared much earlier in the draining lymph nodes. After injecting bacterial antigens into rabbits and mice, White (1954) reported most of the antibody-containing cells in the spleens, as shown by the fluorescence technique, to be mature plasma cells with Russell bodies.

The Coons technique has been applied by Berenbaum (1956) to touch-preparations from the spleens and lungs of rabbits immunized with *Klebsiella* organisms. The polysaccharide, extracted from the same strain of organisms grown in acetate-1-C^{14}, was flooded over these slides, and, after washing, autoradiographs were made. Antibody was found, but the cells containing it could not be identified.

In the same year Askonas and White (1956) added glycine labeled with C^{14} to incubated chopped tissues of lymph nodes and other organs taken from guinea pigs previously injected in the foot pads with egg albumin. The radioactivity was incorporated into the γ-globulin fractions of these tissues, and antibody formed. Further, by a combination of the fluorescence method and staining techniques the authors concluded that plasma cells form antibody.

Still more recently Ortega and Mellors (1957), using the methods of Coons, prepared antihuman γ-globulin and studied the cellular sites of γ-globulin formation in lymphoid tissues taken from human subjects at operation. The fluorescent anti-γ-globulin, when flooded over these tissues, disclosed the formation of γ-globulin in the cytoplasm of three types of cells, two of which were immature or mature cells of the plasma cell series, that is to say, those with and those without Russell bodies. The third type consisted of cells in the germinal centers which the authors designated as "intrinsic" cells to differentiate them from medium or large lymphocytes or from the primitive reticular cells that occur elsewhere and do not form γ-globulin. In all instances the fluorescence was exclusively cytoplasmic and the Russell bodies showed a peripheral concentration of γ-globulin.

The "intrinsic" cells of germinal centers seem to form γ-globulin only when they appear as aggregates. These authors regard the germinal centers as minute organs of internal secretion of γ-globulin. They did not detect γ-globulin in mature lymphocytes or eosinophiles.

In contrast with these findings, Gitlin et al. (1953) reported the fluorescence technique capable of demonstrating γ-globulin in the nuclei of lymphocytes, lymphoblasts, and endothelial cells. Ortega and Mellors (1957) consider the nuclear staining, just mentioned above and in other portions of the review, as an artifact the result of an inadequate absorption of the fluorescent stains with the tissue powders employed for this technique.

X. SOME RECENT ATTEMPTS TO TRACE THE STEPS IN ANTIBODY GLOBULIN FORMATION

The newly acquired knowledge concerning the cells that form antibody has raised a number of questions. Are circulating antibodies formed at the same sites and by the same cells that synthesize other plasma proteins? Do antibody-forming cells convert existing plasma globulin into antibody or do they form it *de novo* from protein building stones? Is all γ-globulin of the blood antibody or is there "normal" γ-globulin?

A. *The Liver and γ-Globulins*

Miller and Bale (1954) and Miller *et al.* (1954, 1951) showed that the isolated, perfused rat's liver formed proteins with the mobilities corresponding to all the plasma proteins except γ-globulins. Lysine-e-C^{14} was incorporated into plasma albumin, β-globulin, and fibrinogen, but γ-globulin, aparently formed by extrahepatic tissues, had no radioactivity. The perfusion of the extrahepatic tissues, either in eviscerated, surviving rats or in the caudal half of "carcasses," produced only plasma proteins with the mobilities of γ-globulins. Massive doses of X-radiation, enough to inhibit antibody formation, suppressed the incorpation of C^{14}-labeled lysine into the γ-globulins, but the synthesis of other plasma protein fractions was not depressed. To the extent that antibodies are looked upon as globulins, they are not considered by these authors as formed by the liver. Further, since the absence of the spleen and mesenteric lymph nodes had little effect in these perfusion experiments. Miller and his co-workers feel that the bone marrow and disseminated lymph follicles, perhaps chiefly the bone marrow, are implicated in the formation of antibody.

B. *Antibodies and Preformed Globulin*

It is now generally agreed that antibody does not arise from preformed globulin. Gros *et al.* (1952), using radioactive materials, injected homologous γ-globulin into animals while they were being actively immunized. The injected globulin was not directly converted into antibody. Green and Anker (1954) gave amino acids with carboxyl-labeled C^{14}-glycine to rabbits during synthesis of antibody in the induction period of a secondary response to ovalbumin as an antigen. The C^{14}-glycine content of the antibody, during this period, depended upon the C^{14} concentration in the nonprotein glycine rather than the concentration of the C^{14} of the serum proteins. Next, rabbits, stimulated for a secondary response to ovalbumin, were injected with amino acids labeled with three different isotopes, C^{14} at 10 hours after the second injection of antigen, N^{15} after 41 hours, and H^2 after 89 hours. The induction period endured for 3 days, and the peak titer appeared on the fifth day. The amino acids labeled with C^{14} and N^{15} and given during the induction periods, were present in the antibody at its peak. These authors estimated that 31% of the antibody was formed during the first 60 hours, in the induction period, and the remainder in the next 72 hours.

Taliaferro and Talmage (1955) injected rabbits with bovine serum albumin and 1 month later reinjected them to produce a secondary response. A clear-cut induction period appeared, enduring for 3 days after the day of injection. On the third day minced splenic tissue was taken from these animals, as donors, and injected into normal, recipient rabbits, which formed antibody. Some of the donors got S^{35}-labeled amino acids (yeast hydrolyzate) during the induction period. Splenic tissue from such, transferred to recipients, yielded antibody rapidly, with its peak in 2 days, that is to say, the antibody-producing mechanism was transferred to unlabeled recipients from donors that had been labeled. As a result the antibody that formed in the unlabeled recipients was labeled to a very small degree (see also Taliaferro, 1957b), although the donor rabbits yielded antibody much more highly labeled with S^{35}. When the S^{35}-labeled amino acids were given to the recipients, but not to the donors that got the secondary injection of bovine serum albumin, there appeared much radioactivity in the antibody formed by these recipients. This antibody must have been drawn from the amino acid pool while the antibody titer was rising, but not during the induction period. Accordingly, the induction period did not seem to represent a time in which antibody was being synthesized from precursors containing amino acids, but rather a period during which new enzymes were being developed that later synthesized antibody. Since these authors found almost all of the amino acids present in the antibody of the recipients to be drawn apparently from the amino acid pool during the rise of antibody formation, and after the induction period, whereas Green and Anker (1954) found 31% formed during the induction period, the difference needs explanation. Taliaferro and Talmage (1955), Dixon *et al.* (1956), and Taliaferro (1957b), suggest that the higher figures of Green and Anker (1954) could be explained by a recycling of labeled amino acids and the inclusion of labeled coprecipitating antibody or complement that might have been present.

Taliaferro and Taliaferro (1957) have extended the work of Taliaferro and Talmage (1955), just mentioned, to show even more clearly the lack of uptake of amino acids labeled with S^{35} during the induction period of antibody formation, that is to say, before circulating antibodies appear. They again divided the secondary response to bovine serum albumin between donor and recipient rabbits by transferring, to unlabeled recipients, minced spleen or bone marrow taken near the end of the induction period from donors that had received both antigen and S^{35}-labeled amino acids. As before, the work showed that the antibody formed by the recipients was produced largely from their amino acids. Within a day after the transfers antibody began to rise, indicating that there is but little time between its synthesis and its appearance in the serum. This was also shown by the finding of a very short period, only 40 minutes, for the "transit time" between the injection of S^{35}-labeled amino acids to the donors, undergoing the secondary response, and the appearance of the label in the antibody that formed.

Stavitsky (1955a, b, 1956, 1957) injected radioactive S^{35}-L-methionine into recipients after transfers of splenic and lymphoid cells from immunized donors, and showed that it was incorporated into the antibody formed in the recipients. By contrast, when S^{35}-labeled normal γ-globulin was injected into recipients, after similar cell transfers, the antibody that formed showed no radioactivity. Again, the transfer of cells, the protein of which had become labeled with S^{35}-L-methionine, did not lead to the formation of labeled antibody in the recipients. Further, Wolf and Stavitsky (1956) cultured lymph node fragments, the proteins of which were labeled by injecting S^{35}-labeled yeast hydrolyzate into the animals furnishing the tissue. The antibody formed in the cultures was not radioactive.

Stavitsky (1957) discussed his work and the findings of others indicating that the conversion of precursors of antibody into fully formed antibody does not occur after the transfer of splenic or lymphoid cells from immunized donor rabbits to either normal or irradiated rabbits. Instead the antibody seems to be formed from materials present in the recipients, and not from antibody or precursors in the transferred cells.

Porter (1955) applied partition chromatography to the study of rabbit antibodies and suggested that both the large number of components found and the appearance of one type of antiovalbumin in the earlier stages of immunization and another at later stages might signify the probability that different types of cells, even in the same tisues, may produce different types of globulin. Humphrey and Porter (1956), using chromatography and rabbits immunized to various antigens, obtained independent patterns of response to each antigen suggesting that different cells, capable of forming somewhat different antibodies, may predominate according to the route taken by the antigen and the duration of the antigenic stimulus.

Humphrey and McFarlane (1954) obtained radioactively labeled antipneumococcus antibody (type III) by giving, during immunization, lysine and phenylalanine with carboxyl groups labeled with C^{14}. The antibody, when transferred to normal animals, showed no change in its specific radioactivity. Taliaferro and Taliaferro (1955) obtained a similar absence of amino acid interchange when the S^{35} label was used. Humphrey and McFarlane (1955) found C^{14}-labeled amino acids incorporated into antibody of rabbits, immunzed against type III pneumococci, in as short a time as 30 minutes, and reaching a peak in 4 to 5 hours. This activity seemed to occur chiefly in the lungs.

Askonas and Humphrey (1955), after producing granulomata in rabbits injected with egg albumin and Freund's adjuvant, removed slices of the tissue when the serum antibody titer was high. When incubated with

C^{14}-glycine, these slices incorporated C^{14}-labeled material into antioval-bumin. When incubated also with the specific soluble substance of the pneumococcus (SSS III), the C^{14} appeared only in the antibody-oval-bumin that the tissue was making. Askonas and White (1956), studying the incorporation of C^{14}-labeled glycine into antiovalbumin in granu-lomata and lymph nodes of guinea pigs, stressed again the important technical factors: that the site of injection of antigen, its dosage, and the species of animal used must be considered in judging the findings obtained.

Very recently Askonas and Humphrey (1958) have shown again that glycine labeled with C^{14} was incorporated into antiovalbumin formed in tissues taken from rabbits immunized against ovalbumin, but that it was not incorporated into antipneumococcal antibody which was added to these tissues. Further, the radioactivity of the antibody did not come from nonspecific protein coprecipitated with the antigen-antibody precipitate. They considered the incorporation of C^{14} with the specific proteins a measure of true synthesis.

Two years earlier Askonas et al. (1956), by determining the rate of incorporation of radioactive amino acids into rabbit γ-globulin, concluded that one form of globulin is not converted into another, neither is already existing γ-globulin changed into antibody. In vitro tests suggested that different tissues, lymph node, spleen, or bone marrow, form separate fractions of γ-globulins at unequal rates, and that the various types of γ-globulin are synthesized by different cells. Further, when these workers injected inert radioactive γ-globulin into a rabbit undergoing active im-munization to alum-precipitated ovalbumin, there was no evidence that the injected γ-globulin was converted into antibody, although it was being formed rapidly. Neither did they obtain evidence of simple conversion of a precursor into antibody.

Again, the uniform labeling of different chromatographic fractions of the formed globulins seemed to indicate that the individual cells do not produce various types of globulin at different times; instead they form characteristic types of γ-globulin, and the distribution of these cells differs in various organs. In other words, if one immunizes an animal to three antigens at once, a single plasma cell, let us say, will form anti-bodies to only one of these antigens, not to all.

C. Is All the γ-Globulin of the Blood Antibody Globulin?

As yet the question above has not been answered to the satisfaction of all. Nevertheless, some of the newer work on the problem is of interest. The papers of Björneboe (1939, 1941, 1943; Björneboe and Gormsen, 1941, 1943; Björneboe et al., 1947), and of many clinicians, referred to above in discussing antibody formation by plasma cells, suggested that all plasma globulin might be antibody, since it has been found that patients with many kinds of infectious processes, and also hyperimmunized animals, had very high concentrations of blood globulin.

Fagraeus (1948a) performed plasmaphoresis experiments in rabbits to promote rapid regeneration of plasma proteins. As a result the animals produced much globulin, but no increase in plasma cells appeared either in the spleen or the bone marrow. These rabbits had not been immunized; the nonspecific protein increase was not related to an increase in plasma cells.

Thorbecke and Keuning (1956) tried, by plasmapheresis, to stimulate the production of γ-globulin in normal rabbits to such an extent that it would appear in cultures of tissues taken from these animals. No significant production was found, nor was there any accumulation of plasma cells in the animals. Electrophoretic studies of the sera indicated that γ-globulin production was slower than that of the other blood proteins, and its formation, in tissue cultures, appeared only if antibody formation was in progress. Since there seems to be little evidence of the conversion of already existing γ-globulin to antibody, and since, in their experiments, the loss of blood proteins did not stimulate the regeneration of γ-globulins nearly as fast as other blood proteins, these authors suggest that γ-globulins are all antibodies. Thorbecke (1954) and Thorbecke and Keuning (1956) stress also the triple correspondence between the numbers of plasma cells, antibody formation, and the level of γ-globulin in the serum of newborn animals, as tending to show that all γ-globulin is antibody. Newborn rabbits, like humans, have blood γ-globulin like that of the mother, and derived from her. Its concentration in their blood falls after birth until the animals are 2–4 weeks old, and then rises. These authors, in agreement with Maximow (1907) and with Schridde (1919), found no plasma cells in newborn rabbits, but the cells began to appear between the first and second weeks, suggesting that they were the source of the rising γ-globulin.

By contrast, Richter (1952) produced high concentrations of γ-globulins in the blood of rabbits by injecting them repeatedly with sodium ribonucleate, while the animals were undergoing immunization with horse serum, but there was no corresponding increase in antibody. Instead the antibody titers in these rabbits which had high blood globulin levels, were sometimes lower than the antibody titers of other rabbits, which, having received only horse serum and saline (no ribonucleate), showed much lower globulin levels.

There are exceptions to the rule that high blood γ-globulins are to be found only in infectious and allergic conditions. Cirrhosis of the liver is an important one. The work of Eisenmenger and Slater (1953) and of Havens *et al.* (1953) has shown that the turnover of γ-globulins in patients with cirrhosis of the liver is more rapid than in normal persons, and Havens *et al.* (1951) and Eichman *et al.* (1953) have found antibody formation to diphtheria toxoid increased in liver disease and in acute hepatitis. Much earlier, Henriques and Klausen (1932) suggested that the globulin increase in liver disease may not necessarily be antibody, but Bing (1940b) described later the presence of many plasma cells in the bone marrow, spleens, and livers of patients with cirrhosis of the liver. Accordingly, it would appear that the globulin might well be antibody.

An electrophoretic study (Miller *et al.*, 1952) of the abnormal plasma proteins that appear in patients with multiple myeloma has disclosed a large sharp peak, which distinguishes their pattern, as present also in the patterns of extracts of myeloma cells. Since myeloma cells resemble plasma cells and are apparently the sources of the abnormal plasma protein, a γ-globulin, found in that disease, the finding suggests that plasma cells are capable of forming this γ-globulin as well as the usual antibodies. However, this abnormal γ-globulin may be something of the nature of an antibody.

As this chapter goes to press a paper by Askonas and Humphrey (1958) has reported that tissue slices, removed from rabbits immunized with ovalbumin, produced specific antibody, but they also formed increased amounts of other γ-globulins which were not identified as specific antibody.

D. A Note on the Speed of Antibody Formation

From time to time throughout this chapter remarks on some of the newer work have indicated that antibody may be formed far more rapidly than is generally supposed. In summary, Nunes (1950) mentioned the appearance of antibodies in the sera of 20–40% of guinea pigs within 5 hours after giving them intraperitoneal injections of pneumococci. Stevens and McKenna (1957a, b) and McKenna and Stevens (1957a, b) have described the appearance of antibodies in animals only 4 hours after the introduction of bovine γ-globulin, and the formation of antibodies in culture in even shorter periods. Its intracellular synthesis occurred with no or little apparent induction period. The release into the culture medium occurred later. Wesslen (1952b), in work already cited, found antibody forming rapidly in tissue cultures. Petersen and Campbell (1955) have reported the appearance of antibodies in the milk of cows in 24 hours after injecting the udders with antigen. Campbell *et al.* (1957) instilled *Salmonella pullorum* into the mouth of a calf which was then allowed to nurse from lactating cows. Within 4 to 7 hours antibodies appeared in the milk. The technique has been termed by them "diathelic immunization."

Humphrey and McFarlane (1955), as already mentioned, studied the incorporation of C^{14}-labeled amino acids into antibody. In animals intravenously injected with killed pneumococci, as antigen, labeled antibody became detectable in the blood in 30 minutes.

Taliaferro and Taliaferro (1957) and Taliaferro (1957b) injected S^{35}-labeled amino acids into rabbits, following the induction period but during the rapid secondary response to bovine serum albumin, to determine the "transit time." They found labeled antibody appearing 40 minutes after injection. Similar results were noted by Askonas and White (1956) who added C^{14}-glycine to bits of lymph nodes, which were producing antibody in tissue cultures, and observed the incorporation of C^{14} into the antibody. Space does not permit further discussion of the subject. The mention of other work upon this theme will be found in the papers already reviewed.

XI. CONCLUDING COMMENT

This chapter presents an impartial review of some of the trends of contemporary thought about the sites of antibody formation and the cells involved in the process. No summary can be given, for the whole is only a summary. Criticism has not been attempted since a wealth of contradictory evidence forbids it. Nevertheless some facts are well established. We know many of the situations or sites at which antibody is formed and much about the cells that either carry it or apparently form it. The work with tracer antigens has indicated some of the sites from which antigens must exert their first stimuli to antibody formation. Unfortunately, the time is not yet ripe to speculate too far beyond the data at hand upon the processes which occur between the seizure of antigen and the appearance of antibody.

At first the reticuloendothelial phagocytic cells were considered as the antibody formers. Later this hypothesis gave way to one accepting the lymphocyte, and then, after more critical observations of lymphoid cells, to the plasma cell hypothesis. Certain papers have been cited purposely in this review to bring out the fact that confusion must reign until more is known about hematological cytology and until the genesis and the life cycles of the various cells that have been mentioned are fully established. It would seem that too much effort has been made in the past to fix the responsibility for antibody formation on some single type of cell, and the mechanism of antibody formation as a whole has been lost to view. Clearly the works presented above indicate that, in the process as a whole, many cell types may be involved. In evaluating past work it becomes clear—for future planning—that technical factors force various conclusions upon the experimenter. Always the route by which an antigen enters the body should be considered as having an all-important effect in calling out antibody responses in different tissues. The frequency with which the tissues are challenged brings about different responses, and the dosage of antigen, if large, may bring about responses in many tissues rather than in only one.

As T. N. Harris and Ehrich (1946) and Ehrich (1956) have suggested, the phase of antibody formation in the antibody-forming cell is preceded by degradation of the antigen, since the injection of antigenic organisms into the foot pads of animals leads to the appearance of soluble antigenic products in the draining lymph nodes. These antigenic materials, liberated by the phagocytic cells at the sites of injection, must induce the fixed cells in the nodes to form antibody. Under such circumstances the antigen behaves like an organizer, either within the phagocytic cells themselves or after its escape from them.

The fact that X-ray given at an appropriate interval after injecting antigen will destroy lymphoid tissue but will not destroy antibody formation, suggests that the phagocytic cells have carried out the earlier steps in antibody formation, perhaps by liberating materials for antibody-forming cells which are not affected by the radiation.

In our present state of ignorance it remains possible that the entire mechanism of antibody formation resides in a single type of cell, or that it is carried out in stages by several generations of the same cell. However, it seems to be more likely that the formation of antibody requires the co-operation of several cell types, and the function of the experimenter seems to be that of a detective to find out just what it is that each type of cell contributes to the process. Accordingly it would seem wise to maintain an open mind rather than to attempt to attribute the whole complex mechanism to the one type of cell which can be shown to contain the finished product—antibody!

REFERENCES

Alexander, J. (1931–1932). *Protoplasma* 14, 296.

Amano, S. B., Unno, G., Hanaoka, M., and Tamaki, Y. (1951). *Acta Pathol. Japon.* 1, 117.

Andersen, H. C., and Bing, J. (1944). *Acta Pathol. Microbiol. Scand.* 21, 455.

Anderson, H. D., Kunkel, H. G., and McCarty, M. (1948). *J. Clin. Invest.* 27, 425.

Anderson, R. J. (1929). *J. Biol. Chem.* 83, 505.

Andreasen, E., Bing, J., Gottlieb, O., and Harböe, N. (1948). *Acta Physiol. Scand.* 15, 254.

Antopol, W. (1950). *Proc. Soc. Exptl. Biol. Med.* 73, 262.

Aschoff, L. (1924). *Ergeb. inn. med. u. Kinderheilk* **26**, 1.

Askonas, B. A., and Humphrey, J. H. (1955). *Biochem. J.* **60**, X.

Askonas, B. A., and Humphrey, J. H. (1958). *Biochem. J.* **68**, 252.

Askonas, B. A., and White, R. G. (1956). *Brit. J. Exptl. Pathol.* **37**, 61.

Askonas, B. A., Humphrey, J. H., and Porter, R. R. (1956). *Biochem. J.* **63**, 412.

Batty, I., and Warrack, G. H. (1955). *J. Pathol. Bacteriol.* **70**, 355.

Beard, J. W., and Rous, P. (1938). *J. Exptl. Med.* 67, 883.

Benjamin, E., and Sluka, E. (1908). *Wien klin Wochschr.* **21**, 311.

Berenbaum, M. C. (1956). *Nature* **177**, 46.

Berglund, K. (1956a). *Acta Pathol. Microbiol. Scand.* **38**, 311.

Berglund, K. (1956b). *Acta Pathol. Microbiol. Scand.* **38**, 329.

Berglund, K., and Fagraeus, A. (1956). *Nature* **177**, 233.

Biggart, J. H. (1932). *J. Pathol. Bacteriol.* **35**, 799.

Bing, J. (1940a). *Acta Med. Scand.* **103**, 547.

Bing, J. (1940b). *Acta Med. Scand.* **103**, 565.

Bing, J., and Plum, P. (1937). *Acta Med. Scand.* **92**, 415.

Bing, J., Fagraeus, A., and Theorell, B. (1945). *Acta Physiol. Scand.* **10**, 282.

Björneboe, M. (1939). *J. Immunol.* **37**, 201.

Björneboe, M. (1941). *Z. Immunitätsforsch.* **99**, 245.

Björneboe, M. (1943). *Acta Pathol. Microbiol. Scand.* **20**, 221.

Björneboe, M., and Gormsen, H. (1941). *Klin. Wochschr.* **20**, 314.

Björneboe, M., and Gormsen, H. (1943). *Acta Pathol. Microbiol. Scand.* **20**, 649.

Björneboe, M., Gormsen, H., and Lindquist, T. (1947). *J. Immunol.* **55**, 121.

Björneboe, M., Fischel, E. E., and Stoerk, H. C. (1951). *J. Exptl. Med.* **93**, 37.

Bloom, W. (1928). *Folia Haematol.* **37**, 63.

Bloom, W. (1938a). *In* "Handbook of Hematology" (H. Downey, ed.), Vol. 1, Sect. 5, p. 373, Hoeber, New York.

Bloom, W. (1938b). *In* "Handbook of Hematology" (H. Downey, ed.), Vol. 2, Sect. 19, p. 1427. Hoeber, New York.

Bordet, J. (1909). "Studies in Immunity" (Trans. by F. P. Gay). Wiley, New York.

Brachet, J. (1940). *Compt. rend. soc. biol.* **133-34**, 88.

Brachet, J. (1952). *Symposia Soc. Exptl. Biol. No.* **6**, 173.

Braude, A. I., and Spink, W. W. (1951). *J. Infectious Diseases* **89**, 272.

Braunsteiner, H., Fellinger, K., and Pakesch, F. (1953). *Klin. Wochschr.* **31**, 357.

Breinl, F., and Haurowitz, F. (1930). *Z. physiol. Chem.* **192**, 45.

Bruton, O. C. (1952). *Pediatrics* **9**, 772.

Bruton, O. C., Apt, L., Gitlin, D., and Janeway, C. A. (1952). *Am. J. Diseases Children* **84**, 632.

Buckley, S. M., Whitney, E., and Rapp, F. (1955). *Proc. Soc. Exptl. Biol. Med.* **90**, 226.

Bukantz, S. C., Dammin, G. J., Wilson, K. S., Johnson, M. C., and Alexander, H. L. (1949). *Proc. Soc. Exptl. Biol. Med.* **72**, 21.

Bunting, C. H. (1925). *Wisconsin Med. J.* **24**, 305.

Burnet, F. M. (1956). "Enzyme Antigen and Virus." Cambridge Univ. Press, London and New York.

Burnet, F. M., and Lush, D. (1938). *Australian J. Exptl. Biol. Med. Sci.* **16**, 261.

Burnet, F. M., and Fenner, F. (1949). "The Production of Antibodies," 2nd ed. Macmillan, New York and London.

Burrows, W., and Havens, I. (1948). *J. Infectious Diseases* **82**, 231.

Burrows, W., Elliott, M. E., and Havens, I. (1947a). *J. Infectious Diseases* **81**, 157.

Burrows, W., Mather, A. N. Elliott, M. E., and Havens, I. (1947b). *J. Infectious Diseases* **81**, 261.

Campbell, B., and Good, R. A. (1949). *Ann. Allergy* **7**, 471.

Campbell, B., and Good, R. A. (1950). *A. M.A. Arch. Neurol. Psychiat.* **63**, 298.

Campbell, B., Porter, R. M., and Petersen, W. E. (1950). *Nature* **166**, 913.

Campbell, B., Sarwar, M., and Petersen, W. E. (1957). *Science* **125**, 932.

Campbell, J. H. (1943). *J. Infectious Diseases* **72**, 42.

Cannon, P. R., and Pacheco, G. A. (1930). *Am. J. Pathol.* **6**, 749.

Cannon, P. R., and Sullivan, F. L. (1932). *Proc. Soc. Exptl. Biol. Med.* **29**, 517.

Cannon, P. R., Baer, R. B., Sullivan, F. L., and Webster, J. R. (1929). *J. Immunol.* **17**, 441.

Carrel, A., and Ingebrigsten, R. (1912). *J. Exptl. Med.* **15**, 287.

Casals, F., and Freund, J. (1939). *J. Immunol.* **36**, 399.

Caspersson, T. (1936). *Skand. Arch. Physiol.* **73**, Suppl. 8.

Caspersson, T., and Theorell, B. (1941). *Chromosoma* **2**, 132.

Chase, J. H., White, A., and Dougherty, T. F. (1946). *J. Immunol.* **52**, 101.

Chase, M. W. (1943). *Proc. Soc. Exptl. Biol. Med.* **52**, 238.

Chase, M. W. (1945). *Proc. Exptl. Biol. Med.* **59**, 134.

Chase, M. W. (1946). *J. Bacteriol.* **51**, 643.

Chase, M. W. (1947). *J. Exptl. Med.* **86**, 489.

Chase, M. W. (1948). *In* "Bacterial and Mycotic Infections of Man" (R. Dubos, ed.), Chapt. 6. Lippincott, Philadelphia, Pennsylvania.

Chase, M. W. (1950a). *Federation Proc.* **8**, 402.

Chase, M. W. (1950b). *Federation Proc.* **9**, 379.

Chase, M. W. (1951). *Federation Proc.* **10**, 409.

Chase, M. W. (1953). *In* "The Nature and Significance of the Antibody Response" (A. M. Pappenheimer, Jr., ed.), p. 156. Columbia Univ. Press, New York.

Chase, M. W., and Wager, O. A. (1957). *Federation Proc.* **16**, 639.

Cheng, K. K. (1949). *J. Pathol. Bacteriol.* **61**, 23.

Cohen, S. M., Gordon, I., Rapp, F., Macauley, J. C., and Buckley, S. M. (1955). *Proc. Soc. Exptl. Biol. Med.* **90**, 118.

Conway, E. (1937). *Anat. Record* **69**, 487.

Conway, E. (1938). *A. M. A. Arch. Pathol.* **25**, 200.

Conway, E. (1939). *J. Infectious Diseases* **64**, 217.

Coons, A. H. (1954). *Ann. Rev. Microbiol.* **8**, 333.

Coons, A. H., and Kaplan, M. H. (1950). *J. Exptl. Med.* **91**, 1.

Coons, A. H., Creech, H. J., Jones, R. N., and Berliner, E. (1942). *J. Immunol.* **45**, 159.

Coons, A. H., Leduc, E. H., and Kaplan, M. H. (1951). *J. Exptl. Med.* **93**, 173.

Coons, A. H., Leduc, E. H., and Connolly, J. M. (1955). *J. Exptl. Med.* **102**, 49.

Councilman, W. T. (1906–1907). *Harvey Lectures* **2**, 268 (1908).

Craddock, C. G., and Lawrence, J. S. (1948). *J. Immunol.* **60**, 241.

Craddock, C. G., Valentine, W. N., and Lawrence, J. S. (1949). *J. Lab. Clin. Med.* **34**, 158.

Crampton, C. F., and Haurowitz, F. (1950). *Science* **112**, 300.

Crampton, C. F., and Haurowitz, F. (1952). *J. Immunol.* **69**, 457.

Crampton, C. F., Reller, H. H., and Haurowitz, F. (1952). *Proc. Soc. Exptl. Biol. Med.* **80**, 448.

Crampton, C. F., Reller, H. H., and Haurowitz, F. (1953). *J. Immunol.* **71**, 319.

Cummings, M. M., Hoyt, M., and Gottshall, R. Y. (1947). *Public Health Repts. (U.S.)* **62**, 994.

Davies, A. (1922). *Lancet* **ii**, 1009.

de Gara, P. F., and Angevine, D. M. (1943). *J. Exptl. Med.* **78**, 27.

Deutsch, L. (1899). *Ann. inst. Pasteur* **13**, 689.

Dews, P. B., and Code, C. F. (1953). *J. Immunol.* **70,** 199.

Digby, K. H., and Enticknap, J. B. (1954). *Brit. J. Exptl. Pathol.* **35,** 294.

Dixon, F. J. (1954). *J. Allergy* **25,** 487.

Dixon, F. J. (1957). *J. Cellular Comp. Physiol.* **50,** Suppl. 1, 27.

Dixon, F. J., and Maurer, P. H. (1955). *J. Exptl. Med.* **101,** 245.

Dixon, F. J., and Talmage, D. W. (1951). *Proc. Soc. Exptl. Biol. Med.* **78,** 123.

Dixon, F. J., and Weigle, W. O. (1957). *J. Exptl. Med.* **105,** 75.

Dixon, F. J., Bukantz, S. C., and Dammin, G. J. (1951a). *Science* **113,** 274.

Dixon, F. J., Bukantz, S. C., Dammin, G. J., and Talmage, D. W. (1951b). *Federation Proc.* **10,** 553.

Dixon, F. J., Talmage, D. W., and Maurer, P. H. (1952). *J. Immunol.* **68,** 693.

Dixon, F. J., Bukantz, S. C., Dammin, G. J., and Talmage, D. W. (1953). *In* "The Nature and Significance of the Antibody Response" (A. M. Pappenheimer, Jr., ed.), p. 170. Columbia Univ. Press, New York.

Dixon, F. J., Maurer, P. H., and Weigle, W. O. (1955). *J. Immunol.* **74,** 188.

Dixon, F. J., Maurer, P. H., Weigle, W. O., and Deichmuller, M. P. (1956). *J. Exptl. Med.* **103,** 425.

Dixon, F. J., Weigle, W. O., and Roberts, J. C. (1957a). *J. Immunol.* **78,** 56.

Dixon, F. J., Roberts, J. C., and Weigle, W. O. (1957b). *Federation Proc.* **16,** 649.

Doan, C. A. (1939). *In* "A Symposium on the Blood and Blood Forming Organs," p. 167 (1941). Univ. Wisconsin Press, Madison, Wisconsin.

Doan, C. A., Sabin, F. R., and Forkner, C. E. (1930). *J. Exptl. Med.* **52,** Suppl. 3, 73.

Doan, C. A., Wright, C. S., Wheeler, W. E., Bouroncle, B. A., Houghton, B. C., and Dodd, M. C. (1950). *Trans. Assoc. Am. Physicians* **63,** 172.

Doerr, R. (1950a). "Die Immunitätsforschung," Vol. VI. Springer, Vienna.

Doerr, R. (1950b). "Die Immunitätsforschung," Vol. VII. Springer, Vienna.

Doerr, R. (1950c). "Die Immunitätsforschung," Vol. VIII. Springer, Vienna.

Dougherty, T. F. (1948). *Am. J. Med.* **4,** 618.

Dougherty, T. F. (1953). *In* "The Mechanism of Inflammation" (G. Jasmin and A. Roberts, eds.), p. 217. Acta Inc., Montreal.

Dougherty, T. F., and White, A. (1943). *Proc. Soc. Exptl. Biol. Med.* **53,** 132.

Dougherty, T. F., and White, A. (1944). *Endocrinology* **35,** 1.

Dougherty, T. F., and White, A. (1945). *Am. J. Anat.* **77,** 81.

Dougherty, T. F., and White, A. (1946a). *Anat. Record* **94,** 457.

Dougherty, T. F., and White, A. (1946b). *Endocrinology* **39,** 370.

Dougherty, T. F., Chase, J. H., and White, A. (1944). *Proc. Soc. Exptl. Biol. Med.* **57,** 295.

Dougherty, T. F., Chase, J. H., and White, A. (1945a). *Proc. Soc. Exptl. Biol. Med.* **58,** 135.

Dougherty, T. F., White, A., and Chase, J. H., (1945b). *Proc. Soc. Exptl. Biol.* **59,** 172.

Downey, H. (1911). *Folia Haematol.* **11,** 275.

Dubois-Ferrière, H. (1951). *In* "Actualites Hématologiques" (Series 1, p. 43). p. 43. Doin, Paris.

Duncan, I. B. R., Goudie, L. G., Mackie, L. M., and Howie, J. W. (1954). *Brit. J. Exptl. Pathol.* **35,** 294.

Ehrich, W. E. (1929a). *J. Exptl. Med.* **49,** 347.

Ehrich, W. E. (1929b). *J. Exptl. Med.* **49,** 361.

Ehrich, W. E. (1929c). *Am. J. Anat.* **43,** 347.

Ehrich, W. E. (1929d). *Am. J. Anat.* **43,** 385.

Ehrich, W. E. (1946). *Ann. N.Y. Acad. Sci.* **46,** 823.

Ehrich, W. E. (1953a). *In* "The Mechanism of Inflammation " (G. Jasmin and A. Roberts, eds.), p. 25. Acta Inc., Montreal.

Ehrich, W. E. (1953b). *In* "Blood Cells and Plasma Proteins: Their State in Nature" (J. L. Tullis, ed.), Sect. 3, Chapt. 4, p. 187. Academic Press, New York.

Ehrich, W. E. (1956). *In* "Handbuch der allgemein Pathologie" (F. Buchner, E. Letterer, and F. Roulet, eds.), Vol. 7, Pt. 1. Springer, Berlin.

Ehrich, W. E., and Harris, T. N. (1942). *J. Exptl. Med.* **76,** 335.

Ehrich, W. E., and Harris, T. N. (1945). *Science* **101,** 28.

Ehrich, W. E., and Voigt, W. (1934). *Beitr. pathol. Anat. u. allgern. Pathol.* **93,** 343.

Ehrich, W. E., Halbert, S. P., Mertens, E., and Mudd, S. (1945). *J. Exptl. Med.* **82,** 343.

Ehrich, W. E., Harris, T. N., and Mertens, E. (1946). *J. Exptl. Med.* **83,** 373.

Ehrich, W. E., Drabkin, D. L., and Forman, C. (1949). *J. Exptl. Med.* **90,** 157.

Ehrlich, P. (1910). "Studies in Immunity" (Trans. by C. Bolduan), 2nd ed. Wiley, New York.

Eichman, P. L., Miller, R. W., and Hariens, W. P., Jr. (1953). *J. Immunol.* **70,** 21.

Eisen, H. N., Mayer, M. M., Moore, D. H., Tarr, R., and Stoerk, H. C. (1947). *Proc. Soc. Exptl. Biol. Med.* **76,** 301.

Eisenmenger, W. J., and Slater, R. J. (1953). *J. Clin. Invest.* **32,** 564.

Ellis, J. T., Toolan, H. W., and Kidd, J. G. (1950). *Federation Proc.* **9,** 329.

Erickson, J. O., Armen, D. M., and Libby, R. L. (1953). *J. Immunol.* **71,** 30.

Erickson, J. O., Hensley, T. J., Fields, M., and Libby, R. L. (1957). *J. Immunol.* **78,** 94.

Erslev, A. (1951). *J. Immunol.* **67,** 281.

Fagraeus, A. (1946). *Nord Med.* **30,** 1381.

Fagraeus, A. (1948a). *Acta Med. Scand.* Suppl. 204.

Fagraeus, A. (1948b). *J. Immunol.* **58,** 1.

Fagraeus, A., and Grabar, P. (1953). *Ann. inst. Pasteur* **85,** 34.

Faragó, F. (1935). *Am. J. Hyg.* **22,** 495.

Fastier, L. B. (1948). *J. Immunol.* **60,** 399.

Felton, L. D. (1949). *J. Immunol.* **61,** 107.

Fields, M., and Libby, R. L. (1952), *J Immunol.* **69,** 581.

Fink, M. A., Smith, P., Rothlauf, M. V. (1955). *Proc. Soc. Exptl. Biol. Med.* **90,** 590.

Fischel, E. E. (1953). *In* "The Effect of ACTH and Cortisone upon Infection and Resistance" (G. Shwartzman, ed.), p. 56. Columbia Univ. Press, New York.

Fischel, E. E., LeMay, M., and Kabat, E. A. (1949). *J. Immunol.* **61,** 89.

Fischel, E. E., Stoerk, H. C., and Björneboe, M. (1951). *Proc. Soc. Exptl. Biol. Med.* **77,** 111.

Fitch, F. W., Barker, P., Soule, K. H., and Wissler, R. W. (1953). *J. Lab. Clin. Med.* **42,** 598.

Fitch, F. W., Wissler, R. W., LaVia, M., and Barker, P. (1956). *J. Immunol.* **76,** 151.

Flemming, W. (1885). *Arch. mikroskop. Anat. u. Entwicklungsmech.* **24,** 50.

Freund, J. (1927). *J. Immunol.* **14,** 101.

Freund, J. (1929a). *J. Immunol.* **16,** 275.

Freund, J. (1929b). *J. Immunol.* **16,** 515.

Freund, J. (1947). *Ann. Rev. Microbiol.* **1,** 291.

Freund, J. (1951). *Am. J. Clin. Pathol.* **21,** 645.

Freund, J. (1956). *In* "Advances in Tuberculosis Research (H. Birkhäuser and H. Bloch, eds.), Vol. 7, p. 130. S. Karger, Basel and New York.

Freund, J., and Lipton, M. M. (1955). *J. Immunol.* **75,** 454.

Freund, J., and McDermott, K. (1942). *Proc. Soc. Exptl. Biol. Med.* **49,** 548.

Freund, J., Casals, J., and Hosmer, E. P. (1937). *Proc. Soc. Exptl. Biol. Med.* **37,** 509.

Freund, J., Thomson, K. J., Hough, H. B., Sommer, H. E., and Pisani, T. M. (1948). *J. Immunol.* **60,** 383.

Freund, J., Schryver, E. M., McGuiness, M. B., and Geitner, M. B. (1952). *Proc. Soc. Exptl. Biol. Med.* **81,** 657.

Friedberg, W., Walter, H., and Haurowitz, F. (1955). *J. Immunol.* **75,** 315.

Garvey, J. S., and Campbell, D. H. (1954). *J. Immunol.* **72,** 131.

Garvey, J. S., and Campbell, D. H. (1956). *J. Immunol.* **76,** 36.

Garvey, J. S., and Campbell, D. H. (1957). *J. Exptl. Med.* **105,** 361.

Gavosto, F., and Ficq, A. (1953). *Nature* **172,** 406.

Gavosto, F., and Ficq, A. (1954a). *Ann. inst. Pasteur* **86,** 320.

Gavosto, F., and Ficq, A. (1954b). *Ann. inst. Pasteur* **86,** 425.

Germuth, F. G., Jr., and Ottinger, B. (1950). *Proc. Soc. Exptl. Biol. Med.* **74,** 815.

Germouth, F. G., Jr., Oyama, B., and Ottinger, B. (1951). *Proc. Soc. Exptl. Biol. Med.* **94,** 139.

Giltner, W., Cooledge, L. H., and Huddleson, I. F. (1916). *J. Am. Vet. Med. Assoc.* **50,** 157.

Gitlin, D. (1950). *Proc. Soc. Exptl. Biol. Med.* **74,** 138.

Gitlin, D., Landing, B. H., and Whipple, A. (1953). *J. Exptl. Med.* **97,** 163.

Glenny, A. T., Pope, C. G., Waddington, H., and Wallace, U. (1926). *J. Pathol. Bacteriol.* **29,** 31.

Glenny, A. T., Buttle, G. A. H., and Stevens, M. F. (1931). *J. Pathol. Bacteriol.* **34,** 267.

Glimstedt, G. (1936). *Acta Pathol. Microbiol. Scand. Suppl.* **30.**

Goldzieher, M. A. (1927). *A. M.A. Arch. Pathol.* **3,** 42.

Good, R. A. (1947). *Anat. Record* **97,** 21.

Good, R. A. (1950). *J. Neuropathol. Exptl. Neurol.* **9,** 78.

Good, R. A. (1954). *Bull. Univ. Minn. Hosp. Minn. Med. Assoc.* **26,** 1.

Good, R. A. (1955). *Am. J. Diseases Children* **90,** 577.

Good, R. A., and Campbell, B. (1950). *Am. J. Med.* **9,** 330.

Good, R. A., and Good, T. A. (1949). *Anat. Record* **103,** 123.

Good, R. A., and Varco, H. L. (1955a). *Ann. Surg.* **142,** 334.

Good, R. A., and Varco, H. L. (1955b). *J. Lancet* **75,** 245.

Good, R. A. and Varco, R. L. (1955c). *J. Clin. Invest.* **34,** 910.

Good, R. A., Varco, R. L., Aust, J. B., and Zak, S. J. (1957). *Ann. N. Y. Acad. Sci.* **64,** 882.

Graham, J. B., and Leskowitz, S. (1956). *J. Immunol.* **76,** 110.

Graham, J. B., Graham, R. M., Neri, L., and Wright, K. (1956). *J. Immunol.* **76,** 103.

Green, H., and Anker, H. S. (1954). *Biochim. et Biophys. Acta* **13,** 365.

Grégoire, C. (1932). *Krankheitsforsch.* **9,** 97.

Gros, P., Coursaget, J., and Macheboeuf, M. (1952). *Bull. soc. chim. biol.* **34,** 1070.

Habel, K., Endicott, K. M., Bell, F., and Spear, F. (1949). *J. Immunol.* **61,** 131.

Halbert, S. P., Mudd, S., and Smolens, J. (1946). *J. Immunol.* **53,** 291.

Hale, W. M., and Stoner, R. D. (1953). *Yale J. Biol. and Med.* **26,** 46.

Hale, W. M., and Stoner, R. D. (1954). *Radiol. Research* **1,** 459.

Hammond, C. W., and Novak, M. (1950). *Proc. Soc. Exptl. Biol. Med.* **74,** 155.

Hanaoka, M., Nakayashiki, T., Tamaki, Y., and Hosomi, T. (1950). *Acta Haematol. Japan* **13,** 222.

Harris, S., and Harris, T. N. (1949). *J. Immunol.* **61,** 193.

Harris, S., and Harris, T. N. (1950). *Proc. Soc. Exptl. Biol. Med.* **74,** 186.

Harris, S., and Harris, T. N. (1951). *Federation Proc.* **10,** 409.

Harris, S., and Harris, T. N. (1954). *J. Exptl. Med.* **100**, 269.

Harris, S., and Harris, T. N. (1955). *J. Immunol.* **74**, 318.

Harris, S., Harris, T. N., and Farber, M. B. (1954). *J. Immunol.* **72**, 148.

Harris, S., Harris, T. N., Ogburn, C. A., and Farber, M. B. (1956). *J. Exptl. Med.* **104**, 645.

Harris, T. N., and Ehrich, W. E. (1946). *J. Exptl. Med.* **84**, 157.

Harris, T. N., and Harris, S. (1949). *J. Exptl. Med.* **90**, 169.

Harris, T. N., and Harris, S. (1950). *J. Immunol.* **64**, 45.

Harris, T. N., and Harris, S. (1955). *Pediatrics* **16**, 709.

Harris, T. N., and Harris, S. (1956). *Am. J. Med.* **20**, 114.

Harris, T. N., Grimm, E., Mertens, E., and Ehrich, W. E. (1945). *J. Exptl. Med.* **81**, 73.

Harris, T. N., Rhoads, J., and Stokes, J., Jr. (1948). *J. Immunol.* **58**, 27.

Harris, T. N., Harris, S., Beale, H. D., and Smith, J. J. (1954a). *J. Exptl. Med.* **100**, 289.

Harris, T. N., Harris, S., and Farber, M. B. (1954b). *J. Immunol.* **72**, 161.

Harris, T. N., Harris, S., and Farber, M. B. (1955). *J. Immunol.* **75**, 112.

Harris, T. N., Harris, S., and Farber, M. B. (1956). *J. Exptl. Med.* **104**, 663.

Harris, T. N., Harris, S., and Farber, M. B. (1957). *Proc. Soc. Exptl. Biol. Med.* **95**, 26.

Harrison, W. T. (1935). *Am. J. Public Health* **25**, 298.

Hartley, G., Jr. (1940). *J. Infectious Diseases* **66**, 44.

Haurowitz, F. (1952). *Biol. Revs. Cambridge Phil. Soc.* **27**, 247.

Haurowitz, F. (1953). *Ann. Rev. Microbiol.* **7**, 389.

Haurowitz, F., and Breinl, F. (1932). *Z. physiol. Chem.* **205**, 259.

Haurowitz, F., and Crampton, C. F. (1952a). *Exptl. Cell Research* **2**, 45.

Haurowitz, F., and Crampton, C. F. (1952b). *J. Immunol.* **68**, 73.

Haurowitz, F., and Kraus, F. (1936). *Z. physiol. Chem.* **239**, 76.

Havens, W. P., Jr., Shaffer, J. M., and Hopke, C. J., Jr. (1951). *J. Immunol.* **67**, 347.

Havens, W. P., Jr., Dickeinsheets, J., Bierly, J. N., and Eberhard, T. P. (1953). *J. Clin. Invest.* **32**, 573.

Hayes, S. P., and Dougherty, T. F. (1952). *Federation Proc.* **11**, 67.

Hayes, S. P., Dougherty, T. F., and Gebhardt, L. P. (1951). *Proc. Soc. Exptl. Biol. Med.* **76**, 460.

Heidelberger, M. (1939). *Bacteriol. Revs.* **3**, 49.

Heidelberger, M., and Kendall, F. E. (1934). *J. Exptl. Med.* **59**, 519.

Heidelberger, M., Kendall, F. E., and Soo Hoo, C. M. (1933). *J. Exptl. Med.* **58**, 137.

Heineke, H. (1905). *Mitt. Grenzg. Med. Chir.* **14**, 21.

Heinlein, H. (1943). *Z. ges. exptl. Med.* **112**, 535.

Hektoen, L. (1909). *J. Infectious Diseases* **6**, 78.

Hektoen, L. (1911). *J. Infectious Diseases* **9**, 103.

Hektoen, L. (1915). *J. Infectious Diseases* **17**, 415.

Hektoen, L. (1918). *J. Infectious Diseases* **22**, 28.

Hektoen, L. (1920). *J. Infectious Diseases* **27**, 23.

Helber, E., and Linser, P. (1905). *Münch. med. Wochschr.* **52**, 689.

Hellman, T. J. (1921). *Beitr. pathol. Anat. u. allgem. Pathol.* **68**, 333.

Hellman, T. J., and White, G. (1930). *Arch. pathol. Anat. u. Physiol. Virchow's* **278**, 221.

Henriques, V., and Klausen, V. (1932). *Biochem. Z.* **254**, 414.

Herbert, P. H., and de Vries, J. A. (1949). *Endocrinology* **44**, 259.

Hewer, T. F. (1930). *Bristol Med. Chir. J.* **47**, 197.

Hill, A. G. S., Deane, H. W., and Coons, A. H. (1950). *J. Exptl. Med.* **92**, 35.

Holden, M., Seegal, B. C., and Adams, L. B. (1953). *J. Exptl. Med.* **98**, 551.

Howie, J. W., Duncan, I. B. R., and Mackie, L. M. (1953). *Lancet* **ii**, 1018.

Huebschmann, P. (1913). *Verhandl. deut. pathol. Ges.* **16**, 110.

Humphrey, J. H., and McFarlane, A. S. (1954). *Biochem. J.* **57**, 160.

Humphrey, J. H., and McFarlane, A. S. (1955). *Biochem. J.* **60**, XI.

Humphrey, J. H., and Porter, R. R. (1956). *Biochem. J.* **62**, 93.

Ingraham, J. S. (1951a). *J. Infectious Diseases* **89**, 109.

Ingraham, J. S. (1951b). *J. Infectious Diseases* **89**, 117.

Jacobson, L. O., and Robson, M. J. (1952). *J. Lab. Clin. Med.* **39**, 169.

Jacobson, L. O., Robson, M. J., Marks, E. K., and Goldman, M. C. (1949). *J. Lab. Clin. Med.* **34**, 1612.

Jacobson, L. O., Simmonds, E. L., Marks, E. K., Robson, M. J., Bethard, W. F., and Gaston, E. O. (1950a). *J. Lab. Clin. Med.* **35**, 746.

Jacobson, L. O., Robson, M. J., and Marks, E. K. (1950b). *Proc. Soc. Exptl. Biol. Med.* **75**, 145.

Jaffe, R. H. (1931). *Physiol. Revs.* **11**, 277.

Janeway, C. A. (1953). *In* "Blood Cells and Plasma Proteins: Their State in Nature" (J. L. Tullis, ed.), Sect. 3. Academic Press, New York.

Janeway, C. A., Apt, L., and Gitlin, D. (1953). *Trans. Assoc. Am. Physicians* **66**, 200.

Jaroslow, B. N., and Taliaferro, W. H. (1956). *J. Infectious Diseases* **98**, 75.

Jenner, E. (1789). "An Inquiry into the Causes and Effects of the Variolae Vaccinae, or Disease Discovered in Some of the Western Counties of England, Particularly Gloucestershire, and Known by the Name of Cowpox." S. Low, London.

Jeschal, E. (1953). *Plasma (Milan)* **1**, 329.

Johnson, A. G., Gaines, S., and Landy, M. (1956). *J. Exptl. Med.* **103**, 225.

Jones, F. S. (1925). *J. Exptl. Med.* **41**, 767.

Kagan, B. M. (1943). *Am. J. Med. Sci.* **206**, 309.

Kaplan, M. H., Coons, A. H., and Deane, H. W. (1950). *J. Exptl. Med.* **91**, 15.

Kass, E. H. (1945). *Science* **101**, 337.

Kass, E. H., and Finland, M. (1953). *Ann. Rev. Microbiol.* **7**, 361.

Kerr, W. R., and Robertson, M. (1947). *J. Comp Pathol. Therap.* **57**, 301.

Kerr, W. R., and Robertson, M. (1953). *J. Hyg.* **51**, 405.

Keuning, F. J., and van der Slikke, L. B. (1950). *J. Lab. Clin. Med.* **36**, 167.

Kidd, J. G., and Toolan, H. W. (1949). *Federation Proc.* **8**, 360.

Kidd, J. G., and Toolan, H. W. (1950). *Federation Proc.* **9**, 385.

Kimura, R. (1932). *Acta Schol. Med. Univ. Imp. Kiota* **15**, 1.

Kirchheimer, W. R., and Weiser, R. S. (1947). *Proc. Soc. Exptl. Biol. Med.* **66**, 166.

Kohn, H. I. (1951). *J. Immunol.* **66**, 525.

Koluch, F. (1938). *Proc. Soc. Exptl. Biol. Med.* **39**, 147.

Koluch, F., Good, R. A., and Campbell, B. (1947). *J. Lab. Clin. Med.* **32**, 749.

Kraus, R. (1897). *Wien. klin. Wochschr.* **10**, 736.

Kruse, H., and McMaster, P. D. (1949). *J. Exptl. Med.* **90**, 425.

Landsteiner, K. (1945). "The Specificity of Serological Reactions," rev. ed. Harvard Univ. Press, Cambridge, Massachusetts.

Landsteiner, K., and Chase, M. W. (1942). *Proc. Soc. Exptl. Biol. Med.* **49**, 688.

Latta, H. (1951). *J. Immunol.* **66**, 635.

La Via, M. F., Barker, P. A., and Wissler, R. W. (1956). *J. Lab. Clin. Med.* **48**, 237.

Lawrence, H. S. (1949). *Proc. Soc. Exptl. Biol. Med.* **71**, 516.

Lawrence, H. S. (1952). *J. Immunol.* **68**, 159.

Lawrence, H. S. (1954). *In* "Streptococcal Infections" (M. McCarty, ed.), Chapt. 11, p. 143. Columbia Univ. Press, New York.

Lawrence, H. S. (1955). *J. Clin. Invest.* **34**, 219.

Lawrence, H. S. (1957). *Ann. N.Y. Acad. Sci.* 64, 826.

Leduc, E. H., Coons, A. H., and Connolly, J. M. (1955). *J. Exptl. Med.* 102, 61.

Lewis, P., and Loomis, D. (1924). *J. Exptl. Med.* 40, 503.

Lewis, P., and Loomis, D. (1925). *J. Exptl. Med.* 41, 327.

Libby, R. L., and Madison, C. R. (1947). *J. Immunol.* 55, 15.

Liebow, A. A., Warren, S., and Delauney, E. (1949). *Am. J. Pathol.* 25, 953.

Linser, P., and Helber, E. (1905). *Deut. Arch. klin. Med.* 83, 479.

Loosli, C. G. (1942). *J. Exptl. Med.* 75, 657.

Lowenhaupt, E. (1945). *Am. J. Pathol.* 21, 511.

Luckhardt, A., and Becht, F. C. (1911). *Am. J. Physiol.* 28, 257.

Ludke, H. (1912). *Berlin. klin. Wochschr.* 49, 1034.

Lurie, M. B. (1942). *J. Exptl. Med.* 75, 247.

McClintock, L. A., and Friedman, M. M. (1945). *Am. J. Roentgenol. Radium Therapy* 54, 704.

McKenna, J. M., and Stevens, K. M. (1957a). *J. Immunol.* 78, 311.

McKenna, J. M., and Stevens, K. M. (1957b). *Federation Proc.* 16, 424.

McMaster, P. D. (1941–1942). *Harvey Lectures* 37, 227 (1942).

McMaster, P. D. (1946). *Ann. N. Y. Acad. Sci.* 46, 743.

McMaster, P. D. (1953). *In* "The Nature and Significance of the Antibody Response" (A. M. Pappenheimer, Jr., ed.), Chapt. 2. Columbia Univ. Press, New York.

McMaster, P. D., and Edwards, J. L. (1957a). *Proc. Nat. Acad. Sci. U.S.* 43, 380.

McMaster, P. D., and Edwards, J. L. (1957b). *J. Exptl. Med.* 106, 219.

McMaster, P. D., and Hudack, S. S. (1935). *J. Exptl. Med.* 61, 783.

McMaster, P. D, and Kidd, J. G. (1937). *J. Exptl. Med.* 66, 73.

McMaster, P. D., and Kruse, H. (1951). *J. Exptl. Med.* 94, 323.

McMaster, P. D., Edwards, J. L., and Sturm, E. (1955). *J. Exptl. Med.* 102, 119.

McNeil, C. (1948). *Am. J. Pathol.* 24, 1271.

McNeil, C. (1950). *J. Immunol.* 65, 359.

Mann, F. C., and Higgins, G. M. (1938). *In* "Downey's Handbook of Hematology," Vol. II, Sect. XVIII, p. 1375. Hoeber, New York.

Markoff, N. G. (1937). *Deut. Arch. klin. Med.* 180, 530.

Marschalko, T. V. (1895). *Arch. Dermatol. and Syphilol.* 30, 3.

Marshall, A. H. E., and White, R. G. (1950). *Brit. J. Exptl. Pathol.* 31, 157.

Martin, C. M., Waite, J. B., and McCullough, N. B. (1957). *J. Clin. Invest.* 36, 405.

Matko, J. (1917–1918). *Z. exptl. Pathol. Therap.* 19, 437.

Matsumura, T. (1949a). *Acta Schol. Med. Univ. Kioto* 27, 1.

Matsumura, T. (1949b). *Acta Schol. Med. Univ. Kioto* 27, 103.

Matsumura, T., Tanaka, T., and Takenaka, Z. (1952). *Acta Schol. Med. Univ. Kioto* 29, 155.

Maurer, P. H., Dixon, F. J., and Talmage, D. W. (1953). *Proc. Soc. Exptl. Biol. Med.* 83, 163.

Maximow, A. A. (1902). *Beitr. pathol. Anat. u. allgem. Pathol.* Suppl. 5, 1.

Maximow, A. A. (1907). *Folia Haematol.* 4, 611.

Maximow, A. A. (1923). *Arch. mikroskop. Anat. u. Entwicklungsmech.* 97, 623.

Maximow, A. A., and Bloom, W. (1957). "A Text Book of Histology," 7th ed. pp. 73, 79. Saunders, Philadelphia, Pennsylvania.

Mayersbach, H. (1957). *Z. Zellforsch. u. mikroskop. Anat.* 45, 483.

Metaxas, M. N., and Metaxas-Bühler, M. (1948). *Proc. Soc. Exptl. Biol. Med.* 69, 163.

Metchnikoff, E. (1887). *Ann. inst. Pasteur* 1, 321.

Metchnikoff, E. (1888). *Arch. pathol. Anat. u. Physiol.* 113, 63.

Metchnikoff, E. (1901). "L'Immunité dans les maladies infectieuses." Masson, Paris.

Meyer, K., and Loewenthal, H. (1927–1928). Z. Immunitätsforsch. 54, 409.

Miller, F. R. (1931). J. Exptl. Med. 54, 333.

Miller, G. L., Brown, C. E., Miller, E. E., and Eitelman, E. S. (1952). Cancer Research 12, 716.

Miller, J. M., and Favour, C. B. (1951). J. Exptl. Med. 93, 1.

Miller, L. L., and Bale, W. F. (1954). J. Exptl. Med. 99, 125.

Miller, L. L., Bly, C. G., Watson, M. C., and Bale, W. F. (1951). J. Exptl. Med. 94, 431.

Miller, L. L., Bly, C. G., and Bale, W. F. (1954). J. Exptl. Med. 99, 133.

Mirsky, A. E., (1951–1952). Harvey Lectures 46, 98 (1953).

Mitchell, C. A., Walker, R. V. L., and Bannister, G. L. (1953). Can. J. Comp. Med. Vet.. Sci. 17, 104.

Moeschlin, S. (1940-1941). Helv. Med. Acta 7, 227.

Moeschlin, S. (1941). Folia Haematol. 65, 181.

Moeschlin, S. (1947). "Die Milzpunktion." Benno Schwabe, Basel.

Moeschlin, S. (1949). Acta Haematol. 2, 399.

Moeschlin, S. (1951). "Spleen Puncture." William Heinemann, London.

Moeschlin, S., and Demiral, B. (1952). Klin. Wochschr. 30, 827.

Moeschlin, S., Pelaez, J. R., and Hugentobler, F. (1951). Acta Haematol. 6, 321.

Moll, F. C., and Hawn, C. von Z. (1953). J. Immunol. 70, 441.

Morgan, I. M., and Olitsky, P. K. (1941). J. Immunol. 42, 445.

Morgan, I. M., Schlesinger, R. W., and Olitsky, P. K. (1942). J. Exptl. Med. 76, 357.

Motohashi, S. (1922). J. Med. Research 43, 473.

Mountain, I. M. (1955a). J. Immunol. 74, 270.

Mountain, I. M. (1955b). J. Immunol. 74, 278.

Mudd, S. (1932). J. Immunol. 23, 423.

Müller, I. (1932). Zentr. allgem. Pathol. u. pathol. Anat. 55, 180.

Murphy, J. B. (1914). J. Am. Med. Assoc. 62, 1459.

Murphy, J. B. (1926). Monographs Rockefeller Inst. Med. Research No. 26.

Murphy, J. B., and Ellis, A. W. M. (1914). J. Exptl. Med. 20, 397.

Murphy, J. B., and Sturm, E. (1919). J. Exptl. Med. 29, 1.

Murphy, J. B., and Sturm, E. (1925). J. Exptl. Med. 41, 245.

Murphy, J. B., and Sturm, E. (1947). Proc. Soc. Exptl. Biol. Med. 66, 303.

Murphy, J. B., and Taylor, H. D. (1918). J. Exptl. Med. 28, 1.

Nakahara, W. (1919). J. Exptl. Med. 29, 17.

Nakano, H. (1955a). Ann. Paediat. 1, 81.

Nakano, H. (1955b). Ann. Paediat. 1, 181.

Nakano, H. (1955c). Ann. Paediat. 1, 300.

Nakano, H. (1956). Ann. Paediat. 2, 65.

Naylor, G. R. E., and Caldwell, R. A. (1953). J. Hyg. 51, 245.

Neil, A. L., and Dixon, F. J. (1959). Arch. Pathol. 67, 643.

Newsom, S. E., and Darrach, M. (1954). Can. J. Biochem. and Physiol. 32, 372.

Nunes, D. S. (1950). Can. J. Research 28, 298.

Oakley, C. L., Warrack, G. H., and Batty, I. (1949). J. Pathol. Bacteriol. 61, 179.

Oakley, C. L., Batty, I., and Warrack, G. H. (1951). J. Pathol. Bacteriol. 63, 33.

Oakley, C. L., Warrack, G. H., and Batty, I. (1954). J. Pathol. Bacteriol. 67, 485.

Oertel, M. J. (1890). Deut. med. Wochschr. 16, 985.

Olitsky, P. K., Schlesinger, R. W., and Morgan, I. M. (1943). J. Exptl. Med. 77, 359.

Ørskov, J. (1956). Acta Pathol. Microbiol. Scand. 38, 375.

Ørskov, J., and Andersen, E. K. (1938a). Z. Immunitätsforsch. 92, 487.

Ørskov, J., and Andersen, E. K. (1938b). *Acta Pathol. Microbiol. Scand.* Suppl. 37, p. 621.

Ortega, L. G., and Mellors, R. C. (1957). *J. Exptl. Med.* 106, 627.

Osgood, E. E., and Hunter, W. C. (1934). *Folia Haematol.* 52, 369.

Osterlind, G. (1938). *Acta Pathol. Microbiol. Scand.* Suppl. 34.

Pappenheim, A. (1901). *Arch. pathol. Anat. u. Physiol. Virchow's* 166, 425.

Pappenheim, A. (1902). *Arch. pathol. Anat. u. Physiol. Virchow's* 169, 372.

Parker, R. C. (1937). *Science* 85, 292.

Parker, R. C. (1950). "Methods of Tissue Culture," 2nd ed. Hoeber, New York.

Parsons, L. D. (1943). *J. Pathol. Bacteriol.* 55, 397.

Pasteur, L. (1880). *Bull. acad. méd. (Paris)* [2] 9, 121, 390, 1119.

Pauling, L. (1940). *J. Am. Chem. Soc.* 62, 2643.

Perla, D., and Marmorston, J. (1941). "Natural Resistance and Clinical Medicine." Little, Brown, Boston, Massachusetts.

Perlzweig, W. A., Delrue, G., and Geschickter, C. (1928). *J. Am. Med. Assoc.* 90, 755.

Petersen, W. E., and Campbell, B. (1955). *J. Lancet* 75, 494.

Pfeiffer, R., and Marx, D. (1898a). *Z. Hyg. Infektionskrankh.* 27, 272.

Pfeiffer, R., and Marx, D. (1898b). *Deut. med. Wochschr.* 24, 47.

Pfeiffer, R., and Proskauer, B. (1896). *Centr. Bakteriol. Parasitenk. Abt. I* 19, 191.

Phillips, F. S., Hopkins, F. H., Freeman, M. L. H. (1947). *J. Immunol.* 55, 289.

Pierce, A. E. (1947). *J. Comp. Pathol. Therap.* 57, 84.

Pierce, A. E. (1949). *Brit. Vet. J.* 105, 286.

Poleff, L. (1928). *Arch. Augenheilk.* 99, 515.

Porter, R. R. (1955). *Biochem. J.* 59, 405.

Pryzgode, P. (1913). *Wein. klin. Wochschr.* 26, 841.

Raffel, S., Arnaud, L. E., Dukes, C. D., and Huang, J. S. (1949). *J. Exptl. Med.* 90, 53.

Ramon y Cajal, S. (1896). "Manual de Anatomie Patologica General," 2nd ed. N. Moya, Madrid.

Ramon y Cajal, S. (1906). *Anat. Anz.* 29, 666.

Ranney, H. M., and London, J. M. (1951). *Federation Proc.* 10, 562.

Rath, D. (1899). *Centr. Bakteriol. Parasitenk. I. Abt.* 25, 549.

Rebuck, J. W. (1947). *Am. J. Clin. Pathol.* 17, 614.

Rebuck, J. W., and Crowley, J. H. (1955). *Ann. N. Y. Acad. Sci.* 59, 757.

Redd, L., and Vaughan, J. H. (1955). *Proc. Soc. Exptl. Biol. Med.* 90, 317.

Reiss, E., Mertens, E., and Ehrich, W. E. (1950). *Proc. Soc. Exptl. Biol. Med.* 74, 732.

Reiter, H. (1913). *Z. Immunitätsforsch.* 18, 5.

Renn, P. (1912). *Beitr. pathol. Anat. u. allgem. Pathol.* 53, 1.

Rich, A. R., Lewis, M. R., and Wintrobe, M. M. (1939). *Bull. Johns Hopkins Hosp.* 65, 311.

Richter, G. W. (1952). *J. Exptl. Med.* 96, 331.

Ringertz, N., and Adamson, C. A. (1948). *Acta Pathol. Microbiol. Scand.* 25, 192.

Ringertz, N., and Adamson, C. A. (1950). *Acta Pathol. Microbiol. Scand.* Suppl. 86.

Ringoen, A. R. (1921). *Folia Haematol.* 27, 10.

Roberts, J. C., and Dixon, F. J. (1955). *J. Exptl. Med.* 102, 379.

Roberts, J. C., Dixon, F. J., and Weigle, W. O. (1957). *A. M. A. Arch. Pathol.* 64, 324.

Roberts, K. B. (1955). *Brit. J. Exptl. Pathol.* 36, 199.

Roberts, S., Adams, E., and White, A. (1948). *J. Biol. Chem.* 174, 379.

Roberts, S., Adams, E., and White, A. (1949). *J. Immunol.* 62, 155.

Römer, P. (1901). *Arch. Ophthalmol. Graefe's* 52, 73.

Ross, J. M. (1932). *J. Pathol. Bacteriol.* 35, 899.

Rous, P., and Beard, J. W. (1934). *J. Exptl. Med.* 59, 577.

Rowley, D. A. (1950a). *J. Immunol.* **65**, 515.

Rowley, D. A. (1950b). *J. Immunol.* **64**, 289.

Sabin, F. R. (1939). *J. Exptl. Med.* **70**, 67.

Scamburow, H. B. (1932). *Z. Hyg. Infektionskrankh.* **114**, 456.

Schiller, A. A., Schayer, R. W., and Hess, E. L. (1953). *J. Gen. Physiol.* **36**, 489.

Schlesinger, R. W. (1949a). *J. Exptl. Med.* **89**, 491.

Schlesinger, R. W. (1949b). *J. Exptl. Med.* **89**, 507.

Schlesinger, R. W., Olitsky, P. K., and Morgan, I. M. (1944). *J. Exptl. Med.* **80**, 147.

Schridde, H. (1919). *In* "Pathologische Anatomie" (L. Aschoff, ed.), 4th ed. Vol. 2, Chapt. 3, p. 118. Gustav Fischer, Jena.

Schwab, L., Moll, F. C., Hall, T., Brean, H., Kirk, M., Hawn, C. v. Z., and Janeway, C. A. (1950). *J. Exptl. Med.* **91**, 505.

Schweet, R. S., and Owen, R. D. (1957). *J. Cellular Comp. Physiol.* **50**, Suppl. 1, p. 199 (see p. 212).

Sédallian, P., Jourdan, F., and Clavel, C. (1939a). *Rev. immunol.* **5**, 34.

Sédallian, P., Jourdan, F., and Clavel, C. (1939b). *Rev. immunol.* **5**, 138.

Seegal, D., and Seegal, B. C. (1931). *J. Exptl.* **54**, 249.

Seegal, B. C., and Seegal, D. (1934). *Proc. Soc. Exptl. Biol. Med.* **31**, 437.

Seegal, B. C., and Seegal, D. (1935). *J. Immunol.* **25**, 221.

Seegal, B. C., and Wilcox, H. B. (1940). *A. M. A. Arch. Pathol.* **30**, 416.

Seegal, B. C., Seegal, D., and Khorazo, D. (1933). *J. Immunol.* **25**, 207.

Selye, H. (1941–1942). *J. Anat.* **76**, 94.

Selye, H. (1950). "Stress." Acta Inc. Montreal.

Selye, H., Harlow, C. M., and Collip, J. B. (1936). *Endokrinologie* **18**, 81.

Shechmeister, I. L., Paris, W. H., Krause, F. T., Paulissen, L. J., and Yunker, R. (1955). *Proc. Soc. Exptl. Biol. Med.* **89**, 228.

Sherwood, N. P. (1951). "Immunology," 3rd ed. Mosby, St. Louis, Missouri.

Shimura, J., Takahashi, K., Miyazawa, Y., Ito, M., Kiuchi, K., Shinoguka, T., and Okamoto, Y. (1950). *Japan. J. Exptl. Med.* **20**, 443.

Silverman, M. S., and Chin, P. H. (1955). *J. Immunol.* **75**, 321.

Sjövall, H. (1936). *Acta Pathol. Microbiol. Scand.* Suppl. **27**.

Sjövall, A., and Sjövall, H. (1930). *Arch. pathol. Anat. u. Physiol.* **278**, 258.

Smetana, H. (1947). *Am. J. Pathol.* **23**, 255.

Smetana, H., and Johnson, F. H. (1942). *Am. J. Pathol.* **18**, 1029.

Smith, F., and Ruth, H. J. (1956). *Proc. Soc. Exptl. Biol. Med.* **90**, 187.

Smith, R. O., and Wood, W. B., Jr. (1949a). *J. Exptl. Med.* **90**, 555.

Smith, R. O., and Wood, W. B., Jr. (1949b). *J. Exptl. Med.* **90**, 567.

Speirs, R. S. (1955). *Ann. N.Y. Acad. Sci.* **59**, 706.

Speirs, R. S. (1956). *J. Immunol.* **77**, 437.

Speirs, R. S. (1957). *RES. Bull.* **3**, 19.

Speirs, R. S. (1958). *Ann. N.Y. Acad. Sci.* **73**, 283.

Speirs, R. S., and Dreisbach, M. E. (1956). *J. Hematol.* **11**, 44.

Speirs, R. S., and Wenck, U. (1955). *Proc. Soc. Exptl. Biol. Med.* **90**, 571.

Speirs, R. S., and Wenck, U. (1957). *Acta Haematol.* **17**, 271.

Speirs, R. S., Wenck, U., and Dreisbach, M. E. (1956). *J. Hematol.* **11**, 56.

Spurr, C. L. (1947). *Proc. Soc. Exptl. Biol. Med.* **64**, 259.

Stavitsky, A. B. (1948). *Proc. Soc. Exptl. Biol. Med.* **67**, 225.

Stavitsky, A. B. (1952). *Federation Proc.* **11**, 482.

Stavitsky, A. B. (1953). *J. Infectious Diseases* **93**, 130.

Stavitsky, A. B. (1954). *J. Infectious Diseases* **94**, 306.

Stavitsky, A. B. (1955a). *Ann. N.Y. Acad. Sci.* **63** (see discussion on pp. 211–213).

Stavitsky, A. B. (1955b). *J. Immunol.* **75**, 214.
Stavitsky, A. B. (1956). *Federation Proc.* **15**, 615.
Stavitsky, A. B. (1957). *Federation Proc.* **16**, 652.
Steiner, D. F., and Anker, H. S. (1956). *Proc. Natl. Acad. Sci. U. S.* **42**, 580.
Sterzl, J. (1955). *Folia Biol. (Prague)* **1**, 193.
Sterzl, J., and Hrubešová, M. (1956). *Folia Biol. (Prague)* **2**, 21.
Stevens, K. M., and McKenna, J. M. (1957a). *Federation Proc.* **16**, 434.
Stevens, K. M., and McKenna, J. M. (1957b). *Nature* **179**, 870.
Stoner, R. D., and Hale, W. M. (1955). *J. Immunol.* **75**, 203.
Süssdorf, D. H., and Draper, L. R. (1956). *J. Infectious Diseases* **99**, 129.
Taliaferro, W. H. (1949). *Ann. Rev. Microbiol.* **3**, 159.
Taliaferro, W. H. (1956). *Am. J. Trop. Med.* **5**, 391.
Taliaferro, W. H. (1957a). *Ann. N. Y. Acad. Sci.* **64**, 745.
Taliaferro, W. H. (1957b). *J. Cellular Comp. Physiol.* **50**, Suppl. **1**, 1.
Taliaferro, W. H., and Mulligan, H. W. (1937). *Indian Med. Research Mem.* **29**, 138.
Taliaferro, W. H., and Taliaferro, L. G. (1950). *J. Infectious Diseases* **87**, 37.
Taliaferro, W. H., and Taliaferro, L. G. (1951a). *J. Immunol.* **66**, 181.
Taliaferro, W. H., and Taliaferro, L. G. (1951b). *J. Infectious Diseases* **89**, 143.
Taliaferro, W. H., and Taliaferro, L. G. (1952). *J. Infectious Diseases* **90**, 205.
Taliaferro, W. H., and Taliaferro, L. G. (1954). *J. Infectious Diseases* **95**, 134.
Taliaferro, W. H., and Taliaferro, L. G. (1955). *J. Infectious Diseases* **97**, 99.
Taliaferro, W. H., and Taliaferro, L. G. (1956). *J. Infectious Diseases* **99**, 109.
Taliaferro, W. H., and Taliaferro, L. G. (1957). *J. Infectious Diseases* **101**, 252.
Taliaferro, W. H., and Talmage, D. W. (1955). *J. Infectious Diseases* **97**, 88.
Taliaferro, W. H., Taliaferro, L. G., and Jannsen, E. F. (1952). *J. Infectious Diseases* **91**, 105.
Talmage, D. W. (1955). *Ann. Rev. Microbiol.* **9**, 335.
Taylor, H. D., Witherbee, W. D., and Murphy, J. B. (1919). *J. Exptl. Med.* **29**, 53.
Theorell, B. (1944). *Acta Med. Scand.* **117**, 334.
Thomas, H. E., and Brüner, F. H. (1933). *Am. J. Roentgenol. Radium Therapy* **29**, 641.
Thompson, R., and Harrison, V. M. (1954). *Federation Proc.* **13**, 515.
Thompson, R., and Olsen, H. (1950). *J. Immunol.* **65**, 633.
Thompson, R., Gallardo, E., Jr., and Khorazo, D. (1936). *Am. J. Ophthalmol.* **19**, 852.
Thompson, R., Pfeiffer, R., and Gallardo, E., Jr. (1937). *Proc. Soc. Exptl. Biol. Med.* **36**, 179.
Thompson, R., Olson, V. H., and Sibal, L. R. (1957). *J. Immunol.* **79**, 508.
Thorbecke, G. J. (1954). "Over de Vorming van Antilichamen en Gamma-Globuline 'in vitro' in Bloedvormende Organen." Dijkstra's Drukkerij N. Y., Groningen.
Thorbecke, G. J., and Keuning, F. J. (1953). *J. Immunol.* **70**, 129.
Thorbecke, G. J., and Keuning, F. J. (1956). *J. Infectious Diseases* **98**, 157.
Thorbecke, G. J., Gordon, H. A., Wostman, B., Wagner, M., and Reyniers, J. A. (1957). *J. Infectious Diseases* **101**, 237.
Tiselius, A. (1937). *Biochem. J.* **31**, 1464.
Tizzoni, G., and Cattani, G. (1892). *Zentr. Bakteriol. Parasitenk. I Abt.* **11**, 325.
Toolan, H. W., and Kidd, J. G. (1949). *Federation Proc.* **8**, 373.
Topley, W. W. C. (1930). *J. Pathol. Bacteriol.* **33**, 339.
Topley, W. W. C., and Wilson, G. S. (1955). "Principles of Bacteriology and Immunity" (G. S. Wilson and A. A. Miles, eds.), 4th ed. Vol. 2, Chapt. 50. Williams and Wilkins, Baltimore, Maryland.
Trowell, O. A. (1952a). *Exptl. Cell Research* **3**, 79.
Trowell, O. A. (1952b). *J. Pathol. Bacteriol.* **64**, 687.

Trowell, O. A. (1953). *J. Physiol. (London)* **119**, 274.

Unna, P. G. (1891). *Monatsh. prakt. Dermatol.* **12**, 296.

Valentine, W. N., Craddock, C. G., and Lawrence, J. S. (1948). *Blood* **3**, 729.

von Behring, E., and Kitasats, S. (1890). *Deut. med. Wochschr.* **16**, 1113.

Wager, O. A., and Chase, M. W. (1952). *Federation Proc.* **11**, 485.

Waldeyer, W. (1875). *Arch. mikroskop. Anat. u. Entwicklungsmech.* **11**, 176.

Walsh, T. E., and Cannon, P. R. (1938). *J. Immunol.* **35**, 31.

Walsh, T. E., Sullivan, F. L., and Cannon, P. R. (1932). *Proc. Soc. Exptl. Biol. Med.* **29**, 675.

Waltzer, M., and Glazer, I. (1950). *Proc. Soc. Exptl. Biol. Med.* **74**, 872.

Warthin, A. J. (1906). *J. Interim. Clin. Ser.* **15**, 4, 243.

Wasserman, A. (1898). *Berlin. klin. Wochschr.* **35**, 209.

Wasserman, M. (1899). *Deut. med. Wochschr.* **25**, 141.

Welch, W. H., and Flexner, S. (1891). *Bull. Johns Hopkins Hosp.* **2**, 107.

Wenck, U. and Speirs, R. S. (1957). *Acta Haematol.* **17**, 193.

Wesslen, T. (1952a). *Acta Dermato-Venereol.* **32**, 195.

Wesslen, T. (1952b). *Acta Dermato-Venereol.* **32**, 265.

Westwater, J. O. (1940a). *J. Exptl. Med.* **71**, 455.

Westwater, J. O. (1940b). *J. Immunol.* **38**, 267.

White, A. (1947–1948). *Harvey Lectures* **43**, 43 (1950).

White, A. (1948). *Bull. N. Y. Acad. Med.* **24**, 26.

White, A., and Dougherty, T. F. (1946). *Ann. N. Y. Acad. Sci.* **46**, 859.

White, R. G. (1954). *Brit. J. Exptl. Pathol.* **35**, 365.

White, R. G. Coons, A. H., and Connolly, J. M. (1955a). *J. Exptl. Med.* **102**, 73.

White, R. G., Coons, A. H., and Connolly, J. M. (1955b). *J. Exptl. Med.* **102**, 83.

Williams, W. L., Stoner, R. D., and Hale, W. M. (1956). *Yale J. Biol. and Med.* **28**, 615.

Wissler, R. W., Robson, M. J., Fitch, F. W., Nelson, W., and Jacobson, L. O. (1953). *J. Immunol.* **70**, 379.

Wissler, R. W., Fitch, F. W., LaVia, M. F., and Gunderson, C. H. (1957). *J. Cellular Comp. Physiol.* **50**, Suppl. **1**.

Wolf, B., and Stavitsky, A. B. (1956). *Federation Proc.* **15**, 623.

Wolfe, H. R., Norton, S., Springer, E., Goodman, M., and Herrick, C. A. (1950). *J. Immunol.* **64**, 179.

Wright, C. S., and Doan, C. A. (1953). In "Blood Cells and Plasma Proteins: Their State in Nature" (J. L. Tullis, ed.), Sect. 5, Chapt. 2, p. 281. Academic Press, New York.

Wyssokowitsch, W. (1886). *Z. Hyg. Infektionskrankh.* **1**, 3.

Many relevant publications have appeared since this chapter was completed for the press. The reader can find references covering practically all this recent material in the following: BURNET, F. M. (1959). "The Clonal Selection Theory of Acquired Immunity." Vanderbilt Univ. Press, Nashville and Cambridge Univ. Press, London and New York. FISHMAN, M. (1959). *Nature* **183**, 1200. GOOD, R. A. (1957). In "Host Parasite Relationships in Living Cells" (H. M. Felton, ed.), p. 78, C. C Thomas, Springfield, Illinois. HUMPHREY, J. H. (1959). *Lectures Sci. Basis Med.* **7**, 239. STERZL, J. (1959). *Experientia* **15**, 62. STERZL, J., and HOLUB, M. (1958). *Folia Biol. (Prague)* **4**, 59. YOFFEY, J. M. (1959). *Lectures Sci. Basis Med.* **7**, 127.

Collections of papers by various authors can be found in the following symposia: *Physiopathology of the Reticulo-Endothelial System* (B. N. Halpern, ed.). Blackwell Sci. Publ. Oxford, 1957. *Mechanisms of Hypersensitivity* (J. H. Shaffer, G. A. Lo Grippo, and M. W. Chase, eds.). Little, Brown, Boston, Massachusetts, 1959. *Ciba Foundation Symposium on Cellular Aspects of Immunity* (G. E. W. Wolstenholme and M. O'Connor, eds.). Little, Brown, Boston, Massachusetts, 1959. *Mechanisms of Antibody Formation* (M. Holub and L. Jarošková, eds.). Publishing House of the Czechoslovak Academy of Sciences, Prague, 1960.

CHAPTER 7

The Morphology of the Cancer Cells

By Ch. OBERLING† and W. BERNHARD
† Died 11th March, 1960

> No single individual will ever be able to present
> adequately what is known about cancer cells.
> Cowdry

In 1837, Gluge in Paris, while examining under a microscope scrapings obtained from a mammary tumor, was struck by an abundance of globules of spherical shape, about $8\,\mu$ in diameter, which covered some three-quarters of the optical field. He reported this finding in a paper read before the Academy of Sciences, and this observation doubtless constitutes the first description of cancer cells. But not until the publications of Johannes Müller and of Virchow was the fundamental fact clearly recognized that cancer, like other tissues and anything else that lives, is composed of cells. The work done by Ribbert (1914) contributed greatly to a clarification of this concept; he demonstrated that cancer, to a certain extent, represents an autonomous race of cells, appropriate to each tissue and which, once established, grows exclusively by proliferation of its own elements. This demonstration inevitably led to the conclusion that the cancer cell is the true substratum of the neoplastic process and, consequently, the bearer of malignancy. This conception, based on purely morphological considerations, was soon to be verified by irrefutable experiments.

Carrel and Ebeling, A. Fischer, Lewis, and others have shown that cancer cells, maintained outside the organism in tissue cultures, retain their pathological properties. When inoculated into a normal organism they reproduce a malignant tumor identical with that from which they have been obtained. The most conclusive facts in this field were found by Fischer and Davidson (1939), who cultivated mouse cancer cells through a period of twelve years in a completely heterologous medium consisting of rat serum, fowl plasma, and chick embryo extract. These cells, when inoculated back into mice after 800 transplantations, produced a tumor identical to the original neoplasm.

On the other hand, it has been proved possible to transfer a neoplastic process by means of a single cell. This experiment was performed for the first time by Furth and Kahn (1937) in mouse leukemia. Ishibashi (1950) and subsequently Hosokawa (1950, 1951) obtained similar results with an ascites sarcoma (Yoshida tumor). Single-cell transmissions of various neoplasms have since been effected on numerous occasions, particularly by Hauschka (1953) and by Makino and Kano (1955).

These facts are to be kept in mind when the question is raised whether it is permissible to speak of "the" cancer cell. Since authentic cancers are formed by cells, each of which bears the essence of malignancy, it would seem to be logical to refer in general terms to "cancer cells." But by so doing and by defending, thus, to some extent the anonymity of the cancer cell, we in no way wish to imply that all cancer cells are identical. This would be an absurd thought, and it is beyond doubt that cancer cells show as much variety as do tumors. In view of the extraordinary complexity of each individual cell it can even be concluded that each cancer cell is probably different from any other. This does not, however, prevent each cancer cell from showing characteristics by which it is distinguished from a normal cell; and the investigation of which is worth while, as we hope that some day it will help to disclose the nature of the process responsible for malignant transformation.

I. The Nucleus

In the classic sense, the nucleus of a cancer cell is characterized by an increase in volume and an irregularity in shape; it contains an abundance of sharply outlined and intensively staining chromatin lumps and possesses one or several large nucleoli. The nuclear hypertrophy, which may be of enormous extent, up to the point of monstrosity, is not compensated by a proportionate increase in cytoplasm; it is striking even in so-called "dwarf" cancer cells in certain carcinomas and sarcomas. Hertwig's (1903, 1908) nucleocytoplasmic ratio is raised to a level found otherwise only in embryonic tissues. When we have added to these characteristics the anomalies of mitosis, the increase in mitotic frequency, and the irregularity of chromosomes in shape, number, and division, we have listed the chief elements of neoplastic karyology, on the basis of which hundreds and probably thousands of cancers are diagnosed every day. These diagnostic signs are generally verified by the clinical behavior of the lesion. This means that these nuclear changes are important and that they are in some way connected with the neoplastic transformation of the cell.

Each of these changes, however, when submitted to more detailed investigation, offers complex and often controversial problems which merit separate discussion if the significance and value to be attached to them is to be understood and correctly interpreted.

A. *The Intermitotic Nucleus*

1. *Shape*

The nucleus of the cancer cell is characterized by the irregularity of its contours and the presence of deep fissures—the *Kernspalten* described by Altmann (1952), which sometimes result in a markedly lobulated appearance. Many neoplastic nuclei appear to be sprouting, deformed by bulby protrusions, constricted at their base or even completely separated from the central nuclear mass. The electron microscope has confirmed these findings and clarified them on two points. The irregularities in contours are often more marked than can be visualized with the aid of the ordinary microscope (Löblich and Landschütz, 1960). It is by no means uncommon to see nuclei so strangely patterned as to appear grotesque (Fig. 1). Such invaginations may lead to pseudo inclusions in the nucleus containing cytoplasmic organelles (Wessel, 1958; Leduc and Wilson, 1959a, b). On the other hand, the frequently deep indentations of the nucleus naturally increase its surface and, therefore, the possibility of nucleo-cytoplasmic interchanges. The incisions may be of very short duration (Leone *et al.*, 1955). Their topography seems to be governed by the position of the nucleolus, and they appear therefore to be best suited to establish the most extensive contact possible between this organelle and the nuclear membrane, a problem which will be subsequently discussed.

2. *Volume*

"These peculiar formations, furnished with large nuclei and nucleolar bodies . . . which have been described as the specific polymorphous cancer cells." This phrase, cited from Virchow (1858), clearly demonstrates that the increase in nuclear and nucleolar volume, has been regarded, from the beginning of the histological era, as characteristic of cancer cells. Since the 1850's this conception has been the subject of numerous investigations. Despite the disappointment of all the authors who, following the footsteps of Lebert (1851), tried to express this nuclear hypertrophy in a more precise and concrete formula, the investigations in this field have never been discontinued; for, as Hoffman (1953) states, "there seems to be an intuitive and persistent notion that the sizes of cell parts seen in the microscope should be made to tell something useful about cancer cells." Heiberg is an example in point. In numerous papers published between 1908 and 1957, he has not ceased to emphasize the importance of quantitative changes of the nucleus and the cancer cell which, according to him, are fundamental not only as diagnostic aids but also as factors in the prognosis of various types of malignant tumors. According to

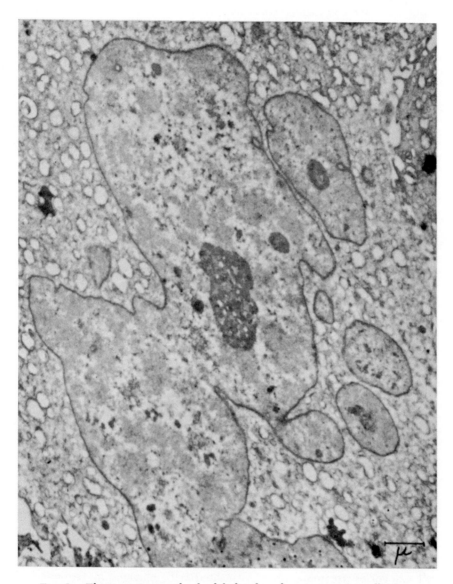

FIG. 1. Electron micrograph of a lobulated nucleus in a cancer cell (mammary tumor of the rat). Marked irregular contours of the nuclear membrane. Magnification: ×11,000.

Heiberg the increase in volume is not confined to the nucleus, but extends to other constituent parts as well such as the nucleolus, mitotic figures, and the cell in its totality. For technical reasons, however, he has confined himself chiefly to measuring nuclei, for which he found a stepwise increase of volumes progressing from the normal nucleus to nuclei in regenerating tissues, adenomata, and cancers. This nuclear hypertrophy can be due to a variety of causes, as will be seen later, but Heiberg maintains that the decisive element in the cancerous transformation is to be found in the multiplication of chromosomes, the number of which is a multiple of the original set; in other words, tetraploidy and polyploidy.

Having thus found a concrete formula for his conception, Heiberg has taken a view which has been defended by other investigators such as those of the Rostock school: Schmitz (1934), Arndt (1935), Stapel (1935), Ehrich (1935, 1936a, b). These authors' opinions have in turn been influenced by the fundamental investigations of Jacobj (1925, 1935, 1942), which are a landmark in the history of karyometry.

When investigating the nuclei of an organ, Jacobj did not content himself, as did all his predecessors, with establishing an over-all average size; but he *classified* the nuclei according to size, recording the respective frequency of each of the classes thus established. He then saw that in organs containing nuclei of varying size, such as the liver, certain sizes appeared at maximal frequency: those especially which showed a proportional interrelationship of $1:2:4:8$, in the manner of a geometrical progression. This law not only applies to the nuclei of a single organ but is applicable to the organism as a whole. Referring to hepatic nuclei as a unit, those of the nucleate erythrocytes can be described as 1/8, of the small lymphocytes as 1/4, of the small round cells of the cerebellum as 1/2, and of the spinal ganglion cells as 8, 16, 32.

According to Jacobj, the regularity of these variations—this "rhythmic growth"—can be explained only as a function of the chromosomes. This notion has been confirmed. It has been established that the volume of a nucleus is largely dependent on the *volume* and *number* of its chromosomes. It is the former factor, i.e., the *volume* of the chromosomes, which chiefly determines the physiological differences in nuclear sizes between the various organs of the same individual; the variations within the same cell categories are more often associated with changes in the *number* of chromosomes (haploidy and polyploidy). In other words, if the nuclei of the lymphocytes are small, compared to those of the liver, it is because their chromosomes are smaller, but if the nuclei of the liver show differences in size between themselves, it is because they are a mixture of diploids and tetraploids. By placing the accent exclusively on the phenomenon of polyploidy in explaining nuclear growth of the cancer cell, Heiberg and the investigators of the Rostock school have oversimplified the problem. On these lines the problem is easily solved by counting the chromosomes; and what is known about the number of these elements in cancer cells will be subsequently discussed.

The volume of the nucleus, however, does not depend only on the chromosomes. The degree of aqueous imbibition (nuclear edema, Benninghoff, 1950) certainly plays a role. On the other hand, graphs of nuclear dimensions are still a function of the rapidity with which the nuclei complete their growth by attaining their definitive volume. This time is short for the differentiated cells, and it is for this reason that the graph of nuclear volumes concerning cells with an important somatic function as a rule shows a steep peak because there are few intermediates between maximal and minimal values; whereas the curve of undifferentiated cells is more widely stretched (Wermel and Ignatjewa, 1932; Jacobj, 1935, 1942; Wilflingseder, 1947). Finally, the appraisal of nuclear volumes in itself entails an enormous source of error inherent in the techniques used, which Hoffman (1953) has submitted to a very judicious and sometimes humorous critical analysis. Many references may furthermore be found in Rather's review article (1958).

From the bulk of often contradictory data on the volume of the nucleus in cancer cells, the following conclusions can be made:

(1) The conception of a constant increase in nuclear size of cancer cells governed by the law of geometrical progression, implying polyploidy, does not stand up to the criticism which has been expressed in the papers of Schairer (1935a, b), Deuticke (1935), Heinkele (1936), Strodtbeck (1937), and Hamperl (1956) (Figs. 2 and 3). All intermediates can be found between duplication and quadruplication of nuclear sizes. This fact has been demonstrated for tumors of the bladder by Kloos and Steffen (1951), for tumors of the thyroid by Wilflingseder (1947), for the nuclei of hepatic rat tumors following administration of butter yellow by Langer (1942). On the other hand, the karyogram sometimes reveals entirely different features in different regions of the same tumor (Schairer, 1935a,b).

(2) Nuclear hypertrophy is no special characteristic of the cancer cell. There are cancers in which the nuclei are of normal size, as demonstrated by Strodtbeck (1937) for cancers of the cervix, and by Schairer (1937) for experimental tumors (tar cancers). On the other hand, nuclear hypertrophy may be observed in the absence of any neoplastic process. The monstrous nuclei of chorionic cells are familiar to all cytologists, and this example should always be borne in mind by those who wish to regard nuclear characteristics as a distinctive sign of cellular cancerization. The chorionic cells are nothing but a particular instance of a very

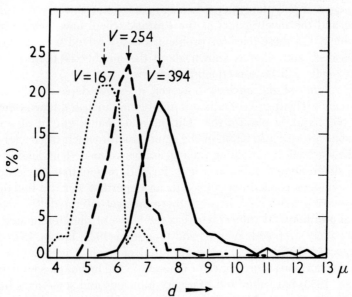

FIG. 2.   Nuclear volume in a case of cirrhosis with carcinoma of the liver (human). Curves: - - - -, bile duct epithelium; – – – –, liver cells; ———, carcinoma. From Schairer (1935).

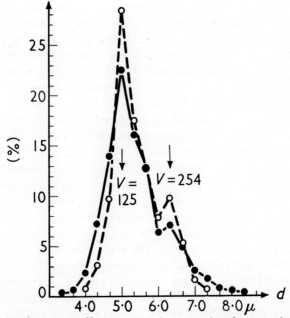

FIG. 3.   Nuclear size in fibrosis cystica (o – – – – o) and in a cirrhotic mammary carcinoma (●———●) similar curves indicate rhythmic increase of the nuclear volume. From Schairer (1935).

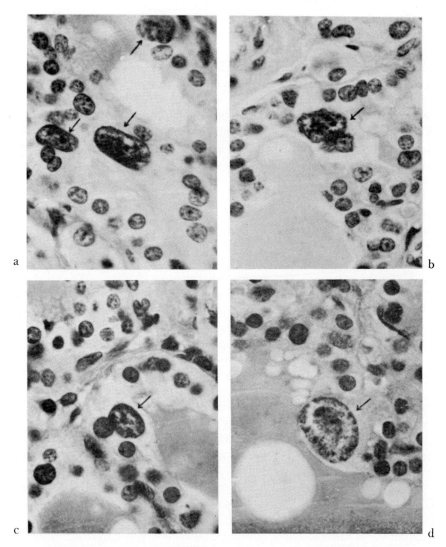

FIG. 4. Giant nuclei (→) in the thyroid of Graves' disease.

general observation illustrating *the influence of hormonal factors on the nuclear volume.* Mention can be made in this respect of the considerable nuclear hypertrophy of myometrial and endometrial cells in the course of pregnancy (Arias-Stella, 1954), and personal observations), the monstrous cells of the thyroid in Graves' disease (Askanazy, 1898) (Fig. 4), and those in parathyroid adenomas.

Generally speaking, *the nuclear volume increases in cells with a very active metabolism.* This explains why, in the normal liver, the cells with large nuclei are found particularly in the intermediate part of the lobule (Jacobj, 1925), where circulation takes place under optimal conditions (Clara, 1930). It also explains why, in cultures of fowl fibroblasts, the cells with nuclei often double the normal size are found especially in the area of growth (Bucher and Gattiker, 1950; Bucher, 1951). This phenomenon is seen in transplanted tumors, in which the nuclei of cells localized in the cortical growth zone, are often larger than the nuclei found at the centre of the tumor (Grynfeltt, 1937).

The multiplicity of phenomena involved in the mechanism of nuclear hypertrophy in cancer cells clearly demonstrates its capricious nature and the variability of its intensity according to the type of tumor involved. *In fact, every neoplasm has its own particular nuclear behavior, and it is not possible to establish a direct relation between neoplastic transformation of a cell and hypertrophy of its nucleus.*

This does not mean that nuclear hypertrophy of the cancer cell is an unimportant phenomenon, and we agree with Bucher (1953) that this question affords a vast and as yet hardly cultivated field of quantitative karyological investigation. It is probable that karyometry with a systematic and more scientific approach, using perfected techniques supported by statistical methods, may be applicable in fields in which classical histology has so far failed: e.g., in revealing tendencies of certain tumors to become malignant, such as polyps of the bladder and the rectum, or in the prognostic evaluation of cancers for which the diagnosis is already established.

In concluding this section, it should be pointed out that nuclear size has been investigated not only in cancer cells, but also in cells of other organs in cancer-bearing animals.

The investigations made by Carminati (1934), Masserini (1935), and Carcupino (1935) showed that, in tumor-bearing animals, there was nuclear hypertrophy in the adrenal medulla, the kidney, and the liver and a diminution in nuclear size in the adrenal cortex. Kasten (1955), who investigated the same subject in animals bearing spontaneous and transplanted tumors, saw no change in the nuclei of various organs in mice with spontaneous tumors. In mice bearing transplanted tumors, however, he found nuclear size increased in the adrenal cortex, the kidney, and the liver.

3. Contents

a. Heterochromatin. It has become customary to apply the term heterochromatin (Heitz, 1935) to the portions of the chromosome that retain their affinity to basic stains even during the interphase, thus representing the "chromatin lumps" of the resting nucleus described by the classic authors. It is not within the scope of this publication to study the genic significance of heterochromatin, its relations with the problem of heteropyknosis in general (White, 1935), nor its still much discussed chemical structure. With Ris (1957), it may be thought that heterochromatin is simply a region of the chromosome in which the elementary fibers have retained a very close spiralization and are very condensed, whereas in the euchromatin the filaments are more dispersed and considerably less coiled.

The heterochromatic regions of the chromosome show a tendency to intertwine, thus forming masses which are microscopically visible as chromatin lumps or chromocenters.

It is certain that the number and shape of these chromocenters vary in different cellular varieties and even in cells of the same type (Barigozzi, 1950).

Koller (1943), Cusmano (1947), Barigozzi and Casabona (1948) confirm that the nucleus of the cancer cell contains more chromocenters than that of the normal cell. This is the classic conception of the cancer cell nucleus having numerous heterochromatin masses, either dense and voluminous, or minute, giving the nucleus a dusty appearance. The systematic investigation of heterochromatin and its variations in certain tumor types or within the same tumor is still to be done. The nucleotypes established in tentative efforts by Castelain and Castelain (1957) are interesting but have as yet not been put on a sufficiently firm basis.

Since heterochromatin constitutes a part of the chromosomes, its irregularities may be at least partly explained by chromosomal anomalies. It is beyond doubt that none of the facts observed in this field can at present be considered as specific of cancer.

The investigations made by Caspersson and Santesson (1942, 1944), based on ultraviolet (UV) microspectrophotometry and essentially devoted to protein synthesis in normal and tumor cells, should be mentioned here because they raise the problems of the nuclear structure of the cancer cell.

Caspersson and Santesson distinguished, in cancer tissue, two types of cells which they refer to as A and B cells. The nucleus of the A cells is described as normal or slightly enlarged, rich in chromatin dust, with a distinct but not markedly enlarged nucleolus. The nucleus of the B cells is considerably enlarged, vesiculated, less rich in chromatin and deoxyribonucleic acid (DNA), with a big nucleolus and a thickened nuclear membrane showing irregular contours. The findings of the Scandinavian

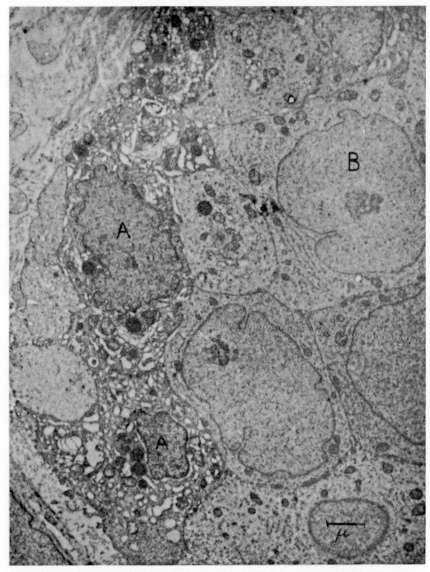

FIG. 5.   Mammary cancer of a woman. Electron micrograph showing tumor tissue with Caspersson's A and B cells. Magnification: ×10,000. Courtesy of Dr. Haguenau.

authors have been verified by numerous investigators, particularly Italians (for the literature, see the paper by Montella, 1954).

It is certain that the cellular types described by Caspersson and Santesson exist and, in this respect, electron microscopy offers observation of undeniable importance. The elements corresponding to A cells (see Fig. 5) are often seen, with their typical nuclear features and a cytoplasm rich in ergastoplasm. The B cells, generally localized in the interior of the neoplastic foci, clearly show a nucleus larger and less rich in chromatin and a very hypertrophic nucleolus. However, these may be changes associated with the blood circulation and in no way characteristic of cancer tissues. We must conclude then that the exact significance of the A and B cells has not been established so far.

Special mention must be made of the *sex chromatin* which has been the subject of a large number of investigations since the publications of Barr and Bertram (1949). This heterochromatin is seen in the form of a small, slightly planoconvex body, with a diameter of $1\,\mu$, its flat surface being in contact with the nuclear membrane. Sometimes it assumes a triangular shape, but it may also appear as a disk or annular element. It is known to be formed by the junction of the two characteristic X chromosomes of the female sex. The Y chromosome in males is generally invisible because it is attached to the nucleolus (Reitalu, 1957). This sex chromatin has been found in tumor cells (Hunter and Lennox, 1954; Moore, 1955; Cruikshank, 1955; Tavares, 1955; Moore and Barr, 1955; Rivière, 1956; Rodermund, 1956). These investigations have shown that, in benign tumors, it is generally possible to differentiate between male and female nuclei. In malignant tumors, the demonstration of sex chromatin is more difficult, as it may be confused with the often larger chromocenters. Identification offers no difficulty in certain types of tumors; e.g., squamous cell carcinomas. All in all, it can be detected in about 75% of cases (Fig. 6). The behavior of the sex chromatin is of special interest in teratomas and in tumors of the genital glands. In females, teratomas and tumors of the genital glands are invariably of the female sex. This holds true even for arrhenoblastomas: i.e., for virilizing tumors formed by Berger's (1923) sympathicotropic cells in the ovarian hilus which, on the basis of the evidence available, correspond to the Leydig cells in males. In other words, these cells which secrete the male hormone are, in females, of the female sex. This demonstrates complete independence between cellular sex and secretory activity.

In male genital glands, on the other hand, undoubtedly female tumors may be found, a logical finding since the male germinal cells possess the elements of both sexes. It is therefore not surprising to see that in males not only testicular teratomas but even a *mediastinal teratoma* of

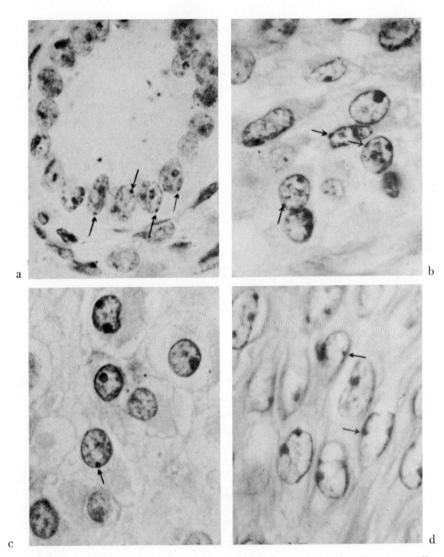

Fig. 6.   Sex chromatin (→) in interphase nuclei of various human tumor cells.
a. Dysembryoma of the ovary. b. Arrhenoblastoma   c. Berger's sympaticotropic cells
in the ovarian hilus. d. Squamous cell carcinoma.

the female sex (Sohval and Gaines, 1955) have been observed. This may be an argument in favor of a parthenogenic against a blastomeric origin of these tumors.

b. *Cytochemical note.* In view of the tremendous difficulties involved in the accurate estimation of nuclear sizes, volumes, and weights, investigations have long been made into the possibility of replacing these data by a histochemical assay of the chief nuclear constituents. Since the fundamental investigations of Caspersson (1950), Brachet (1950), Boivin *et al.* (1948), Ris and Mirsky (1949), Swift (1950), Pollister *et al.* (1951), Leuchtenberger *et al.* (1951), DNA has retained its prominence in this field.

Photometric determinations in Feulgen-stained sections of normal cells showed that there are variations from cell to cell; these variations are very limited, however, and graphs grouping cells of the same DNA value show the most common figures closely centered about one or two peaks. The behavior in cancer cells is different. The investigations of Stowell (1947), Mellors *et al.* (1952), and Leuchtenberger *et al.* (1954a, b), based on microspectrophotometry (micro absorption analysis of individual nuclei stained with Feulgen reaction), have shown that DNA values are often increased in neoplastic cells (Figs. 7, 8). The colorimetric reactions are similar to the staining reactions currently used in cytology for the diagnosis of nuclear basophilia (Caspersson and Santesson, 1942; Michaelis, 1947; Brachet, 1950). The increase in DNA, so frequently seen in the nuclei of cancer cells, is thus a cytochemical manifestation of the older conception of the hyperchromaticity of the nuclei of cancer cells. Mellors (1956) is certainly right in stating that: "A substantial amount of applied morphological cytology, perhaps more than is generally realized, utilizes in one way or another visual estimations of the cellular and, particularly, the nuclear content and distribution of nucleic acids."

The increase in DNA, meanwhile, is extremely variable. First, it differs in intensity with different tumors. The amount of aneuploid values has no relation with the degree of malignancy (Stich *et al.*, 1960). Secondly, it varies within the same tumor in correlation with changes in nuclear volume and chromosome count (Leuchtenberger, 1956). However, these variations often differ conspicuously from those in normal tissues. The karyograms grouping the nuclei according to their DNA level do not show one or two peaks as they do in normal tissues but present a more stretched-out curve owing to the great number of intermediate DNA levels between the extremes (Swift, 1950; Leuchtenberger *et al.*, 1954; Mellors, 1955). This particular type of curve is not specific for cancer but can also be observed in tissues engaged in active multiplication and particularly in embryonic tissues (Vendrely and Vendrely, 1957).

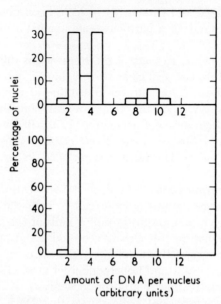

Fig. 7.   Cytophotometric measurements of DNA in nuclei from normal liver of a 68-year-old man (*above*) and a 6-year-old child (*below*). From Leuchtenberger

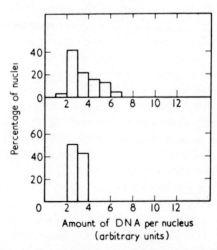

Fig. 8.   Cytophotometric measurements of DNA in nuclei from a gastric adeno-carcinoma (*above*) and adjacent normal tissue (*below*) of a 65-year-old man. From Leuchtenberger

Thirdly, this increase in nuclear DNA is not constant (Metais and Mandel, 1950; Klein and Klein, 1950; Klein, 1951; Petermann and Schneider, 1951). Fourthly, it may become apparent *before* the cancer is disclosed, as shown by the investigations of Grundmann (1955) in the liver of rats treated with butter yellow. Moreover, histopathologists have long been familiar with the exaggerated chromaticity of nuclei found in "cancer *in situ*," or in villous tumors of the rectum, lesions which *may* turn into cancers but are not cancerous by themselves, at the time where exaggerated basophilia is already highly evident.

DNA constitutes but a fraction of the nuclear contents. Together with the histones with which it is always associated and of which the concentration is always proportionate, it constitutes about 50% of the nuclear mass (Alfert and Geschwind, 1953). The remainder consists of nuclear proteins (Pollister and Ris, 1947; Rasch and Swift, 1953), ribonucleic acid (RNA), and water.

Nuclear proteins play a predominant role in the determination of nuclear size (Leuchtenberger, 1954). Quantitative appreciation of their mass as obtained by interferometry (Davies and Engfeldt, 1954) or by absorption of X-rays (Moberger, 1954) in squamous cell carcinomas has demonstrated that their increase shows geometrical progression (two to four times). Consequently nuclear proteins show some degree of proportional relationship with the number of chromosomes, in the constitution of which they are probably involved. The RNA content of neoplastic nuclei is as a rule increased (Goldberg *et al.*, 1950; Klein, 1951; Petermann and Schneider, 1951).

In summary, none of the changes in the intermitotic nucleus of cancer cells has been found to be typical of cancer. The present state of our knowledge in this respect cannot be better formulated than by Ludford's conclusion (1954): "It would be more compatible with the results of the present study if hypertrophy of the nuclear system and a hyperchromatinic condition of the nucleus were the morphologic expression of an increased growth rate rather than being indicative of malignant growth."

4. The Nucleolus

It is a classic statement that cancer cells contain a *very hypertrophic nucleolus*, among other signs, a characteristic aid in the diagnosis of cancer. Pianese (1896) was the first to draw attention to this hypertrophy. Subsequently Quensel (1928) attempted to establish a correlation between nuclear hypertrophy and nucleolar hypertrophy; in this respect he compared cancer cells and inflammatory cells. But it was McCarty in particular (1923, 1925, 1928, 1936, 1937; McCarty and Haumeder, 1934) who emphasized the increase in nucleolar size in tumors of all types and considered it to be pathognomonic of the cancer cell. He stressed the importance of the examination of fresh, nonfixed tissues, thus refuting criticisms (Guttman and Halpern, 1935) based on routine histological techniques. Other authors, however, have expressed doubt as to the specificity of this phenomenon. Heiberg (1921), although impressed by the nucleolar hypertrophy in cancers, believed it to be due to accelerated cellular metabolism rather than to the cancer process as such. Subsequently, Cowdry and Paletta (1941) and Stowell (1949) formally rejected the theory that nucleolar hypertrophy might constitute a reliable criterion of malignancy. Stowell showed that intensive synthesis in the regenerating liver cells following hepatectomy, may cause the nucleolar size to exceed that of certain malignant hepatomas. All cells involved in an intensive process of protein synthesis are equipped with a hyperactive nucleolar system.

Page *et al.* (1938) have drawn attention to the very high frequency of *nucleolar inclusions* (argentophile vacuoles and granules)—"nucleolini" —in malignant tumors. Yet these findings cannot be regarded as pathognomonic of cancer growth: intranucleolar vacuoles are known to have been often seen in normal and pathological tissues (Lewis, 1943; Yokoyama and Stowell, 1951; Peters, 1956) and particularly in nerve cells (Vogt and Vogt, 1947; Höpker, 1953; Seite, 1955). The presence of argentophile granules in the nucleolus is not surprising since Estable and Sotelo (1951, 1955) have demonstrated reticular argentophile structures in normal nucleoli of varying origin, which they called "nucleolonema."

The *number of nucleoli* in cancer cells is frequently increased as compared with normal homologous tissues (Quensel, 1928; Cowdry and Paletta, 1941). Perhaps this is related to the polyploidy frequently seen in cancers. However, there is no constant relation between malignancy and the number of nucleoli. The *position* of this organite in the nucleus is likewise indifferent. It is frequently found to be attached to the surface of the nuclear membrane, but this is also observed in normal tissues (see in particular the films made by Frédéric, 1954, 1958; Lettré and Siebs, 1954; Lettré, 1955. The *shape* is more irregular than normally (Reitalu, 1957).

Caspersson and Santesson (1942) have contributed greatly to a better understanding of the nucleolar apparatus. The heterochromatin system (nucleolus-associated chromatin) together with the nucleolus in the strictest sense (plasmasome) is believed to undergo extreme stimulation for some unknown reason, and to play a decisive role both in *carcinogenesis* and in the *growth* of cancer cells. In view of the primary importance of the nucleolus in protein synthesis, derangement of its normal activity might be followed by serious chemical changes and consequently

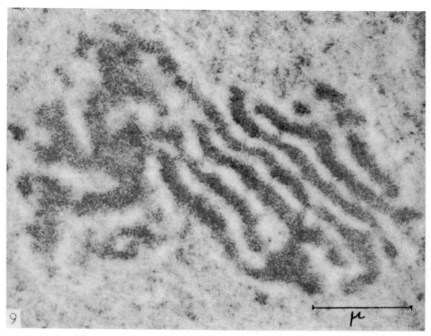

FIG. 9. Electron micrograph of a giant nucleolus in a tumor cell (human primary hepatoma). Granular structure of the 'nucleolonema' well visible. Magnification: × 27,000.

by disturbances in the biological behavior of the cancer cells. In the A cells, but especially in the B cells, nucleolar hyperactivity is found. The highly basophilic cytoplasm of the A cells responds to the stimulation of the nucleolus. In the B cells, the products elaborated by the nucleolar apparatus seem to be arrested by the nuclear membrane. The very voluminous cytoplasm contains little RNA and no longer takes up basic stains.

Although the investigations made by Caspersson (1950) and his school, and those of Brachet (1950) have had the merit of definitely demonstrating the importance of the nucleolus in the synthetic activity of the cell, it is today rather doubtful that the hyperfunction of the heterochromatin-nucleolus system is a primary phenomenon specifically related to the cancer process (see Ludford, 1954; Brachet, 1957).

Electron microscopy has made an interesting contribution to our knowledge of the normal nucleolus. Borysko and Bang (1951) and Bernhard *et al.* (1952) have reported the filamentous or reticulate appearance of nucleoli examined in ultrathin sections, structures which are probably identical with the "nucleolonema" described by Estable (Fig. 9).

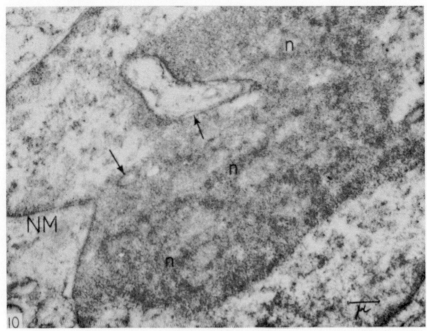

FIG. 10.   Electron micrograph of an Ehrlich ascites tumor cell. Giant nucleolus whose "nucleolonema" (*nu*) is embedded in a diffuse, granular matrix. Cross-sections of infoldings of the nuclear membrane (→) in close contact with the nucleolar substance. *NM*, nuclear membrane. Magnification: × 8,500.

In a certain number of nucleoli this organized part is surrounded by a finely granular mass without visible organization, doubtless corresponding to Estable's "pars amorpha" or to Caspersson's "nucleolus-associated chromatin" (Fig. 10). At higher magnifications, the nucleolar network and the amorphous region can be demonstrated to consist of two types of elements: on the one hand, rounded, very dense granules with a diameter of 100–200 A.; on the other, filaments of very variable length but with fairly constant thickness, averaging 80–100 A. When examined with the aid of a powerful electron microscope, these elongated elements do not appear to be distinct from those found throughout the chromatin masses of the nucleus, and the round granules seem to be morphologically very much like the RNP granules described in the cytoplasm by Palade (1955a, b). But there is no constant ultrastructural difference between "nucleolonema" and "pars amorpha." Granules and fibrillae are densely packed in the reticulate part and very scattered in the diffuse part. The proportion of granules and fibrillae may markedly change in the "nucleolonema" or in the diffuse part. This change in the ultrastructural composition may explain the cytochemical differences mentioned in the literature. The diffuse mass appears to be Feulgen positive, while the "nucleolonema" seems to be pyroninophilic. Lettré and Siebs (1954) demonstrated, however, that the latter also enclosed Feulgen-positive substances. No constant difference has been visible between the nucleolar ultrastructure of normal and cancer cells (Bernhard *et al.*, 1955a; Bolognari, 1959). Cancer cells as a rule show an increase in the size of this organelle, but the ultramicroscopic structure is not altered. As in normal cells, the nucleolus of the cancer cell consists of a filamentous or clustered element which may occur alone or in combination with a diffuse granular mass. Figs. 9 and 10 show two extreme types of nucleolus, between which all the intermediates can be found without difficulty. In a given type of tumor, either a reticulate nucleolus (Yoshida's sarcoma) or a nucleolus almost entirely composed of a diffuse substance (Ehrlich's ascites tumor, see Wessel and Bernhard, 1957) is predominant. It seems, however, that the latter appearance predominates throughout the series of tumors hitherto examined with the aid of the electron microscope.

The *mechanism of transfer* of nucleolar material through the nuclear membrane into the cytoplasm has not yet been elucidated. It seems probable that this material passes the barrier of the membrane at the level of the pores (see Section I, A, 5). Complete expulsion of the entire nucleolus into the cytoplasm, as has been repeatedly seen in normal ovogenesis (see Brachet, 1957) does not seem to occur in well-preserved cancer tissue. Here the nucleolus appears attached to the internal surface of the nuclear membrane, an observation well established by micro-cinematographic examination of normal tissues (Frédéric, 1954; Lettré, 1955). The ultrastructure of this membrane moreover shows no visible change at this point. The extrusion of whole nucleoli into the cytoplasm, in various human tumor cells, reported by Kopac and Mateyko (1958), seems rather doubtful, as the smear technique, used by these authors, may damage the nuclear membrane. Certain tumors (Ehrlich's ascites tumor, for instance) show deep invaginations or diverticular penetrations of the nuclear membrane into the interior of the nucleus where they are surrounded by nucleolar substance (Fig. 10). One can assume that transportation of nucleolar products toward the cytoplasm is highly facilitated by this arrangement (Schulz, 1957; Wessel and Bernhard, 1957). On the other hand, no ultrastructural element in the nuclei of cancer cells has made it possible to confirm the existence of intrachromosomal canalicular structures (Altmann, 1952; Homann, 1954; Lettré and Siebs, 1954), used as pathways to convey the nucleolar products toward the exterior.

It can be concluded that the nucleolus of the cancer cell shows no specific characters associated with neoplastic transformation as such, and

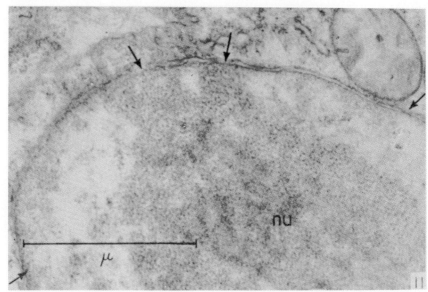

FIG. 11. Electron micrograph showing a double nuclear membrane (→) in a sarcoma cell of the Murray-Begg chicken tumor; *nu*, nucleolus. Magnification: × 47,000. Courtesy of Dr. Rouiller.

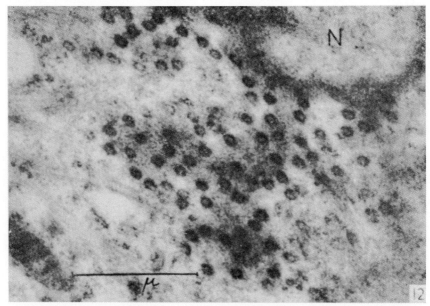

FIG. 12. Tangential section across the double nuclear membrane of a cancer cell (human primary hepatoma). Porelike structures well visible in this electron micrograph. *N*, nucleus. Magnification: × 34,000.

that all its changes and particularly its hypertrophy are merely a mani-festation of metabolic disturbances to which the cell is subject.

5. The Nuclear Membrane

The nuclear membrane of cancer cells is often described as thickened and irregularly folded; Caspersson and Santesson (1944), notably in describing their B cells, have emphasized the thickening of the membrane. Electron microscopy has not confirmed this. It may therefore be believed that the exaggerated outline of the membrane so common in ordinary microscopy is due to precipitation of the ergastoplasmic structures on the surface of the membrane. In ultrathin sections the membrane of a neoplastic nucleus appears to have two sheets, and thus resembles that of a normal nucleus (Figs. 11, 12). The only difference is that the highly lobulated shape of the neoplastic nucleus frequently realizes tangential orientations of the sections with respect to the nuclear surfaces, thus affording excellent opportunities for the study of its ultra-structural behavior (Haguenau *et al.*, 1955; Haguenau and Bernhard, 1955b). These studies have confirmed all the details described by Watson (1954, 1955), in particular, the existence of the pores (Figs. 12, 13), recognized in *Triton* and *Xenopus* (Callan and Tomlin, 1950; Gall, 1954), in *Amoeba proteus* (Bairati and Lehmann, 1952; Harris and James, 1952, in *Chironomus* (Bahr and Beermann, 1954), and in sea urchin and starfish oöcytes (Afzelius, 1955).

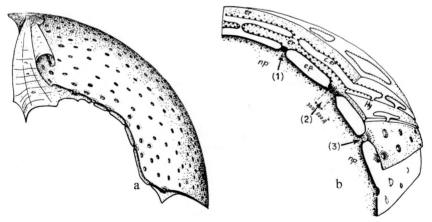

FIG. 13. Schematic reconstruction of the nuclear membrane as shown in electron micrographs of various normal and cancer tissues. a. Small segment of the two sheets with numerous "pores." b. Higher magnification showing relationship of the nuclear membrane with nucleoplasm (*np*), perinuclear space (*e.p.*), ergastoplasm (*er*), and hyaloplasm (*Hy*); (1), (2), (3), porelike structures closed or apparently open.

These formations evidently establish direct communication between the nuclear contents and the cytoplasm (Fig. 13); they are not simple orifices, however, and their very complex structure suggests regulatory systems and modalities of function of which the details are still completely unknown. In the Yoshida sarcoma, mitochondria were found to be in close connection with the nuclear membrane (Yasuzumi, 1959).

B. The Mitotic Nucleus

1. The Chromosomes

Hansemann (1890) was the first to draw attention to the abnormal behavior of chromosomes in cancer cells which, he thought, might be responsible for the neoplastic transformation. Since then, the interest in these structures has been greatly stimulated by the theory of carcinogenesis developed by Boveri (1914) and by the mutation theories (Whitmann, 1919; Bauer, 1928; Strong, 1929; Ludford, 1930; Lockhart-Mummery, 1934; Sutton, 1938; Fardon, 1953; see Burdette, 1955, for important bibliography).

a. Number. There are, at present, precise estimations of the number of chromosomes in malignant tumors. Without pretending to be complete, the following investigations can be listed: human tumors (Heiberg and Kemp, 1929; Ortiz Picon, 1930; Andres, 1932; Schairer, 1935a, b; Deckner, 1939; Regen, 1951; Hsu, 1954a, b; Levan, 1956a, b; Hansen-Melander *et al.*, 1956; Koller, 1947a, b, 1956; Fritz-Niggli, 1955, 1956; Ising and Levan, 1957; these papers have been published before the exact number of human chromosomes (46) was known (Tjio and Levan, 1956a; Ford *et al.*, 1958); mouse tumors (Winge, 1930; Levine, 1931; Bayreuther, 1952; Hauschka and Levan, 1953; Levan and Hauschka, 1953a; Schairer, 1955; Hellström, 1959; Hsu and Klatt, 1958); rabbit tumors (Biesele, 1945; Palmer, 1959); rat tumors (Lewis and Lockwood, 1929; Yoshida, 1952a, b; Makino, 1952a, b); a hamster sarcoma (Yerganian, 1955); fowl tumors (Levine, 1931); tumors *in vitro* (Hirschfeld and Klee-Rawidowicz, 1929; Goldschmidt and Fischer, 1929; Hsu, 1954a, b; Levan, 1956a, b; Hsu and Moorhead, 1957; Chu and Giles, 1958); plant tumors (Winge, 1927).

These investigations have shown that there are tumors with a practically normal diploid chromosomal number although they may show considerable clinical malignancy, Timonen and Therman, 1950). Bayreuther and Theorell (1959) found a normal diploid macrochromosome constitution in the virus-induced erythroleukemia of chickens, and Nowell *et al.* (1958) observed normal diploid values in acute human myeloblastic leukemias. Observations have even been made on tumors with a reduced number of chromosomes, approaching the haploid figure (Winge, 1930; Ortiz Picon,

1930; Koller, 1947a, b; Anghelesco, 1948; Makino, 1952). These observations, however, are rare. *As a rule, the number of chromosomes in cancer cells, and in normal cells as well, deviates from the diploid figures characteristic of the species; it is aneuploid.* These deviations, however, are invariably more marked and more frequent than those in corresponding normal cell populations. In normal tissues the karyographic chart, which indicates the frequency of cells according to the number of their chromosomes, shows sharp peaks corresponding to the diploid or tetraploid chromosome numbers. In tumor tissues, the curve is more irregular owing to increased aneuploid values interpolated between the maximal diploid or tetraploid figures. A similar flattening of the maxima is seen not only in cancer tissues, but also in the case of precancerous lesions: e.g., liver when treated with butter yellow (Marquardt and Gläss, 1957a; Gläss, 1960), skin when painted with carcinogenic hydrocarbons (Biesele and Cowdry, 1944), or tissues cultured *in vitro* (Moore, 1956; Levan, 1956a, b). *In cancer cells the number of chromosomes is most frequently increased:* hyperdiploid, hypotetraploid, hypertetraploid, or polyploid values are found. Thus the chromosomal values of cancer cells can be extremely irregular, the range being from 4 to 5000 (Timonen and Therman, 1950; Levan, 1956); yet even in this apparent chaos there is a certain order.

Gläss (1956, 1957a, b) and Marquardt and Gläss (1957b) have demonstrated that certain chromosomal groups may segregate themselves from others (Genomsonderung), and this phenomenon may even manifest itself in the morphological features of the equatorial plate by a separate grouping of the chromosomes involved. As a result of their elimination, aneuploid figures or their multiple values appear in the course of subsequent mitoses.

The reverse may occur when some chromosomes multiply on their own account by endoreduplication and then appear four- or eightfold. It may then occur that a given chromosome set is hypodiploid for some chromosomes and hypertetraploid for others (Levan, 1956a, b).

Makino (1952, 1956, 1957) and Makino and Kano (1953), examining ascites tumors in rats, made morphological and statistical analyses which showed that the apparent incoherence of the chromosomal number can be explained on the basis of coexistence of several cellular lines (*stem lines*), each with its own particular chromosomal equipment or chromosome pattern, characterized by number, shape, orientation at the time of the metaphase, V-chromosome number, etc. The value of this concept of stem lines or "fundamental lines" has been experimentally demonstrated by the technique of monocellular ascites tumor grafts in rats as well as in mice by Hauschka (1953), Hauschka and Levan (1953), and Makino and Kano (1955) (Fig. 14). By this procedure it is possible to obtain cellular clones

Fig. 14. Metaphasic chromosomes of stem-line cells of single-cell clones from the Hirosaki sarcoma, an ascites tumor in rats. Each clone is characterized by a distinguishable stem-line chromosome pattern. Presence of V-shaped chromosomes: 4–6, 3V-type cell clone; 7–9, 2V-type; 10–12, 3V-type; 13–15, 2V-type. Courtesy of Dr. Makino.

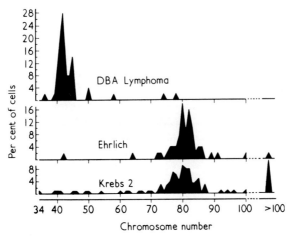

FIG. 15. Graph showing the distribution of chromosome numbers in three ascites tumors. From Levan and Hauschka (1953).

with the distinct properties of stem lines, which are maintained throughout numerous generations and persist even after heterotransplantation or chemotherapy. Lettré (1953) believes that, apart from the process of isolating stem lines by unicellular transfer, it is possible to select them by application of certain cytotoxic substances such as colchicine. These substances act as chemical filters and produce resistant lines with the behavior of stem lines.

It appears then that each tumor is a population with a very variable number of chromosomes, but including one or several lines with a relatively stable chromosome pattern. Deviations may occur at any time as a result of accidents in regrouping of pluripolar mitoses, but these are merely ephemeral repercussions, without lasting consequences for the chief line or lines which maintain the continuity of the neoplastic process (Fig. 15).

The stem line concept was subsequently corroborated by the chromosomal analyses of Tonomura (1953), Tjio and Levan (1954), Sachs and Gallily (1955, 1956), and by Richards et al. (1956).

The results obtained with ascites tumors are completely in agreement with the chromosomal analyses of other tumors, particularly human neoplasms. Fritz–Niggli (1956), investigating mammary tumors, demonstrated that every tumor, independently of its histological structure, shows a special chromosomal pattern with reference to number, shape, and certain peculiarities of division, setting off characteristic cellular lines.

In order to understand the significance of the facts given above, it should be borne in mind that these irregularities in the number of chromosomes are not restricted to cancer. Makino and Tanaka (1953) and Gläss (1957a, b) have observed that heteroploidy is marked in regenerating tissues, particularly in the liver. Hsu and Moorhead (1957) and Hsu and associates (1957) observed heteroploid transformation of human cells cultured *in vitro;* they believe this to be a phenomenon of adaptation which may well be connected with the "spontaneous" neoplastic transformation sometimes seen in certain strains of cells kept *in vitro* for a considerable time. The dwarf mice studied by Leuchtenberger *et al.* (1954a), show purely diploid hepatic nuclei. Administration of pituitary hormone causes these mice to grow, and at the same time tetraploid and octoploid nuclei appear. Thus polyploidy accompanies increased metabolism and growth, a fact long established in plants. Phytophysiologists well know that tetraploidy enhances resistance toward unfavorable environmental conditions. The frequency of polyploids is highest in extreme climates (e.g., the arctic flora).

This leads to the most interesting and unexpected development in this field: the relationships between heteroploidy and adaptation of cancer cells to the environment.

Genetic instability is one of the outstanding characteristics of cancer cells, and chromosomal changes occur constantly during the evolution of a tumor strain, even if it is clonal from the onset as are probably most spontaneously recurring tumors (Levan, 1956a, b). A tumor of diploid predominance, for instance, containing 5% of polyploids, may show a shift of this formula to polyploidy with only 5% of diploids (Hauschka, 1953). This reversal is accompanied by a loss of the tumor's selective affinity to a determined strain of animals.

These changes may appear spontaneously, without any apparent cause, sometimes on the occasion of a change in strain or environment. This explains why Ehrlich's ascites tumor may be hyperdiploid in some German laboratories, tetraploid in others—in Scandinavian countries, for instance, or in the United States (Lettré, 1953; Ising, 1957).

Thus the passage of an ascites tumor to a less favorable environment (e.g., to strains of refractory animals) (Sachs and Gallily, 1955; Kaziwara, 1954), or to animals of another species (Yoshida, 1952a, b; Ising, 1957) often changes the number of chromosomes in the direction of polyploidy or heteroploidy. Hauschka and Levan (1953), observing that such deviations in chromosome number involve increased histocompatibility, have forwarded the hypothesis of *immunoselection.* Hauschka *et al.* (1956) found a particularly pertinent example in favor of this hypothesis in the behavior of the ascitic lymphosarcoma 6C3HED, which is normally diploid. When grafted to strain DBA 2, the tetraploid form becomes predominant, although the growth of this form is not accelerated and its invasiveness is even diminished. The only selective advantage lies in its greater resistance to its new host. Retransmission to C3H mice is followed by reappearance of the diploid form. In order to explain this correlation between heteroploidy and increased resistance, Hauschka *et al.* (1956), Klein and Klein (1958), incriminated the loss of specific antigens (antigens of histocompatibility) situated at the H2 locus.

This hypothesis has been confirmed by the investigations of Amos (1956). Certain diploid forms of ascites lymphosarcoma in fact absorb more H2 antibodies and provoke the formation of these antibodies in higher concentrations than do polyploïd forms. In summary, polyploidy and heteroploidy seem to be related to antigenic dedifferentiation and, therefore, with increased resistance of cancer cells to adverse conditions of the environment.

We may conclude, then, that many of the chromosomal changes occurring constantly in cancer cells are the expression of a genetic adjustment to modifications of the environment and that they may have positive selective value in forming new stem lines with enhanced growth capacity and invasiveness (Hauschka and Levan, 1958; Hauschka, 1958). *But the abnormal chromosome pattern is probably of secondary importance for the acquirement of malignant properties* (Bayreuther, 1960).

b. Form. The form of the chromosomes, in turn, is extremely variable in cancer cells, showing an appearance which is often characteristic of the stem line of the tumor. The chromosomes may be very short, squat, and contracted (contraction chromosomes). These minute, "dwarf" chromosomes are found in many stem lines, and they are therefore likely to be the vehicle for important genetic material.

Other types of chromosomes are elongated, filamentous, thin, and very irregular in their response to Feulgen staining (Polli, 1951; Biesele, 1945) or, on the contrary, thick and of a giant type (Ludford, 1954). Generally speaking, the range between extreme chromosomal sizes is wider in cancer cells than in normal cells.

The viscous consistency of the cancer chromosomes, as demonstrated with the aid of a microdissection needle, is manifested morphologically by adherence of chromosomes or failure of chromosomes to separate at the time of anaphase, when they form chromosomal bridges.

Dicentric and annular chromosomes are also observed (Tjio and Levan, 1954). These highly anomalous forms, in spite of the considerable difficulties which they offer at the time of mitosis, are often perpetuated in the stem lines, particularly in tetraploid or polyploid cells (Ising and Levan, 1957).

Tjio and Levan (1956) in complete idiogram analyses of Yoshida rat sarcomas and normal rat cells have insisted on the great number of structural arrangements which have taken place, so that the morphological identity within the originally homologous chromosome pairs may be totally blurred during the development of the tumor.

Special mention should be made of abnormal chromosomes encountered in rodents in which, normally, chromosomes are telocentric. Makino (1952a, b) and Makino and Kano (1953) were the first to describe in rat ascites sarcomas (Yoshida's tumor and two chemically induced sarcomas, MTK 1 and 2), chromosomes with a central or metacentral kinetochore, in a V-shaped or J-shaped pattern (Fig. 14). Similar chromosomes have been described by Bayreuther (1952) in Ehrlich's carcinoma in mice. These chromosomes were first regarded as specific for certain varieties of cancer cells. But Makino and Hsu (1954) at last discovered similar but smaller chromosomes in the tissues of normal rats.

On the other hand, Sachs and Gallily (1955), Schairer (1955), and Schümmelfeder and Wessel (1956) have pointed out that the V forms are not constant and that in some chemically induced and in various ascites tumors they may even be completely lacking. This phenomenon, therefore, has to be considered as a secondary chromosomal mutation without causal correlation with cancer.

Bayreuther (1952) described in Ehrlich's sarcoma A-chromosomes provided with a square achromic zone, without connection with the fusorial insertion or with the zone of nucleolus-associated chromatin.

The B-chromosomes described by the same author show a median centromere and, thus, correspond to a variety of V forms. These facts have been confirmed by Querner (1955).

Mention should also be made of chromosomal fragmentations, elongation with spherical swellings, occurrence of club- or dumbbell-shaped elements, tetrades, and important structural remodelings inside the chromosomes, characterized by the appearance of negative heteropyknotic zones.

As far as the ultrastructure of cancer chromosomes is concerned, no special morphological feature could hitherto be revealed with the electron microscope (Fig. 16).

To conclude this section on chromosomal anomalies, so important as an aid in understanding carcinogenesis, particularly in connection with the theory of mutation, it can be stated that *modifications in number,* in the sense of polyploidy or heteroploidy (*genome mutations*), or in *form* (*chromosomal mutations*) are frequently observed in cancer cells. None of these, however, is in actual fact specific for the cancer process.

The only change which, according to current belief, should be taken into consideration as responsible for cancer, is a "fatal" combination of

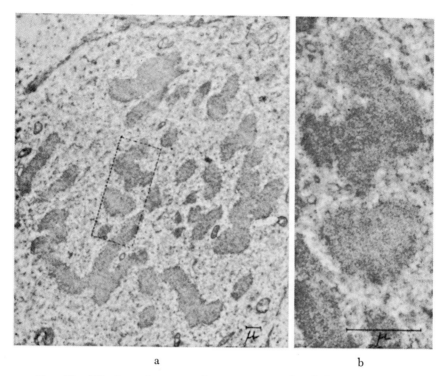

a   b

FIG. 16. Mitosis as shown on electron micrographs of thin sections of tumor cells (Ehrlich ascites). a. Longitudinal and cross sections of chromosomes. b. Enlarged area of (a), showing granular and apparently unorganized structure of the chromosomes. Magnifications: ×4000 and 20,000.

genes in the sense introduced by Boveri, i.e., a specific *gene mutation.* The existence of gene mutations has become highly probable in view of the constancy of chromosomal changes in the cancer cell. This "chromosomal disease" manifests itself by marked modifications affecting consistency, chromaticity, and texture and which may often result in a loss of genes, as has been pointed out in some typical cases of immunoselection. No investigator, however, has hitherto been able to demonstrate (1) *that the changes observed correspond with conditions such as are required by the theory of mutation;* (2) *that they are a cause rather than a consequence of malignant transformation.*

2. Mitosis

The anomalies of mitosis to be found in cancer cells are of extraordinary variety. They are due to different mechanisms and affect both the chromosomes and the achromatic apparatus. In many cases mitosis is abortive, although the mechanism involved cannot as yet be explained.

a. Anomalous Behavior of Chromosomes. Sometimes the chromosomes of the equatorial plate spread away from the center, thus forming a pattern known as *hollow metaphase.*

At the time of anaphase certain chromosomes fail to follow the movement of others toward the poles but remain behind (laggards). They usually end up by rejoining the other chromosomes, but some are ultimately excluded (*aberrant chromosomes*). If an entire group of chromosomes is involved in this anomaly, it may produce one or several small accessory nuclei (micronuclei). Very often, the chromosomes, owing to their stickiness, do not completely separate, and, at the time of anaphase, the two adherent chromosomes, with their extremities drawn to opposed poles, form *chromosomal bridges.*

Another anomaly connected with the abnormal consistency of chromosomes has been described by Tjio and Levan (1954) in certain mouse ascites tumors; it is referred to as *D-mitosis* (deviating type of mitosis). This type of mitosis may be artificially produced by means of quinoline. These are mitoses with large spindles and flat poles, which occur as a rule in voluminous cells. The equatorial plate is spread out and the anaphasic groups move toward their respective poles in a widely deployed front. At this moment, an anomalous cohesion of chromosome pairs delays their separation. The terminal parts of the chromosomes, dragged by the fusorial filament, thin out and lose part of their chromatin. Finally the separation of chromosomes takes place, but at different times (heterochronism).

b. Anomalous behavior of the spindle fibers. This anomaly may manifest itself in the absence of spindle fibers as seen in *colchicine mitoses* (C-mitoses). In this condition the arrangement of chromosomes at the equatorial plate is absolutely irregular. The chromosomes divide but do not move toward the opposite poles. The nucleus is reconstituted after a certain time and then contains a double quantity of chromosomes. This is one of the mechanisms often involved in polyploidy of the neoplastic cell (Ludford, 1930).

The restoration of nuclei does not always occur; in certain transplantable tumors with very rapid proliferation, Ludford (1930) saw mitoses without spindles terminate by dispersion of doubled chromosomes along the cellular membrane, followed by degeneration of the cell (abortive mitosis).

The lack of spindle may be partial and thus involve very unequal division of chromosomes among the daughter cells.

Overproduction of spindles can be explained by heterochronism in various phases of mitosis, particularly by prolongation of the metaphase, which permits additional multiplication of centrosomes (Therman and Timonen, 1950). Multipolar mitoses thus occur, which are very common in neoplastic cells (Makino and Yoshida, 1951; Nakahara, 1953).

This point of view has been experimentally confirmed. It has been possible to produce tripolar mitoses in normal tissues of the cornea in the triton (Peters, 1946) or in fibroblasts cultured *in vitro,* and to increase the number of multipolar mitoses in ascites tumors (Lettré and Lettré, 1954) by means of colchicine, its derivatives, or other products such as di-β-trimethoxybenzoic acid.

The *three-group metaphase* presents very characteristic features (Parmentier and Dustin, 1948, 1951, 1953) (Fig. 17). From the beginning of the metaphase there are, apart from the group of chromosomes which form the equatorial plate, two groups of juxtapolar chromosomes. To explain this anomaly, one may assume an accelerated migration of certain chromosomes toward the poles, a hypothesis forwarded by Moricard (1951). Parmentier (1953) and Dustin and Parmentier (1953) have accepted the opposite theory according to which the juxtapolar chromosomes are whole chromosomes, often longitudinally fissured, which from the beginning have lingered in the centrosomal region instead of migrating to the equatorial zone of the cell. The anomaly involves, in short, a premetaphasic rather than an anaphasic disturbance. The Belgian authors found verification of their interpretation in the experimental observations made earlier by von Möllendorff (1940), who found *chromosomal dispersion* in mitoses of fibroblasts under the influence of sex hormones. A metaphase in three groups, as experimentally produced by Parmentier and Dustin (1948) with quinoline, is especially frequent in cancers of the cervix uteri (Hamperl *et al.,* 1954; Scarpelli and von Haam, 1957), i.e., in cells that have certainly been under the influence of estrogens.

c. Uncompleted mitoses. The usual mitosis shows a series of stages which may be summarily characterized as follows: duplication of chromosomes, separation of chromosomes, division of the nucleus, division of the cytoplasm. The process may be interrupted at any of these stages, and accordingly we may describe the following abnormalities:

(1) *Lack of cytodieresis.* This is frequently a sequel of cytoplasmic disturbance and leads to the formation of multinucleated giant cells (Wilson and Leduc, 1948), a characteristic finding in cancer tissues.

(2) *Lack of karyodieresis.* The *endomitosis* defined by Geitler (1939, 1941) and described by Biesele *et al.* (1942) and by Levan and Hauschka (1953) involves a division of chromosomes within the intact nuclear membrane. Its various stages are completed normally, and distinction can be made between an endoprophase, an endometaphase, and an endotelophase. No equatorial plate is formed. The chromosomes are scattered

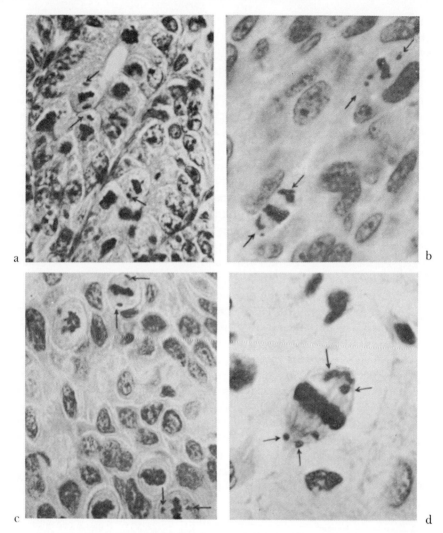

Fig. 17. Three-group metaphases in various tissues. a. Mouse intestinal tract, 1 hour after hydrochinone treatment. b. Bone marrow of the rat, 2 hours after hydrochinone treatment. c. Epithelioma *in situ* of a human cervix. d. Human glioblastoma. Courtesy of Dr. Dustin.

throughout the nuclear area. The achromatic figure is absent. The nucleolus disappears and reappears at the end of the process. Endomitosis finally produces a nucleus with a double number of chromosomes. Its frequency is very variable according to the tumor type; it may amount to 50 % of the over-all number of mitoses (Homann, 1955).

(3) *Lack of chromosome separation. Endoreduplication* (Levan and

Hauschka, 1953) involves duplication without separation of the chromosomes. As it manifests itself simply by a nuclear hypertrophy, it is not recognized at the time of its occurrence and its result does not become apparent until the next complete mitosis. Each duplicated chromosome gives rise then to quadruple chromosomes which appear at the equatorial plate. In some insect cells, endoreduplication may occur many times in succession, so that a great number of regular mitoses is needed to separate these multifold split chromosomes.

Hsu and Moorhead (1957) suggested a slightly different nomenclature based on the time at which polyploidization takes place. Since in endoreduplication this phenomenon takes place apparently during the interphase, they referred to it as *interreduplication*. Endomitosis, regarded as a separation of chromosomes taking place in the prophase, the nuclear membrane still being present, thus becomes the *proreduplication*. The C-mitosis, which is a metaphase without spindle, becomes *metareduplication*. Hsu and Moorhead described as *anareduplication* the anomalous behavior of chromosomes which, after having joined their respective poles, return to the equatorial zone and finish up by integrating to form a single nucleus. *Telereduplication* is a fusion of two cells already separated. In this case, however, the two nuclei probably always remain separate, the result being the equivalent of abortive cytodieresis.

Before ending this section it should be emphasized that *none of the mitotic irregularities described so far is specific for cancer cells.* All of them may be occasionally seen in normal tissues or can be provoked by cold, distilled water (von Möllendorff, 1938), X-rays, sex hormones, colchicine, urethane, mustard oil, quinoline, and other cellular toxins (Biesele, 1958).

Ludford (1953) lists some fifty substances and influences of various types which are capable of inhibiting spindle formation, prolonging metaphase producing anomalous separation of chromosomes, stickiness, chromosome bridges, etc. This paper makes very useful reading for all those who think that mitotic anomalies are a specific property of cancer cells.

Other changes in the mitosis of cancer cells have been described but have hardly been further investigated. Warren (1933), returning to the investigations of Ellermann (1923), measured the angle of mitotic spindles and confirmed marked differences between sarcomas and carcinomas.

Fabre-Domergue (1898) considered *disorientation of the axis of mitosis* to be a very striking feature in carcinomas of the surface linings. Catsaras (1911) showed, however, that this anomaly is shared by benign tumors and is consequently in no way specific for cancer, an opinion largely confirmed by our personal observations.

d. Chronological behavior of cancer mitoses. Since the first publications of Lambert (1913, 1914), several authors have studied the duration of mitosis and its different phases in normal and cancer cells. The investigations were made either by direct observation of mitosis in cells cultured *in vitro* or by proportional evaluation of the different mitotic phases in fixed and stained preparations. The main difficulties met with in this field of research are due to the fact that nearly every author has his own definition of the beginning and the end of mitosis and the exact limits of its different phases. A study of fixed tissues of course permits only an appreciation of the relative duration of the various phases. According to Moorhead and Hsu (1956), moreover, the first half of the prophase is not visible by this method; it has the advantage, on the other hand, of showing the tissues as they were in the organism.

An investigation of living cells cultured *in vitro* entails considerable sources of error in as much as variations in temperature and in the composition of the milieu may decisively influence the duration of mitosis (Makino and Nakahara, 1953).

As early an investigator as Lambert (1913) saw that mitoses in rat sarcoma cells are of much longer duration than those in normal corresponding fibroblasts. Lewis (1951) made the same observation. Moorhead and Hsu (1956), investigating HeLa cells, found a total mitosis time of 192 minutes, with variations considerably more marked than in normal tissues, ranging from 80 to 230 minutes. In comparative terms, the duration of a mitosis in the human epidermis has been indicated by Thuringer (1939) as being 15–30 minutes. The duration of mitoses in general is prolonged in anomalous divisions with aberrant chromosomes in multipolar mitoses.

Timonen and Therman (1950), comparing the mitotic figures of the uterine epithelium with those of endometrial cancer, noticed that in cancer tissues prophase figures appear to be less numerous than metaphases, a fact previously observed by Biesele (1945) in rabbit tumors. This conclusion has been corroborated by Schairer (1955) and particularly by Scarpelli and von Haam (1957). According to the latter, the correlation between metaphase and prophase as expressed in the prophase index according to Timonen and Therman (1950) is 1 in normal cells; it progressively increases in the case of inflammatory lesions, precancerous conditions, and carcinoma, to a point where figures of 2.56 and higher can be regarded as characteristic of malignancy, with a margin of error no wider than 0.0005 (Fig. 18). David (1958), however, could not confirm these observations.

e. Frequency of mitoses. The frequency of mitoses is always a striking feature in neoplastic tissues. In order to appreciate its true value, it should

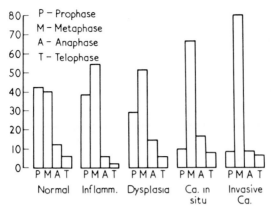

FIG. 18. The average frequency of mitotic stages in normal animals and in four types of cervical lesions during experimental inflammation and carcinogenesis. From Scarpelli and von Haam (1957).

be borne in mind that no spontaneous tumor grows at the rate of a fetus and that many tissues in regeneration, e.g., uterine mucosa after menstruation, may present a mitotic index as high as that of malignant tumors. In spite of this, histologists frequently are basing their evaluation of the malignancy of a tumor, at least partly, on the frequency of mitotic figures (Broders, 1926). The *karyokinetic index* suggested by De Nabias and Forestier (1923) was meant by its authors to furnish data important to the radiotherapist. Regaud (1923) and Roussy *et al.* (1923) have discussed the criticisms aroused by this opinion: the mitoses can be very unequally distributed within a tumor and the development of a tumor is not always dependent on the frequency of its mitoses. Investigations into the mitotic index have nevertheless been continued, and those interested will find very well documented indications in Hoffman's book (1953). Casey (1937) believes that the mitotic index, applied on the basis of statistical techniques, may furnish interesting prognostic data on certain tumors such as lymphosarcomas. Furthermore, the frequency of mitoses provides a valuable indication in the study of certain mechanisms of carcinogenesis, as shown in papers by Cooper and Reller (1942) and Reller and Cooper (1944). On the whole, it can be agreed with Bauer (1949) that *the frequency of mitoses is a symptom but not a specific sign of cancer.*

C. Amitosis

For a detailed description of what was classically called amitosis we refer to the article of Wassermann (1929), who gives a survey of the chief errors of interpretation concerning this mode of cellular division, which according to the majority of modern cytologists is insignificant as compared with that of mitosis.

All those familiar with the morphological irregularity of the neoplastic nucleus, the frequency of deep invaginations of the nuclear membrane, the folds which are often in contact with the nucleolus, the lobulations, the formations of accessory nuclei as a result of chromosomal aberrations or nuclear buddings (Howard and Schultz, 1911; Ghon and Roman, 1916), will understand that the margin of error in the chapter on amitosis is at its widest with regard to cancer cells. In our opinion the greatest restraint should be observed in an evaluation of amitosis as a mode of multiplication of cancer cells.

We believe that the pictures produced by static histology are not convincing in this matter and that only cinematography will be able to provide definite proof. It is especially valuable in that it teaches us how easily postmitotic figures (late telophase) may simulate amitoses (Wilson and Leduc, 1948; Levan and Hauschka, 1953b; and Nagata, 1957). It is not astonishing then to see that those precisely who are most familiar with the observation of living cells generally deny formally the existence of amitosis. Perhaps the frequent association of amitosis and endomitosis mentioned by Homann (1955) is not merely a coincidence, and amitosis could be merely a karyodieresis preceded by endomitotic polyploidization.

One may agree with Ris (1953), who contends that the ill-defined term amitosis should be definitely replaced by the term nuclear fragmentation.

In summary, we believe that amitosis in the classic sense, if it exists, is very rare and does not play the important role which certain authors such as Grynfeltt (1932, 1934, 1935) have ascribed to it in the multiplication of neoplastic cells.

II. The Cytoplasm

Until recently the interest aroused by investigations on the cytoplasm has been very moderate as compared with the paramount importance attached, since Boveri (1914), to investigations of the nucleus. This is due to considerations based on the chromosomal or mutational theory of cancer and to the fact that classic histological techniques, particularly methods of fixation and staining, greatly facilitated observations on the nucleus and the chromosomes but were less adequate for studying the cytoplasm. The cytoplasmic organelles are too small in size, and their form is too readily variable following changes in the environment, to allow as many incontestable data as have been collected on the nuclear structures. A clear example of this is the controversy which has lasted half a century with regard to the existence of the Golgi apparatus. This situation has changed drastically only since the introduction of phase contrast microscopy and, more particularly, of electron microscopy. Since then there has been a preference for cytoplasmic investigations, as demonstrated by the observations made by phase contrast microcinematography and by the impressive quantity of electron microscopic cytological findings collected in the course of the past four years.

The classic descriptions concerning the cytoplasm of cancer cells essentially apply to two apparently contradictory phenomena, viz: *dedifferentiation* or *anaplasia* (Hansemann, 1920), which characterizes the majority of malignant tumors, and the increased cytoplasmic *structural diversity* in other cases of cancer.

Anaplasia is both a histological and a cytological conception. The term denotes variable degrees of diminution of the structural organization of the tissue in general, and of the cell in particular; thus, the cancerous tissue approaches the *embryonic state* of the original tissue. It is subsequently to be seen that this dedifferentiation is not confined to the level of the structures visible by ordinary microscopy but can be observed throughout the ultrastructural organization of the tumor cell (Oberling and Bernhard, 1955), sometimes even down to the level of the protein molecules, which then lose part of their specific antigenic power (Weiler, 1952, 1956). This phenomenon of anaplasia tends to lend a more *uniform* aspect to the cytoplasm of all cancer cells. Whatever their origin— epithelial or mesenchymal—they are usually remarkable for their *paucity of organized structure*, even to the point of a complete disappearance of cytoplasmic organelles (Albertini, 1951) (Fig. 19).

It should be immediately emphasized, however, that this merely constitutes a *rule, which implies exceptions*. It is consequently necessary to mention in each chapter the *diversity* of cytoplasmic elements which may be encountered in specific cases. Ordinary microscopy has for a long time permitted the observation of certain cancer cells which show more advanced differentiation than that seen in the homologous normal tissue (*progressive metaplasia: prosoplasia*); it could hardly have been expected, however, that electron microscopy would reveal such *ultrastructural polymorphism* as it has already shown. In the following pages, a detailed description will be given of each organelle or particular cytoplasmic structure observed in a wide variety of tumors. A subsequent section will deal with the cytoplasmic changes found in virus tumors. The reader who requires supplementary references to older works dealing with this subject should consult the important reviews of Ludford (1952), Hamperl (1956), and Cowdry's valuable book (1955).

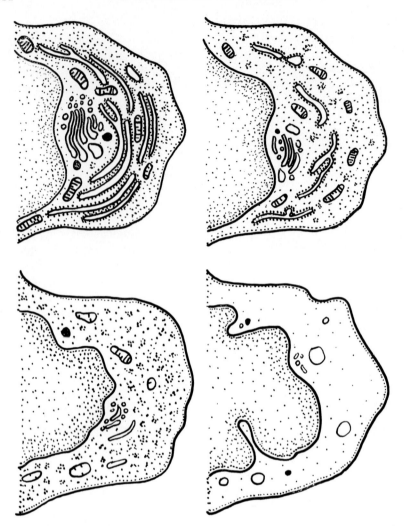

Fig. 19.   Diagram showing gradual dedifferentiation (anaplasia) of the cytoplasm from tumor cells.

A. The Mitochondria

1. The Question of Quantity

The indications found in the literature—based on a great variety of material—are contradictory. Mention is sometimes made of an increase of the number of mitochondria in cancer cells, whereas other investigators report a diminution. In his attempt to summarize the findings reported by numerous authors, Cowdry (1955) was forced to speak of "conflicting data." Ludford (1952), too, is hesitant in his conclusion but believes that there is frequently an increase in the number of mitochondria in cancer cells. The distinction between mitochondria and other cellular structures has been simplified by the use of the electron microscope. It can be assumed that in many of the cases investigated merely optically in the past, the techniques of fixation and staining used in the demonstration of mitochondria have not been entirely satisfactory since these organelles in the malignant cell would seem to be more fragile than normally and since their internal structure, and therefore their affinity to certain stains, may not remain the same. Such studies have been facilitated by the use of tissue culture; it is doubtful, however, whether tumors cultured *in vitro* for some time retain the same quantity of mitochondria. The macrophages of the Rous sarcoma, when cultured *in vitro,* may show an abundant chondrioma, whereas the same tumor, obtained from an animal and studied in ultrathin sections, possesses but few mitochondria (personal observation).

It must be admitted that the number of mitochondria may vary, not only from one type of tumor to another, but in each specimen obtained from the same tumor, and particularly from one cell to another in the same tumor tissue. An investigator using the electron microscope will find it easy to present examples of cancer cells packed with mitochondria (Fig. 20) and of other cancer cells almost completely devoid of them (Figs. 26 and 28). *On the whole, however, there is a more or less marked decrease in the number of mitochondria,* associated with parallel cytoplasmic dedifferentiation. In this respect it is interesting to note that the approximate number of mitochondria has been quantitatively evaluated by direct counting (Allard *et al.,* 1952; Shelton *et al.,* 1953). A decrease of the mitochondrial fraction in hepatomas obtained by ultracentrifugation has been reported (Allard *et al.,* 1952).

In order to detect exceptions to what is merely a general rule, each tumor variety should be very carefully examined.

2. *Shape and Size*

The above criticism with regard to the classic techniques of demonstrating chondriosomes apply even more strongly to studies of the shape and size of these organelles which have an average diameter at the extreme limit of optical resolution. Numerous authors working with various tumors have observed rod-shaped chondrioconts or rounded granular mitochondria, with a complete range of intermediate forms: no one form seems to be characteristic of the cancer cell. Thus critical observers such as Carrel and Ebeling (1928a), on the one hand, and Ludford (1934a) on the other, have described contradictory facts when observing

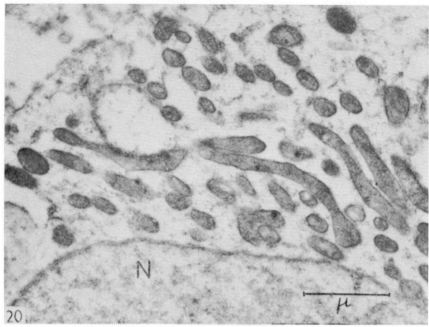

FIG. 20. Human myxosarcoma. Very numerous dense, but normal looking, mitochondria; *N*, Nucleus. Magnification: × 23,500.

the same tumor in tissue culture. According to the former investigators, the mitochondria of Jensen's sarcoma are shorter and thicker, whereas the latter author contends that they are smaller than those of normal fibroblasts. Lewis (1939), on the basis of a study of tissue cultures of various types of sarcoma, concludes that these mitochondria are generally more numerous and smaller than those found in normal tissue. Long, thick rod-shaped chondrioconts have been found in clusters in tissue cultures of Rous sarcoma (personal observations) and small rounded mitochondria in the same type of tumor, fixed immediately after having been obtained from animals and examined in ultrathin sections. *Therefore, it seems inadvisable to attach any great importance to the exterior form of mitochondria* (Lewis and Lewis, 1914–1915), particularly since movies made by Chèvremont and Frédéric (1952), Frédéric (1954, 1958), and Gey *et al.* (1955) have revealed the extremely dynamic behavior of the chondriosomes in tissue culture. Electron microscopy gives further examples (Fogh and Edwards, 1959).

With regard to the *size* of the organelles, electron microscopic observations have shown that cancer cells often contain smaller mitochondria than normal cells (Howatson and Ham, 1955; Novikoff, 1957), whereas in other cancer cells, they are on the contrary perfectly normal in size or have a larger diameter because they are swollen (see below). Very occasionally, cancer cells may contain such giant mitochondria as have been reported in a peritoneal metastasis of an Ehrlich ascites tumor (Wessel and Bernhard, 1957). In the latter case their diameter may attain $4\,\mu$.

3. Density and Internal Structure

In most cancers the internal appearance of the mitochondria corresponds to the general description given by Palade (1953a), Sjöstrand (1953), and many other authors: the organelle is surrounded by a double membrane; a variable number of "cristae" (infoldings of the inner membrane) form partitions in the mitochondrial body which contains a homogeneous substance of medium density. This matrix may occasionally include rounded, highly osmiophilic corpuscles with a diameter of about $40\,\mathrm{m}\mu$. In the cells of a hepatoma in a rat, Novikoff (1957), Howatson and Ham (1955) noticed the presence of highly developed mitochondria with many cristae, resembling those found in the embryonic liver. This observation has not been made in other neoplastic tissues.

Although the ultrastructure of the chondrioma in numerous cancer cells shows no particular features, a striking fact is the frequency of two phenomena encountered in tumors, viz., mitochondrial *tumefaction* on the one hand, and *densification* of the organelles on the other.

At present it can be regarded as established that mitochondrial tumefaction is the essential morphological basis of *cloudy swelling* (Zollinger, 1948, 1950; Rouiller and Gansler, 1955; Rouiller et al., 1956), a lesion found in numerous cases of cell damage due to a variety of causes. *In cancer cells, this alteration is extremely frequent.* The mitochondria become rounded and more or less swollen, progressively lose the majority of their internal crests, become clear, and are finally transformed into apparently empty sacs, the mitochondrial origin of which can only be guessed on the basis of the persistence of a double external membrane (Fig. 21). In extreme cases, this double membrane may be reduced to one. Thus, the origin of such a vesicle surrounded by a single membrane can no longer be established. The cytoplasm of cancer cell which has undergone such a transformation may seem to be entirely filled with tightly packed vesicles, obscuring any other cytoplasmic structure (Fig. 22). This process of mitochondrial degeneration in tumor cells has been reported by Rouiller et al. (1956), Bernhard and Oberling (1957), Weissenfels (1957), Edwards et al. (1959), Gusek (1959b), and Frajola et al. (1958). It is very likely that this change is often due to secondary causes which have nothing to do with the cancer phenomenon as such, e.g., unsatis-

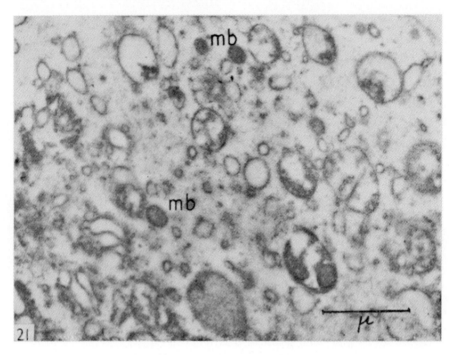

FIG. 21. Brown-Pierce tumor of the rabbit Swollen and partially damaged mitochondria; *mb*, microbodies. Magnification: ×23,500.

factory vascularization of the tumor tissue, which gives rise to respiratory and trophic disturbances and thus forms more or less extensive necrobiotic zones. Unsatisfactory fixation alone has also been found responsible for creating such swelling artificially. However, in view of the fact that mitochondria may exhibit the same transformations in tumor cells in which other cellular organelles *are perfectly fixed, it must be admitted that the cancer cell often has more fragile mitochondria which are defective from the structural point of view.*

The second observation would seem to contradict the above, but it might be an indirect sequel to the degenerative process: *sometimes the mitochondria are not swollen and empty but, on the contrary, appear denser than normal* and of somewhat subnormal size. The internal cristae are normal or rarefied, and osmiophilic corpuscles are frequently seen. Two tumor specimens examined by the authors were particularly rich in very dense mitochondria; the specimens were a human myxosarcoma (Leplus *et al.*, 1957) (Fig. 20) and some hepatomas induced by means of butter yellow in rats (Bernhard and Bauer, 1955; Heine *et al*, 1957); other

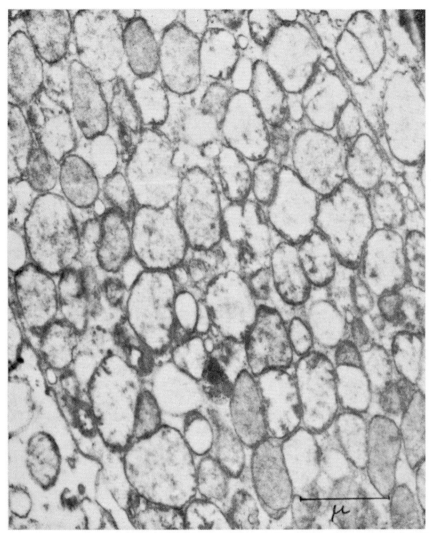

FIG. 22. Human hypernephroma. Swelling of all mitochondria. "Cloudy swelling" of the cytoplasm. Magnification: ×23,500.

tumors showed only occasional mitochondria of this type, scattered among the normal mitochondria in the cytoplasm. The hypothesis that these are *young regenerating mitochondria* seems probable. Gansler and Rouiller (1956) and Rouiller and Bernhard (1956) suggest that the renal and hepatic chondrioma regenerates on the basis of the microbodies described by Rhodin (1954). In these tissues, the whole range of transitional forms can be found between the dense bodies and normal mitochondria.

1. Growth Granules, Ultrachondrioma, Microbodies and their Relationship with Mitochondrial Regeneration

Porter and Kallman (1952) drew attention to granular or filamentous elements of sinuous appearance, found in rat sarcoma cells; they were smaller than mitochondria and were found scattered among the latter in the cytoplasm. The authors called these corpuscles growth granules; as they seemed to contain ribonucleic acid, they were assumed to be related to protein synthesis and growth. Oberling *et al.* (1950, 1951) presented electron micrographs of human leukemic cells in which the cytoplasm included *granulofilamentous* corpuscles, smaller than mitochondria and highly similar to growth granules. These elements, dumbbell-shaped or ellipsoid, were arranged in small chains; some features may evoke virus morphology. The technique of ultrathin sections was in its earliest stage and did not, at the time, make it possible to explore their internal structure. Oberling connected these structures with the chondrioma, as there were transitional forms between the two, and he therefore called them *"ultrachondrioma."*

Selby and Berger (1952) found similar formations in other cancers cultured *in vitro*, but not in normal cells. Harel and Oberling (1954), meanwhile, demonstrated them in noncancerous cells in inflammatory effusions. Their specificity in the neoplastic process has not been demonstrated. The growth granules or ultrachondrioma were thus forgotten for a few years until the technique of ultrathin sections made it possible to describe the microbodies, very dense corpuscles, smaller than mitochondria, without internal cristae and enveloped in a single membrane (Rouiller and Bernhard, 1956) (Fig. 23). In the light of these investigations, it seems likely that there is a very close connection between growth granules, ultrachondrioma and microbodies—terms which may well refer to the same cellular organelle, viz., promitochondria—precursors of the mitochondria and a manifestation of intensive regeneration of the mitochondria. If this hypothesis can be verified, it will not be difficult to understand the significance of their presence in cancer cells, the chondrioma of which is defective. The affected cell attempts unceasingly to replace the mitochondria lost by cloudy swelling. Certain cancers may be capable, thus, of compensation for their respiratory deficiency, whereas other tumors with a cytoplasm already too dedifferentiated would be unable to do so.

Mention should be made, in this respect, of another important observation by Ludford and Smiles (1953) on living sarcoma cells. According to these investigators, these cells enclose a large proportion of very small mitochondria characterized by marked absorption of UV-light and containing RNA which, according to the latest biochemical findings,

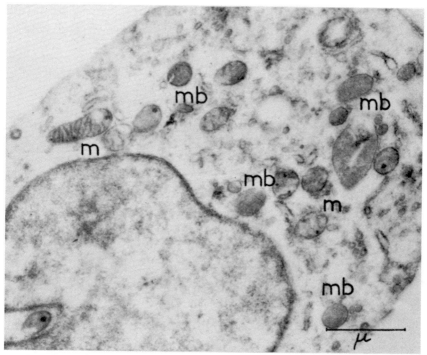

FIG. 23. Chemically induced chicken sarcoma. Cytoplasm with mitochondria (*m*) and many microbodies (*mb*). Magnification: ×21,500.

hardly ever seems to be present in the mitochondria of the normal tissue. The biochemical behavior of the very small mitochondria, ultrachondrioma, or microbodies, however, is likely to differ from typical mitochondria.

5. Storage

It is an established fact that normal mitochondria are capable of storing various kinds of substances in their interior. The first electron microscope studies on this subject were published by Rhodin (1954), and later by Gansler and Rouiller (1956), Miller and Sitte (1955), and Bessis and Breton–Gorius (1957b). Few facts are available with regard to this process in the mitochondria of tumor cells. Figure 24 shows a good example: a large number of mitochondria in ovarian mouse tumors contain a very homogeneous, dense mass which obscures the majority of the cristae and gives rise to considerable swelling of the mitochondrial bodies.

The mitochondria in rat hepatoma (Bernhard and Bauer, 1955) and in

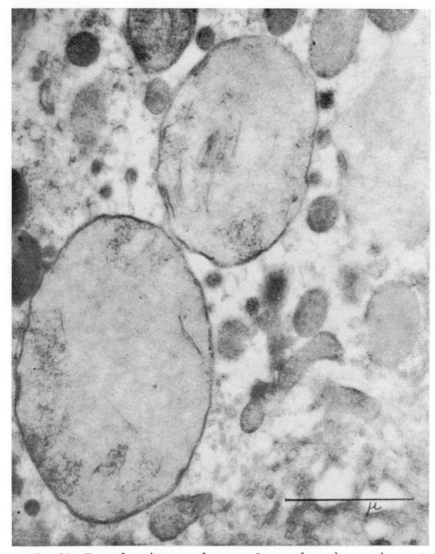

Fig. 24.   Tumor from the ovary of a mouse. Storage of an unknown substance in mitochondria which appear considerably enlarged. Magnification: ×35,000.

primary human hepatic cancer (Rouiller, 1957) can store substances of unknown chemical nature. In the virus-induced chicken erythroblastosis, the microbodies present in the erythroblasts may contain hemosiderin or ferritin granules (Benedetti and Leplus, 1958).

6. Relationships between Mitochondrial Structures and Certain Cancer Theories

The importance attributed by Warburg (1956) (see Le Breton and Moulé, Chap. 8, Vol. V) to the respiratory deficiency of cancer cells is well known. Normal cellular oxidation is essentially connected with the chondrioma; it is therefore tempting to correlate the relative rarefaction and the more or less pronounced structural lesions of the mitochondria with the biochemical observations. It may be supposed that the submicroscopic mitochondrial lesions are connected with the cancer process itself rather than that they should be attributable to secondary phenomena as discussed. There are no ways at present, however, which would enable the morphologist to settle this question.

Furthermore, interesting observations were made with the aid of the fluorescence microscope by Graffi (1940), who showed that certain carcinogenic substances concentrate in the mitochondria—the possible site of attack of these carcinogens. If examined with the electron microscope, these same mitochondria probably would show simple nonspecific swelling, a phenomenon which cannot be further analyzed by the current techniques of electron microscopy.

Finally, the presence of virus particles inside the mitochondria of reticular cells in leukemic chicken spleen and bone marrow (Benedetti *et al.*, 1956) has for the first time illustrated in a simple way the site at which the virus tumor agents could attack to transform a normal cell into a tumor cell. These findings are undeniable, but they seem to be somewhat rare; further, it is still uncertain whether the carcinogenic virus is capable of producing regular and extensive lesions at the level of the chondrioma.

B. Basophilia and Its Relationship to the Ergastoplasm

The affinity of the cytoplasm of cancer cells to basic stains is well known; it is one of the characteristics which constitute its "embryonic" aspect. This basophilia, generally more evident in cancer tissue than in the homologous normal material, is known to vary considerably from one species of tumor to another, and even within a limited region of the same specimen. Caspersson and Santesson (1942) have emphasized this variability of cytoplasmic basophilia, which is very marked at the periphery of tumor nodules and near the vessels (A cells) (Fig. 5). Cytoplasmic basophilia has since been submitted to systematic investigation in various tumors (Amistani, 1951). At the same time biochemists have attempted to determine, in pellets of centrifuged microsomes, the RNP content of the cytoplasm (see Le Breton and Moulé, Chap. 8, Vol. V). To date, remarkably varied results have been obtained.

Studies on basophilia which had formerly been described in terms of color only, have been transformed into investigations of structure (ergastoplasm), as result of the introduction of the electron microscope, and very considerable progress has been made (Oberling et al., 1953). The history of the ergastoplasm, first described by Garnier (1900) at the end of the last century with a definition of its functional significance, has been summarized elsewhere (Haguenau, 1958). The ergastoplasm consists of a system of membranes which form canaliculi or flat lamellae, Porter's endoplasmic reticulum (1953), as well as of osmiophilic granules, with a diameter of about $150 m\mu$, containing a very high concentration of RNP (Palade, 1953, 1955a, b; Palade and Porter, 1954). The cytoplasmic basophilia is obviously dependent on the presence of these granules (Palade and Siekewitz, 1956). In the presence of granules and membranes (rough membranes), the term "organized ergastoplasm" may be used. If only granules are visible, as is often the case in embryonic cells, the term "unorganized ergastoplasm" seems acceptable (Howatson and Ham, 1955, 1957a, b), or one may simply speak of RNP granules.

It was, thus, of importance to study the features of ergastoplasm in various tumors. Again, as in the case of the mitochondria, there were considerable quantitative and qualitative variations in these structures. However, it became apparent that there is *a trend in the cancer cell toward a decrease in organized ergastoplasm*. It is not difficult to demonstrate in almost all malignant cells at least remnants of ergastoplasmic lamellae or vesicles covered with RNP granules, but in the majority of these tumors lamellar structures are rare. On the other hand, in certain benign tumors, such as experimental pituitary adenomas in rats, ergastoplasmic lamellae are quite dense, as dense as in normal glandular cells (Haguenau and Lacour, 1955) (Fig. 25).

In a mouse hepatoma (Fawcett and Wilson, 1955) and in certain primary cancers of the human liver (Rouiller, 1957), some of the tumor cells show an ergastoplasm far more developed than in normal liver cells. Some of these cases may involve cells of the A type described by Caspersson, but it is also possible that such tumor cells belong to an adenomatous and still benign region. A highly developed canalicular ergastoplasmic system was also found in the Rous sarcoma (Fig. 26) in a human osteosarcoma (Gusek, 1959a) and in human myeloma cells where the disease was accompanied by globulin production (Braunsteiner et al., 1957). The higher the degree of dedifferentiation of a tumor cell, the greater its loss of membranous ergastoplasm. It can still remain unmistakably basophilic, but in that case the basophilia is represented by

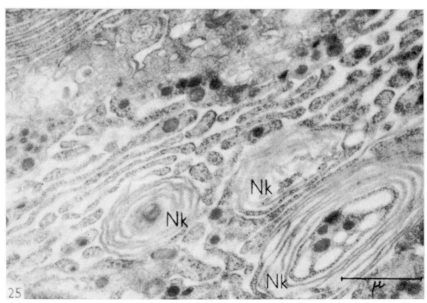

FIG. 25. Benign pituitary adenoma of the rat; *Nk, Nebenkerne*. Magnification: ×22,000.

FIGS. 25–28. Variable densities of ergastoplasmic structures in tumor cells.

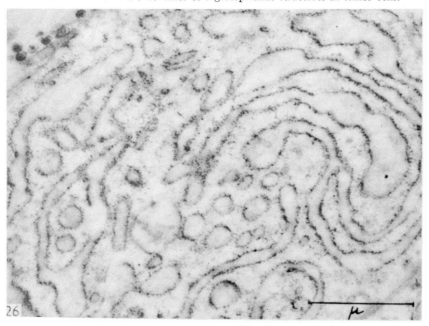

FIG. 26. Rous sarcoma, chicken. Magnification: ×28,000.

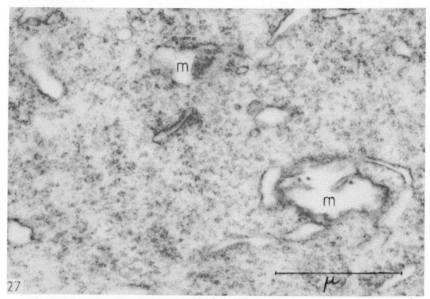

FIG. 27. Murray-Begg sarcoma, chicken. RNP granules without membranes dispersed in the cytoplasm; *m*, swollen mitochondria. Magnification: ×35,000.

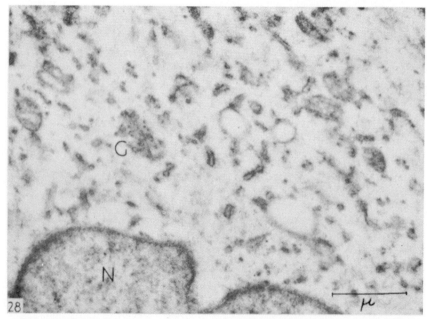

FIG. 28. Brown-Pierce carcinoma, rabbit. Basophilic structures practically disappeared. Magnification: ×20,000.

numerous granules of the RNP type, dispersed without membranous support throughout the hyaloplasm (Fig. 27). Howatson and Ham (1955, 1957a) have pointed out the existence of this disorganized, embryonic form of structural support of basophilia in certain cancer cells. As in the case of very young cells (Palade, 1955a, b), these RNP granules are single and distributed homogenously within the cell; in other cells, they are arranged in small clusters, which may show rosettes or spiral patterns. This observation has been confirmed by Rouiller *et al.* (1956), Selby *et al.* (1956), Bernhard and Oberling (1957), and Novikoff (1957). Finally, in other very dedifferentiated cells, neither membranes nor granules can be seen; their cytoplasm is assumed to have lost its basophilia completely (Fig. 28). Porter and Bruni (1959) have noticed that the first alterations produced in the liver of rats with a butter yellow diet are characterized by the detachment of RNP-particles from the ergastoplasmic membranes, accompanied by a loss of glycogen.

In certain virus tumors (see below) the RNP granules are concentrated in the area of the viroplasm, where the virus corpuscle subsequently develops. This has been observed by Febvre *et al.* (1957) in the Shope fibroma. Moreover, it was possible to demonstrate cytoplasmic inclusions in Rous sarcoma, suspected of representing the matrix in which its agent is formed (Bernhard *et al.*, 1956b). Again, a dense accumulation of RNP granules is observed in this area.

Cancer cells, further, show *pathological features of the ergastoplasm.* These lesions are certainly not specific for the cancer process itself but merely imply that the cytoplasm has in some way been affected. *Vesiculation* of the ergastoplasmic lamellae, a phenomenon which can be experimentally provoked by a hypotonic medium (Weiss, 1953), is very common in cancer cells.

On the other hand, there are ergastoplasmic sacs or small cysts, which are swollen and filled with a homogenous dense substance, probably representing secretory product (Fig. 29). Similar findings have been obtained in a series of tumors (Porter, 1955; Schulz, 1957; Bernhard *et al.*, 1956b), but also in cells involved in an inflammatory process (Policard *et al.*, 1957a).

Summarizing these observations, one can say that the structural variability in cancer cells of the elements forming the substrate of cytoplasmic basophilia is in accordance with the observations made with the ordinary and UV-light microscope and the great variations of staining properties. At the ultrastructural level, the frequently marked basophilia of cancer cells corresponds to an increase in RNP granules freely dispersed in the cytoplasm; less frequently, there may also be an increase in organized membranous ergastoplasm. There is perfect superposition

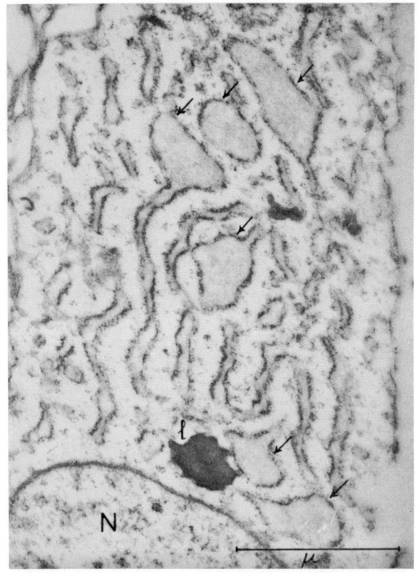

Fig. 29.   Human reticulosarcoma. Swollen ergastoplasmic sacs, containing homogeneous substance (→); *l*, lipids; *N*, nucleus. Magnification: ×44,200.

of pictures of A and B cells according to Caspersson and to electron micrographs of cells rich or poor in ergastoplasm (Haguenau, 1959) (Fig. 5); the true significance of the observations made by the Swedish investigators, however, still remains to be elucidated.

C. The Golgi Apparatus

In view of the controversy only recently settled as to the morphology of this cellular organelle, various authors, having raised this question with regard to cancer cells, confine themselves to mentioning its demonstration in various cancerous tissues by classical impregnation techniques. Ludford (1939) believes the organelle to be more readily demonstrable in tumor cells but mentions that no useful differences have been discovered. According to Cowdry (1955), "the knowledge of the Golgi apparatus in normal cells, let alone malignant cells, is disappointing." Despite numerous light microscopic investigations on cancer cells (Ludford, 1929, 1952; Bagozzi, 1933; Dalton and Edwards, 1942) it was not possible to determine a constant and characteristic feature of the Golgi zone in tumor cells. The localization and extent of such regions were found to be very variable, but in certain tumors a marked hypertrophy was noticed (mouse hepatoma, Dalton and Edwards, 1942; Bothe et al., 1950). A hypertrophy was also found in certain pituitary adenomas experimentally produced by application of hormones (Severinghaus, 1937). The ultrastructural architecture of the organelle remained obscure, however, and therefore a precise morphological definition of the Golgi apparatus could not be given at that time (Worley and Spater, 1952).

It is due to the important work of Dalton and Felix (1953, 1954, 1956a) that the ultrastructure of the Golgi zone has been identified by electron microscope studies and found to consist of a number of lamellae with smooth double membranes, vacuoles, and microvesicles or granules [see also Chapter 7, Volume II, by Dalton; and Pollister and Pollister in their recent review (1957)].

The first description of the Golgi apparatus of a benign tumor as seen with the electron microscope was given by Haguenau and Lacour (1955) with regard to a chromophobe pituitary adenoma, and a comparison of the submicroscopic structure of the Golgi zone in normal and in cancer cells was made by Haguenau and Bernhard (1955a). Electron micrographs of various malignant tumors have since been presented by Dalton and Felix (1956b), Howatson and Ham (1955, 1957a), Selby et al. (1956), Suzuki (1957), Epstein (1957b), and Dalton (1959). On the basis of these investigations, it can be concluded that no constant specific features of these structures have been discovered in tumor cells. The Golgi apparatus seems to be present in all neoplasms. Very hypertrophied in certain

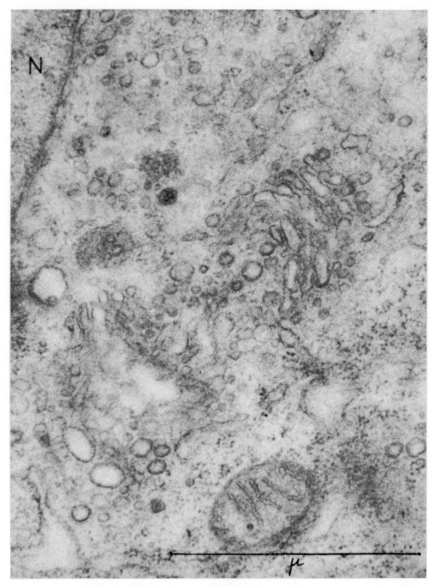

FIG. 30.   Golgi zone in a mouse mammary tumor. Typical aspect after osmic acid fixation; N, nucleus. Magnification: ×68,000.

benign or malignant tumors (Fig. 30), they are only occasionally visible in highly dedifferentiated tumors. The ascites tumors of Yoshida and Ehrlich (Wessel and Bernhard, 1957) have a Golgi apparatus of greatly reduced size, which is also observed in the Brown-Pierce tumor. Briefly, neither hypotrophy nor hypertrophy would seem to be characteristic of this organelle for all tumors, but, in a neoplasm of a given type, one may expect it to show relatively consistent morphological features. Its *ultrastructure*, although having all the characteristics of the Golgi apparatus in the normal cell, is often found to be of an embryonic type: its microvesicular constituent is predominant; large vacuoles and secretory granules are absent (Dalton and Felix, 1956b). This rule, however, has its exceptions.

Special mention should be made of an observation published in 1955 (Bernhard *et al.*, 1955c) and recently completed (Bernhard and Guérin, 1957). In mammary adenocarcinomas of mice which contain the Bittner milk factor, inclusion bodies formed by thousands of viruslike particles often fill the paranuclear region, the site of predilection for Golgi structures in normal cells. This phenomenon is particularly well visible in less dedifferentiated tumors with a pseudotubular structure: all inclusion bodies are found at the secretory pole of the cells, situated toward the lumen of the tubules. The *polarity* of tumor cells is completely intact in these cases. In less advanced stages of the formation of inclusion bodies it is easy to observe their connection with the Golgi elements. The first pathological particles considered as viruses appear in the Golgi region and are in close contact with its normal constituents.

D. The Centriole

Directly associated with the Golgi zone (region of the centrosphere), the *centrosome* is encountered during the interphase. It may be either single or double (diplosome) and less frequently also multiple (megakaryocytes) (Yamada, 1957). The centrosome encloses the *centriole*. Its role during mitosis is well known; it is an established fact that this minute organelle is responsible for certain pathological forms of mitosis in cancer cells (Therman and Timonen, 1950). As it lies at the limit of resolution of the ordinary microscope, no exact morphological data on its structure can be found in the literature. It has been stated that the centrosome is more easily visible or even hypertrophic in certain tumors (Lewis and Lewis, 1932; Zweibaum, 1933; Orsos, 1935), and the entire centrosomal region has been held responsible for the formation of bird's eye inclusions or "Plimmer bodies," cytoplasmic inclusions of unknown origin seen in certain tumors (see below) (Borrel, 1901; Le Count, 1902). Hypertrophy of the centrosphere, however, is also seen in noncancerous tissue.

It is particularly striking in epithelioid and giant cells in tuberculous tissue (Orsos, 1935; Gedigk, 1954).

The size and ultrastructure of the centriole in normal cells can be revealed with accuracy in ultrathin sections examined with the electron microscope (Bernhard and De Harven, 1956; De Harven and Bernhard, 1956; Amano, 1957; Tanaka *et al.*, 1957; Stoeckenius, 1957a; Bessis and Breton–Gorius, 1957a). It is seen to be a small cylinder with a diameter of about $200\,m\mu$ and a length of 300–$500\,m\mu$ (Fig. 31). This cylinder consists of nine parallel very thin tubules, arranged in a circle at its periphery; some of these tubules may appear to be divided into two or three units. They are embedded in a dense homogeneous centrosomal mass. This organelle appears to be the same whether it occurs in normal or in cancer cells (De Harven and Bernhard, 1956). Antimitotics such as sodium cacodylate or colchicine do not visibly alter these structures, but, owing to their action, spindle fibers disappear and mitoses remain blocked.

The *position of the Golgi apparatus and, therefore, of the centriole determines the polarity of cells* in the interphase. This phenomenon is marked in cells where the functional activity is oriented. It is easily demonstrable in single layers of epithelial cells surrounding a lumen (ducts) and displaying absorptive or secretory activity (glands). The centrosphere is found beside the nucleus in these tissues, directed toward the lumen of the tubules. The polarity is gradually lost in proportion to the degree of anaplasia. Cowdry (1955) particularly emphasized the importance of this loss of structural and functional orientation of cancer cells. It is associated with general dedifferentiation of the cytoplasm. Before disappearing completely, the polarity of the tumor cells in certain types of glandular cancer may simply be reversed—the secretory pole then being on the side of the basement membrane (Masson, 1922, 1956). A similar reversal can exceptionally be seen in normal glandular tissue (Cowdry, 1922).

In cancers showing relatively little dedifferentiation, and retaining some functional activity, polarity may persist. This is the case, for instance, in mouse mammary tumors, some of which may still secrete milk (Foulds, 1956; Bernhard *et al.*, 1956a) (Fig. 35).

E. Other Cytoplasmic Structures

1. Fenestrated Membranes

With the aid of the electron microscope, a hitherto entirely unknown organelle was discovered simultaneously by Afzelius (1955), by Palade (1955b) who called it "cysternae fenestratae," and by Swift (1956) who described it as "annulate lamellae." These structures were found in normal cells of invertebrates, batrachians, and rodents, but they have also been

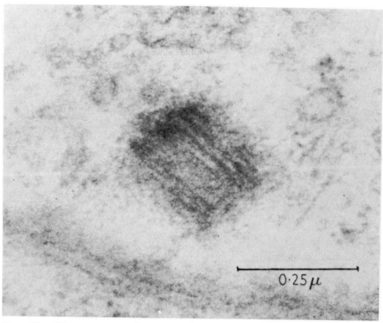

Fig. 31. Ultrastructure of a centriole. Mammary tumor of mice. Magnification: × 130,000.

demonstrated in certain tumor cells (Schulz, 1957; Wessel and Bernhard, 1957), where the term "fenestrated membranes" was used. Generally grouped in bundles of three to twenty, straight or curved, they appear anywhere in the cytoplasm. These membranes have an ultrastructure similar to that of the nuclear membrane: in transverse sections they seem to be double, showing a periodic structure formed by fusion of the two leaflets at regular intervals. In tangential sections annular formations of a diameter of 400–500 A. become visible (Figs. 32, 33).

Fenestrated membranes are sporadically seen in the Yoshida's sarcoma, Ehrlich's ascites tumor (Wessel and Bernhard, 1957), mammary rat tumors (Schulz, 1957), and an avian sarcoma (Friedlaender, 1959), but they are probably present also in other neoplasms. These elements are identical with those described in normal cells. They may show structural continuity with ergastoplasmic membranes (Pasteels et al., 1959; Friedlaender, 1959). As to their significance, it has been suggested that they play a role in the elaboration of the basophilic components of the cytoplasm (Swift, 1956), that they represent remnants of the nuclear membranes (Afzelius, 1955) probably originating from anomalous telophases,

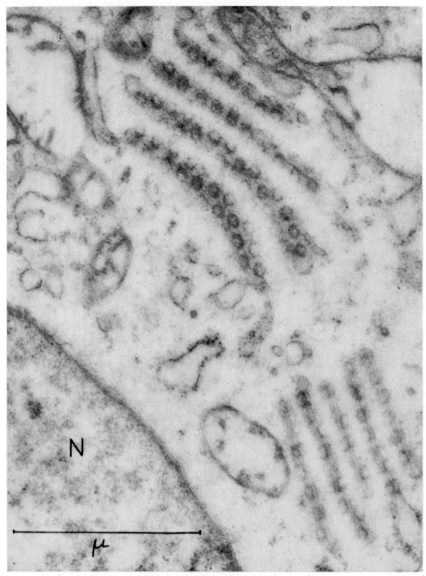

Fig. 32. "Membranae fenestratae" in the cytoplasm of a mammary tumor of the rat; N, nucleus. Magnification: ×51,000.

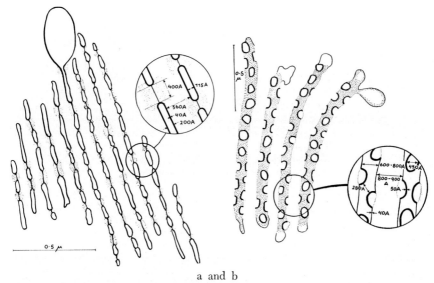

a and b

FIG. 33.   Diagram showing bundles of "membranae fenestratae" in transversal (a) and tangential sections (b). From Schulz (1957).

or that they may form a genuine cellular organelle whose function is still obscure (Schulz, 1957).

2. The Ciliate Border

Very occasionally, the tumor cells are furnished with *cilia*. These are encountered in a variety of benign polyps (nasal, bronchial, and uterine) (Hamperl, 1950), and in dysembryomas (see Masson, 1956). The cell which has become malignant, however, is as a rule incapable of elaborating so differentiated a cytoplasmic structure. A ciliate border has been found in renal tumors in hamsters (Mannweiler and Bernhard, 1957), which, while dependent on a prolonged estrogen effect, invades the normal tissue and even forms metastases (Matthews *et al.*, 1947; Horning and Whittick, 1954) (Fig. 34). These cells, otherwise highly dedifferentiated, may have cilia inserted in the cytoplasm by numerous basal corpuscles, the ultrastructure of which is found to be similar to that of the centriole. This would seem to involve a particular type of metaplasia, as cilia have not been found either in the adult or in the embryonic kidney in these animals.

F. Various Cytoplasmic Inclusions

In addition to the cellular organelles described above, the cytoplasm of tumor cells may contain inclusions of variable size and origin. Some

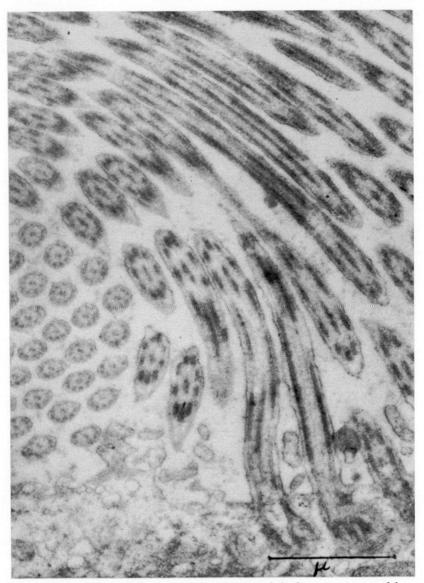

Fig. 34.   Ciliated border in a renal carcinoma of the hamster. Cross and longitudinal sections of the cilia. Magnification: ×33,000.

of these have for many years attracted the attention of investigators searching for a specific cancer agent (bacteria, virus). In the course of their investigations the "bird's eye inclusions" and "Plimmer bodies" were discovered; they are believed to be related to hyperplasia or multiplication of the centrosome (Borrel, 1901; Le Count, 1902; see also Ewing, 1942; Masson, 1956). Meanwhile, no definite statement can be made as long as the ultrastructure of such bodies remains unknown. As Cowdry (1955) and Sträuli (1959) suggest, they may be of very diverse origin.

Other inclusions are more clearly defined. They may be divided into three groups.

1. Inclusions Due to Persistence of Cellular Activity

These are exclusively encountered in tumors with little dedifferentiation: vesicles filled with bile in malignant hepatomas, milk in adenocarcinomas of the breast (Fig. 35), colloid droplets in the cytoplasm of thyroid cancers (Dalton, 1959), secretory products in various other endocrine tumors (Boyd et al., 1959), etc. Also included in this group are melanomas in which the cytoplasm is overcharged with pigment (Braunsteiner, 1958; Dalton, 1959; Wellings and Siegel, 1960), and the erythroblasts of chicken erythroblastosis, which sometimes include microbodies overcharged with hemosiderin or ferritin granules (Benedetti and Leplus, 1958).

2. Deposits Due to Degenerative Phenomena

The most common finding in this respect is the deposit of fat in cancer cells. It is prevalent in the perinuclear region, and electron microscopically visible in the form of very opaque bodies with an irregular outline. They increase in number with an increase in the cell's age and are very frequent in necrobiotic zones (Fig. 36). Other lipid inclusions show a concentric lamellar structure resembling the myelin figures described in noncancerous cells by Stoeckenius (1957b) and Policard et al. (1957b). The accumulation of *myxoid* substance in the cytoplasm probably always indicates actual cellular activity although, if this substance becomes abundant, the cells may degenerate and perish. The product concentrates in small isolated cytoplasmic vesicles (Leplus et al., 1957) (Fig. 37), or it may fill a large single vacuole compressing the nucleus which flattens and then appears in the form of a thin crescent (signet ring cell). Certain avian sarcomas, epitheliomas, and various myxosarcomas in man may produce a considerable quantity of mucus.

Another inclusion of unknown but probably degenerative nature was described by Friedlaender (1959) in the cells of an avian sarcoma produced by subcutaneous injection of dibenzanthracene. It involves limited

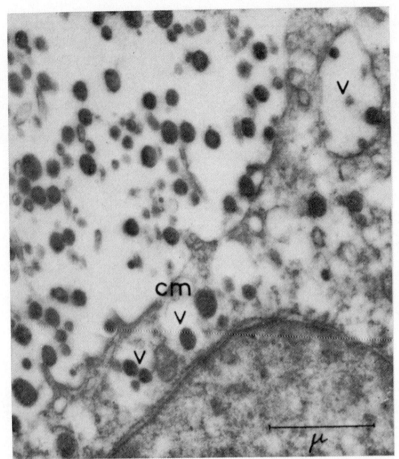

Fig. 35.   Mammary carcinoma of mice. Secretion of milk. Cytoplasmic vesicles
(*v*) containing milk droplets; *CM*, cell membrane. Magnification: × 28,500.

zones of cytoplasm marked by the presence of *tubular or annular structures* with an osmiophilic double membrane. Mention should also be
made of *erythrocytic* debris visible in the cells of the Yoshida sarcoma.
This tumor may originate from Kupffer cells, a fact which would explain
its considerable phagocytic activity (Wessel and Bernhard, 1957). Fragments of erythrocytes have also been noticed in Sternberg cells of Hodg-
kin's disease (André *et al.*, 1955).

3. Inclusions of a Viral Nature

These inclusions, formed either by a viroplasm or by complete or
incomplete virus particles, are visible both microscopically and electron

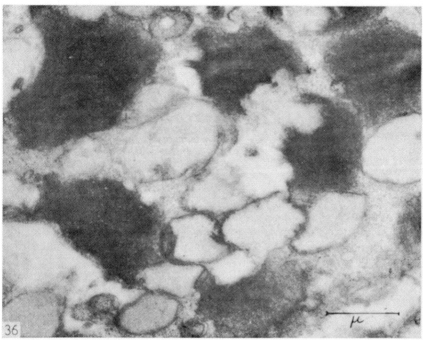

FIG. 36. Lipid deposits in the cytoplasm of a hypernephroma cell (human). Magnification: ×20,000.

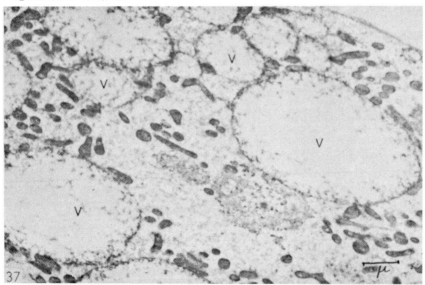

FIG. 37. Human myxosarcoma. Myxoid substance in cytoplasmic vesicles. Magnification: ×10,000.

microscopically in virus tumor cells in animals. They have not been established with certainty in human tumors. Details are to be found in Section J.

G. The Hyaloplasm (Ground Substance)

No constant difference in the consistency of the ground substance is discernible between normal and cancer cells, although it may seem to be less dense in tumors, a fact doubtless due to their frequently reduced mineral, and increased water, content. Fixed in osmic acid and examined with the electron microscope, the ground substance shows a very homogeneous aspect without any clear structure. Occasionally, however, *extremely thin fibrils* with a diameter of 40–80 A., forming a very loose network may be discovered. Where these fibrils form bundles in some areas of the cell, well-defined zones may appear in the ground substance of the cytoplasm. Such fibrillar formations without visible periodicity are relatively often seen in human leukemic cells (Bessis and Breton–Gorius, 1955), in avian sarcomas (Rouiller *et al.*, 1956) and in a human bronchogenic carcinoma (Edwards *et al.*, 1959). The chemical nature and significance of these elements is still obscure.

When the hyaloplasm is loaded with *glycogen*, as it is the case in the hypernephroma (Oberling, Rivière and Haguenau, 1959), electron microscopic examination reveals a fluffy low-contrast aspect of the substance, which resembles the glycogenic features of the normal liver as described by Fawcett (1955).

H. The Cell Membrane

Ludford (1934b), discussing his observations on the rapid penetration of fat-soluble substances such as Sudan III and Sudan black in the cells of Crocker's sarcoma, believed that the plasma membrane of certain tumor cells must be richer in lipids than that of the normal controls, an assumption which could explain its modified permeability to certain substances. It is hardly necessary to discuss the contradictory data presented in the literature on this subject, and reported by Cowdry (1955), since many microcinematographic investigations have been made on *phagocytosis* and *pinocytosis* (Lewis, 1931, 1937) and have shown that the cellular membrane as well as the ectoplasm are actively involved in transportation of material. Esterase and phosphatase activity was shown to be located in the cell surface of HeLa cells (Gropp and Hellweg, 1959). An interpretation of phenomena such as permeability and absorption should not be exclusively based on the structure of the cell membrane. Current electron microscopic investigations have not revealed any ultrastructural differences between plasma membranes in normal and cancer cells. In all cases its thickness is about 75 A., although values up to 200 A. have been found. These data are undoubtedly affected by the manner of fixation (Robertson, 1957).

According to the origin and structure of cancers, their cells may show considerable, poor, or no *phagocytotic* and *pinocytotic activity*. It is mainly confined to free cells, primarily macrophages, which can best be observed in tissue cultures or in ascites tumors. However, phagocytosis in particular is also a property inherent in a given strain of cells, as illustrated by the differences in the phagocytic activity of Yoshida's and Ehrlich's ascites tumor cells. As mentioned above, the cells of Yoshida's sarcoma often contain phagocytosed erythrocytes, rarely observed in Ehrlich's ascites. The former phagocytose a considerable quantity of particles of India ink, whereas the Ehrlich tumor cells retain considerably less, although they are of comparable size and ultrastructure.

Ranvier's *clasmatosis* (1900), another phenomenon partially based on the activity of the cellular membrane, and observed normally in megakaryocytes and also in macrophages in tissue cultures, is characterized by continuous detachment of cytoplasmic strips from the cell periphery which may or may not contain some cellular organelles. This process is also observed in cancer cells; it has been seen in macrophages of the Rous sarcoma (personal observation), but in this latter case, it probably indicated a process of necrobiosis.

The cell membrane may constitute a *smooth surface* or form microvilli, which are particularly well developed in the epithelia lining the bile ducts and the excretory ducts of various glands (Dalton *et al.*, 1951; Yamada, 1955). Although less numerous, microvilli can equally be found on endothelial surfaces. Cancers of such tissues show identical formations, which vary more markedly in number and size (Glatthaar and Vogel, 1959) (Fig. 38).

The superficial extensions of epithelial cells play an important role in *cohesion of the tissues*. They are often entangled with adjacent cells, thus ensuring close continuity (Fig. 39). It can be assumed that, when the entanglement is less marked or completely lacking owing to anaplastic growth, cells can be more readily separated; this could at least partly explain the decrease in cohesion and the tendency to fray shown by certain cancer tissues (Coman, 1944, 1953). Other types of connections similar to those existing in normal cells are also found in epithelial cancer cells. The one is characterized by intercellular bridges interrupted by *desmosomes* (Figs. 40, 41), the other, by *terminal bars* whose ultrastructure is similar. A cytoplasmic continuity across the intercellular bridges has not been observed. These connections were found disrupted in a case of a cervix carcinoma (Vogel, 1958) and in chemically induced skin

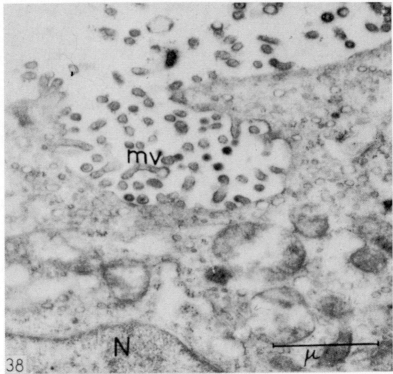

Fig. 38.   Microvilli (*mv*) on the border of a tumor cell (mammary carcinoma of the mouse); *N*, nucleus. Magnification: ×28,500.

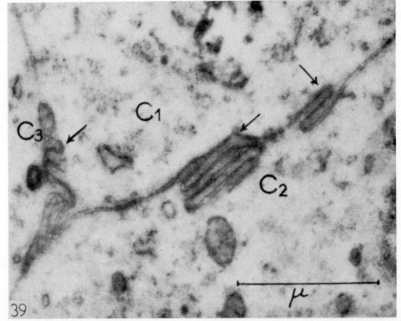

Fig. 39.   Entangled cell processes (→) of three mammary tumor cells. (*C₁*, *C₂*, *C₃*). Magnification: ×38,000.

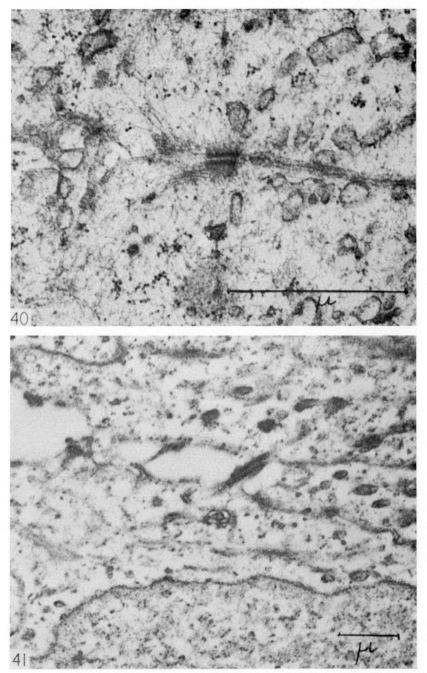

FIGS. 40 and 41. "Desmosomes" in a human cervical carcinoma. Longitudinal and tangential sections. Magnifications: ×48,000 and 16,000.

tumors of the mouse (v. Albertini, 1959). In the Vx2 carcinoma of the rabbit, disappearance of prickles at the cell surface was observed by Berwick, 1959).

It has also to be mentioned that the cell membrane plays a role in the *formation of collagen*. Fibrosarcomas are known to be capable of forming an abundant quantity. The collagen fibers would seem to form in contact with the cellular membrane, always outside the cell. No difference between normal and sarcomatous tissue has been observed.

The *contact between the cancer cells and the normal tissue cells* which they invade is probably conditioned by the surface properties of their cellular membrane (Ambrose, 1958). A microcinematographic study of the invasion of normal fibroblasts by sarcomatous cells has shown the absence of "contact inhibition" of cancer cells when in touch with normal cells, a fact which could favor the propagation of malignant cells in the organism (Abercrombie and Heaysman, 1954). It should be borne in mind, however, that normal polynuclear leucocytes also possess an extraordinary invasive power.

The "contact inhibition" phenomenon may be related to the *electrical charge* of the free cells which, according to their origin, move at various rates of speed in the electrical field. Thus, the electrophoretic mobility of certain cancer cells would seem to be considerably enhanced (Ambrose *et al.*, 1956). Human leukemia cells, however, have shown a decreased mobility when compared to normal leucocytes (Robineaux and Bazin, 1958).

The invasiveness of tumor cells is also conditioned by the extent of connective tissue reaction and collagen formation (Leighton *et al.*, 1959).

I. The Problem of the Specific Staining Affinities of Cancer Cells

It is evident that cells showing such a variable ultrastructure cannot be uniform in their staining properties. According to the nature of the tumor tissue, chemical reactions may be more or less marked in relation to the original tissue. An enumeration of all these variations is not within the scope of this publication. Although cytoplasmic basophilia is among the more constant signs, it is not necessarily always present (see above). From a practical point of view Papanicolaou's stain (1954) is doubtless a very important diagnostic aid in *exfoliative cytology*, but it does not pretend to be based on a specific cytochemical reaction.

Lipschütz' *plastin reaction* (1931) has been much discussed. According to the author, a modification of the Giemsa reaction can be used to demonstrate, in the cytoplasm of most of the cancer cells, a special zone containing "stegosomes." Liegeois (1933) demonstrated that no definite distinction can be made between basophilic and plastin-positive zones, and that this staining technique has no specificity whatever. The staining technic of Roskin (1938), based on *rongalite white* as a reagent believed to be specific for malignant cells, has also been found to be misleading. The use of *triphenyltetrazole chloride* to distinguish malignant cells from normal cells by simple staining (Cheronis *et al.*, 1948) has not recently yielded the results which these authors seem to have obtained (Calcutt, 1952; Hsu and Hoch-Ligeti, 1953).

According to Ludford (1929) there is a very marked difference between normal fibroblasts and sarcomatous cells in the intracytoplasmic segregation of trypan red and trypan blue. The vital stain is not concentrated in living sarcoma cells. The author correlates their functional deficiency with a change in the cytoplasmic ultrastructure; this seems to be highly probable in the light of recent electron microscopic observations on changes in the Golgi apparatus in certain strains of cancer.

Louis (1958) uses a fluorescein-globulin stain to distinguish normal cells from malignant cells. However, this method cannot be used with a high degree of certainty.

J. Virus Tumor Cells

The nuclear and cytoplasmic changes described in the two preceding chapters with regard to the cancer cell in general are also to be seen in virus tumors in animals. Here, the same quantitative and qualitative variability is found in cellular organelles, and the same tendency toward dedifferentiation of the cytoplasm. In addition, however, there are *specific lesions* due to the presence of infectious agents or their precursors in the cells. The results recently obtained in this field have, therefore, been briefly summarized. More detailed reviews with full bibliography are given elsewhere (Bernhard, 1958, 1960; Oberling, 1959). The morphology of viruses in general is reviewed by Williams (1957).

1. Shope's Fibroma

This tumor, benign in adults, but malignant when transplanted to newborn rabbits (Duran-Reynals, 1945; Harel, 1956) is caused by a large virus belonging to the pox group (Fenner, 1953). This agent forms cytoplasmic inclusion bodies, at first Feulgen-positive but subsequently with a pyronine affinity, which are demonstrable by light microscopy (Constantin and Febvre, 1958). Electron microscopic examinations show a series of different aspects, simultaneously present in these tumors but succeeding each other in well-established order when the evolution of the virus is studied *in vitro* (Bernhard *et al.*, 1955c; Lloyd and Kahler, 1955; Febvre *et al.*, 1957). After 4 hours of infection, very dense and finely

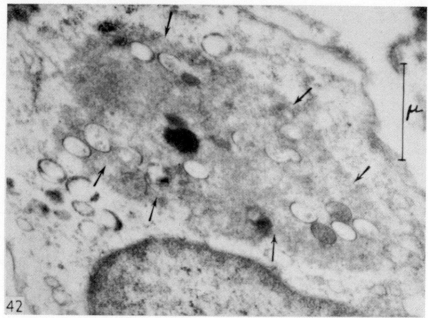

FIG. 42. Shope's fibroma, rabbit. Cytoplasmic inclusion (→) with viroplasm and dense or empty virus bodies. Magnification: × 25,000.

grained corpuscles appear, which subsequently enlarge to become diffuse viroplasm (Fig. 42) within which the first virus corpuscles appear after about 10 hours. After 24 hours' incubation the viroplasm is almost entirely replaced by these elements which are surrounded by a single membrane (Fig. 43); this membrane may subsequently duplicate. Where the cytoplasm is invaded by these corpuscles, the tumor cells often show signs of damage as described above. No particular lesions have yet been observed in the nucleus, but the latter is often hypertrophic and contains a giant nucleolus. The other viruses of the pox group (fowlpox, vaccinia, ectromelia, molluscum contagiosum) reveal a very similar morphology in the cells which they invade (Morgan *et al.*, 1954; Gaylord and Melnick, 1953).

2. *Mammary Adenocarcinomas in Mice*

Since the first investigations made by electron microscopy (Porter and Thompson, 1948; Kinosita *et al.*, 1953; Bang and Andervont, 1953; Dmochowski, 1954), the demonstration of viruslike particles in cancer tissue has made considerable progress (Bernhard *et al.*, 1955b, 1956a, 1957; Dmochowski, 1956; Bang, 1955; Bang *et al.*, 1956a, b; Suzuki, 1957).

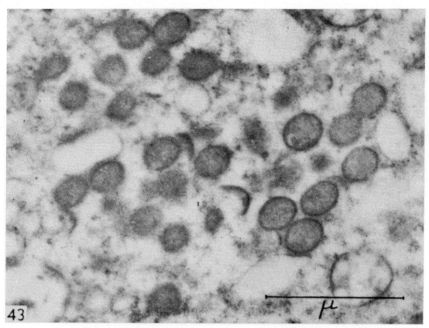

Fig. 43. Shope's Fibroma. Group of cytoplasmic virus bodies, surrounded with a single membrane. Magnification: ×37,000.

Although no absolute proof has so far been presented, it seems nevertheless probable that these particles, present in the cytoplasm, at the cellular surface, and in the intercellular spaces, are related with Bittner's milk factor (Moore *et al.,* 1959). In strains considered rich in viruses there are numerous tumor cells containing large paranuclear inclusion bodies which can be microscopically demonstrated (Guérin, 1955) and which are composed of thousands of homogeneously sized particles about 70 mμ in diameter (Fig. 44, a). These elements are enclosed in a double membrane which can be well demonstrated at higher magnification (Fig. 44, b). The Golgi apparatus would seem to play an important role in the elaboration of these elements. In the intercellular spaces electron microscopic examination shows other viruslike particles about 105 mμ in diameter, with an excentric nucleoid. Their ultrastructure is visible in Fig. 44, c. The frequency of images showing the passage of intracytoplasmic particles through the cellular membrane suggests their transformation into extracellular elements by means of a particular mechanism. However, there are also particles of the extracellular type found in the interior of cytoplasmic vacuoles. These two groups of particles may also

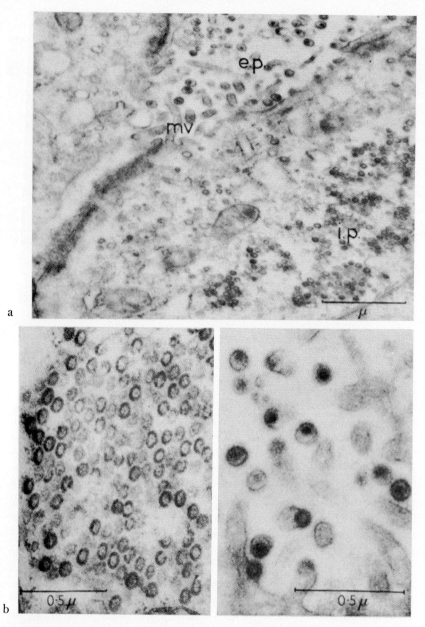

FIG. 44. Mammary carcinoma of mice. a. Intracytoplasmic (*i.p.*) and extra-cytoplasmic (*e.p.*) viruslike particles; *mv*, microvilli. Magnification: × 22,000. b. Intracytoplasmic particles at higher magnification (× 50,000). c. Extracytoplasmic particles at higher magnification (× 60,000).

be found in mammary tumors from strains with a low cancer incidence or even in the apparent biological absence of the Bittner factor.

3. Viruslike Particles in Various Other Tumors of Mice

Cells in other mouse neoplasms contain particles the viral nature of which is also probable but not established with certainty. Interesting facts have been described by Selby *et al.* (1954), who accidentally discovered, in an Ehrlich ascites tumor, cytoplasmic zones of numerous corpuscles of homogeneous size, measuring about 60 mμ in diameter. Each corpuscle is surrounded by a single membrane and contains a nucleoid with a diameter of 30–35 mμ. The elements form a striking geometrical pattern. In addition, tubular structures are found. Wessel and Bernhard (1957) have described the presence of similar corpuscles in one out of thirty-two Ehrlich tumors. Another type of viruslike particle has been described in the Ehrlich ascites tumor by Friedlaender and Moore (1956), and subsequently by Adams and Prince (1957) and Yasuzumi and Sugihara (1958). The particles have a double membrane, no nucleoid, and resemble the intracytoplasmic form as described in mammary tumors above. They are chiefly found within small cytoplasmic vesicles. Dalton and Felix (1956b) found them in a melanoma, and De Harven and Friend (1958), in an undifferentiated leukemia in mice. Dmochowski and Grey (1956, 1957, 1958), Bernhard and Guérin (1958), Bernhard and Gross (1959) described elements suspected of being of viral nature in a granulocytic leukemia and in spontaneous or induced lymphoid leukemia in the same animal.

The discovery of the polyoma-virus of the mouse (Gross, 1953; Stewart *et al.*, 1958) which is capable of inducing many different tumors in the mouse, in rats and in hamsters after a short latency, is of fundamental importance for cancer research. This virus induces extensive nuclear lesions in embryonic cells cultivated *in vitro* (Love and Rabson, 1959). The electron microscope has revealed an agent of 30 mμ in diameter, which may aggregate and form crystal-like inclusions in the nuclei (Banfield *et al.*, 1959; Bernhard *et al.*, 1959; Howatson *et al.*, 1960).

4. Avian Tumors

Since the investigations of Claude *et al.* (1947) on the *Rous sarcoma*, this most frequently studied chicken neoplasm has often been examined by electron microscopy (Bernhard *et al.*, 1953; Epstein, 1956). Thanks to the introduction of ultrathin sectioning, it has been possible to identify particles suspected of being the pathogenic agents of this tumor (Gaylord, 1955; Bernhard *et al.*, 1956b; Epstein, 1957a). A recent investigation revealed a remarkable correlation between infectivity of these sarcomas

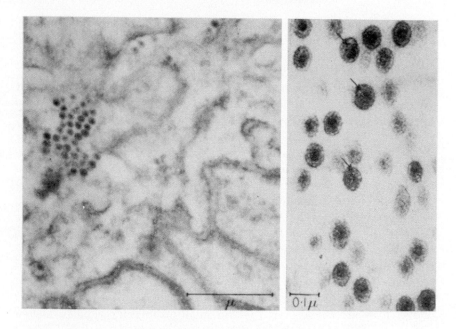

Fig. 45. Rous sarcoma, chicken. a. Group of viruses at the cell surface. Magnification: × 23,000. Same particles at higher magnification (× 80,000). Second membrane (→) well visible.

and their particle content, thus transforming what had been a hypothesis into certainty (Haguenau *et al.*, 1958; Epstein and Holt (1958). The occurrence of viruses in the tissues varies; there are relatively few in cytoplasmic vacuoles, yet more frequently they adhere to the cellular surface (Fig. 45a). High-resolution micrographs reveal internal structure in the virus, i.e., a central nucleoid, surrounded by an internal and external membrane (Fig. 45b) (Oberling *et al.*, 1957). The external diameter of these particles is about 70–80 mμ. In a small percentage of sarcomas there are granular areas which are regarded as viroplasm and in the center of which viruses are found. Morphologically identical particles have been found in other neoplastic diseases in chickens, viz., Murray-Begg's tumor (Rouiller *et al.*, 1956), the spleen and bone marrow of leukemic animals (erythroblastosis) (Benedetti *et al.*, 1958), purified plasma in myeloblastosis and erythroblastosis (Bernhard *et al.*, 1958), in myeloblasts of myeloblastosis (Bonar *et al.*, 1959), in Fujinami's tumor (Mannweiler and Bernhard, 1958), in the spleen of animals suffering from *neurolymphomatosis* (Hollmann, 1957), and in the liver in visceral lymphomatoses (Dmochowski and Grey, 1957). Benedetti (1957) has demonstrated

NUCLEAR   CYTOPLASMIC

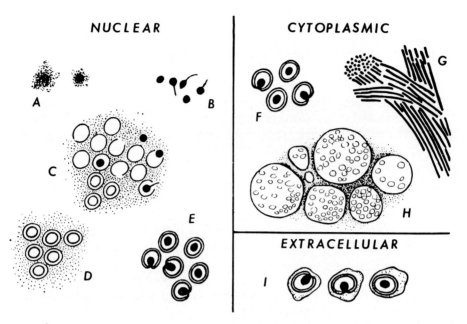

EXTRACELLULAR

Fɪɢ. 46. Diagram representing viruslike particles as found in the nucleus, the cytoplasm, and outside the cells of renal carcinomas of frogs. From Fawcett (1956).

that apparently normal control tissues may also be found to contain a certain number of these particles. From a theoretical point of view, these observations are important in that they suggest the existence of an extremely widespread family of viruses in fowl. Often innocuous and latent, these viruses may become carcinogenic under certain well-defined conditions.

5. Lucké's Tumor

A detailed electron microscopic investigation made by Fawcett (1956) in renal adenocarcinoma in leopard frogs showed that the nuclear inclusion bodies microscopically studied by Lucké (1934) probably correspond to virus particles suspected of being the carcinogenic agent. The infected cells show a marked increase of both DNA and RNA-synthesis (Duryee et al., 1960). These agents are extremely polymorphous in size and structure and seem to be part of a developmental cycle as yet only poorly established. Fawcett found these elements not only in the nucleus, but also in the cytoplasm. The schematic reconstruction of Fig. 46 gives an idea of their appearance in various phases of their development which is analogous to the evolution of the Herpes virus.

6. Human Virus Tumors

So far only small benign tumors are known to be transmissible by filtrates, viz., molluscum contagiosum, verrucae, venereal acanthomas, and the larynx papilloma.

The elementary bodies of molluscum contagiosum correspond to viruses belonging to the pox group. They are microscopically visible, but a detailed study requires electron microscopy. Melnick *et al.* (1952), Peters and Stoeckenius (1954), and Dourmashkin and Duperrat (1958), Dourmashkin and Bernhard (1959); Banfield and Brindley (1959) have been able to determine all the developmental stages of this virus, whose size is of the order of 220–250 mμ.

The nuclei of the stratum corneum of warts may contain inclusion bodies which have been shown to consist of thousands of virus particles in crystalloid formations (Melnick *et al.*, 1952). The particles measure 50–60 mμ in diameter. Their ultrastructure and developmental cycle are still unknown.

A study on the larynx papilloma has revealed the existence of cytoplasmic inclusions whose relationship to viral infection cannot yet be considered proved (Meessen and Schulz, 1957).

It has so far been impossible to demonstrate viral inclusions in malignant human tumors with certainty. Four reports have dealt so far with this problem. Leuchtenberger *et al.* (1956) found Feulgen-positive cytoplasmic inclusions in benign and malignantly degenerated rectal polyps. These might be correlated with the presence of a virus. On the other hand, for a human myxosarcoma, Leplus *et al.*, (1957) described cytoplasmic areas of fibrillar or granular aspect, in some way connected with the production of myxoid substance. The possibility of a viral origin comes to mind, but no definite conclusions have so far been reached. Dmochowski and Grey (1958) as well as Braunsteiner *et al.* (1959) have observed virus particles in acute human leukemias. Nothing is known about the significance of these findings. The presence of a non-specific, contaminating agent is not discarded.

III. Conclusions

Classical and modern investigations have failed to reveal a single morphological sign which is truly specific of cancer cells; however, it is possible to enumerate certain phenomena which, when taken together, characterize tumor cells. Although tumor tissue presents on the whole a more homogeneous appearance than do the numerous varieties of normal differentiated tissue, it does show considerable diversity as compared with the normal homologous tissue from which it originates. Cancer cells tend to have a more simple structural organization (dedifferentiation, anaplasia); their nucleocytoplasmic ratio is often changed in favor of the nucleus. The principal observations can be summarized as follows:

(A) *The nucleus*

(1) The nucleus in interphase often has a more irregular form than normal, and its size may be considerably increased. Every neoplasm, however, has its own pattern of nuclear behavior, and it is impossible to determine direct relationships between neoplastic transformation of a cell and hypertrophy of its nucleus.

(2) Heterochromatin anomalies in the resting nucleus are very marked in tumors but have not shown specific features, as yet.

(3) Cancer cells often show an increased nuclear deoxyribonucleic acid content, but this increase varies in different tumors and even within the same tumor, along with changes of the nuclear volume and chromosome number.

(4) The nucleolus is as a rule hypertrophic. There is no constant ultrastructural difference as compared with normal nucleoli.

(5) The nuclear membrane frequently shows deep invaginations; its ultrastructure does not seem to be changed as compared with that of the normal nucleus.

(6) The chromosomes show the most marked irregularities as to number and shape. Cancer cells are often heteroploid and the chromosomes may show considerable structural changes. There has never been any demonstration, so far, that these anomalies are a cause rather than a consequence of malignant transformation.

(7) Mitotic anomalies are very frequent; but all such irregularities can be experimentally reproduced in noncancerous tissues.

(8) Changes in the duration of mitosis are considerable. The metaphase seems to be relatively longer than in homologous normal tissues; the mitotic frequency is not increased in cancer tissue as compared with embryonic tissue in full growth.

(9) Amitosis, in the classic sense of the word, does not seem to play a role in the multiplication of cancer cells.

(B) *The cytoplasm*

(1) The *number* of mitochondria varies considerably from tumor to tumor. All the intermediate states between extreme richness and almost complete absence of chondriosomes are found; often, however, tumor cells tend to show fewer, smaller, and denser mitochondria. Phenomena of degeneration are also found, such as defective ultrastructure and swelling.

(2) Cytoplasmic basophilia is as a rule more marked, but may also be absent. On the ultrastructural level, this basophilia corresponds to an increase of RNP granules, freely dispersed in the cytoplasm, and rarely to an increase of membranous ergastoplasm. The latter, on the contrary, tends to disappear in the cancer cell.

(3) The Golgi apparatus can be very hypertrophic in certain tumors and may be hypotrophic in others. Its ultrastructure does not differ from that of the Golgi zone in normal cells.

(4) The ultrastructure of the centriole in cancer cells does not seem to be abnormal. The position of this organelle and of the Golgi apparatus determines the polarity of the cells, which may be reversed or completely lost in dedifferentiated cancer tissue.

(5) The cytoplasm of cancer cells may contain numerous inclusions of variable origin, resulting from degenerative processes, from partly preserved functional activity, from the phagocytic properties of certain cancers, or from the presence of virus corpuscles in filterable tumors.

(6) The ground substance shows no electron microscopic alterations.

(7) The ultrastructure of the cell membrane of cancer cells also seems to be identical to that of normal cells.

(8) So far it has been impossible to distinguish between normal and cancer cells on the basis of differences in staining dependent on any specific cytochemical reaction.

(9) In filterable tumors, the electron microscope has made it possible to demonstrate virus particles contained in the nucleus or the cytoplasm. Highly polymorphous inclusions reveal a complex development of oncogenic viruses. A study of their formation would seem to be particularly promising in future morphological investigations on cancer cells.

References

Abercrombie, M., and Heaysman, J. E. M. (1954). *Nature* **174**, 697–698.

Adams, W. R., and Prince, A. M. (1957). *J. Biophys. Biochem. Cytol.* **3**, 161–170.

Afzelius, B. A. (1955). *Exptl. Cell Research* **8**, 147–158.

Albertini, v. A. A. (1951). *Verhandl. deut. Ges. Pathol.* **35**, 54–70.

Albertini, v. A. (1958). *Schweiz. Z. Pathol. u. Bakteriol.* **21**, 773–820.

Alfert, M., and Geschwind, I. I. (1953). *Proc. Natl. Acad. Sci. U. S.* **39**, 991.

Allard, C., Mathieu, R., De Lamirande, G., and Cantero, A. (1952). *Cancer Research* **12**, 407–412.

Altmann, H. W. (1952). *Z. Krebsforsch.* **58**, 632–645.

Amano, S. (1957). *Symposia Soc. Cellular Chem.* **6**, 43–73.

Ambrose, E. J., James, A. M., and Lowich, J. H. B. (1956). *Nature* **177**, 576–577.

Ambrose, E. J. (1958). *Nature* **182**, 1419–1421.

Amistani, B. (1951). *Tumori* **35**, 1–23.

Amos, D. B. (1956). *Ann. N. Y. Acad. Sci.* **63**, 706–710.

André, R., Dreyfus, B., and Bessis, M. (1955). *Presse Méd.* **63**, 967–970.

Andres, A. H. (1932). Z. Zellforsch. u. mikroskop. Anat. 16, 88–122.
Anghelesco, V. (1948). Bull. Cancer 35, 217–221.
Arias-Stella, J. (1954). A.M.A. Arch. Pathol. 58, 112–128.
Askanazy, M. (1898). Arch klin. Med. 61, 118–186.
Arndt, G. (1935). Z. Krebsforsch. 41, 393–444.
Bagozzi, U. C. (1933). Tumori 19, III, Nota 1, 1–16.
Bahr, G. F., and Beermann, W. (1954). Exptl. Cell Research 6, 519–522.
Bairati, A., and Lehmann, F. E. (1952). Experientia 8, 60–61.
Banfield, W. G., and Brindley, D. C. (1959). Ann. N.Y. Acad. Sci. 81, 145–163.
Banfield, W. G., Dawe, C. J., and Brindley, D. C. (1959). J. Natl. Cancer Inst. 23, 1123–1135.
Bang, F. B. (1955). Federation Proc. 14, 619–632.
Bang, F. B., and Andervont, H. B. (1953). J. Appl. Phys. 24, 1418.
Bang, F. B., Vellisto, I., and Libert, R. (1956a). Bull. Johns Hopkins Hosp. 98, 255–285.
Bang, F. B., Andervont, H. B., and Vellisto, I. (1956b). Bull. Johns Hopkins Hosp. 98, 287–308.
Barigozzi, C. (1950). Portugaliae Acta Biol. Sér. A. 594–620.
Barigozzi, C., and Casabona, U. (1948). Experientia 4, 156–157.
Barr, M. L., and Bertram, E. G. (1949). Nature 163, 676–677.
Bauer, K. H. (1928). "Mutationstheorie der Geschwulstentstehung." Springer, Berlin.
Bauer, K. H. (1949). "Das Krebsproblem." Springer, Berlin.
Bayreuther, K. (1952). Z. Naturforsch. 7b, 554–557.
Bayreuther, K., and Theorell, B. (1959). Exptl. Cell Research 18, 370–373.
Bayreuther, K. (1960). Nature 186, 6–9.
Benedetti, E. L. (1957). Bull. Cancer 44, 473–482.
Benedetti, E. L., and Leplus, R. (1958). Rev. hématol. 13, 199–230.
Benedetti, E. L., Bernhard, W., and Oberling, C. (1956). Compt. rend. 242, 2891–2894.
Benninghoff, A. (1950). Anat. Nachr. 1, 50–52.
Berger, L. (1923). Arch. anat. Strasbourg 2, 255–306.
Bernhard, W. (1958). Cancer Research 18, 497–509.
Bernhard, W. (1960). Cancer Research 20, No. 5, 712–727.
Bernhard, W., and Bauer, A. (1955). In "Symposium on the Fine Structure of Cells," pp. 294–304. Interscience, New York.
Bernhard, W., and De Harven, E. (1956). Compt. rend. 242, 288–290.
Bernhard, W., Febvre, H. L., and Cramer, R. (1959). Compt. rend. 249, 483–485.
Bernhard, W., and Guérin, M. (1957). Proc. 2nd Int. Symposium Mammary on Tumors Perugia, 627–639.
Bernhard, W., and Guérin, M. (1959). Compt. rend. 247, 1802–1805.
Bernhard, W., and Gross, L. (1959). Compt. rend. 248, 160–163.
Bernhard, W., and Oberling, C. (1957). Proc. Can. Cancer Research Conf. 2nd Conf. Honey Harbour, Ontario, pp. 59–81.
Bernhard, W., Haguenau, F., and Oberling, C. (1952). Experientia 8, 58–59.
Bernhard, W., Dontcheff, A., Oberling, C., and Vigier, P. (1953). Bull. Cancer 40, 311–321.
Bernhard, W., Bauer, A., Gropp, A., Haguenau, F., and Oberling, C. (1955a). Exptl. Cell Research 9, 88–100.
Bernhard, W., Bauer, A., Guérin, M., and Oberling, C. (1955b). Bull. Cancer 42, 163–178.
Bernhard, W., Bauer, A., Harel, J., and Oberling, C. (1955c). Bull. Cancer 41, 423–444.
Bernhard, W., Guérin, M., and Oberling, C. (1956a). Acta Unio Intern. contra Cancrum 12, 544–557.

Bernhard, W., Oberling, C., and Vigier, P. (1956b). *Bull. Cancer* **43**, 407–422.

Bernhard, W., Bonar, R. A., Beard, D., and Beard, J. W. (1958). *Proc. Soc. Exptl. Biol. Med.* **97**, 48–52.

Berwick, L. (1959). *Cancer Research* **19**, 853–855.

Bessis, M., and Breton-Gorius, J. (1955). *Presse méd.* **63**, 189–192.

Bessis, M., and Breton–Gorius, J. (1957a). *Bull. microscop. appl.* **7**, 54–56.

Bessis, M., and Breton–Gorius, J. (1957b). *Compt. rend.* **244**, 2846–2847.

Biesele, J. J. (1945). *Cancer Research* **5**, 179–182.

Biesele, J. J. (1958). "Mitotic Poisons and the Cancer Problem." Elsevier Publishing Company.

Biesele, J. J., and Cowdry, E. V. (1944). *J. Natl. Cancer Inst.* **4**, 373–384.

Biesele, J. J., Poyner, M., and Painter, T. S. (1942). *Univ. Texas Publ. No.* **4243**, 1–68.

Boivin, A., Vendrely, R., and Vendrely, C. (1948). *Compt. rend.* **226**, 1061–1063.

Bolognari, A. (1959). *Biologica Latina.* **XII**, fasc. 4, 1015–1040.

Bonar, R. A., Parsons, D. F., Beaudreau, G. S., Becker, C., and Beard, J. W. (1959). *J. Natl. Cancer Inst.* **23**, 199–225.

Borrel, A. (1901). *Ann. Inst. Pasteur* **15**, 49–67.

Borysko, E., and Bang, F. B. (1951). *Bull. Johns Hopkins Hosp.* **89**, 468–473.

Bothe, A. E., Dalton, A. J., Hastings, W. S., and Zillesen, F. O. (1950). *J. Natl. Cancer Inst.* **11**, 239–243.

Boveri, T. (1914). "Zur Frage der Entstehung maligner Tumoren." Fischer, Jena.

Boyd, J. D., Lever, J. D., and Griffiths, A. N. (1959). *Ann. Otol. Rhinol. & Laringol.* **68**, 273–283.

Brachet, J. (1950). "Chemical Embryology." Interscience, New York.

Brachet, J. (1957). "Biochemical Cytology." Academic Press, New York.

Braunsteiner, H. (1958). *Klin. Wochschr.* **36**, 262–263.

Braunsteiner, H., Fellinger, K., and Pakesch, F. (1957). *Blood* **12**, 278–294.

Braunsteiner, H., Fellinger, K., and Pakesch, F. (1959). *Wien. Z. inn. Med. u. Grenzg.* **40**, 384–388.

Broders, A. C. (1926). *Arch. Pathol.* **2**, 376–380.

Bucher, O. (1951). *Arch. Julius Klaus–Stift. Vererbungsforsch. Soziale Anthropol. u. Rassenhyg.* **26**, 177–186.

Bucher, O. (1953). *Bull. microscop. appl.* **3**, 114–127.

Bucher, O., and Gattiker, R. (1950). *Rev. suisse zool.* **57**, 769–788.

Burdette, W. J. (1955). *Cancer Research* **15**, 201–226.

Calcutt, G. (1952). *Brit. J. Cancer* **6**, 197–199.

Callan, H. G., and Tomlin, S. G. (1950). *Proc. Roy. Soc.* **B137**, 367–378.

Carcupino, F. (1935). *Tumori* **9**, 561–568.

Carminati, V. (1935). *Biol. Abstr.* **9**, 14036.

Carrel, A., and Ebeling, A. H. (1928a). *J. Exptl. Med.* **48**, 105–124.

Carrel, A., and Ebeling, A. H. (1928b). *J. Exptl. Med.* **48**, 285–298.

Casey, A. E. (1937). *Am. J. Cancer* **29**, 47–56.

Caspersson, T. O. (1950). "Cell Growth and Cell Function—A Cytochemical Study." Norton, New York.

Caspersson, T. O., and Santesson, L. (1942). *Acta Radiol. (Suppl.)* **46**, 1–105.

Caspersson, T. O., and Santesson, L. (1944). *Acta Radiol.* **25**, 113–120.

Castelain, G., and Castelain, C. (1957). *Presse méd.* **65**, 1946–1947.

Catsaras, J. (1911). *Arch. pathol. Anat. u. Physiol. Virchow's* **204**, 105–116.

Cheronis, N. D., Straus, F. H., and Straus, E. (1948). *Science* **108**, 113–115.

Chèvremont, M., and Frédéric, J. (1952). *Arch. Biol. Paris* **63**, 259.

Chu, E. H. Y., and Giles, N. H. (1958). *J. Natl. Cancer Inst.* **20**, 383–396.

Clara, M. (1930). *Z. mikroskop. anat. Forsch.* **22,** 145–219.

Claude, A., Porter, K. R., and Pickels, E. G. (1947). *Cancer Research* **7,** 421–430.

Coman, D. R. (1944). *Cancer Research* **4,** 625–629.

Coman, D. R. (1953). *Cancer Research* **13,** 397–404.

Constantin, T., and Febvre, H. L. (1958). *Compt. rend.* **246,** 332–334.

Cooper, Z. K., and Reller, H. C. (1942). *J. Natl. Cancer Inst.* **2,** 335–344.

Cowdry, E. V. (1922). *Am. J. Anat.* **30,** 25–38.

Cowdry, E. V. (1955). "Cancer Cells." Saunders, Philadelphia. Pennsylvania.

Cowdry, E. V., and Paletta, F. X. (1941). *J. Natl. Cancer Inst.* **1,** 745–759.

Cruickshank, D. B. (1955). *Lancet* **i,** 253.

Cusmano, L. (1947). *Tumori* **33,** 10–19.

Dalton, A. J. (1959). *Lab. Invest.* **8,** 510–537.

Dalton, A. J., and Edwards, J. E. (1942). *J. Natl. Cancer Inst.* **2,** 565–575.

Dalton, A. J., and Felix, M. D. (1953). *Am. J. Anat.* **92,** 277–305.

Dalton, A. J., and Felix, M. D. (1954). *Am. J. Anat.* **94,** 171–208.

Dalton, A. J., and Felix, M. D. (1956a). *J. Biophys. Biochem. Cytol.* **2,** Suppl. 4, 79–83.

Dalton, A. J., and Felix, M. D. (1956b). *Ann. N.Y. Acad. Sci.* **63,** 1117–1140.

Dalton, A. J., Kahler, H., and Lloyd, B. J., Jr. (1951). *Anat. Record* **111,** 67.

David, H. (1958). *Arch. Geschwulstforsch.* **13,** 258–263.

Davies, H. G., and Engfeldt, B. (1954). *Lab. Invest.* **3,** 277–284.

Deckner, K. (1939). *Z. Krebsforsch.* **48,** 129–148.

De Harven, E., and Bernhard, W. (1956). *Z. Zellforsch. u. mikroskop. Anat.* **45,** 378–398.

De Harven, E., and Friend, C. (1958). *J. Biophys. Biochem. Cytol.* **4,** 151–156.

De Nabias, S., and Forestier, J. J. (1923). *Compt. rend. soc. biol.* **88,** 83–84.

Deuticke, K. (1935). *Z. Krebsforsch.* **43,** 39–53.

Dmochowski, L. (1954). *J. Natl. Cancer Inst.* **15,** 785–787.

Dmochowski, L. (1956). *Acta Unis. Intern. contra Cancrum* **12,** 582–618.

Dmochowski, L., and Grey, C. E. (1956). *Texas Repts. Biol. and Med.* **15,** 704–756.

Dmochowski, L., and Grey, C. E. (1957). *Ann. N.Y. Acad. Sci.* **68,** 559–615.

Dmochowski, L., and Grey, C. E. (1958). *Blood* **13,** 1017–1042.

Dmochowski, L., Haagensen, G. D., and Moore, D. H. (1955). *Acta Unio Intern. contra Cancrum* **11,** 640–645.

Dourmashkin, R., and Duperrat, B. (1958). *Compt. rend.* **246,** 313–335.

Dourmashkin, R., and Bernhard, W. (1959). *J. Ultrastructure Research* **3,** 11–38.

Duran-Reynals, F. (1945). *Cancer Research* **5,** 25–39.

Duryee, W. R., Long, M. E., Taylor, H. C., Jr., McKelway, W. P., and Ehrmann, R. L. (1960). *Science* **131,** 276–280.

Dustin, P., and Parmentier, R. (1953). *Gynécol. et obstét.* **42,** 258–265.

Edwards, G. A., Ruska, C., Ruska, H., and Skiff, J. V. (1959). *Cancer* **12,** 982–1002.

Ehrich, W. (1935). *Centr. Pathol.* **43,** 277–282.

Ehrich, W. (1936a). *Z. Krebsforsch.* **44,** 308–324.

Ehrich, W. (1936b). *Am. J. Med. Sci.* **192,** 772–789.

Ellermann, V. (1923). *Folia Haematol.* **28,** 207–216.

Epstein, M. A. (1956). *Brit. J. Cancer* **10,** 33–40.

Epstein, M. A. (1957a). *Brit. J. Cancer* **11,** 268–273.

Epstein, M. A. (1957b). *J. Biophys. Biochem. Cytol.* **3,** 851–858.

Epstein, M. A., Holt, S. J. (1958). *Brit. J. Cancer* **12,** 363–369.

Estable, C., and Sotelo, J. R. (1951). *Invest. Sci. Biol.* **1,** 105–126.

Estable, C., and Sotelo, J. R. (1955). *In* "Symposium on the Fine Structure of Cells," pp. 170–190. Interscience, New York.

Ewing, J. (1942). "Neoplastic Diseases," 4th ed. Saunders, Philadelphia, Pennsylvania.

Fabre-Domergue, F. (1898). "Les cancers épithéliaux." Carre and Naud, Paris.

Fardon, J. C. (1953). Science 177, 441–445.

Fawcett, D. W. (1955). J. Natl. Cancer Inst. 15, Suppl. 5, 1475–1503.

Fawcett, D. W. (1956). J. Biophys. Biochem. Cytol. 2, 725–742.

Fawcett, D. W., and Wilson, J. W. (1955). J. Natl. Cancer Inst. 15, 1505–1512.

Febvre, H., Harel, J., and Arnoult, J. (1957). Bull. Cancer 44, 92–105.

Fenner, F. R. (1953). Nature 171, 562–563.

Fischer, A. (1930). "Gewebszüchtung," 3rd ed. Müller & Skinick, Munich.

Fischer, A., and Davidson, F. (1939). Nature 143, 436–437.

Fogh, J., and Edwards, G. A. (1959). J. Natl. Cancer Inst. 23, 893–923.

Ford, C. E., Jacobs, P. A., and Lajtha, L. G. (1958). Nature 181, 1565–1568.

Foulds, L. (1956). Acta Unio. Intern. contra Cancrum 12, 619–622.

Frajola, W. J., Greider, M. H., and Bouroncle, B. A. (1958). Ann. N.Y. Acad. Sci. 73, 221–236.

Frédéric, J. (1954). Ann. N.Y. Acad. Sci. 58, 1246–1263.

Frédéric, J. (1958). Thèse agrégation, Univ. Liège, Impr. H. Vaillant Carmanne.

Friedlaender, M. (1959). J. Biophys. Biochem. Cytol. 5, 143–152.

Friedlaender, M., and Moore, D. H. (1956). Proc. Soc. Exptl. Biol. Med. 92, 828–831.

Fritz-Niggli, H. (1955). Oncologia 8, 121–135.

Fritz-Niggli, H. (1956). Acta Unio Intern. contra Cancrum 12, 623–637.

Furth, J., and Kahn, M. C. (1957). Am. J. Cancer 31, 276–282.

Gall, J. G. (1954). Exptl. Cell Research 7, 197–200.

Gansler, H., and Rouiller, C. (1956). Schweiz. Z. allgem. Pathol. u. Bakteriol. 19, 217–248.

Garnier, C. J. (1900). J. Anat. Physiol. 36, 22–98.

Gaylord, W. H. (1955). Cancer Research 15, 80–83.

Gaylord, W. H., and Melnick, J. L. (1953). J. Exptl. Med. 98, 157–171.

Gedigk, P. (1954). Arch. pathol. Anat. u. Physiol. Virchow's 325, 366–378.

Geitler, L. (1939). Chromosoma 1, 1–22.

Geitler, L. (1941). Ergeb. Biol. 18, 1–54.

Gey, G. O., Shapras, P., Bang, F. B., and Gey, M. K. (1955). In "Symposium on the Fine Structure of Cells," pp. 38–54. Interscience, New York.

Ghon, A., and Roman, B. (1916). Frankfurt. Z. Pathol. 19, 1–138.

Gläss, E. (1956). Chromosoma 7, 655–669.

Gläss, E. (1957a). Chromosoma 8, 468–492.

Gläss, E. (1957b). Naturwissenschaften 44, 639–640.

Gläss, E. (1960). Z. Krebsforsch. 63, 294–310.

Glatthaar, E., and Vogel, A. (1959). Gynaecologia 148, 1–8.

Gluge, G. (1837). Compt. rend. 4, 20–21.

Goldberg, L., Klein, E., and Klein, G. (1950). Exptl. Cell Research 1, 543–570.

Goldschmidt, R., and Fischer, A. (1929). Z. Krebsforsch. 30, 281–285.

Graffi, A. (1940). Z. Krebsforsch. 49, 477–495.

Gropp, A., and Hellweg, H. R. (1959). Z. Zellforsch. u. Mikroskop. Anat. 50, 315–331.

Gross, L. (1953). Proc. Soc. Expd. Biol. Med. 83, 414–421.

Grynfeltt, E. (1932). Compt. rend. Assoc. anat. 27, 343–351.

Grynfeltt, E. (1934). Arch. soc. sci. méd. biol. Montpellier et Languedoc 12, 665–668.

Grynfeltt, E. (1935). Biol. méd. (Paris) 25, 353–454.

Grynfeltt, E. (1937). Bull. Cancer 26, 354–360.

Grundmann, E. (1955). Verhandl. deut. Ges. Pathol. 38, 362–370.

Guérin, M. (1955). Bull. Cancer 42, 14–28.

Gusek, W. (1959a). *Beitr. Pathol. Anat. u. allgem. Pathol.* **120**, 302–318.
Gusek, W. (1959b). *Arch. ital. patol. clin. tumori* **3**, 259–277.
Guttman, P. H., and Halpern, S. (1935). *Am. J. Cancer* **25**, 802–806.
Haguenau, F. (1958). *Intern. Rev. Cytol.* **7**, 425–478.
Haguenau, F. (1959). *Bull. Cancer* **46**, 177–211.
Haguenau, F., and Bernhard, W. (1955a). *Arch. anat. microscop. morphol. exptl.* **44**, 27–55.
Haguenau, F., and Bernhard, W. (1955b). *Bull. Cancer* **42**, 537–544.
Haguenau, F., and Lacour, F. (1955). *In* "Symposium on the Fine Structure of Cells," pp. 316–322. Interscience, New York.
Haguenau, F., Rouiller, C., and Lacour, F. (1955). *Bull. Cancer* **42**, 350–357.
Haguenau, F., Dalton, A. J., and Moloney, J. B. (1958). *J. Natl. Cancer Inst.* **20**, 633–641.
Hamperl, H. (1950). *Arch. pathol. Anat. u. Physiol. Virchow's* **319**, 265–281.
Hamperl, H. (1956). *In* "Handbuch der allgemeinen Pathologie (F. Büchner, E. Letterer, and F. Roulet, eds.), Vol. 6/III, pp. 18–106, Springer, Berlin.
Hamperl, H., Kaufmann, C., and Ober, K. G. (1954). *Arch. Gynäkol.* **184**, 181–280.
Hansemann, D. V. (1890). *Arch. pathol. Anat. u. Physiol. Virchow's* **119**, 299–326.
Hansemann, D. V. (1920). *Z. Krebsforsch.* **17**, 172–191.
Hansen-Melander, E., Kullander, S., and Melander, J. (1956). *J. Natl. Cancer Inst.* **16**, 1067–1081.
Harel, J. (1956). *Compt. rend. soc. biol.* **150**, 139–142.
Harel, J., and Oberling, C. (1954). *Brit. J. Cancer* **8**, 353–360.
Harris, P., and James, T. W. (1952). *Experientia* **8**, 384.
Hauschka, T. S. (1953). *Trans. N.Y. Acad. Sci.* [2], **16**, 64–73.
Hauschka, T. S. (1958). *J. of Cellular Comp. Physiol.* **52**, 197–267.
Hauschka, T. S., and Levan, A. (1953). *Exptl. Cell Research* **4**, 457–467.
Hauschka, T. S., and Levan, A. (1958). *J. Natl. Cancer Inst.* **21**, 77–135.
Hauschka, T. S., Kredar, B. J., Grinnell, S. T., and Arms, B. D. (1956). *Ann. N.Y. Acad. Sci.* **63**, 683–705.
Heiberg, K. A. (1908). *Nord. Med. Arkiv* **8**, 1–20.
Heiberg, K. A. (1921). *Arch. pathol. Anat. u. Physiol. Virchow's* **234**, 469–480.
Heiberg, K. A. (1929). *Z. Krebsforsch.* **30**, 60–65.
Heiberg, K. A. (1934). "Einige Carcinome und Adenome." Kopenhagen and Leipzig.
Heiberg, K. A. (1935). *Acta Cancrol.* **1**, 3.
Heiberg, K. A. (1957). On progress and point of view regarding cancer research (privately printed).
Heiberg, K. A., and Kemp, T. (1929). *Arch. pathol. Anat. u. Physiol. Virchow's* **273**, 692–700.
Heine, U., Graffi, A., Helmcke, H. J., and Randt, A. (1957). *Z. ärztl. Fortbild.* **51**, 648–652.
Heinkele, T. (1936). *Z. Krebsforsch.* **43**, 323–336.
Heitz, E. (1935). *Z. Induktive Abstammungs- u. Vererbungslehre* **70**, 402–447.
Hellström, K. E. (1959). *J. Natl. Cancer Inst.* **23**, 1019–1033.
Hertwig, R. (1903). *Biol. Centr.* **23**, 49–62, 108–119.
Hertwig, R. (1908). *Arch. Zellforsch.* **1**, 1–32.
Hirschfeld, H., and Klee-Rawidowicz, E. (1929). *Z. Krebsforsch.* **30**, 406–427.
Höpker, W. (1953). *Z. Zellforsch. u. mikroskop. Anat.* **38**, 218–229.
Hoffman, J. G. (1953). "The Size and Growth of Tissue Cells." C. C Thomas, Springfield, Illinois.
Hollmann, K. H. (1957). Personal communication.

Homann, W. (1954). Z. *Krebsforsch.* **59**, 673–678.

Homann, W. (1955). Z. *Krebsforsch.* **60**, 283–290.

Horning, E. S., and Whittick, J. W. (1954). *Brit. J. Cancer* **8**, 451–457.

Hosokawa, K. (1950). *Gann* **41**, 236–237.

Hosokawa, K. (1951). *Gann* **42**, 343–345.

Howard, W. T., and Schultz, O. T. (1911). *Monographs Rockefeller Inst. Med. Research No.* **2**.

Howatson, A. F., and Ham, A. W. (1955). *Cancer Research* **15**, 62–69.

Howatson, A. F., and Ham, A. W. (1957a). *Proc. Can. Cancer Research Conf. 2nd Conf. Honey Harbour, Ontario, 1956*, pp. 17–58.

Howatson, A. F., and Ham, A. W. (1957b). *Can. J. Biochem. and Physiol.* **35**, 549–564.

Howatson, A. F., McCulloch, E. A., Almeida, J. D., Siminovitch, L., Axelrad, A. A., and Ham, A. W. (1960). *J. Natl. Cancer Inst.* **24**, 1131–52.

Hsu, T. C. (1954a). *J. Natl. Cancer Inst.* **14**, 905–933.

Hsu, T. C. (1954b). *Texas Repts. Biol. and Med.* **12**, 833–846.

Hsu, T. C., and Hoch-Ligeti, C. (1953). *Am. J. Pathol.* **29**, 105–111.

Hsu, T. C., and Klatt, O. (1958). *J. Natl. Cancer Inst.* **21**, 437–473.

Hsu, T. C., and Moorhead, P. S. (1957). *J. Natl. Cancer Inst.* **18**, 463–470.

Hsu, T. C., Pomerat, C. M., and Moorhead, P. S. (1957). *J. Natl. Cancer Inst.* **19**, 867–872.

Hunter, W. F., and Lennox, B. (1954). *Lancet* **ii**, 633–634.

Ishibashi, K. (1950). *Gann* **41**, 1–14.

Ising, U. (1957). *Acta Pathol. Microbiol. Scand.* **40**, 315–327.

Ising, U., and Levan, A. (1957). *Acta Pathol. Microbiol. Scand.* **40**, 13–24.

Jacobj, W. (1925). *Wilhelm Roux' Arch. Entwicklungsmech. Organ.* **106**, 124–192.

Jacobj, W. (1935). Z. *mikroskop. anat. Forsch.* **38**, 161–240.

Jacobj, W. (1942). *Wilhelm Roux' Arch. Entwicklungsmech. Organ.* **141**, 584–692.

Kahler, H., and Lloyd, B. J., Jr. (1952). *J. Natl. Cancer Inst.* **12**, 1167–1175.

Kasten, F. M. (1955). *J. Natl. Cancer Inst.* **16**, 579–589.

Kaziwara, K. (1954). *Cancer Research* **14**, 795–801.

Kinosita, R., Erickson, J. O., Armen, D. M., Dolch, M. E., and Ward, J. P. (1953). *Exptl. Cell Research* **4**, 353–361.

Klein, E., and Klein, G. (1950). *Nature* **166**, 832.

Klein, G., and Klein, E. (1958). *J. Cellular Comp. Physiol.* **52**, Suppl. I, 125–168.

Klein, G. (1951). *Exptl. Cell Research* **2**, 518–573.

Kloos, K., and Steffen, J. (1951). Z. *Krebsforsch.* **55**, 577–613.

Koller, P. C. (1943). *Nature* **151**, 244.

Koller, P. C. (1947a). *Brit. J. Cancer* **1**, 38–47.

Koller, P. C. (1947b). *Symposia Soc. Exptl. Biol.* **1**, 270–290.

Koller, P. C. (1956). *Ann. N.Y. Acad. Sci.* **63**, 793–816.

Kopac, M. J., and Mateyko, G. M. (1958). *Ann. N.Y. Acad. Sci.* **73**, 237–282.

Lambert, R. A. (1913). *J. Exptl. Med.* **17**, 499–510.

Lambert, R. A. (1914). *Zentr. Bakteriol. Parasitenk.* **60**, 178.

Langer, E. (1942). Z. *Krebsforsch.* **52**, 443–454.

Lebert, H. (1851). "Traité pratique des maladies cancéreuses et des affections curables confondues avec le cancer." Baillière, Paris.

Le Count, E. R. (1902). *J. Med. Research* **7**, 383–393.

Leduc, E. H., and Wilson, J. W. (1959a). *J. Histochem. and Cytochem.* **7**, 8–16.

Leduc, E. H., and Wilson, J. W. (1959b). *J. Biophys. Biochem. Cytol.* **6**, 427–430.

Leighton, J., Kalla, L., Kline, I., and Belkin, M. (1959). *Cancer Research* **19**, 23–27.

Leone, V., Hsu, T. C., and Pomerat, C. M. (1955). Z. Zellforsch. u. mikroskop. Anat. 41, 481–492.

Leplus, R., Bernhard, W., and Oberling, C. (1957). Compt. rend. 244, 2110–2113.

Lettré, H. (1953). Z. Krebsforsch. 59, 568–580.

Lettré, R. (1955). In "Symposium on the Fine Structure of Cells," pp. 141–150. Interscience, New York.

Lettré, H., and Lettré, R. (1954). Z. Krebsforsch. 60, 1–8.

Lettré, R., and Siebs, W. (1954). Z. Krebsforsch. 60, 19–30.

Leuchtenberger, C. (1954). "Statistics and Mathematics in Biology." Iowa State College Press, Ames, Iowa.

Leuchtenberger, C. (1956). Exptl. Cell Research 11, 506–507.

Leuchtenberger, C., Vendrely, R., and Vendrely, C. (1951). Proc. Natl. Acad. Sci. U. S. 37, 33–38.

Leuchtenberger, C., Helweg-Larsen, H. F., and Murmanis, L. (1954a). Lab. Invest. 3, 245–260.

Leuchtenberger, C., Leuchtenberger, R., and Davis, A. M. (1954b). Am. J. Pathol. 30, 65–85.

Leuchtenberger, C., Leuchtenberger, R., and Lieb, E. (1956). Acta Genet. et Statist. Med. 6, 291–297.

Levan, A. (1956a). Exptl. Cell Research 11, 613–629.

Levan, A. (1956b). Cancer 9, 648–663.

Levan, A., and Hauschka, T. S. (1953a). J. Natl. Cancer Inst. 14, 1–43.

Levan, A., and Hauschka, T. S. (1953b). Hereditas 39, 137–148.

Levine, M. (1931). Am. J. Cancer 15, 1410–1494.

Lewis, M. R., and Lewis, W. H. (1914–1915). Am. J. Anat. 17, 339–401.

Lewis, M. R., and Lewis, W. H. (1932). Am. J. Cancer 16, 1153–1183.

Lewis, M. R., and Lockwood, J. (1929). Bull. Johns Hopkins Hosp. 44, 187–198.

Lewis, W. H. (1931). Bull. Johns Hopkins Hosp. 49, 17–27.

Lewis, W. H. (1937). Am. J. Cancer 29, 666–679.

Lewis, W. H. (1939). Arch. exptl. Zellforsch. Gewebezücht. 23, 8–26.

Lewis, W. H. (1943). Cancer Research 3, 531–536.

Lewis, W. H. (1951). Ann. N.Y. Acad. Sci. 51, 1287–1294.

Liegeois, P. (1933). Bull. Cancer 22, 8–50.

Lipschütz, B. (1931). Z. Krebsforsch. 34, 299–312.

Lloyd, B. J., Jr., and Kahler, H. (1955). J. Natl. Cancer Inst. 15, 991–999.

Lockhart-Mummery, J. P. (1934). "The Origin of Cancer." Churchill, London.

Löblich, H. J., and Landschütz, Chr. (1960). Z. Krebsforsch. 63, 269–283.

Louis, C. J. (1958). Brit. J. Cancer 12, 537–546.

Love, R., and Rabson, A. S. (1959). J. Natl. Cancer Inst. 23, 875–891.

Lucké, B. (1934). Am. J. Cancer 20, 352–379.

Ludford, R. J. (1929). Proc. Roy. Soc. B104, 493–512.

Ludford, R. J. (1930). Sci. Repts. Invest. Imp. Cancer Research Fund 9, 109–119.

Ludford, R. J. (1934a). Sci. Repts. Invest. Imp. Cancer Research Fund 11, 147–168.

Ludford, R. J. (1934b). Sci. Repts. Invest. Imp. Cancer Research Fund 11, 169–177.

Ludford, R. J. (1952). In "Cytology and Cell Physiology" (G. Bourne, ed.), 2nd ed., pp. 373–418. Oxford Univ. Press, London and New York.

Ludford, R. J. (1953). J. Roy. Microscop. Soc. 73, 1–23.

Ludford, R. J. (1954). Brit. J. Cancer 8, 112–131.

Ludford, R. J., and Smiles, J. (1953). J. Roy. Microscop. Soc. 73, 173–178.

McCarty, W. C. (1923). J. Am. Med. Assoc. 81, 519–522.

McCarty, W. C. (1925). Arch. Clin. Cancer Research 1, 11.

McCarty, W. C. (1928). *J. Lab. Clin. Med.* **13**, 364–365.

McCarty, W. C. (1936). *Am. J. Cancer* **26**, 529–532.

McCarty, W. C. (1937). *Am. J. Cancer* **31**, 104–106.

McCarty, W. C., and Haumeder, E. (1934). *Am. J. Cancer* **20**, 403.

Makino, S. (1952a). *Chromosoma* **4**, 649–674.

Makino, S. (1952b). *Gann* **43**, 17–34.

Makino, S. (1956). *Ann. N.Y. Acad. Sci.* **63**, 818–830.

Makino, S. (1957). *Inter. Rev. Cytol.* **6**, 26–84.

Makino, S., and Hsu, T. C. (1954). *Cytologia (Tokyo)* **19**, 23–28.

Makino, S., and Kano, K. (1953). *J. Natl. Cancer Inst.* **13**, 1213–1235.

Makino, S., and Kano, K. (1955). *J. Natl. Cancer Inst.* **15**, 1165–1181.

Makino, S., and Nakahara, H. (1953). *Cytologia (Tokyo)* **18**, 128–132.

Makino, S., and Tanaka, H. (1953). *Texas Repts. Biol. and Med.* **11**, 588–592.

Makino, S., and Yoshida, T. H. (1951). *J. Fac. Sci. Hokkaido Univ. Ser. VI* **10**, 225–242.

Mannweiler, K., and Bernhard, W. (1957). *J. Ultrastructure Research* **1**, 158–169.

Mannweiler, K., and Bernhard, W. (1958). *Bull. Cancer* **45**, 223–236.

Marquardt, H., and Gläss, E. (1957a). *Chromosoma* **8**, 617–636.

Marquardt, H., and Gläss, E. (1957b). *Naturwissenschaften* **44**, 640.

Masserini, A. (1935). *Tumori* **9**, 548–560.

Masson, P. (1922). *Bull. Cancer* **11**, 345–365.

Masson, P. (1956). "Tumeurs humaines," 2nd ed. Maloine, Paris.

Matthews, V. S., Kirkman, H., and Bacon, R. L. (1947). *Proc. Soc. Exptl. Biol. Med.* **66**, 195–196.

Meesen, H., and Schulz, H. (1957). *Klin. Wochschr.* **35**, 771–773.

Mellors, R. C. (1955). *Cancer Research* **15**, 557–573.

Mellors, R. C. (1956). *Ann. N.Y. Acad. Sci.* **63**, 1117–1201.

Mellors, R. C., Keane, J. F., Jr., and Papanicolaou, G. N. (1952). *Science* **116**, 265–269.

Melnick, J. L., Bunting, H., Banfield, W. G., Strauss, M. J., and Gaylord, W. H. (1952). *Ann. N.Y. Acad. Sci.* **54**, 1214–1226.

Metais, P., and Mandel, P. (1950). *Compt. rend. soc. biol.* **144**, 277–279.

Michaelis, L. (1947). *Cold Spring Harbor Symposia Quant. Biol.* **12**, 131–141.

Miller, F., and Sitte, H. (1955). *Verhandl. deut. Ges. Pathol.* **39**, 183–189.

Moberger, G. (1954). *Acta Radiol.* (Suppl.) **12**, 1–108.

Montella, G. (1954). *Tumori* **40**, 232–243.

Moore, A. E. (1956). *Science* **124**, 127–129.

Moore, D. H., Lasfargues, E. Y., Murray, M. R., Haagensen, C. D., and Pollard, E. C. (1959). *J. Biophys. Biochem. Cytol.* **5**, 85–92.

Moore, K. L. (1955). *Anat. Record* **121**, 409–410.

Moore, K. L., and Barr, M. L. (1955). *Brit. J. Cancer* **9**, 246–256.

Moorhead, P. S., and Hsu, T. C. (1956). *J. Natl. Cancer Inst.* **16**, 1047–1066.

Morgan, C., Ellison, S. A., Rose, H. M., and Moore, D. H. (1954). *J. Exptl. Med.* **100**, 301–308.

Moricard, R. (1951). *Bull. federation soc. gynécol. et obstét* **4**, 108–132.

Müller, J. For a detailed account of his work with references, see Wolff, J. (1907). "Die Lehre von der Krebskrankheit von der ältesten Zeit bis zur Gegenwart," Pt. 1. Fischer, Jena.

Nagata, T. (1957). *Med. J. Shinshu Univ.* **2**, 199–208.

Nakahara, H. (1953). *J. Fac. Sci. Hokkaido Univ. Ser. II* **11**, 473–480.

Novikoff, A. B. (1957). *Cancer Research* **17**, 1010–1027.

Nowell, P. C., Hungerford, D. A., and Brooks, C. D. (1958). *Proc. Am. Assoc. Cancer Research* **2**, 331–332.

Oberling, Ch. (1959). *Krebsforsch. u. Krebsbekämpfung* **3**, 10–24.

Oberling, Ch., and Bernhard, W. (1955). *In* "Actualités de Biol. Cellulaire" (A. Thomas, ed.), pp. 279–300. Masson, Paris.

Oberling, Ch., Bernhard, W., Braunsteiner, H., and Febvre, H. L. (1950). *Bull. Cancer* **37**, 15–19.

Oberling, Ch., Bernhard, W., Febvre, H. L., and Harel, J. (1951). *Rev. hématol.* **6**, 395–400.

Oberling, Ch., Bernhard, W., Gautier, A., and Haguenau, F. (1953). *Presse méd.* **61**, 719–724.

Oberling, Ch., Bernhard, W., and Vigier, P. (1957). *Nature* **180**, 386–387.

Oberling, Ch., Rivière, M., and Haguenau, Fr. (1959). *Bull. Cancer* **46**, 356–381.

Orsos, F. (1935). *Verhandl. deut. pathol. Ges. Pt.* **3**, 95–155.

Ortiz Picon, J. M. (1930). *Arch. exptl. Oncol.* **1**, 277–296.

Page, R. C., Reagan, J. F., and McCarty, W. C. (1938). *Am. J. Cancer* **32**, 383–394.

Palade, G. E. (1953a). *J. Histochem. and Cytochem.* **1**, 188–211.

Palade, G. E. (1953b). *J. Appl. Phys.* **24**, 1419.

Palade, G. E. (1955a). *J. Biophys. Biochem. Cytol.* **1**, 59–67.

Palade, G. E. (1955b). *J. Biophys. Biochem. Cytol.* **1**, 567–582.

Palade, G. E., and Porter, K. R. (1954). *J. Exptl. Med.* **100**, 641–655.

Palade, G. E., and Siekevitz, P. (1956). *J. Biophys. Biochem. Cytol.* **2**, 671–690.

Palmer, C. G. (1959). *J. Natl. Cancer Inst.* **23**, 241–249.

Papanicolaou, G. N. (1954). "Atlas of Exfoliative Cytology." Harvard Univ. Press, Cambridge, Massachusetts.

Parmentier, R. (1953). *Nature* **171**, 1029.

Parmentier, R., and Dustin, P. Jr. (1948). *Nature* **161**, 527–528.

Parmentier, R., and Dustin, P., Jr. (1951). *Caryologia* **4**, 98–109.

Parmentier, R., and Dustin, P., Jr. (1953). *Rev. belge pathol. et méd. exptl.* **23**, 20–30.

Pasteels, J., Castiaux, P., and Vandermeerssche, G. (1959). *Arch. Biol.* **59**, 627–643.

Petermann, M. L., and Schneider, R. M. (1951). *Cancer Research* **11**, 485–489.

Peters, D., and Stoeckenius, W. (1954). *Z. Tropenmed. u. Parasitol.* **5**, 329–341.

Peters, J. J. (1946). *J. Exptl. Zool.* **103**, 33.

Peters, K. (1956). *Z. Zellforsch. u. mikroskop. Anat.* **44**, 14–26.

Pianese, G. (1896). *Ziegler's Beitr.* **142**, Suppl. 1, 1–193.

Policard, A., Collet, A., and Prégermain, S. (1957a). *Z. Zellforsch. u. mikroskop. Anat.* **46**, 147–154.

Policard, A., Collet, A., and Prégermain, S. (1957b). *Compt. rend.* **244**, 979–984.

Polli, E. (1951). *Experientia* **7**, 138.

Pollister, A. W., and Ris, H. (1947). *Cold Spring Harbor Symposia Quant. Biol.* **12**, 147–157.

Pollister, A. W., Swift, H. H., and Alfert, M. (1951). *J. Cellular Comp. Physiol.* **38**, 101–119.

Pollister, A. W., and Pollister, P. F. (1957). *Intern. Rev. Cytol.* **6**, 85–106.

Porter, K. R. (1953). *J. Exptl. Med.* **97**, 727–750.

Porter, K. R. (1955). *In* "Symposium on the Fine Structure of Cells," pp. 236–250. Interscience, New York.

Porter, K. R., and Bruni, C. (1959). *Cancer Research* **19**, 997–1009.

Porter, K. R., and Kallman, F. L. (1952). *Ann. N.Y. Acad. Sci.* **54**, 882–891.

Porter, K. R., and Thompson, H. P. (1948). *J. Exptl. Med.* **88**, 15–25.

Quensel, U. (1928). *Acta Med. Scand.* (Suppl.) **23**, 1–190.

Querner, H. (1955). *Z. Krebsforsch.* **60**, 307–315.

Ranvier, L. (1900). *Arch. anat. microscop.* **3**, 122–139.

Rasch, E., and Swift, H. (1953). *J. Histochem. and Cytochem.* **1**, 392.

Rather, L. J. (1958). *Ergeb. Allgem. Pathol. u. Pathol. Anat.* **38**, 127–199.

Regaud, C. (1923). *Bull. Cancer* **12**, 482–487.

Regen, H. J. (1951). Chromosomenzählungen der menschlichen Carcinomen zur Klärung des Wesens der Polyploidie, I. Dissertation, Kiel.

Reitalu, J. (1957). *Acta Genet. Med. et Gemellol.* **6**, 393–401.

Reitalu, J. (1957). *Acta Pathol. Microbiol. Scand.* **41**, 257–266.

Reller, H. C., and Cooper, Z. K. (1944). *Cancer Research* **4**, 236–239.

Rhodin, J. (1954). Thesis. Karolinska Inst. Stockholm.

Ribbert, H. (1914). "Geschwulstlehre," 2nd ed. Cohen, Bonn.

Richards, B. M., Walker, P. M. B., and Deeley, E. M. (1956). *Ann. N.Y. Acad. Sci.* **63**, 831–846.

Ris, H. (1953). *In* "Cell Division" (B. H. Willier, P. A. Weiss, and V. Hamburger, eds.), pp. 91–125. Saunders, Philadelphia, Pennsylvania.

Ris, H. (1957). *In* "The Chemical Basis of Heredity" (W. D. McElroy and B. Glass, eds.), pp. 23–62. Johns Hopkins Press, Baltimore, Maryland.

Ris, H., and Mirsky, A. E. (1949). *J. Gen. Physiol.* **33**, 125–146.

Rivière, M. (1956). *Bull. Cancer* **43**, 37–50.

Robertson, J. D. (1957). *J. Physiol.* **140**, 58–59.

Robineaux, R., and Bazin, S. (1958). Personal communication.

Rodermund, O. E. (1956). *Z. Krebsforsch.* **61**, 259–262.

Roskin, G. (1938). *Bull. histol. appl. physiol. et pathol. et tech. microscop.* **15**, 20–23.

Rouiller, C. (1957). *Ann. anat. pathol.* **2**, 548–562.

Rouiller, C., and Bernhard, W. (1956). *J. Biophys. Biochem. Cytology* **4**, Suppl. 2, 355–360.

Rouiller, C., and Gansler, H. (1955). *In* "Symposium on the Fine Structure of Cells," pp. 82–86, Interscience, New York.

Rouiller, C., Haguenau, F., Goldé, A., and Lacour, F. (1956). *Bull. Cancer* **43**, 10–22.

Roussy, G., Laborde, S., and Leroux, R. (1923). *Bull. Cancer* **12**, 467–481.

Sachs, L., and Gallily, R. (1955). *J. Natl. Cancer Inst.* **15**, 1267–1286.

Sachs, L., and Gallily, R. (1956). *J. Natl. Cancer Inst.* **16**, 803–834.

Scarpelli, D. G., and von Haam, E. (1957). *Cancer Research* **17**, 880–885.

Schairer, E. (1935a). *Z. Krebsforsch.* **43**, 1–38.

Schairer, E. (1935b). *Verhandl. deut. pathol. Ges.* **28**, 109–112.

Schairer, E. (1937). *Z. Krebsforsch.* **45**, 279–297.

Schairer, E. (1955). *Z. Krebsforsch.* **60**, 460–469.

Schmitz, W. (1934). *Z. Krebsforsch.* **41**, 372–381.

Schümmelfeder, N., and Wessel, W. (1956). *Z. Krebsforsch.* **61**, 214–226.

Schulz, H. (1957). *Oncologia* **10**, 307–329.

Seite, R. (1955). *Arch. anat. microscop. et morphol. exptl.* **44**, 89–139.

Selby, C. C., and Berger, R. E. (1952). *Cancer* **5**, 770–786.

Selby, C. C., Grey, C. E., Lichtenberg, S., Friend, C., Moore, A. E., and Biesele, J. J. (1954). *Cancer Research* **14**, 790–794.

Selby, C. C., Biesele, J. J., and Grey, C. E. (1956). *Ann. N.Y. Acad. Sci.* **63**, 748–767.

Severinghaus, A. E. (1937). *Physiol. Revs.* **17**, 556–588.

Shelton, E., Schneider, W. C., and Striebich, M. J. (1953). *Exptl. Cell Research* **4**, 32–41.

Sjöstrand, F. S. (1953). *Nature* **171**, 30–31.

Sohval, A., and Gaines, J. (1955). *Cancer* **8**, 896–902.

Stapel, E. (1935). Z. Krebsforsch. 42, 488–496.

Stewart, S. E., Eddy, E. B., and Borgese, N. (1958). J. Natl. Cancer Inst. 20, 1223–1243.

Stich, H. F., Florian, S. F., and Emson, H. E. (1960). J. Natl. Cancer Inst. 24, 471–482.

Stoeckenius, W. (1957a). Frankfurt. Z. Pathol. 68, 404–409.

Stoeckenius, W. (1957b). Exptl. Cell Research 13, 410–414.

Stowell, R. E. (1947). Symposia Soc. Exptl. Biol. 1, 190–205.

Stowell, R. E. (1949). Cancer 2, 121–131.

Sträuli, P. (1959). Schweiz. Z. allgem. Pathol. u. Bakteriol. 22, 368–375.

Strodtbeck, W. (1937). Z. Krebsforsch. 45, 268–278.

Strong, L. C. (1929). Am. J. Cancer Research 13, 103–115.

Sutton, R. L. (1938). Arch. Dermatol. u. Syphilis 37, 737–780.

Suzuki, T. (1957). Gann 48, 39–56.

Swift, H. H. (1950). Physiol. Zool. 13, 169–198.

Swift, H. H. (1956). J. Biophys. Biochem. Cytol. 2, 415–418.

Tanaka, H., Hanaoka, M., and Amano, S. (1957). Acta Haematol. Japon. 20, 85–98.

Tavares, A. S. (1955). Lancet 268, 948–949.

Therman, E., and Timonen, S. (1950). Hereditas 36, 343–405.

Thuringer, J. M. (1939). J. Invest. Dermatol. 2, 313–326.

Timonen, S., and Therman, E. (1950). Cancer Research 10, 431–439.

Tjio, J. H., and Levan, A. (1954). Lunds Universitets Arskrift 50, 1–39.

Tjio, J. H., and Levan, A. (1956a). Hereditas 42, 1–6.

Tjio, J. H., and Levan, A. (1956b). Hereditas 42, 218–234.

Tonomura, A. (1953). Dobytsugaku Zasshi 62, 411–415.

Vendrely, R., and Vendrely, C. (1957). "L'acide désoxyribonucléique (DNA), substance fondamentale de la cellule vivante." Legrand, Paris.

Virchow, R. (1858). "Die Cellularpathologie." Hirschwald, Berlin.

Virchow, R. (1907). In Wolff, J. (1907). "Die Lehre in der Krebskrankheit von der ältesten Zeit bis zur Gegenwart," Pt. 1. Fischer, Jena.

Vogel, A. (1958). Verhandl. deut. Ges. Pathol. 41 Tagung 285–295.

Vogt, C., and Vogt, O. (1947). Arztl. Forsch. 1, 8–14, 43–50.

von Möllendorff, W. (1938). Z. Zellforsch. u. mikroskop. Anat. 28, 512–546.

von Möllendorff, W. (1940). Z. Zellforsch. u. mikroskop. Anat. 29, 706–749.

Warburg, O. (1956). Oncologia 9, 75–84.

Warren, S. (1933). Am. J. Pathol. 9, 781–788.

Wassermann, F. (1929). In "Handbuch der Mikroskopischen Anatomie des Menschen" (W. von Möllendorff, ed.), Vol. 1, pp. 549–583. Springer, Berlin.

Watson, M. L. (1954). Biochem. et Biophys. Acta 15, 475–479.

Watson, M. L. (1955). J. Biophys. Biochem. Cytol. 1, 257–270.

Weiler, E. (1952). Z. Naturforsch. 7b, 324.

Weiler, E. (1956). Z. Naturforsch. 11b, 31–38.

Weiss, J. M. (1953). J. Exptl. Med. 98, 607–618.

Weissenfels, N. (1957). Krebsforsch. u. Krebsbekämpf. 2, 102–108.

Wellings, S. R., and Siegel, B. V. (1960). J. Natl. Cancer Inst. 24, 437–462.

Wermel, E. M., and Ignatjewa, Z. (1932). Z. Zellforsch. u. mikroskop. Anat. 16, 674–688.

Wessel, W. (1958). Virchows Arch. pathol. Anat. u. Physiol. 331, 314–328.

Wessel, W., and Bernhard, W. (1957). Z. Krebsforsch. 62, 140–162.

White, M. J. D. (1935). Proc. Roy. Soc. B119, 61–84.

Whitmann, R. C. (1919). Am. J. Cancer Research 5, 155–197.

Wilflingseder, P. (1947). Krebsarzt 2, 249–209.

Williams, R. C. (1957). *Intern. Rev. Cytol.* **6,** 130–191.

Wilson, J. W., and Leduc, E. H. (1948). *Am. J. Anat.* **82,** 353–391.

Winge, O. (1927). *Z. Zellforsch. u. mikroskop. Anat.* **6,** 397–423.

Winge, O. (1930). *Z. Zellforsch. u. mikroskop. Anat.* **10,** 683–735.

Worley, L. G., and Spater, H. W. (1952). *Quart. J. Microscop. Sci.* **93,** 413–425.

Yamada, E. (1955). *J. Biophys. Biochem. Cytol.* **1,** 445–458.

Yamada, E. (1957). *Acta Anat.* **29,** 267–290.

Yasuzumi, G. (1959). *Z. Zellforsch. u. mikroskop. Anat.* **50,** 110–120.

Yasuzumi, G., and Sugihara, R. (1958). *Cancer Research* **18,** 1167–1170.

Yerganian, G. (1955). *Proc. Am. Assoc. Cancer Research* **2,** 56.

Yokoyama, H. O., and Stowell, R. E. (1951). *J. Natl. Cancer Inst.* **11,** 939–945.

Yoshida, T. H. (1952a). *Gann* **43,** 35–43.

Yoshida, T. H. (1952b). *J. Natl. Cancer Inst.* **12,** 947–969.

Zollinger, H. U. (1948). *Schweiz. Z. Pathol. u. Bakteriol.* **11,** 42–634.

Zollinger, H. U. (1950). *Rev. hématol.* **5,** 696–745.

Zweibaum, J. (1933). *Arch. exptl. Zellforsch. Gewebezücht.* **14,** 358–390.

CHAPTER 8

Biochemistry and Physiology of the Cancer Cell

By ELIANE LE BRETON and YVONNE MOULÉ

I. Introduction[*]

Morphological studies, made both with the optical and the electron microscopes, have shown that although there is no specific character common to all cancer cells, a certain number of properties can nevertheless be attributed to them. These are a large nucleus, abnormally big or multiple nucleoli, chromatin unequally scattered in dense masses, an irregular nuclear membrane, and a high nucleoplasmic ratio (this volume, Chapter 7). Let us now see whether a general characteristic of the cancer cell is emerging from the innumerable results of biochemists' and physiologists' investigations.

Methods of studying the cell and its mode of functioning that have been developed in the last fifteen years have thrown new light on many aspects of cellular physiology. It was therefore legitimate to suppose that research on cancer, a disease of the cell, would profit by this work and that the study of the composition and of the enzymatic activities of the different cell structures would have unveiled specific characteristics of cancer cells. In spite, however, of numerous and interesting advances in various fields, a critical study of original papers, of books, and of recent reviews devoted to the subject shows that today the biochemist, like the cytologist, can give as a general definition of the cancer cell only its capacity for multiplication and invasion, leading to the death of the host. Although a certain number of qualitative and quantitative differences between a cancer cell and its normal counterpart have been brought to light, it is still impossible to supply answers to two really fundamental questions: (1) What is *the initial and specific lesion that gives rise to the tumor?* (2) What are *the characteristics of a cell that is becoming cancerous?* It is evident that difficulties of many sorts, encountered by the workers, are in part responsible for this situation, and we shall deal with the extent of these difficulties before giving an account of results obtained. Nevertheless, when one considers the accomplishments of the last few years and the progress attained one gets the impression that research is now following hopeful roads.

The study of the cancer cell has been approached from many different

---

[*] The following abbreviations are used: DNA, deoxyribonucleic acid; RNA, ribonucleic acid; PL, phospholipids; ATP, adenosine triphosphate; ADP, adenosine diphosphate; DPN, diphosphopyridine nucleotide; DPNH, diphosphopyridine nucleotide reduced form; UMP, uridine monophosphate; TMP, thymidine monophosphate; DAB, 4-dimethylaminoazobenzene; 3'-Me-DAB, 3'-methyl-4-dimethylaminoazobenzene; Pi, inorganic phosphorus; Q_{O_2}, microliters oxygen consumption per milligram dry weight per hour; Q_LN_2, micromoles anaerobic lactic acid per milligram dry weight per hour; Q_LO_2, micromoles aerobic lactic acid per milligram dry weight per hour.

points of view, but whatever the theoretical attitude may be, the bio-chemist at once encounters a major difficulty, namely the heterogeneity of the cancerous tissues. This heterogeneity consists not only in an unequal distribution of the various histological categories (connective tissue, par-enchyma, etc.) in different parts of the tumor, but more especially in a heterogeneity of the cancer cells themselves, which have been classified by Caspersson (1950) as A cells (with very active syntheses), B cells, and necrotic cells. Cells situated at the periphery of a cancer nodule have a high rate of division, while as one leaves these, the supply of oxygen and nutritive material become less and less good. The cells in general lose their potential of multiplication, they are deficient in nucleic acids and may be regarded as "sickly" cancer cells, becoming more and more necrosed toward the center of the tumor.*

The obstacle which the heterogeneity of the tissues presents has been emphasized by many workers. Greenstein (1954) discards the necrotic center of primary cancers and considers that the transplanted tumors, richer in active cells, furnish more favorable material. Likewise Khouvine and co-workers, for their preparations, separate the living from the necrosed cancer tissue in Guérin's epithelioma (Khouvine et al., 1953; Khouvine and Grégoire, 1953). The results of Zamecnik et al. (1951) clearly illustrate the consequences of heterogeneity and show that the problem is not solved by discarding the necrosed portion of the tumor. On making serial sections of a rat hepatoma cancer nodule, after in-jecting a labeled amino acid into the animal, they find that the specific activity as measured by autoradiography decreases progressively from the periphery toward the center of the tumor, thus emphasizing the im-portance of circulation. To overcome these obstacles certain biochemists have recently made use of ascites tumors, which correspond to a culture of cancer cells multiplying in the peritoneal cavity. The ascites liquid obtained by puncture is almost homogeneous if taken during the logarith-mic phase of growth.

There is, it appears to us, another type of heterogeneity, depending upon the conditions of experimentation and in particular on the use of different carcinogenic agents. In effect, it is conceivable that a carcinogen may act on sensitive cells in producing several categories of disorders: (1) specific cellular lesions that are causes or consequences of the carcino-genesis—it is these that the worker hopes to identify—and (2) accessory concomitant lesions, due to the toxicity of the agent and which would not alone be able to bring about a cancer. When (1) act together with (2) they

---

* The necrosis of the central tumor tissue is, in general, the more pronounced the greater is the rate of growth of the cancer.

interfere with the metabolism of the cancer cell and can mask the fundamental cellular lesions that are specific for the cancerous condition.

Before giving an account of research results, the concept of the
precancerous state should be considered. This concept has appealed to
biologists for practical, and particularly for theoretical, reasons. Undoubtedly if among cellular disequilibria one could point to those that
are irreversible and lead inevitably to specific anarchy of cell division,
a great advance would be made in understanding the mechanism of
cancer formation.

The long latent period which separates the action of the carcinogen
from its perceptible results, namely the multiplication of the affected cells,
permits a priori of all sorts of investigations. But it must be realized that
the biochemical significance of the precancerous state may be very
different according to the initial number of cells affected by the carcinogenesis. If it happens that only a few cells have at the start undergone
the premalignant transformation, then an analysis of the whole tissue
during the latent period is without significance because it will relate to
a whole population of cells among which only a few are of interest. It
could, of course, happen that the altered metabolism of these precancerous
islets would influence the organ as a whole, but it is above all the specific
lesion that is sought, not its repercussions.

If, on the contrary, it happens that the majority of the cells have been
initially altered by the carcinogen, but only a few of them become precociously cancerous, then an examination made at regular intervals will
supply information on the course of the phenomenon. In this alternative,
it is essential to set aside additional modifications not connected with the
carcinogenesis.

Apart from the initial number of cells affected, one may inquire into
the individual characteristics of these cells. It would indeed be of great
importance to be able to decide between the two following possibilities:
(1) the malignant change has the same probability of affecting all cells
in a population, or (2) some cells are more likely than others to undergo
carcinogenesis, being for some reason exceptions in the population. In
relation to this problem the work of Weiler (1956) is particularly interesting. Whereas all normal parenchymal cells are very uniformly shown up
by an antiserum tagged with a fluorescent dye, slices of precancerous
liver, like slices of hepatoma, show islands where the fluorescence is
markedly reduced. These results suggest that, in liver, cancerous cells
derive from parenchyma cells. It should be added that the irregularities
in fluorescence can be observed even before histological abnormalities
are visible.

A very large number of original publications have been devoted to the biochemistry of cancer cells; in addition, a certain number of books (Stern and Willheim, 1943; Potter, 1950; Greenstein, 1954), of symposia, and of reviews have regularly dealt with new results in this field (Potter, 1944; Zamecnik, 1950; Heidelberger, 1953, 1956; Begg, 1955, 1957; Haddow, 1955; Greenberg, 1955; Rondoni, 1955; Greenstein, 1956; Haven and Bloor, 1956; Weinhouse, 1955, 1956a, b; Kit and Griffin, 1958; Skipper and Bennett, 1958); in these the reader will find the very numerous original papers.

In the present article we have made no attempt at completeness. The aim has been quite a different one. Our purpose is to give an account of essential and significant facts emerging in the different fields explored, with emphasis on those giving rise to interesting working hypotheses. More space has been devoted to a discussion of the problem of glycolysis and respiration, not only because it has again come to the fore of recent years, but because it appears to be one of the most important aspects of the metabolism of the cancer cell. The article will conclude by summarizing, from among the theories that seek to explain carcinogenesis, those which are based on biochemical mechanisms.

II. COMPOSITION OF THE CANCER CELL

The majority of investigations on the biochemistry and the metabolism of the cancer cell have dealt with spontaneous or experimental tumors of the mouse, the rat, or more rarely the fowl. Among these cancers, hepatoma occupies the chief place. It can easily be produced in the two first-mentioned animals by a whole series of agents, or by transplantation (Novikoff, 1957); moreover, the liver is the most favorable organ for work at the cellular level. It was obvious that biochemists would rapidly apply to cancer, a cellular disease, the methods of differential centrifugation which permit work on the various isolated cell structures. By employing at the same time other recent techniques, such as chromatography, the use of radioisotopes, etc., it became possible to study the composition of these structures and to determine the part they play in the cell.

A. The Nucleus of the Cancer Cell

1. Deoxyribonucleic Acid

Among observations on the mechanism of carcinogenesis, the transmission of the specific characters of the neoplastic cell through successive generations focused the attention of biochemists on the nucleus and particularly on deoxyribonucleic acid, the main constituent of the genes. The investigations of biochemists became superimposed on those of cytologists on the nuclear anomalies of the cancerous cells.

Qualitative variations. There are few results to mention concerning the nucleotide composition of DNA in cancer tissues. It has been noted by Chargaff (1951a, b) that the DNA of a human hepatoma has the same composition as that of a normal liver. Identical results have been obtained by Uzman and Desoer (1954) for the highly polymerized DNA from leukemic, diseased but not leukemic, and normal human spleens, and by Butler *et al.* (1956) from tumors of rat and mouse. On the other hand, in liver tumors induced by azo dyes Griffin and Rhein (1951) found the DNA to show an increased concentration of guanylic acid. Likewise Khouvine and Grégoire (1953) were able to find differences of composition in the case of an atypical epithelioma in the rat: the DNA of the cancer tissue and that of ganglionic metastases contained less adenine and more thymine than DNA from the testes. To draw firm conclusions it would be necessary to have analyses of the DNA of numerous normal tissues and their cancerous counterparts.

In addition to the nucleotide composition of the DNA, carcinogenesis might affect its physicochemical properties. Thus Polli and Shooter (1958) find that differences exist between the sedimentation characteristics of DNA from normal leucocytes and from those of leukemia.

b. Quantitative variations. During the development of a tumor the successive mitoses evidently imply an increase in the *total quantity* of DNA; nevertheless, even in fully developed cancer tissues, the DNA content per gram is higher than in homologous normal tissues. These facts call for no comment; they are bound up with the phenomenon of cell division. On the other hand, the question may be put whether carcinogenesis involves variation in the *DNA content per nucleus* in the affected cells. This working hypothesis has given rise to numerous investigations, and an attempt will now be made to sort out the principal facts derived from them.

One of the chief characteristics emerging is the high frequency of polyploidy in cancer cells. It has been seen that often they have an abnormally large nucleus, and cytology has shown that this property corresponds to a polyploid or heteroploid condition of the chromosomes. In Chapter 7 of this volume will be found the references to the cytochemical and spectrophotometric data. The biochemical investigations express these facts in terms of quantities of DNA. Thus, in the mouse Menten *et al.* (1953) have shown that a lymphocyte from a normal spleen contains 4.6×10^{-9} mg. of DNA, whereas in the case of leukemia the figure is 6.6 to 7.2×10^{-9} mg. Similar values have been given by Mizen and Petermann (1952). Other instances of polyploidy or of increase in DNA content per nucleus in tumor cells have been published by a series of authors, in particular by Mirsky and Ris (1949), Klein (1951), Mac Indoe and Davidson (1952), Barer (1952), and White *et al.* (1953). It has been stated further by Leuchtenberger *et al.* (1958) that in the case of the tracheobronchial tree and lungs of mice exposed to cigarette smoke, increases in nuclear volume and in DNA are seen only when cells show a histological picture indicating cancer formation.

If polyploidy is relatively common in the nuclei of cancer cells, it cannot be considered as a specific characteristic inseparable from carcinogenesis. Indeed it can be induced by noncarcinogenic agents, and further there are many cases where tumors have a normal amount of DNA per nucleus. In the rat Price *et al.* (1950) find no difference in DNA content between nuclei of normal liver cells and of a hepatoma induced by DAB. Again, in rat hepatoma Mark and Ris (1949) find a DNA content of 6×10^{-9} mg. for the nuclei, a value similar to that of diploid nuclei in this animal. This result agrees with that of Thomson and Frazer (1954), who find a very high proportion of class I diploid nuclei. Carcinogenesis of liver parenchyma is said even to result in reducing the polyploidy that exists in the normal liver (Swift, 1950).

As regards bone marrow, Davidson *et al.* (1951) find no difference between the amounts of DNA in normal and leukemic bone marrow cells; Metais and Mandel (1950) had recorded analogous results. Mellors (1955) points out that polyploidy can be more or less pronounced according to the type of tumor: whereas carcinomas and adenomas frequently show increased DNA, this is very rare in lymphomas.

It is natural to suppose that this frequent, but not constant, increase in amount of DNA in the nucleus of cancer cells could result from the mitotic activity of these cells. Nevertheless, there is a question the theoretical importance of which cannot be denied: Is the *quantity of DNA the same per diploid set of chromosomes in a cancer cell and in a normal cell,* or does carcinogenesis result in variation in the amount of DNA for a given number of chromosomes? Unfortunately, this is a very difficult problem to solve, and one about which few precise data are available. Cunningham *et al.* (1950) have found that if polyploidy is taken into account, the DNA content per diploid set of chromosomes is normal in rat hepatoma. Leuchtenberger *et al.* (1954) have determined the amount of DNA in 2500 normal and cancerous cells of man, all studied at interphase. They find that for normal diploid nuclei the DNA contents have a low standard deviation around 5.6×10^{-9} mg.; for precancerous and cancerous nuclei the dispersion of the values is much greater in comparison with corresponding normal tissues, but this deviation, even in malignant cases, may be taken as a result of DNA synthesis in relation to mitotic activity. The example cited emphasizes the difficulties encountered in the determination of this characteristic of the cancer cell, fundamental as it is. To solve the problem, the determination should be made immediately after the end of mitosis, in the first stages of interphase, during which there is no DNA synthesis in normal cells. But a question arises at the outset: in the case of the cancer cell, does the synthesis of new molecules of DNA not recommence as soon as mitosis ends, at a rate determining the duration of interphase? On the other hand, it is evident that the biochemical methods available at present are not sufficiently sensitive to settle problems of this nature.

2. Ribonucleic Acid

Although only a few scattered data are available on the RNA content of the nuclei of cancer cells, it seems that their amount is in general greater than normal. Thus, Petermann and Schneider (1951) state that nuclei isolated from spleen of mice bearing transplanted leukemia have an abnormally high RNA:DNA ratio. The same has been found for Ehrlich ascites cells (Leuchtenberger and Klein, 1952).

The increase of nuclear RNA appears to be connected with the degree of activity of the tumor cells, a fact related without any doubt to the participation of RNA in synthetic reactions.

3. Proteins

In mitotic cells the synthesis of DNA is accompanied by a synthesis of various nuclear proteins, but there are few data on the quantities or nature of these proteins in cancer cells. Cruft et al. (1954) state that malignant cells may contain a smaller quantity of histones than the normal ones: thus the nuclei of a human breast carcinoma have 4% of their dry weight in the form of histones, whereas normal nuclei have 25–30%. But here again this is not a general rule and the authors interpret the phenomenon as a consequence, rather than a cause, of carcinogenesis. Apart from these quantitative variations, the same authors show that the histones of various malignant tumors (Walker rat carcinoma, mouse carcinoma 2146, human breast carcinoma, etc.) are less soluble and have lower electrophoretic mobilities. In the case of histones from rat hepatoma, there are even differences in lysine and arginine content compared with normal histones. More recently, Davison (1957) could find no difference between the histones from normal and malignant cells.

Furthermore, Moulé (1959) has studied the ratio between nuclear proteins and DNA in normal, precancerous, and cancerous liver. In the last two cases the author found a wide dispersion of the values compared to normal, indicating that the balance between nuclear proteins and DNA is disturbed in the course of carcinogenesis.

4. Lipids

Nuclear lipids have been studied particularly in liver cells. The low lipid content of the nucleus makes determinations difficult and above all necessitates special precautions being taken in the separation of this structure (Chauveau *et al.*, 1956).

In cancerous and precancerous cells the nuclear lipids, particularly the unsaponifiable matter, increase: the quantity of free cholesterol may be as much as forty-five times the normal (Clément *et al.*, 1954). The authors show that there is an effect, on the one hand, of the diet and, on the other, of the DAB used as carcinogen: diets inducing steatosis lead principally to an increase in cholesterol esters, whereas the addition of DAB results in a rise in free cholesterol, whether or not the cancer is established.

B. The Cytoplasm of the Cancer Cell

It appears that an increase in the nucleo-cytoplasmic ratio is a more or less general property of cancer cells. Obviously, variations in the size of the nucleus and of the cell sometimes make an exact estimation of the proportion of cytoplasm difficult. Taking this into account, however, it seems that the nucleus of a neoplastic cell exercises control on a more limited territory. Thus, calculations made from the figures of Laird and Miller (1953), Price *et al.* (1949a, b), and Petermann *et al.* (1956) show that there is much less cytoplasmic nitrogen *per cell* in an induced hepatoma than in normal liver; the same is true of hepatoma cells transplanted into the intraperitoneal cavity of the rat by Novikoff (1957; Allard *et al.*, 1957). This mode of expressing *amounts of constituents per cell* is more precise, for it brings out differences that are not evident when the results are given per unit wet weight.

The data relating to cytoplasm are both qualitative and quantitative. The most recent and interesting results concern the proportions and composition of the various cellular structures isolated by differential centrifugation.

1. Proteins

The proteins of cancer tissues have been much studied. It is clear that the discovery of an abnormal protein, specific for cancer would be of prime interest. The problem has been approached by different ways and some of the results obtained are contradictory.

The work of Kögl and Erxleben (1939), on the presence of *d*-glutamic acid in tumors, has been much criticized, and it seems that there is no undisputed fact in support of the existence of this isomer. In many instances the amino acid composition of cancer proteins is very close to that of their normal counterparts (Miller, 1950; Wiltshire, 1953; Bassi and Bernelli–Zayzera, 1954), but nevertheless modifications in composition have been reported (Schweigert *et al.*, 1949). At present, however, no general difference between proteins from normal and malignant tissues has been discovered.

It has been shown that the electrophoretic patterns of soluble proteins from hepatoma induced by azo dyes show differences compared with those from normal livers (Sorof and Cohen, 1951; Sorof *et al.*, 1951b; Eldredge and Luck, 1952), but it is impossible to demonstrate them for precancerous stages or for regenerating tissue after partial hepatectomy (Sorof *et al.*, 1951a). Likewise Miller *et al.* (1950a, b) have found that electrophoretic analysis of a muscle extract from the mouse reveals three components while electrophoresis of rhabdomyosarcoma shows seven; myosin is always present, but some of its properties are modified in the tumor.

As regards liver, Miller and Miller (1947) have shown that the ingestion of azo dyes leads to the formation of a protein-bound dye which appears very rapidly after the carcinogen is given. At first the amount of this complex increases with the time of ingestion and then decreases, to disappear completely from the tumor while the apparently healthy tissues surrounding the nodule still contain it. It has been shown by Hultin (1957) that the protein-dye complex appears first in the microsomal fraction; after a certain time the protein-bound dye is present in all the cell structures but more than half of it is found in the soluble cytoplasmic phase. These results have been confirmed by Gelboin *et al.* (1958), who add that "the dye is bound to certain liver proteins during their synthesis rather than to proteins found previously." Sorof *et al.* (1958) have reported that the h proteins contain the bulk of the soluble azoproteins and that in the course of the first weeks of ingestion of azocarcinogens the amount of slow-h_2 proteins increases. We shall see below what interpretation Miller and Miller (1953, 1955) have given to these facts.

Petermann *et al.* (1953) have studied the ultracentrifuge and electrophoretic patterns of proteins from normal and cancerous tissues. In both cases, about the same gradients are found but the relative proportions of the different components may vary. Thus, for submicroscopic ribonucleoprotein particles containing 50% RNA the authors have emphasized that component C, with a sedimentation constant of 40 S, is quantitatively very important in tumors and in growing tissues, as if it were connected with cell division (Petermann *et al.*, 1954, 1956; Petermann, 1954). This point is interesting in relation to the preponderant part known to be played in protein synthesis by microsomal structures rich in RNA (see the review by Brachet, 1958).

At this point, mention ought to be made of the immunological investigations on cancer tissues. It is understandable that the sensitivity and specificity of these methods should have been utilized with the aim of revealing specific antigens in malignant tissues. It is, however, impossible to discuss this topic here, owing to the complexity of the problem. The reader is referred to the review of Zilber (1958).

2. Ribonucleic acid

As far back as 1947 Schneider had showed that tumors of different origins have RNA contents per unit weight very like one another and that the values are fairly near to those of the corresponding normal tissues (Schneider, 1947). The figures collected by Leslie (1955) confirm these conclusions.

If the results are expressed *per cell* it is seen that, in general, *carcinogenesis produces a fall in ribonucleic acid* (Price et al., 1949a; Laird and Miller, 1953). Calculations of the RNA:DNA ratio lead to the same conclusion: thus, in a primary or transplanted hepatoma the value is close to 1, whereas for a normal liver cell 2.7–3.2 is usually found (Leslie, 1955; Novikoff, 1957).

For the Landschütz ascites tumor it has been found by Ledoux and Revell (1955) that when the mitotic index decreases (that is to say, when multiplication no longer follows an exponential curve), there is a marked fall in the RNA content per cell, without it being possible to attribute this fall to a disappearance of RNA precursors in the medium.

Moreover, it has been shown by De Lamirande and co-workers that in the rat the RNA of DAB-induced hepatoma differs from that of normal liver. In healthy tissue the RNA is said to have a nucleotide composition characteristic of each cellular structure (De Lamirande et al., 1955a); on the contrary, according to these authors, "in tumor cells, the nucleotide composition of RNA is similar in all fractions, but different from any fraction of the normal liver. This relative homogeneity is characterized by a high guanylic acid content" (De Lamirande and Allard, 1954; De Lamirande et al., 1955b). It should be noted that Butler et al. (1956) also found a high content of guanylic acid in the RNA of various tumors of the rat and the mouse. These results seem to show a tendency to uniformity of the RNA in hepatomas, but it must be pointed out that in normal liver Crosbie et al. (1953) then Elson et al. (1955) did not find any significant differences in the RNA of different cytoplasmic structures.

3. Lipids

A review by Haven and Bloor (1956) deals very adequately with the various aspects of the problem of lipids in cancer, and underlines the importance of these constituents in carcinogenesis. This present discussion will be limited to a few particular topics, with special emphasis on results obtained for cancerous and precancerous hepatic tissue.

As early as 1914 Bullock and Cramer came to the conclusion that various rat and mouse tumors contain more phospholipids and cholesterol than normal tissues (Bullock and Cramer, 1914); more recent results confirm these older findings (Lewis, 1927; Jowett, 1931; Willheim and Fuchs 1932; Bierich and Lang, 1933).

Other authors, however, have recorded results at variance with these. Randall (1940) found that in malignant neuromas the content of phospholipids and cholesterol is lower than in nervous tissue. Kandutsch and Baumann (1954) analyzed eleven types of neoplasms and found the Δ^7-cholestenol content to be identical in normal and cancerous tissues despite the fact that quantitative variations occur during the period preceding the appearance of the tumor. It should be noted that the amount of phospholipids in Novikoff's hepatoma is definitely lower than that in normal hepatic tissue (Novikoff, 1957; Weber and Cantero, 1957b). In addition, the ratio of the different phospholipids seems also to be affected by carcinogenesis (Albert and Johnson, 1954; Jablonski and Olson, 1955).

It follows from these investigations that although quantitative changes often do occur in neoplastic tissues, as compared with homologous normal tissues, the direction of these changes cannot be predicted. It is possible that these variations reflect real differences in the behavior of malignant cells in various tissues, but it is also possible that experimental difficulties are responsible for them.

As an example of the influence of carcinogenesis on lipids, the experiments of Wicks and Suntzeff (1942) on the mouse have often been quoted: in the course of carcinogenesis of the skin by methylcholanthrene they found a decrease of 40% in total lipids. But the investigations of Carruthers (1950) and later of Cambel (1951) on the rat show that this decrease is in reality due to the destruction of sebaceous glands by the action of the carcinogen and that this destruction is itself responsible for the cancer formation. After the carcinoma has developed it contains twice as much lipid phosphorus than in normal skin (Costello *et al.*, 1947). In this instance there is a superposition of injuries masking the true phenomenon.

An animal's diet has a direct effect on the lipid content of tissues and on the composition of the lipids. It is certain that in studies of carcinogenesis in rat liver, the nutritional factor is of prime importance. The diets normally used for inducing hepatoma in the rat by azo dyes are generally hypoproteic and deficient in certain vitamins (Copeland and Salmon, 1946;

Miller *et al.*, 1948). Whether or not these diets are deficient in choline, they rapidly cause profound changes in the liver tissue (the more profound, the more carcinogenic and toxic is the azo compound added) which result in phenomena such as steatosis, fibrosis, and cirrhosis (Miller and Miller, 1948; Striebich *et al.*, 1953; Daoust, 1955). Although it is known that *these histological disorders are not inevitable stages in carcinogenesis* (Dyer, 1951; Firminger, 1955) *but simply favorable factors,* these diets continue to be used. From the biochemical point of view, however, it is only the resultant of several reactions that is being studied, and it is evident that investigations carried out on lipids under these conditions must be regarded with caution.

Insofar as it is desirable to strip the phenomenon of cancer formation of concomitant lesions, the work of Le Breton and co-workers deserves mention. These authors adopted a semisynthetic balanced diet (20% proteins, 10% lard, and all the vitamins) deficient only in choline (Le Breton, 1954, 1955). In the presence of DAB they obtained hepatomas, not only without steatosis of the parenchyma, but also without fibrosis or cirrhosis of the hepatic tissue. Research on the lipids made under these experimental conditions would appear to be more valid.

As regards the nature of the phospholipids, Haven (1940) finds an increase of cephalin relative to lecithin in Walker 256 rat carcinoma; Johnson and Dutch (1952) get the same result with mammary carcinoma. On the other hand, the presence of acetalphospholipids has been reported in various carcinomas (Haven, 1940), but these compounds are perhaps merely evidence of mitotic activity, since they have also been demonstrated by Yarbro and Anderson (1954) in regenerating liver of the rat.

Note should be taken of an interesting fact published by Clément (1955). In the liver of rats fed on a complete diet, no phosphoric esters are found to accompany the phospholipids, but when carcinogenesis of the liver is induced by azo dyes these esters are always present in the tumor (where they may furnish 50% of the total lipid phosphorus), while the hepatic tissue surrounding the nodule may contain none.

Recently the presence of a new phospholipid has been reported in cancer tissues of man and animals (Kosaki *et al.*, 1958; Kosaki, 1958). This "malignolipin" is also present in the blood of cancer patients and according to the Japanese authors should have diagnostic possibilities.

There can be no doubt that, as Haven and Bloor (1956) have already pointed out, investigations on cell structures in liver tumor induced by DAB are perhaps the most interesting. Unfortunately they are up to now only fragmentary owing to the difficulties of quantitatively isolating the different phospholipids and their constituent fatty acids.

C. Changes in the Relative Proportions of Cell Structures during Carcinogenesis

We have seen that carcinogenesis often involves a decrease of cytoplasm in the cancer cell, but the question may now be asked: to what extent does this decrease affect the various cell structures?

In this field more than in any other, research has been concentrated on experimental hepatomas transplanted or induced by carcinogens; the systematic investigations of several groups of workers may here be quoted (Cantero *et al.*, Griffin *et al.*, Laird, Le Breton *et al.*, Novikoff, Price and the Millers, Schneider *et al.*).

One fact is at once clear, namely that per unit weight of tumor tissue there is a diminution in the quantity of mitochondria and microsomes, while the supernatant increases and likewise the nuclear fraction (Schneider, 1946; Price *et al.*, 1949a; Schneider and Hogeboom, 1950; Allard *et al.*, 1953, 1957; Laird and Miller, 1953; Fiala *et al.*, 1955; Novikoff, 1957). It appears to Laird and Barton (1956) that these variations correspond to a biochemical pattern common to many tumors.

But what are quantitative changes in the mitochondria and microsomes *per cell*? And do there exist *qualitative* differences in these components in neoplastic tissues?

1. Microsomes

Their importance in the process of protein synthesis is well known. The particular interest of the microsomes in cancer cells is thus comprehensible, when one considers the intensity of syntheses occurring in tumor.

It has been shown in Chapter 7 of this review that a decrease of ergastoplasm whose relationship with cytoplasmic basophilia and RNA has been demonstrated, is frequently revealed by the electron microscope. It is of interest that the biochemical analyses performed on a given type of cancer are in agreement with the observations of morphologists. It has been found by Price *et al.* (1949a, b, 1950) that in rat hepatoma the quantity of microsomes per gram of tissue is close to the normal, but *per cell carcinogenesis involves a definite decrease in hepatic microsomes* (Laird, 1954). In the Novikoff hepatoma, the decrease in microsomes appears to be particularly marked: Allard *et al.* (1957) have shown that per cancer cell the microsomal material undergoes a diminution of 80% in comparison with normal liver. Similar results had been reported by Fiala *et al.* (1955) for mouse hepatoma.

As regards the *composition* of the microsomes, Laird and Miller (1953) find that the RNA concentration in relation to the nitrogen is unchanged in hepatoma. On the other hand, in the small microsomes obtained from

Novikoff hepatoma transplants this concentration is greatly increased (110 instead of 50). This may be explained by the presence in this fraction of innumerable ribonucleoprotein particles, whereas microsomal membranes are extremely rare (Novikoff, 1957). The essential part played by these particles in the incorporation of amino acids and protein synthesis is well known (Littlefield *et al.*, 1955). It is probable that their increase is connected with the phenomena of rapid growth found in Novikoff tumors.

On the other hand, Charlot (1957) has shown that in tissue surrounding a DAB-induced hepatoma the amounts of polyethylenic acids, of phospholipids, and of neutral fats show significant variations compared with a normal liver. In the microsomes tetraenes decrease, whereas hexaenes and pentaenes disappear in the mitochondria.

There is no doubt that the discovery of any modification in nature, permeability (Arcos and Arcos, 1958), or activity of the microsomal membranes in the course of carcinogenesis are of prime importance in understanding the process. It would be desirable to make such investigations on very small nodules engaged in active proliferation.

2. Mitochondria

It is well known that Warburg considers a change in cell respiration to be the determining cause of malignancy. "If the respiration of a growing cell is disturbed, as a rule the cell dies. If it does not die, a tumor cell results" (Warburg, 1930). Knowing the nature and importance of mitochondria in these oxidation reactions, it is understandable that research has been directed to the obtaining of evidence for qualitative and quantitative variations in the mitochondria both in tumors and in precancerous tissues.

a. Cancerous tissues. All authors agree on the fact that *per gram of tissue, the quantity of mitochondria expressed as nitrogen is much smaller in a hepatoma than in the liver,* the decrease sometimes reaching 50–75% of the normal figure (Schneider, 1946; Price *et al.*, 1949a; Schneider and Hogeboom, 1950; Fiala *et al.*, 1955; Allard *et al.*, 1953, 1957; Novikoff, 1957). But, considering the variations in size of tumor cells, the only significant mode of expression of the results is the estimation of the mitochondrial material per cell.

In rat hepatoma induced by 2-acetylaminofluorine, Laird and Miller (1953) find that the quantity of mitochondria expressed as nitrogen is about one-sixth that of the normal cells; similar figures are obtained when using DAB as carcinogen. Some authors have attempted to estimate *the number* of mitochondria per nucleus by counting under the phase contrast microscope (Striebich *et al.*, 1953; Allard *et al.*, 1952; Klein, 1957):

the counts are made either on the total homogenate or on a suspension of isolated mitochondria.

In the mouse Shelton *et al.* (1952) find the same number of mitochondria per cell in hepatoma and in normal liver. In the rat Allard *et al.* (1952, 1953, 1957) arrive at an opposite conclusion: in DAB-induced primary tumor, they always find definitely fewer mitochondria per liver cancer cell than per normal cell. The results in their most recent paper give about 800 mitochondria per tumor cell as against 1400 in normal liver cell. As regards Novikoff hepatoma transplants, each cell appears to contain only a very small number of mitochondria (Allard *et al.*, 1957; Novikoff, 1957). It must be remarked that comparison of the results of different authors is made difficult by the fact that the number of mitochondria found for the normal liver varies much according to the workers: thus Striebich *et al.* give 500 mitochondria per cell (1953), whereas Allard *et al.* (1952, 1957) count 2500 or 1400 according to the cellularity of the livers of their animals.

In addition to quantitative variations in cancer cells, it seems well established that the mitochondria show *qualitative* modifications. Hogeboom and Schneider (1951) have shown that the soluble protein fraction released upon disruption of mitochondria from 98/15 hepatoma by sonic oscillations or high pressure lacks one of the sedimentation boundaries that is consistently observed in corresponding mitochondrial extracts from normal or regenerating liver. In addition, the mitochondria from a hepatoma are more fragile (Emmelot and Bos, 1957a) and their permeability to ions and coenzymes is altered; the experiments of Emmelot and Bos (1957b), of Clerici and Cudkowicz (1956), and of Mutolo and Abrignani (1957) give clear proof of this. It is evident that such modifications must influence the metabolic activity of the cell and the relative rates of enzymatic reactions taking place inside or outside of the mitochondrion.

Moreover, authors are on the whole in agreement that the mitochondria of tumors contain a lesser quantity of pyridine nucleotides than those of normal tissues (Carruthers and Suntzeff, 1952, 1954; Carruthers *et al.*, 1954). In general, the total amount of pyridine nucleotides of neoplastic cells appears to be much less than in corresponding normal cells (Jedeikin and Weinhouse, 1955; Jedeikin *et al.*, 1956; Emmelot, 1957a; Glock and Mac Lean, 1957; Sketol *et al.*, 1958). In the opinion of Wenner and Weinhouse (1953) the link uniting DPN to mitochondria in tumors is not so strong as in normal tissues, which would allow an easier passage of this enzyme toward the cytoplasm. Carruthers and Suntzeff (1954) attribute the low DPN concentration in tumor mitochondria to the particular susceptibility of the DPN to nucleotidases. The demonstration of nucleotide-splitting activity in the mitochondria of certain tumors would seem to support this opinion (Emmelot and Bos, 1956; Emmelot and Brombacher, 1956).

b. Precancerous tissues. When an azo dye is ingested for 28 days, a decrease is found in mitochondrial material per gram of liver, the magnitude of the decrease being proportional to the carcinogenic potencies of the various azo dyes in a given series (Price *et al.*, 1949b, 1950). Identical results have been obtained by Allard *et al.* (1957). But when the primary hepatoma has appeared, the quantity of mitochondria returns to a normal value in the apparently healthy tissue, whereas the decrease is accentuated in the tumor (Price *et al.*, 1949a).

It is known that the ingestion of azo dyes results, during carcinogenesis, in considerable tissue alterations (steatosis, fibrosis, biliary hyperplasia, cirrhosis), which are not the cause of hepatoma but are merely *additional* lesions. In order to make clear what it is that affects the parenchymatous cell in the precancerous stages, Striebich *et al.* (1953) have evaluated the proportion of the different cell types by the method of Chalkley (1943). Their results show that the ingestion of 3′-Me-DAB (which has a very high carcinogenic potency and is very toxic), produces after 28 days a severe diminution in the number of mitochondria per cell. But, taking into account the morphological changes, it becomes apparent that the volume of cytoplasm in a cell has decreased by about 50%, so that the number of mitochondria (or the amount of mitochondrial nitrogen) per unit of cytoplasm is little modified. Their results are summarized thus: compared with a normal hepatic cell "the preneoplastic liver cell contains fewer mitochondria of approximately normal biochemical composition and at normal cytoplasmic concentration, while the hepatoma cell contains the same number of mitochondria of greatly altered biochemical properties and at reduced cytoplasmic concentration" (Schneider *et al.*, 1953).

From the theoretical point of view it would be important to know if this decrease in mitochondria is really a specific lesion of carcinogenesis, but it does not seem to us possible to decide this at present. Nevertheless we think it worth remarking that some noncarcinogenic compounds, such as cortisone, can also induce considerable decreases of mitochondria in the liver of the rat (Lowe, 1955). But, on the other hand, when DAB is added to a complete diet which does not cause tissue alterations, no fall in mitochondria is observed after 4 weeks of ingestion (Le Breton *et al.*, 1958).

III. Metabolism of the Cancer Cell

In a normal cell, the constituents undergo a continuous renewal leading to a steady state. In a growing cell or in a cell with mitotic or secretory activity, the metabolic system performing "net" syntheses of cellular constituents, is no longer in the steady state. A cancer cell will necessarily have a considerable part of its metabolic activity engaged in these syntheses, the intensity of which depends directly upon the rate of growth of the tumor.

Taking these circumstances into account, it is understandable that for the biochemist a valuable method of approach is a *comparison, for the same tissue, of normal regulated growth with anarchic cancerous multiplication.* Skin epithelioma has been and continues to be used for cancer research, but the liver is a particularly suitable material for study, for the following reasons: (1) There are hardly any mitosis in the hepatic tissue of an adult animal (Brues and Marble, 1937). (2) Partial hepatectomy leads to intense cell multiplication, which remains perfectly controlled, coming to an end when the DNA content has returned to about its initial value (Harkness, 1957). (3) The development of a cancerous growth can be followed either in the primary tumor or in a transplanted one.

It may be desirable at this point to recall the conditions peculiar to malignancy. The circulation in tumors is frequently disturbed and irregular; it follows not only that the supply of materials for synthesis must be very different according to the topographic region of the neoplasm, but that the oxygenation will also vary, imposing on the cells all intermediate states between aerobiosis and anaerobiosis. The experiments of Zamecnik *et al.* (1951) are particularly suggestive in this connection.

These facts explain, partially at least, the difficulties encountered in comparing the results of different authors. Thus, after giving glycine-C^{14} to rats bearing a DAB-induced hepatoma, Griffin *et al.* (1950) find that the incorporation of the amino acid is more rapid in the noncancerous part of the liver than in the tumor; on the contrary, with tyrosine Kremen *et al.* (1949) obtain the opposite result. This may mean, as Greenstein (1954) has pointed out, that tyrosine and glycine have a different metabolism, but it could also be due to inherent differences in the experimental conditions of the two groups of workers (such as size of nodules, circulation, degree of necrosis, etc.). In the reviewers' opinion, the development of a tumor should never be considered without taking into account the biological conditions which accompany its growth.

On the other hand, it has often been emphasized that results obtained *in vitro* do not correspond to those obtained *in vivo* (Farber *et al.*, 1951; Greenstein, 1954; Greenberg, 1955). Thus, the rate of uptake of glycine-C^{14}

is greater in slices of hepatoma than in liver slices (Zamecnik *et al.*, 1948), whereas *in vivo* the opposite occurs (Griffin *et al.*, 1950; Norberg and Greenberg, 1951; Tyner *et al.*, 1952). These contradictions can be explained by the fact that in slices the cancerous and the normal tissues are in the same conditions relative to the isotope introduced into the medium, while *in situ* the supply of precursor to the tumor is deficient because of the bad circulation. Investigations conducted *in vitro* make the inequalities of supply between the tumor and the normal tissue disappear to a certain extent, thus bringing out the potentialities of the cells. This is very well illustrated by the differences in utilization of guanine *in vitro* and *in vivo* by normal and cancer tissues (Balis *et al.*, 1955; Balis, 1957; Bennett and Skipper, 1957).

If one makes an exception of the quantitative variations which are due to the synthetic activity of the cancer cell, certain fundamental problems still continue to present their complexity to the biochemist:

(1) Are there specific synthetic pathways, qualitatively different, in the cancer cell?

(2) What is the evolution of the enzymatic equipment of the cancer cell?

(3) What are its energy sources?

A. Synthetic Pathways in the Cancer Cell

1. Nucleic Acids

a. Deoxyribonucleic acid. It is usually considered that DNA is *metabolically stable* and that the incorporation of a precursor is a sign of a "net" synthesis connected with mitosis (Barnum *et al.*, 1957; Hecht and Potter, 1956, 1958; Sibatani, 1957). It is evident that in neoplasms there will be a synthesis of DNA, the greater, the more rapid is the growth. Thus, it has been observed that when a tumor grows very quickly, DNA has a specific activity greater than that found for RNA with the same precursor (Carlo and Mandel, 1953).

It is none the less true that DNA presents to the biologist a crucial question which is perhaps the key to the problem of carcinogenesis: *Why and how does DNA synthesis escape the control to which it is subjected in a normal cell?* Are the pathways of DNA synthesis in tumors the same as those of normal cell multiplication? We shall return to these questions in the last part of this chapter, in connection with mechanisms of carcinogenesis, but without further delay certain results must be considered.

In the absence of DNA synthesis the amount of acid-soluble deoxyribosidic compounds is very small (Potter *et al.*, 1957) and the major component found in normal liver is deoxycytidine (Schneider, 1955, 1957); in growing tissues and regenerating liver their amount is much increased (Schneider and Brownell, 1957; Rotherham and Schneider, 1958). Moreover, in the Novikoff hepatoma the deoxynucleotides are different from those found in normal tissues (Potter *et al.*, 1958; Schneider and Rotherham, 1958).

The presence of these compounds assumes at the present time a prime interest: not only do we know that they may serve as precursors to the pyrimidine portions of DNA (Bollum and Potter, 1957; Friedkin and Kornberg, 1957; Kit, 1957; Kornberg, 1957), but in addition it is in the alternative pathways of their metabolism that may consist the bifurcation between normal and uncontrolled growth. Indeed, Potter (1957) puts forward the suggestion that the equilibrium of a reaction depends directly on "two opposing influences which may be considered in terms of a positive factor, the inducer, which in most cases is probably the substrate for the enzyme, and a negative feedback mechanism which in most cases is probably a distal product of the enzyme being synthesized" (Potter, 1957). It is evident that if, for some reason, the negative feedback mechanism does not function any more, there will be no brake on the reaction and "the biosynthetic pathway might be stimulated by the deletion of a competing catabolic pathway" (Potter, 1958b). And this American biochemist indeed puts forward the hypothesis that carcinogenesis could consist of the loss of specific enzyme-forming systems which insure the control of cell multiplication through a negative feedback mechanism. Concerning DNA synthesis, "it appears that an important bottleneck may be the conversion of UMP to TMP or of uridine to thymidine. . . . The problem is to find a feedback mechanism that will suppress TMP formation when DNA is not needed and increase its formation when it is needed" (Potter, 1957).

b. Ribonucleic acid. The synthesis of RNA can be followed experimentally with the aid of various precursors introduced into the medium, and the results obtained with normal tissues have formed the subject of very complete reviews (Bendich, 1952; Brown and Roll, 1955; Schlenk, 1955; Smellie, 1955).

Cancerous tissues, too, are able to make use of these compounds in their anabolic stages (see Heidelberger, 1953, 1956; Greenberg, 1955; Skipper and Bennett, 1958). Thus, glycine is very rapidly incorporated into ascites tumor cells (Eliasson *et al.*, 1951; Le Page, 1953; Le Page *et al.*, 1956; Williams and Le Page, 1958a, b, c), into uterine sarcoma cells (Pileri and Ledoux, 1957), into sarcoma 180 (Barclay and Garfinkel, 1955), etc. Adenine and formate are also good precursors (Forssberg *et al.*, 1958), and so too glucose (Schmitz *et al.*, 1954a, b, c; Kit *et al.*, 1957), phosphate (Mortreuil and Khouvine, 1957; Yagi and Khouvine, 1957), and orotic acid, a specific precursor of pyrimidine bases. In the case of human uterus carcinoma (strain HeLa), Pileri and Ledoux (1957) have shown that the cells are capable of directly utilizing preformed bases, as well as synthesizing them from smaller molecules. Many results suggest that, in general, the intermediate stages in the synthesis of RNA in cancer tissues are the same as in normal tissues (see Skipper and Bennett, 1958). Nevertheless, certain interesting points must be emphasized here.

The first point concerns the part played by glucose as precursor of the ribose portion of nucleotides. It is known, on the one hand, that the oxidative shunt of glucose exists in tumors (Schmitz et al., 1954a, b, c); not only has the mechanism been demonstrated, but the results obtained, using different tumors, show that the oxidative pathway of glucose toward gluconic acid and pentoses is quantitatively more important in cancer tissues than in corresponding normal tissues. Thus, Kit (1956) finds that lymphatic tumor cells produce from glucose a quantity of pentoses three to five times greater than that formed by normal lymphatic cells. Similar results have been obtained by other authors (Abraham et al., 1955, 1956; Emmelot, 1955; Villavicencio and Barron, 1955; Wenner and Weinhouse, 1956). It has recently been shown that the transketolase-transaldolase pathway occurs in malignant lymphatic cells (Kit et al., 1957), in human carcinoma cells (Hiatt, 1957) and in the Ehrlich ascites tumor (Wenner et al., 1957). The participation of this mechanism may be quite important.

Another interesting point concerns the metabolism of uracil in cancer tissues. Uracil is degraded by normal liver and is poorly incorporated into RNA; in contrast to this, hepatoma and the Flexner-Jobling carcinoma are able to use free uracil provided by the medium (Rutman et al., 1954; Canellakis, 1957; Heidelberger et al., 1957; Melnick et al., 1958). These results have been confirmed by the observations of Reichard and Skold (1958); they studied enzymes concerned with uracil metabolism and found that there are "much higher levels of enzymes involved in anabolic reactions in ascites tumour extracts than in liver, whereas, the enzymes of uracil catabolism which are present in rat and mouse liver, are absent from the tumour." We must mention, in conclusion, that Weed (1951) had already found that orotic acid is incorporated more rapidly into tumor slices than into slices of normal tissue, and that the ratio of the amounts of uridylic and cytidylic acids taken up, which is from 2 to 4 for normal and regenerating liver, varies from 7 to 10 for hepatoma.

In another connection, Auerbach and Waisman (1958) and Calva et al. (1959) have stressed the importance of the de novo synthesis of uracil derivatives by malignant tissues. Thus, in primary and transplanted hepatoma, and in Ehrlich ascites cells, there are increased levels of carbamyl phosphate aspartate transcarbamylase, an enzyme that catalyzes one of the steps in the formation of uridine-5′-phosphate. These experimental results have led Potter to formulate the hypothesis that here also carcinogenesis might involve the suppression of a negative feedback mechanism controlling the metabolism of uracil derivatives (Potter, 1958a).

Observations of this kind are very important inasmuch as they can direct the use of antimetabolites in chemotherapy (see the publications of Heidelberger and his co-workers, and of Skipper and his co-workers).

2. Proteins and Amino Acids

Malignant growth and multiplication imply the existence of a "net" protein synthesis in the tumors. It is known that cancer tissues are capable of building up their proteins de novo from small molecules supplied by the medium, and this incorporation can be followed experimentally with the aid of radioactive precursors (see the recent review of Campbell, 1958). But jointly with these syntheses, does a tumor live at the expense of its host, taking up complex molecules already elaborated?

This possibility is supported by certain experimental data (Griffin et al., 1950; Yuile et al., 1951; Le Page et al., 1952; Tyner et al., 1952), as well as by results published by Babson and Winnick (1954), who found that the activity of a plasma protein previously labeled by leucine-C^{14} is taken over by tumors without any decrease being detected in the activity of the free leucine present. This would mean that tumors make use of preformed polypeptides to build up their own proteins. The findings of Busch et al. (1956; Busch and Greene, 1955) are also in agreement with these conclusions. In contrast to this, more recent experiments of Campbell and Stone (1957) lead to an opposite conclusion, for after the injection of homologous labeled serum albumin into a rat bearing a primary hepatoma, "there is no evidence that serum albumin is utilized for the synthesis of soluble tissue proteins without first being broken down to free amino acids. In this respect, liver tumour did not appear to differ from the other tissues studied."

Regarding the mechanism of protein anabolism, many arguments plead in favor of the direct participation of microsomes, and more exactly of ribonucleoprotein particles rich in RNA (Littlefield et al., 1955), the energy of the reaction being supplied by the mitochondria (Siekevitz and Zamecnik, 1951). Only a few results are concerned with the incorporation of amino acids into the microsomes of cancer tissues. It has been shown by Littlefield and Keller (1957) that for Ehrlich ascites cells "most of the amino acids incorporated into whole cell proteins pass through the ribonucleoprotein particles." On the other hand, Emmelot (1957b), using an in vitro technique, has found that the ability of microsomes from rat and mouse cancer tissues to incorporate leucine-C^{14} varies considerably with the tumor. Campbell and Greengard (1957) found too that the microsomes of primary hepatoma incorporate only very little of the amino acids present in the medium. The significance of these facts might be, in our opinion, that the tumor microsomes are deficient in certain coenzymes or other factors, and if the system were adequately supplemented, better incorporation could be obtained.

It is known, on the other hand, that in cancer tissues protein catabolism may be subject to certain deviations compared to normal mechanisms. Thus, whereas in the liver and the kidney glutamic acid is transformed into arginine through an intermediary stage of ornithine, in tumors this stage is said not to exist (Greenberg, 1955). It has also been stated by Claudatus and Ginors (1957) that kynurenine is not metabolized by hepatoma, contrary to what occurs in normal liver. Finally, it has been found that rats bearing sarcomas excrete in the urine up to nine to twelve times more creatine than normal rats, which is a proof of considerable catabolism of certain amino acids in these animals (Bach and Maw, 1953).

There is another aspect of metabolism which would certainly be very interesting to explore: this is *the turnover of the different cell constituents before malignant growth actually commences,* during a period in which cells do not yet multiply, but *are already on the road leading with certainty to carcinogenesis.* Undoubtedly, modifications in the turnover of the membrane phospholipids, of the nucleic acids and of certain proteins (structural proteins, enzymes?) would be of great interest in understanding the phenomenon of carcinogenesis. It seems likely that such investigations could at the present time be undertaken in the case of the liver.

B. *The Enzyme Equipment of the Cancer Cell and the Problem of Dedifferentiation*

The interest attached to the study of the enzyme characteristics of cancer cells is undoubted. Indeed, the mode of functioning of cells is really the algebraic sum of regulated and coordinated activities of enzymatic systems present in the diverse structures and different tissues of an individual. Although the presence of an apoenzyme is determined by the genetic pattern of the cell, the degree of activity of the system is a function of the particular type of cell and of the physiological conditions of the animal. It is understandable then that biochemists have tried to grasp in different cancer cells an enzymatic behavior specific to "the" cancer cell, and capable perhaps of explaining its biological properties.

It is appropriate here to quote Greenstein, who, with his co-workers, has attempted to codify the enzymatic characteristics of cancer cells and to trace their general lines (Greenstein, 1945, 1954, 1956). The essential points of his concept may be summarized as follows:

(1) If the enzymatic equipment of a cancer tissue is compared with that of its normal counterpart, the disappearance of a certain number of functional activities becomes apparent; as a result of this, characteristics differentiating between tissues from which the tumors originate also tend to disappear.

(2) A given type of tumor tends to have the same enzymatic equipment, whatever be its age, its rate of growth, and the animal species from which it comes.

(3) For a given enzymatic system, the range of variation between the different types of tumors is always smaller than that found among the healthy original tissues.

This double tendency in neoplasms, toward *a leveling of their enzymatic activities* and a *disappearance of some of their functional systems,* is reflected on the biochemical plane by the appearance of a common enzymatic pattern, to which the different tumors converge.

In his book Greenstein (1954) has developed experimental arguments that support these conclusions; some of them are the following: carcinogenesis in the gastric mucosa involves the disappearance of pepsin and rennin secretion (Greenstein and Stewart, 1942); at the same time it is difficult to distinguish by their enzymatic equipment sarcomas, carcinomas, adenocarcinomas, and lymphomas, even when the tissues of origin of these tumors are functionally as different as the gastric mucosa and the liver (Greenstein, 1954).

The wealth of the hepatic cell in differentiated enzymatic systems has given rise to particularly significant findings in the case of hepatomas, namely the disappearance of the activity of cystine desulfurase (Hirschberg *et al.*, 1952), of dehydropeptidase II (Greenstein and Leuthardt, 1946; Levintow *et al.*, 1950; Greenstein *et al.*, 1949), of aminooxidase (Shack, 1943), and of glucose-6-phosphatase (Weber and Cantero, 1957a); as well as a marked lowering of the activity of arginase (Greenstein *et al.*, 1941) and of the enzyme that metabolizes *p*-aminohippuric acid (Tung and Cohen, 1950). It is known too that the catalase content of the liver decreases progressively in the course of DAB ingestion (Mori and Manoki, 1952; Seabra and Deutsch, 1955; Adams and Burgess, 1957), and that the failure of a hepatoma to store glycogen (Olson, 1951; Goranson, 1955) is due in part to the low activity of its phosphorylase (Hadjiolov and Dancheva, 1958). The fat metabolism of a hepatoma offers a particularly good example of the loss of functional systems. Normal liver tissue is distinguished from other tissues by the fact that it is the site of a marked ketogenesis; hepatomas, on the contrary, only produce small quantities of ketonic bodies, although possessing a high capacity for the oxidation of fats (Dickens and Simer, 1931; Chapman *et al.*, 1954). The work of Weinhouse and his co-workers (1951, 1953) has shown more precisely that this result was attained by a diminution of ketogenesis and an increase of ketolysis, which tends to bring hepatoma closer to other tissues of the organism.

Novikoff's transplanted hepatoma (1957) has been very much studied in the last few years by several groups of workers. From an enzymatic point of view, it shows itself to be particularly dedifferentiated, with a loss of uricase, of xanthine oxidase (De Lamirande and Allard, 1957), of glucose-6-phosphatase (Weber and Cantero, 1957a), of glutamic dehydrogenase (Allard *et al.*, 1957), and of transhydrogenase (Reynafarje and Potter, 1957), with a feeble activity of esterase and phosphoglucomutase (Weber and Cantero, 1957a). Also, the work of Auerbach and Waisman (1958) has shown that this tumor is lacking in about twelve enzymes involved in the amino acid metabolism.

At the conclusion of this summary of the topic of enzymatic dedifferentiation (studied mainly in the case of hepatomas but found also in other tumors, such as HeLa and ascites cells, carcinoma, etc.), there is yet another fundamental question to put to the biochemist: Among these lesions, which are those that are *specifically* bound up with carcinogenesis and those which appear in the course of successive cell multiplications but *after* the initiation of the cancer? An experimental result will help to define the query. In a rat hepatoma induced by DAB, Allard *et al.* (1957) find no variation in acid phosphatase, whereas in the Novikoff hepatoma transplants the activity of this enzyme is always very low. It thus seems to us impossible to consider this fall in activity as a factor determining carcinogenesis; for us, it is a secondary disturbance brought on by the cancerous process.

Considering the importance of the problem raised, it is worth while to make several comments at this point concerning enzymatic dedifferentiation in tumors.

(1) The rapid succession of cell divisions cannot be considered to be the only factor responsible for the loss of enzymatic systems, for the liver, in the course of regeneration after a partial hepatectomy, is the seat of an "explosion" of mitoses, much more intense than in any neoplasm, but in spite of this the different enzymes (like the cell structures) show a normal rhythm of duplication.

(2) Although enzymatic dedifferentiation is a phenomenon that is relatively frequent in tumors, yet one cannot generalize since examples are known of cancer tissues where the functional characters of the tissue of origin are retained even in the metastases. Thus, epidermal cancer cells

produce keratin, melanomas always form melanin, and the metastases of thyroid tumors secrete the thyroid hormone.

(3) There are instances where a decrease in activity of an enzymatic system is due not to a parallel diminution of the apoenzyme, but to the abnormally low concentration of coenzymes, which then play the part of limiting factors. In this manner the poverty in DPN of the mitochondria of certain cancer cells explains how, in order to obtain the reducing amination of α-ketoglutaric acid to glutamic acid in presence of ammonia, it is essential to supplement the system with this cofactor (Emmelot, 1957a). But since transamination does not require the presence of this coenzyme, the mitochondria of these tumors can always transform α-ketoglutaric acid into glutamic acid in the presence of valine without the addition of DPN (Kit and Awapara, 1953; Emmelot, 1957a).

Another example of the same nature has recently been published by Sketol et al. (1958); for them the apparently low activity of cystine-desulfurase in tumour tissue was found to be due to low concentration of available DPNH and pyridoxal phosphate, the cofactors of cystine reductase and cystein desulfurase, respectively. Analogous observations have been made by Morton (1958).

One would also think that the very low concentration of TPN, and more especially of TPNH, in tumors would explain experimental data of the same order (Glock and Mac Lean, 1957).

(4) Finally, on the theoretical plane a last remark seems called for. It is well known that tissue culture results in morphological dedifferentiation for certain types of cell, but it is very interesting that it has recently been shown that this phenomenon may sometimes be accompanied by a loss of enzymatic systems, even in normal cells. Thus Perske et al. (1957) have found that when normal hepatic cells (Chang line) are cultured in vitro the activity of glucose-6-phosphatase disappears (as it vanishes in hepatoma). In the same way Reynafarje and Potter (1957) find that the same cells show a tumor-like behavior inasmuch as, like the Novikoff hepatoma transplants, they lose their transhydrogenase activity. Since the liver cells grown in tissue culture remain apparently normal, it seems to us that observations of this sort should lead biochemists to reconsider the meaning of enzymatic dedifferentiation.

It is none the less true that a demonstration of the loss of an enzymatic activity, or of its regulating mechanism, which would specifically determine the process of carcinogenesis remains the objective of biochemists working in this field. In this connection the investigations of Potter (1957, 1958a, b) and his co-workers on the loss of the regulation of certain enzymatic systems are of the greatest interest.

C. Glycolysis and Respiration

There are several reasons for the particular interest attached to the study of the glycolysis and respiration of cancer cells. These are two all-important processes that supply energy to the cells in a form which can be utilized for maintaining the steady state and to allow of growth and multiplication. Glycolysis and respiration consist of a sequence of reactions depending upon multiple enzymatic systems, the localization and functioning of which are fairly well known, even if the mechanisms that normally ensure their coordination still escape us.

In a series of publications, Warburg and co-workers (Warburg et al., 1924; Warburg, 1930, 1955, 1956a, b; Warburg and Hiepler, 1952) has given values for the glycolysis and the respiration in vitro of a large number of tumor tissues, and in spite of criticisms and reserves that have been made, it seems to be established that intense anaerobic glycolysis, remaining high in aerobic conditions, is a general and fundamental characteristic of all types of cancer cells. As regards respiration, the general conclusion from the data (Weinhouse, 1955) is that the Q_{o_2} (microliters of oxygen consumed by milligram dry weight of tissue per hour) measured by the Warburg technique is of the same order in normal and cancerous tissues.

From the year 1930 onward, Warburg has definitely expressed the idea that the injury of the respiratory systems is the prime cause of carcinogenesis in normal cells; this injury obliges the cell to adopt an anaerobic metabolism in order to satisfy its energy requirements. If the cell is able to adapt itself to these new conditions it can survive, but then it is transformed slowly, but surely, into a cancer cell (Warburg, 1930). In 1956 Warburg's theory remained unchanged: all carcinogens act by injuring respiration and this results in an increase in glycolysis. "For cancer formation, there is necessary not only an irreversible damaging of respiration but also an increase in the fermentation" (Warburg, 1956a). In experiments made on ascites tumor cells Warburg finds new arguments in favor of his concept.

Since 1930 a series of reviews have been devoted to this question and in them will be found the results obtained on neoplastic tissues compared with various normal tissues (Burk, 1939, 1942, 1956; Elliott, 1942; Dickens, 1951; Schmidt, 1951; Weinhouse, 1955, 1956a; Emmelot, 1957a).

The review by Weinhouse (1955) is of prime importance, since it systematically collects the data obtained for the various stages of glycolysis and respiration and gives a critical examination of the results. His conclusions are undoubtedly responsible for the renewal of interest in this question; the numerous investigations, publications, and symposia on the problem are a proof of this,

In this review we shall confine ourselves to pointing out the essentials of the subject of glycolysis and respiration in cancer cells, emphasizing the problems still open to the biochemist.

1. Intracellular Localization and Energy Yield of Glycolysis and of Respiration

It is now generally admitted that the multiple system of glycolysis is situated in the cytoplasm and in the nucleus (Le Page and Schneider, 1948; Siebert, 1958), whereas the respiratory enzymes are localized in the mitochondria. We have seen already that the microsomes intervene directly in protein synthesis; this being so, they are dependent on the glycolyzing and respiratory systems, which by the intermediary of certain molecules such as ATP provide them with the energy necessary for their syntheses.

As regards the yield of available energy, the results differ according to the mode of degradation of the glucose: In glycolysis, 1 molecule of glucose is transformed into 2 molecules of lactic acid with the formation of 2 molecules of ATP and a change in free energy of $\Delta F'$ (pH 7.5) $= -57$ kcal. In respiration, according to the Embden-Meyerhof pathway, 1 molecule of glucose is oxidized with the formation of 38 molecules of ATP and a change in free energy of $\Delta F'$ (pH 7.5) $= -690$ kcal.

It is clear that the yield is greater by the oxidative pathway: to get the same number of molecules of ATP it would theoretically be necessary to use nineteen times more glucose by glycolysis than by way of phosphorylating oxidations.

In a cell the quantity of available ATP is known to be a direct expression of the available energy. It contains an energy-rich bond $(P\sim)$ and it has long been considered that it represents a change in free energy of $+11$ to $+12$ kcal. (which corresponds to the reaction: ATP→ADP+Pi). There is now a tendency to consider that the energy liberated is no more than 8 kcal., which would lower by one-third the yield of both processes. Actually, the true yield should take into account the possible losses in the process of energy transfer from energy-rich bonds to endergonic reactions, these losses being dependent upon the degree of adjustment between coupled reactions.

2. Comparison between the Glycolytic and Respiratory Systems in Normal Cells and Cancer Cells

Whereas, according to Warburg, there is an *injury to the respiratory systems,* Weinhouse (1955) considers that these systems remain normal, so that respiration retains the same characteristics, qualitatively and quantitatively, in cancer cells. However, the cells would not be able to eliminate by respiration the lactic acid formed, so high is the level of the glycolysis. In the opinion of Weinhouse the characteristic *injury concerns the glycolysis*. Let us briefly examine what is known up to the present about the constituents of these two great systems in the tumor cell.

The experiments on which the conclusions are based have for the most part been made on tissue slices or homogenates. The tumors most frequently used have been hepatomas of rat and mouse, Walker and Flexner-Jobling carcinomas of the rat, and Jensen sarcoma; the state of development of these tumors is rarely stated. Since 1951, cells from rat ascites tumors of various origins (hepatoma, Ehrlich carcinoma, rhabdomyosarcoma) have sometimes been utilized. The great advantage of these tumors is that they furnish a material which is homogeneous in cancer cells if taken in the logarithmic growth phase. The media in which Q_{O_2} has been measured, and the anaerobic and aerobic lactic acid production (Q_LN_2, Q_LO_2), vary with authors and problems tackled.

a. Glycolysis. This requires the presence of a series of specific enzymes for each of the intermediate stages; it requires, moreover, at the start the presence of two cofactors, namely ATP and DPN, located in the cytoplasm.

The experiments of Novikoff *et al.* (1948) have demonstrated in tumors the existence of all the glycolytic enzymes; they also showed that the intermediary stages are the same as in normal tissues. The crucial experiments of Le Page (1950) brought to light two important facts.

(1) On a complete medium, containing both substrates (glucose and hexose diphosphate) and coenzymes (DPN and ATP), the *potential* glycolytic power of normal cells is of the same order as that of cancer cells. This means that in tumors the amount of apoenzymes is not higher than in normal cells. But *in vitro*, in a Ringer medium, only the *actual* glycolysis can be measured, and in these conditions it is indeed greater in cancer tissues than in normal tissues.

(2) Tumor cells *in vitro* are able to utilize glucose just as well as hexose diphosphate. This property is found also in brain tissue, the anaerobic glycolysis of which is high and partly persists in aerobiosis. It shows the presence of a high content of hexokinase, an enzyme phosphorylating glucose, to glucose-6-phosphate. We shall return to this point below. This fact explains the higher glycolysis *in vitro* of cancer cells as compared with normal tissues in media containing glucose (and no hexose phosphate). On the other hand, experiments of Le Page and Schneider (1948) show that the glycolytic activity of nuclei if expressed as a percentage of the glycolytic activity of the whole cell is the same in normal and in tumor cells.

b. Pasteur effect. When glycolysis is evaluated first in nitrogen and then in oxygen, diminution of lactic acid formation is observed: this is termed the Pasteur Effect. In the majority of normal tissues, in presence of oxygen, lactic acid disappears almost completely from the medium, thus aerobic glycolysis is very weak. On the other hand, in cancerous tissues, aerobic glycolysis is very important whatever the oxygen consumption may be. According to Warburg, in tumors, "respiration is always disturbed, inasmuch as it is incapable of causing the disappearance of fermentation." Nevertheless, each molecule of oxygen breathed causes the same amount of lactic acid to disappear in all cases. In other words, the Meyerhof quotient[*] is identical for normal and cancerous tissues (Warburg, 1930).

In his review, Weinhouse (1955) emphasizes the fact that Warburg does not postulate "quantitatively disturbed respiration." But in view of the high aerobic glycolysis of cancerous tissues "the decrease in glycolysis due to oxygen is lower percentagewise in tumors than in most normal tissues. For this reason a low Pasteur Effect has been mistakenly attributed to cancer cells." So, Weinhouse states that the cancer cell is characterized by a *high glycolysis persisting in an aerobic medium, the respiration remaining normal:* a bottleneck would thus be formed.

c. Respiration. Here, too, both qualitative and quantitative aspects of the problem have been examined. As it is generally admitted that respiratory systems are located in mitochondria, these cellular components have been most thoroughly investigated in this respect.

The main problems raised are the following:

(1) Do mitochondria of cancer cells contain all the enzymes and coenzymes present in normal mitochondria?

(2) Are they able to make use of the same initial substrates as mitochondria of corresponding normal tissues?

(3) Are the amounts of various constituents of the respiratory systems of the same order in normal and cancer cells?

In normal mitochondria most of the enzyme systems allowing the formation of acetyl CoA from pyruvic acid or fatty acids are present; in addition they contain the enzymes of the Krebs cycle, intervening subsequently in the reactions of oxidative phosphorylation.

In view of the low values of the respiratory quotient of tumor slices, Dickens and Simer as early as 1931, postulated an oxidation of fatty acids. Their hypothesis was corroborated by experiments performed with labeled fatty acids (see the review of Weinhouse, 1955). Afterwards,

---

[*] Meyerhof quotient $= \dfrac{(Q_L N_2 - Q_L O_2)}{Q_{O_2}}$.

Elliott *et al.* (1935) demonstrated oxidation of lactate and pyruvate, results confirmed subsequently by numerous authors (Meister, 1950; Weinhouse *et al.*, 1951; Wenner *et al.*, 1952; Emmelot *et al.*, 1955; Van Vals *et al.*, 1958). Furthermore, recent experiments (in particular determination of the P : G ratio) on ascites tumor cells have shown that the oxidation takes place via the tricarboxylic cycle (Weinhouse, 1955). Thus the problem does not relate to the existence of these enzymatic systems but to their quantity and availability in the cancer cell.

Mitochondria. The problem of the number of mitochondria per cancer cell has already been dealt with in this review.

Pyridine nucleotides. We have already quoted work proving that tumor tissues contain less of these coenzymes than their normal homologs. In addition, in order to obtain a normal level of oxidative phosphorylation of the intermediates in the Krebs cycle, it is necessary to supply greater amounts of DPN (Kielley, 1952; Williams–Ashman and Lehninger, 1951; Kennedy and Williams–Ashman, 1952; Weinhouse, 1955); this could be due, according to Weinhouse, to a modified permeability of the mitochondrial membrane to DPN and TPN.

Cytochromes and cytochrome oxidase. According to Greenstein (1956), a characteristic of tumors is their low content in cytochrome c and cytochrome oxidase. On the contrary, Chance and Castor (1952), using a spectrophotometric method of estimation *in situ,* find a normal concentration of the cytochromes as a whole, but a higher content of cytochrome c in the cancer cells. Schlief and Schmidt (1955; Schmidt and Schlief, 1955) arrive at similar conclusions.

The oxygen consumption of cancer cells cannot be attributed wholly to oxidative phosphorylation of the Krebs cycle, as it is known that the oxidative shunt is operative in tumor cells.

Measurements of the P : O ratio supplying information on the coupling of oxidations to phosphorylations have been made on tumor mitochondria. A certain number of experiments done with tumors give yields inferior to the normal values (Williams–Ashman and Kennedy, 1952; Siekevitz and Potter, 1953). In contrast to this, Kielley (1952) finds for the mouse hepatoma 98/15, and Andrejew *et al.* (1956) observe for the ascites hepatoma cells during the logarithmic phase, P : O ratios very nearly equal to those of normal mitochondria. It is of interest that Andrejew *et al.* (1956) demonstrated that β-hydroxybutyrate alone is not a substrate for oxidative phosphorylation in their system. Furthermore, Kielley (1957) studying *in vitro* the effect of carcinogens on normal rat liver mitochondria finds no decrease in the P : O ratio.

The low values of P : O ratios observed in cancerous tissues could perhaps be partially explained by the heterogeneity of the material used for the experiments. In addition other factors may also be responsible, such as morphological changes in the mitochondria or tissue rearrangements independent of carcinogenesis (Emmelot, 1957a).

Finally, one may ask, as Weinhouse (1955) and Skipper and Bennett (1958) have done, to what extent do these results obtained *in vitro* correspond to the phenomena taking place *in vivo*? On the one hand, the oxygenation brought about in the Warburg apparatus is certainly superior to that existing in tumors, but, on the other hand, the supply of substrate and of coenzymes may be very different according to the media used for the slices and homogenates: sometimes conditions of maximum activity (Le Page medium) are adopted, at others the levels of activity in the tissues at the time of their isolation (Ringer glucose medium) are measured. Thus, *in vivo* Potter and Busch (1950) found that, in contrast to normal tissues, tumors do not accumulate fluorocitrate after injection of fluoroacetate, and Busch and Baltrush (1954) observed a very slow disappearance of labeled acetate in tumors. However *in vitro*, Kit and Greenberg (1951) and then Wenner *et al.* (1952) obtained evidence for the condensation of acetyl CoA with oxalate, using analogous tumors. From the preceding, the conclusion may be drawn that the substrate concentrations and the oxygen pressure present *in vitro* favor oxidation by the Krebs cycle (Busch *et al.*, 1957).

3. Significance of the Increase of Aerobic Glycolysis

To explain the coexistence of a high aerobic glycolysis and a normal respiration in cancer tissues, some authors have thought that the high glycolysis was inseparable from the mitotic activity of the cancer cell (Holzer *et al.*, 1955). This would be a particularity of the neoplastic cell, since in regenerating liver, where the mitotic rhythm is more rapid than in any tumor, Burk (1942) found an aerobic glycolysis similar to that in normal liver. In the latter case the glycolysis is adjusted to the intensity of respiration by a mechanism which seems to have disappeared in neoplastic tissues.

There remains nevertheless a crucial question: at what moment does this characteristic disturbance arise? Is it actually bound up with carcinogenesis, or is it merely a consequence of this?

If it is an acquired characteristic, one might suppose that it consists of an enzymatic adaptation to the low oxygen pressures existing, either inside solid tumors or in the ascites liquid in which the cells float. Such an adaptation would explain how tumor slices can continue to glycolyze *in vitro* for long periods of time (Warburg, 1930). The phenomenon would be similar to what takes place in yeast cells, which manifest a much higher aerobic glycolysis when multiplying in a medium lacking oxygen

(Meyerhof, 1925; Lipmann, 1942). This mechanism of adaptation to partial anaerobiosis would involve an increase in hexokinase content and an impossibility for the cancer cell to store glycogen. No experimental evidence can at present be produced in support of this hypothesis.

If, on the contrary, as Warburg thinks, the initial cause of carcinogenesis consists in an alteration of respiration which would itself involve an increase in glycolysis, an attempt could be made to solve the problem experimentally; a study of the development of glycolysis and of respiration in the precancerous stages should supply valuable information.

Le Breton *et al.* have measured the glycolysis and the respiration of rat liver tissue during the months preceding carcinogenesis induced by a diet containing DAB. The experimental conditions were such that the hepatoma appeared after 12–14 months of diet; no anatomical pathological lesions were visible in the specimen taken for measurement of glycolysis after 6–9 months of the diet. The main results obtained are as follows.

(1) The increase in glycolysis, detectable after 6 months of diet, precedes the fall in respiration. After 9 months, the respiration is approximately 15% lower than that of control rats and the aerobic glycolysis has increased by 85%, both measurements being made in Tyrode medium.

(2) Thyroxine, activated by CoA, which increases the respiration of slices of normal liver in Tyrode (Le Breton and Van Hung, 1956), has no effect on that of precancerous liver.

This lack of action on the oxidations in the slices of precancerous liver cannot be attributed to an absence of activation of the thyroxine in presence of CoA, as in the same conditions aerobic glycolysis is increased by 400% in slices both in normal and precancerous livers: thus the activation of the thyroxine has taken place. If nevertheless the respiration is not increased, it is perhaps because the activated thyroxine was unable to penetrate into the mitochondria of the precancerous liver. In this case, after 9 months of diet containing DAB, the permeability of the mitochondrial membranes seems to have been modified and the mechanism which controls the adjustment between respiration and glycolysis badly damaged.

These findings suggest that during carcinogenesis, together with the disappearance of certain functional enzyme systems, the permeability of cellular membranes, particularly those of mitochondria become modified. The resulting changes should be important for the permeability of these membranes and are undoubtedly among the factors coordinating the metabolism of the various cell components.

IV. Carcinogenesis of Cells Cultured in Vitro

We shall briefly review the results obtained from the culture of cells *in vitro* concerning the transformation of the normal cell into a cancer cell and the characters of the latter. During the last twenty years numerous strains of cells derived from various tissues, normal and cancerous, have been isolated and cultivated. A certain number of these strains have been kept indefinitely under continuous cultivation without modification appearing in their characters in spite of a wide variety of media; this constitutes "a remarkable record of stability" (Syverton, 1957). Nevertheless—and this is the point that interests us—there is an increasing body of evidence that a pure line, or even a single cell, can give rise to lines with very different characteristics in the course of repeated transfers.

A. Modifications of Normal Cells in Vitro

Starting from a single cell or from the same initial clone, one can obtain diverse cell lines, characterized by their behavior with regard to the nutritive medium and/or their requirements of essential substances (Chang, 1957). Little by little the tissue culture experts are forced to admit that besides visible morphological changes there also exist changes of an enzymatic nature which often escape observation. Thus, Perske *et al.* (1957) find that after a hundred transfers of liver cells (Chang line) three of the four enzymatic systems tested have disappeared, among them glucose-6-phosphatase; alone glucose-6-dehydrogenase, an enzyme operative in the oxidative shunt, is found.

Parker *et al.* (1957) have recently summarized their investigations on alteration of strains kept in continuous culture, and have shown that from normal cells of very various origins, or even from cancer cells (HeLa), "altered" cells are obtained having the same general characteristics. Such "altered" cells turn up in cultures which may be young or old and with very different mitotic activities. These cells are able to multiply and their descendants to survive indefinitely by repeated transfers, while retaining the same morphological type of cell. A selected stable line has arisen, capable of multiplying at a rapid rate, even in conditions that are unfavorable for the original cell type. These cells react differently toward viruses and toxic substances. In spite of chromosome changes, the authors do not state positively that a mutation has taken place, in the sense in which the word is generally understood. Nevertheless, their observations, together with similar ones by many others, make this probable.

B. Malignant Transformations

Only in rare cases, has it been possible to prove that the "alteration" occurring during the transfers leads to *cancer cells,* since the only valid test for such a transformation is the malignant character of these cells when transplanted into a suitable host, i.e., multiplication and formation of metastases leading to the death of the animal. These tests have become easier to carry out for certain cells, e.g., those of human origin, since it was shown by Toolan (1951, 1954) that animals which have been irradiated and/or treated with cortisone can accept heterologous cells. In other words, such animals provide an excellent culture medium for cells of the invasive type (Leighton, 1957).

Nevertheless, cancer cells have been produced *in vitro* by several workers.

Gey and his co-workers (Firor and Gey, 1945; Gey *et al.*, 1949, 1952; Gey, 1956) have observed important morphological modifications in cell forms and mitotic characteristics in two different lines of rat fibroblasts (Ansatt 21 and 14p) which have been cultured in roller tubes for several months. Polynucleolar giant cells possessing heteroploidy appeared, and the rhythm of mitoses was four- to fivefold that of normal cells. When these cells were injected subcutaneously into inbred rats from which the fibroblasts had been obtained, metastasizing fibrosarcomas appeared which could be grafted; intraperitoneal injections produced ascites tumor cells.

Earle *et al.* (1943, 1950) have obtained cancerous transformation of strain L mice fibroblasts whether methylcholanthrene was present or absent in the culture medium. From a single cell, they got eight cell lines, six of which produce sarcomas when injected into mice. Cytological abnormalities of these cells are similar to those described by Gey. These experiments prove that cancerous or noncancerous strains may be derived from a single cell.

Goldblatt and Cameron (1953) obtained from rat fibroblasts cultured under nitrogen for 30 months, subcultures which when injected to rats give rise to fibrosarcomas. Injection of control subcultures which have grown in presence of oxygen do not produce tumors. These authors consider that their results confirm Warburg's theory.

Sanford (1958) got from a single cell two cell lines which possess different degrees of malignancy in the mouse. It has been established by the author that the arginase content of the two cell lines are also very different.

Experiments on the conversion of normal human cells into cancer cells have been made by Coriell *et al.* (1957), Leighton (1957), and Moore (1957). In spite of the difficulties of defining precisely the malignant character in the case of human cells, the conclusions of Moore in favor of a cancerous transformation are generally accepted, although not without certain reservations (Syverton, 1957).

Thus, it may be stated that the *transformation of normal cells to cancer cells in vitro can be obtained.*

C. Physiological Characters of Cancer Cells Obtained in Vitro

Gey (1956) has made a comparative study of the behavior of two cell lines of isogenic origin, which have been referred to above: 14p normal and T333 cancerous issued from 14p.

He found a more rapid multiplication of T333. Moreover, these cells have a definitely higher metabolic activity; they consume greater quantities of fluid medium and exhibit a pinocytosis ten times greater than that of the line 14p.

The anaerobic and aerobic glycolysis of the T333 cells is much higher than that of the line 14p, and so is the oxygen consumption (see results of Gey and Hellerman reported in Gey, 1956, p. 210). Moore (1957), too, found that cultured human cancer cells have a high glycolysis compared with that of normal cells (fibroblasts).

Gey (1956) has noted another interesting fact: the sensitivity to viruses of normal strains is different from that of homologous cancer cells. Certain viruses rapidly destroy neoplastic cells, without markedly injuring normal cells. In conclusion, we wish to refer to observations by Gey concerning differences between the 14p and T333 colonies as regards cohesion in identical media; he notes "an increased production of microfibrils of the normal cell line with a concomitant increase in colony stability through cellular cohesiveness as compared to the poorer cohesiveness of its tumor cell derivative." It is well known that the decreased adhesiveness of cancer cells is one of their important biological characteristics; it is the basis of their ameboid mobility.

It appears from this short account that a normal cell can be transformed into a cancer cell *in vitro.* This transformation is, indeed, only a particular case among all the modifications that can arise in normal cells in the course of successive transfers. It is an interesting fact that cancer cells in culture are themselves able to undergo modification: thus, the recent work of Hsu and Klatt (1959) shows that Novikoff hepatoma cells cultured *in vitro* develop into cells which have *lost their malignancy.*

Unfortunately it has not been possible to determine factors that are responsible for these transformations. Nutrition must certainly play an important part, as shown by the experiments of Puck (1957). Systematic investigations along these lines should be facilitated if the cancer cell possesses a characteristic easier to test than its malignancy. The problem is extremely complex; there is, however, no doubt that coordinated *in vitro* and *in vivo* experiments are of the greatest value in attempting to understand the cancer cell.

V. MECHANISMS OF CARCINOGENESIS AND CONCLUSIONS

It is not our intention to present in this review the various theories that have been put forward in connection with carcinogenesis. We merely wish to remark that most of them involve a modification, direct or indirect, of the DNA in the cell. Nevertheless, in concluding we should like to show how at the present time certain scientists have drawn a picture of the biochemical mechanisms of carcinogenesis which may lead to precise working hypotheses.

Leaving on one side the problem of the etiology of cancer, we may note the fact that carcinogenic agents are of very diverse nature (viruses, ionizing radiations, chemicals, hormonal and nutritional imbalance). Moreover, the most various animal species and almost all the different types of tissues are capable of developing a cancer, which emphasizes the generality of the mechanisms involved. One exception must, however, be made and that is the nerve cell (neuron) of the adult; as is well known, this cellular type has lost its faculty of division.

The majority of biologists tend to agree with Burnet (1958) that cancer seems to be due to a series of successive somatic mutations which result in endowing the cancer cells with their own transmissible characters. Sinsheimer (1957) speaks of "molecular parasitism of DNA," a concept which includes modifications by mutation and by introduction of exogenous DNA.[*] Indeed, Mellors (1958) considers that the two hypotheses relating to etiology, namely viruses and somatic mutations, appear sometimes to be separable only with difficulty.

What do these mutations consist of, and by what mechanisms may they be induced?

It is the present-day opinion that genes may be identified as DNA molecules occupying a definite place on the chromosome. The hypothesis of Watson and Crick permits a representation of the specificity of genes (the genetic information consisting in a specific sequence of the bases in DNA), of their replication, and of diverse types of modifications which lead to mutation (Beadle, 1957).

The function of the gene is to determine the specificity of a protein molecule, an apoenzyme which acts as template or pattern for the synthesis of various enzyme-forming systems (EFS). The transmission or transfer of the genetic informations of DNA occurs through the RNA directly involved in a template reproduction mechanism. It has not yet been unequivocally demonstrated that a specific RNA does function as

---

[*] As examples where exogenous DNA or RNA seems to act as transforming principles, work by Hays et al. (1957), Harel et al. (1958), Hewer and Meek (1958), and Latarjet et al. (1958) may be quoted.

a model in the formation of a specific protein, but it is certain that RNA is a key substance in protein synthesis.

The activity of the enzyme forming systems or synthetases will be controlled by cellular metabolism. The rate of an enzymatic reaction is function of a series of factors. It depends on the presence of coenzymes and on that of substrates; it may be modified by the hormones, the chemical mediators of the nervous system, and by other enzymatic reactions. It is also very important to underline that the normal functioning of certain enzymatic systems necessitates the integrity of the selective permeability of cellular membranes (the oxidative phosphorylations are a good example).

The rate of a reaction determines the turnover of the apoenzyme which catalyzes this reaction and through the phenomena of induction or repression influences the biogenesis of this enzyme; other synthetases (EFS) might however not be influenced (Vogel, 1957). In other words, there is metabolic adaptation according to the conditions of the cell. This mechanism has been much studied in bacteria; during the last few years, it has also given rise to many lines of research on mammalian tissues (see the review of Knox *et al.*, 1956). Under normal conditions, the synthetases (or EFS) are maintained at characteristic levels by mechanisms of induction or repression responsible for homeostasis. It is thus understandable that the permanent suppression of an enzymatic activity could involve the deletion of an EFS pattern connected with a gene.

We have seen that a more or less marked dedifferentiation is one of the characteristics of the cancer cell. It appears in the light of present data that this phenomenon may be connected with carcinogenesis; yet the conditions of life to which tumors are subjected might result in new enzymatic adaptations. Many biologists, including recently Burnet (1958), have expressed the idea that a precise knowledge of the mechanisms of embryonic differentiation might well throw light on the dedifferentiation of cancer cells. It appears to us that the conception of the enzyme forming system subjected to inductive or repressive mechanisms connects these two processes; we shall briefly deal with this question.

Various theories have been developed to explain how the cells of different lineages possess an identical genetic equipment but very different physiological activities (Brachet, 1957). We shall refer here only to the theory of Goldschmidt, as stated by Ephrussi (1952).

According to Goldschmidt, there appears in the course of embryonic development a methodical, orderly, and specific activation of the sets of genes in the cells of origin of the different lineages. It is the specific properties of the cytoplasm of each cell which, after cleavage of the egg, intervenes and allows of the activation and manifestation of particular genes. The characters thus acquired would be transmitted in the course of subsequent cell divisions.

Thus the molecules of DNA would not fulfill their template function of manufacturing EFS until they had undergone this activation. The fundamental genes corresponding to enzyme systems common to all types of cells would be activated before the functional genes: the phenomenon of activation would take place by stages in the course of embryonic differentiation.

One could then suppose that some of the characters acquired during this differentiation would be reversible: in the process of carcinogenesis certain genes would be "deactivated" as a result of feedback mechanisms. On the other hand, for other genes the activation would be irreversible whatever were the cellular conditions. In this way it is understandable that anaplasia can be very different according to the type of tumor.

As regards the mechanisms of mitosis itself, it is likely that this is identical in normal and cancer cells (Swann, 1957, 1958). Yet one fact is certain: it is not the high rate of mitosis which can explain the absence of certain enzyme systems; in regenerating liver all the genes are duplicated and the synthesis of fundamental and of functional enzyme systems is simultaneous.

The deletion of certain enzyme systems leading to the production of cancer could find its place in the theory of Miller and Miller (1953, 1955). In the opinion of these authors the protein which controls the growth and division of cells in a normal liver is inactivated by its union with a derivative of the carcinogenic azo dyes. But cancer of the liver can be obtained without the use of an azo dye by a disequilibrium of the diet fed to the rat (Copeland and Salmon, 1946; Le Breton, 1955). It appears to us that to explain the suppression of EFS (and the mutations which might result from this) another more general mechanism may be put forward.

The different types of membranes present in the cell (mitochondrial, microsomal, etc.) are composed of complex molecules. Many factors may modify their nature, their structure, and their degree of dispersion and cohesion; in this manner, the selective permeability of these membranes would be altered. We have seen how in precancerous states a modification of mitochondrial permeability led to an increase in glycolysis and an alteration of the effect of the thyroid hormone. These considerations on the mechanisms of carcinogenesis lead us to return to ideas expressed in the introduction of this review.

The extremely large number of biochemical and physiological investigations of the last few years has not provided an explanation of the biological behavior of the cancer cell. Only a few general characteristics have been demonstrated for some types of cancers particularly well studied (e.g., high aerobic glycolysis, enzymatic dedifferentiation). However, when reading recent publications, the rather stimulating impression results that as biochemistry penetrates more and more into all branches of biology, certain general conceptions and working hypotheses emerge whose incidence on cellular physiology is largely recognized by scientists engaged in cancer research. It is also beyond argument that progress is a function of knowledge concerning the normal cell; it would be therefore of prime interest to obtain separately the various cellular types of a given tissue, to establish the mechanisms coordinating metabolic reactions between the different cellular structures, and to determine the influence, and to determine the influence upon the fundamental enzyme systems of the activity levels of the various functional enzyme systems. All these problems have been set and good progress is made to resolve them.

On the other hand, it seems to be necessary to study the precancerous stages and to follow the evolution of the normal cell to a cancer cell. To grasp the cellular modifications which are responsible for carcinogenesis, it would be of interest to compare biochemical alterations of the same cellular type as a function of the carcinogenic agent; such experiments would make it possible to distinguish between *true* carcinogenic action and *concomitant* lesions caused by the carcinogens. Whether the mutations which are bound to carcinogenesis originate in primary effect on the cytoplasm, on mitochondria, on membranes, or on the nucleus, in every case there is *a moment when cellular modifications converge and the enzymatic alterations lead irreversibly to the cancerous state*. It is precisely *these* specific modifications which the biochemist tries to identify.

The survey of the literature pertaining to this review was completed in January, 1959.

REFERENCES

Abraham, S., Hill, R., and Chaikoff, I. L. (1955). *Cancer Research* **15**, 177–180.
Abraham, S., Cady, P., and Chaikoff, I. L. (1956). *Proc. Am. Assoc. Cancer Research* **2**, 89.
Adams, D. H., and Burgess, A. (1957). *Brit. J. Cancer* **11**, 310–325.
Albert, S., and Johnson, R. M. (1954). *Cancer Research* **14**, 271–276.
Allard, C., Mathieu, R., De Lamirande, G., and Cantero, A. (1952). *Cancer Research* **12**, 407–412.
Allard, C., De Lamirande, G., and Cantero, A. (1953). *Can. J. Med. Sci.* **31**, 103–108.
Allard, C., De Lamirande, G., and Cantero, A. (1957). *Cancer Research* **17**, 862–879.
Andrejew, A., Rosenberg, A. J., and Zajdela, F. (1956). *Compt. rend. soc. biol.* **150**, 1855–1861.
Arcos, J. C., and Arcos, M. (1958). *Biochim. et Biophys. Acta* **28**, 9–21.

Auerbach, V. H., and Waisman, H. A. (1958). *Cancer Research* 18, 543–547.

Babson, A. L., and Winnick, T. (1954). *Cancer Research* 14, 606–611.

Bach, S. J., and Maw, G. A. (1953). *Biochim. et Biophys. Acta* 11, 69–77.

Balis, M. E. (1957). *Proc. Am. Assoc. Cancer Research* 2, 186.

Balis, M. E., Van Praag, D., and Brown, G. B. (1955). *Cancer Research* 15, 673–678.

Barclay, R. K., and Garfinkel, E. (1955). *J. Biol. Chem.* 212, 397–401.

Barer, R. (1952). *Nature* 169, 366–367

Barnum, C. P., Jardetzky, C. D., and Halberg, F. (1957). *Texas Repts. Biol. and Med.* 15, 134–147.

Bassi, M., and Bernelli–Zayzera, A. (1954). *Tumori* 40, 21–41.

Beadle, G. W. (1957). *In* "The Chemical Basis of Heredity" (W. D. McElroy and B. Glass, eds.), pp. 3–22. Johns Hopkins Press, Baltimore, Maryland.

Begg, R. W., ed. (1955). *Proc. Can. Cancer Research Conf. 1st Conf. Honey Harbour, Ontario, 1954.*

Begg, R. W., ed. (1957). *Proc. Can. Cancer Research Conf. 2nd Conf. Honey Harbour, Ontario, 1956.*

Bendich, A. (1952). *Exptl. Cell Research* Suppl. 2, 82–86.

Bennett, L. L., Jr., and Skipper, H. E. (1957). *Cancer Research* 17, 370–373.

Bierich, R., and Lang, A. (1933). *Z. physiol. Chem.* 216, 217–223.

Bollum, F. J., and Potter, V. R. (1957). *J. Am. Chem. Soc.* 79, 3603–3604.

Brachet, J. (1957). "Biochemical Cytology." Academic Press, New York.

Brachet, J. (1958). *Bull. soc. chim. biol.* 40, 1387–1416.

Brown, G. B., and Roll, P. M. (1955). *In* "The Nucleic Acids" (E. Chargaff and J. N. Davidson, eds.), Vol. 2, pp. 341–392. Academic Press, New York.

Brues, A., and Marble, B. B. (1937). *J. Exptl. Med.* 65, 15–27.

Bullock, W. E., and Cramer, W. (1914). *Proc. Roy. Soc.* B87, 236–239.

Burk, D. (1939). *Cold Spring Harbor Symposia Quant. Biol.* 7, 420–459.

Burk, D. (1942). *Symposium on Respiratory Enzymes*, pp. 235–245.

Burk, D. (1956). *Science* 124, 269–270.

Burnet, F. M. (1958). *Federation Proc.* 17, 687–690.

Busch, H., and Baltrush, H. (1954). *Cancer Research* 14, 448–455.

Busch, H., and Greene, H. S. N. (1955). *Yale J. Biol. and Med.* 27, 339–349.

Busch, H., Simbonis, S., Anderson, D., and Greene, H. S. N. (1956). *Yale J. Biol. and Med.* 29, 105–116.

Busch, H., Davis, J. R., and Olle, E. W. (1957). *Cancer Research* 17, 711–716.

Butler, J. A., Johns, E. W., Lucy, J. A., and Simson, P. (1956). *Brit. J. Cancer* 10, 202–208.

Calva, E., Lowenstein, J. M., and Cohen, P. P. (1959). *Cancer Research* 19, 101–103.

Cambel, P. (1951). *Cancer Research* 11, 370–375.

Campbell, P. N. (1958). *Advances in Cancer Research* 5, 97–155.

Campbell, P. N., and Greengard, O. (1957). *Biochem. J.* 66, 47–48.

Campbell, P. N., and Stone, N. E. (1957). *Biochem. J.* 66, 669–677.

Canellakis, E. S. (1957). *J. Biol. Chem.* 227, 701–709.

Carlo, P. E., and Mandel, H. G. (1953). *J. Biol. Chem.* 201, 343–347.

Carruthers, C. (1950). *Cancer Research* 10, 255–265.

Carruthers, C., and Suntzeff, V. (1952). *Cancer Research* 12, 879–885.

Carruthers, C., and Suntzeff, V. (1954). *Cancer Research* 14, 29–33.

Carruthers, C., Suntzeff, V., and Harris, P. N. (1954). *Cancer Research* 14, 845–847.

Caspersson, T. O. (1950). "Cell Growth and Cell Function." Norton, New York.

Chalkley, H. W. (1943). *J. Natl. Cancer Inst.* 4, 47–53.

Chance, B., and Castor, L. N. (1952). *Science* 116, 200–202.

Chang, R. S. (1957). *Spec. Publ. N.Y. Acad. Sci.* **5**, 315–320.

Chapman, D. D., Brown, O. N., Chaikoff, I. L., Dauben, W. O., and Fanash, N. O. (1954). *Cancer Research* **14**, 372–376.

Chargaff, E. (1951a). *J. Gen. Physiol.* **38**, Suppl., 1–41.

Chargaff, E. (1951b). *Federation Proc.* **10**, 654.

Charlot, D. (1957). *Arch. sci. physiol.* **11**, 169–178.

Chauveau, J., Moulé, Y., and Rouiller, C. (1956). *Exptl. Cell Research* **11**, 317–321.

Claudatus, J., and Ginors, S. (1957). *Science* **125**, 394–395.

Clément, J. (1955). *Arch. sci. physiol.* **9**, 51–62.

Clément, G., Clément, J., and Le Breton, E. (1954). *Arch. sci. physiol.* **8**, 259–277.

Clerici, E., and Cudkowicz, G. (1956). *J. Natl. Cancer Inst.* **16**, 1459–1470.

Copeland, D. H., and Salmon, W. H. (1946). *Am. J. Pathol.* **22**, 1059–1079.

Coriell, L. L., Mac Allister, R. M., and Wagner, B. M. (1957). *Spec. Publ. N.Y. Acad. Sci.* **5**, 343–350.

Costello, C. J., Carruthers, C., Kamen, M. D., and Simoes, R. L. (1947). *Cancer Research* **7**, 642–646.

Crosbie, G. N., Smellie, R. M. S., and Davidson, J. N. (1953). *Biochem. J.* **54**, 287–292.

Cruft, H. J., Mauritzen, C. M., and Stedman, E. (1954). *Nature* **174**, 580–585.

Cunningham, L., Griffin, A. C., and Luck, J. M. (1950). *J. Gen. Physiol.* **34**, 59–63.

Daoust, R. (1955). *J. Natl. Cancer Inst.* **15**, 1447–1449.

Davidson, J. N., Leslie, I., and White, J. C. (1951). *Lancet* **i**, 1287.

Davison, P. F. (1957). *Biochem. J.* **66**, 703–707.

De Lamirande, G., and Allard, C. (1954). *Proc. Am. Assoc. Cancer Research* **1**, 27.

De Lamirande, G., and Allard, C. (1957). *Proc. Am. Assoc. Cancer Research* **2**, 224.

De Lamirande, G., Allard, C., and Cantero, A. (1955a). *J. Biol. Chem.* **214**, 519–524.

De Lamirande, G., Allard, C., and Cantero, A. (1955b). *Cancer Research* **15**, 329–332.

Dickens, F. (1951). *In* "The Enzymes" (J. B. Sumner and K. Myrbäck, eds.), Vol. 2, Pt. 1, p. 624. Academic Press, New York.

Dickens, F., and Simer, F. (1931). *Biochem. J.* **25**, 985–993.

Dyer, H. M. (1951). *J. Natl. Cancer Inst.* **11**, 1073–1080.

Earle, W. R., Nettleship, H., Shilling, E. L., Stark, T. H., Straus, N. R., Brown, M. F., and Shelton, E. (1943). *J. Natl. Cancer Inst.* **4**, 165–212.

Earle, W. R., Shelton, E., and Shilling, E. L. (1950). *J. Natl. Cancer Inst.* **10**, 1105–1113.

Eldredge, N. T., and Luck, J. M. (1952). *Cancer Research* **12**, 801–806.

Eliasson, N. A., Hammersten, E., Reichard, P., Aqvist, S. E. G., Theorell, B., and Ehrensvard, G. (1951). *Acta Chem. Scand.* **5**, 431–444.

Elliott, K. A. C. (1942). *Symposium on Respiratory Enzymes*, pp. 229–233.

Elliott, K. A. C., Benoy, M. P., and Baker, Z. (1935). *Biochem. J.* **29**, 1937–1950.

Elson, D., Trent, L. W., and Chargaff, E. (1955). *Biochim. et Biophys. Acta* **17**, 362–366.

Emmelot, P. (1957a). *Chem. Weekblad* **53**, 255–263.

Emmelot, P. (1957b). *Exptl. Cell Research* **13**, 601–604.

Emmelot, P., and Bos, C. J. (1956). *Biochim. et Biophys. Acta* **19**, 565–566.

Emmelot, P., and Bos, C. J. (1957a). *Enzymologia* **18**, 179–189.

Emmelot, P., and Bos, C. J. (1957b). *Exptl. Cell Research* **12**, 191–195.

Emmelot, P., and Brombacher, P. J. (1956). *Biochim. et Biophys. Acta* **21**, 581–583.

Emmelot, P., Bosch, L., and Van Vals, G. H. (1955). *Biochim. et Biophys. Acta* **17**, 451–452.

Ephrussi, B. (1953). "Nucleo-Cytoplasmic Relations in Microorganism." Oxford Univ. Press, London and New York.

Farber, E., Kit, S., and Greenberg, D. M. (1951). *Cancer Research* 11, 490–494.

Fiala, S., Sproul, E. E., Blutinger, M. E., and Fiala, A. E. (1955). *J. Histochem. and Cytochem.* 3, 212–225.

Firminger, H. I. (1955). *J. Natl. Cancer Inst.* 15, 1427–1442.

Firor, W. M., and Gey, G. O. (1945). *Ann. Surg.* 121, 700–703.

Forssberg, A., Finlayson, J. S., and Dreyfus, G. (1958). *Biochim. et Biophys. Acta* 30, 258–265.

Friedkin, M., and Kornberg, A. (1957). *In* "The Chemical Basis of Heredity" (W. D. McElroy and B. Glass, eds.), pp. 609–614. Johns Hopkins Press, Baltimore, Maryland.

Gelboin, H. V., Miller, J. A., and Miller, E. C. (1958). *Cancer Research* 18, 608–617.

Gey, G. O. (1956). *Harvey Lectures Ser.* 50, 154–229.

Gey, G. O., Gey, M. K., Firor, W. M., and Self, W. O. (1949). *Acta Unio Intern. contra Cancrum* 6, 706–712.

Gey, G. O., Coffman, W. D., and Kubicek, M. T. (1952). *Proc. Am. Assoc. Cancer Research* 2, 264–265.

Glock, G. E., and Mac Lean, P. (1957). *Biochem. J.* 65, 413–416.

Goldblatt, H., and Cameron, G. (1953). *J. Exptl. Med.* 97, 525–552.

Goranson, E. S. (1955). *Proc. Can. Cancer Research Conf. 1st Conf. Honey Harbour, Ontario, 1954*, pp. 330–344.

Greenberg, D. M. (1955). *Cancer Research* 15, 421–436.

Greenstein, J. P. (1945). "Mammary Tumors in Mice." Am. Assoc. Advance. Sci., Washington, D. C.

Greenstein, J. P. (1954). "Biochemistry of Cancer," 2nd ed. Academic Press, New York.

Greenstein, J. P. (1956). *Cancer Research* 16, 641–653.

Greenstein, J. P., and Leuthardt, F. M. (1946). *J. Natl. Cancer Inst.* 6, 317.

Greenstein, J. P., and Stewart, H. L. (1942). *J. Natl. Cancer Inst.* 2, 631–633.

Greenstein, J. P., Jenrette, W. V., Mider, G. B., and White, J. (1941). *J. Natl. Cancer Inst.* 1, 687–706.

Greenstein, J. P., Werne, J., Eschenbrenner, A. B., and Leuthardt, F. M. (1949). *J. Natl. Cancer Inst.* 9, 389–390.

Griffin, A. C., and Rhein, A. (1951). *Acta Unio Intern. contra Cancrum* 7, 763–767.

Griffin, A. C., Bloom, S., Cunningham, S., Teresi, J. D., and Luck, J. M. (1950). *Cancer* 3, 316–320.

Haddow, A. (1955). *Ann. Rev. Biochem.* 24, 689–742.

Hadjiolov, A. A., and Dancheva, K. I. (1958). *Nature* 181, 547–548.

Harel, J., Huppert, J., Lacour, F., and Lacour, J. (1958). *Compt. rend.* 247, 795–796.

Harkness, R. D. (1957). *Brit. Med. Bull.* 13, 87–93.

Haven, F. L. (1940). *J. Natl. Cancer Inst.* 1, 205–209.

Haven, F. L., and Bloor, W. R. (1956). *Advances in Cancer Research* 4, 238–314.

Hays, E. F., Simmons, N. S., and Beck, W. S. (1957). *Nature* 180, 1419–1420.

Hecht, L. I., and Potter, V. R. (1956). *Cancer Research* 16, 988–993.

Hecht, L. I., and Potter, V. R. (1958). *Cancer Research* 18, 186–192.

Heidelberger, C. (1953). *Advances in Cancer Research* 1, 273–338.

Heidelberger, C. (1956). *Ann. Rev. Biochem.* 25, 573–600.

Heidelberger, C., Leibman, K. C., Harbers, E., and Bhargava, P. M. (1957). *Cancer Research* 17, 399–404.

Hewer, T. F., and Meek, E. S. (1958). *Nature* 181, 990–991.

Hiatt, H. H. (1957). *Federation Proc.* 16, 251.

Hirschberg, E., Kream, J., and Gellhorn, A. (1952). *Cancer Research* 12, 524–528.

Hogeboom, G. H., and Schneider, W. C. (1951). *Science* **113**, 355–358.

Holzer, H., Haan, J., and Pette, D. (1955). *Biochem. Z.* **327**, 195–201.

Hsu, T. C., and Klatt, O. (1959). *J. Natl. Cancer Inst.* **22**, 313–339.

Hultin, T. (1957). *Exptl. Cell Research* **13**, 47–50.

Jablonski, J. R., and Olson, R. E. (1955). *Proc. Am. Assoc. Cancer Research* **2**, 26.

Jedeikin, L. A., and Weinhouse, S. (1955). *J. Biol. Chem.* **213**, 271–280.

Jedeikin, L. A., Thomas, A., and Weinhouse, S. (1956). *Cancer Research* **16**, 867–872.

Johnson, R. M., and Dutch, P. H. (1952). *Arch. Biochem. Biophys.* **40**, 239–244.

Jowett, M. (1931). *Biochem. J.* **25**, 1991–1998.

Kandutsch, A. A., and Baumann, C. A. (1954). *Cancer Research* **14**, 667–671.

Kennedy, E. P., and Williams–Ashman, H. G. (1952). *Cancer Research* **12**, 274.

Khouvine, Y., and Grégoire, J. (1953). *Bull. soc. chim. biol.* **35**, 603–608.

Khouvine, Y., Grégoire, J., and Zalta, J. P. (1953). *Bull. soc. chim. biol.* **35**, 244–256.

Kielley, R. K. (1952). *Cancer Research* **12**, 124–128.

Kielley, R. K. (1957). *J. Natl. Cancer Inst.* **19**, 1077–1085.

Kit, S. (1956). *Cancer Research* **16**, 70–76.

Kit, S. (1957). *Cancer Research* **17**, 56–63.

Kit, S., and Awapara, J. (1953). *Cancer Research* **13**, 694–698.

Kit, S., and Greenberg, D. M. (1951). *Cancer Research* **11**, 495–499.

Kit, S., and Griffin, A. C. (1958). *Cancer Research* **18**, 621–652.

Kit, S., Klein, J., and Graham, O. L. (1957). *J. Biol. Chem.* **229**, 853–864.

Klein, G. (1951). *Exptl. Cell Research* **2**, 518–573.

Klein, G. (1957). *Zentr. allgem. u. Pathol. u. pathol. Anat.* **96**, 20–24.

Knox, W. E., Auerbach, V. H., and Lin, E. C. C. (1956). *Physiol. Revs.* **36**, 164–254.

Kögl, F., and Erxleben, H. (1939). *Z. physiol. Chem.* **258**, 57–95.

Kornberg, A. (1957). *In* "The Chemical Basis of Heredity" (W. D. McElroy and B. Glass, eds.), pp. 579–608. Johns Hopkins Press, Baltimore, Maryland.

Kosaki, T. (1958). *Science* **158**, 485–486.

Kosaki, T., Ikoda, T., Kotani, Y., Nakagawa, S., and Saka, T. (1958). *Science* **127**, 1176–1177.

Kremen, A. J., Hunter, S. W., Moore, G. E., and Hitchcock, C. R. (1949). *Cancer Research* **9**, 174–176.

Laird, A. K. (1954). *Exptl. Cell Research* **6**, 30–44.

Laird, A. K., and Barton, A. D. (1956). *Science* **124**, 32–34.

Laird, A. K., and Miller, E. C. (1953). *Cancer Research* **13**, 464–470.

Latarjet, R., Rebeyrotte, N., and Moustacchi, E. (1958). *Compt. rend.* **246**, 853–855.

Le Breton, E. (1954). *Compt. rend.* **258**, 2446–2448.

Le Breton, E. (1955). *Voeding* **16**, 373–377.

Le Breton, E., and Van Hung, L. (1956). *Compt rend.* **242**, 1357–1359.

Le Breton, E., Chauveau, J., Jacob, A., and Moulé, Y. (1958). *Acta Unio Intern. contra Cancrum* **14**, 48–49.

Le Breton, E., Chauveau, J., Jacob, A., and Moulé, Y. (Unpublished experiments.)

Ledoux, L., and Revell, S. H. (1955). *Biochim. et Biophys. Acta* **18**, 416–425.

Leighton, J. (1957). *Cancer Research* **17**, 929–941.

Le Page, G. A. (1950). *Cancer Research* **10**, 77–88.

Le Page, G. A. (1953). *Cancer Research* **13**, 178–185.

Le Page, G. A., and Schneider, W. C. (1948). *J. Biol. Chem.* **176**, 1024–1027.

Le Page, G. A., Potter, V. R., Busch, H., Heidelberger, C., and Hurlbert, R. B. (1952). *Cancer Research* **12**, 153–157.

Le Page, G. A., Greenless, J., and Fernandes, J. F. (1956). *Ann. N.Y. Acad. Sci.* **63**, 999–1005.

Leslie, I. (1955). *In* "The Nucleic Acids" (E. Chargaff and J. N. Davidson, eds.), Vol. 2, pp. 1–50. Academic Press, New York.

Leuchtenberger, C., and Klein, E. (1952). *Cancer Research* **12**, 480–483.

Leuchtenberger, C., Leuchtenberger, R., and Davis, A. M. (1954). *Am. J. Pathol.* **30**, 65–85.

Leuchtenberger, C., Leuchtenberger, R., and Doolin, P. F. (1958). *Cancer* **11**, 490–506.

Levintow, L., Fu, S. C., Price, V. E., and Greenstein, J. P. (1950). *J. Biol. Chem.* **184**, 633–640.

Lewis, W. C. M. (1927). *J. Cancer Research* **11**, 16–53.

Lipmann, F. (1942). *Symposium on Respiratory Enzymes*, pp. 48–73.

Littlefield, J. W., and Keller, E. B. (1957). *J. Biol. Chem.* **224**, 13–30.

Littlefield, J. W., Keller, E. B., Gross, J., and Zamecnik, P. C. (1955). *J. Biol. Chem.* **217**, 111–123.

Lowe, C. U. (1955). *J. Natl. Cancer Inst.* **15**, 1619–1621.

Mac Indoe, W. M., and Davidson, J. N. (1952). *Brit. J. Cancer* **6**, 200–214.

Mark, D. D., and Ris, H. (1949). *Proc. Soc. Exptl. Biol. Med.* **71**, 727.

Meister, A. (1950). *J. Natl. Cancer Inst.* **10**, 1263–1271.

Mellors, R. C. (1955). *Cancer Research* **15**, 557–572.

Mellors, R. C. (1958). *Federation Proc.* **17**, 714–723.

Melnick, I., Cantarow, A., and Paschkis, K. E. (1958). *Arch. Biochem. Biophys.* **74**, 281–283.

Menten, M. L., Willms, M., and Wright, W. D. (1953). *Cancer Research* **13**, 729–732.

Metais, P., and Mandel, P. (1950). *Compt. rend. soc. biol.* **144**, 277–279.

Meyerhof, O. (1925). *Biochem. Z.* **162**, 43–86.

Miller, J. A. (1950). *Cancer Research* **10**, 65–72.

Miller, E. C., and Miller, J. A. (1947). *Cancer Research* **7**, 468–480.

Miller, E. C., and Miller, J. A. (1955). *J. Natl. Cancer Inst.* **15**, 1571–1590.

Miller, E. C., Miller, J. A., Kline, B. E., and Rusch, H. P. (1948). *J. Exptl. Med.* **88**, 89–98.

Miller, G. L., Green, E. V., Kolb, J. J., and Miller, E. E. (1950a). *Cancer Research* **10**, 141–147.

Miller, G. L., Green, E. V., Miller, E. E., and Kolb, J. J. (1950b). *Cancer Research* **10**, 148–154.

Miller, J. A., and Miller, E. C. (1948). *J. Exptl. Med.* **87**, 139–156.

Miller, J. A., and Miller, E. C. (1953). *Advances in Cancer Research* **1**, 339–396.

Mirsky, A. E., and Ris, H. (1949). *Nature* **163**, 666–667.

Mizen, N. A., and Petermann, M. L. (1952). *Cancer Research* **12**, 727–730.

Moore, A. C. (1957). *Spec. Publ. N.Y. Acad. Sci.* **5**, 321–329.

Mori, K., and Manoki, K. (1952). *Gann* **43**, 431–436.

Morton, K. R. (1958). *Nature* **181**, 540–542.

Mortreuil, M., and Khouvine, Y. (1957). *Bull. soc. chim. biol.* **39**, 161–168.

Moulé, Y. (1959). *Arch. sci. physiol.* **13**, 379–386.

Mutolo, V., and Abrignani, F. (1957). *Brit. J. Cancer* **11**, 590–596.

Norberg, E., and Greenberg, D. M. (1951). *Cancer* **4**, 383–386.

Novikoff, A. B. (1957). *Cancer Research* **17**, 1010–1027.

Novikoff, A. B., Potter, V. R., and Le Page, G. A. (1948). *Cancer Research* **8**, 203–210.

Olson, R. E. (1951). *Cancer Research* **11**, 571–581.

Parker, R. C., Castor, L. N., and Mac Culloch, E. A. (1957). *Spec. Publ. N.Y. Acad. Sci.* **5**, 303–313.

Perske, W. F., Parks, R. E., Jr., and Walker, D. L. (1957). *Science* **125**, 1290–1291.

Petermann, M. L. (1954). *Texas Repts. Biol. and Med.* **12**, 921–930.

Petermann, M. L., and Schneider, M. R. (1951). *Cancer Research* **11**, 485–489.

Petermann, M. L., Mizen, N. A., and Hamilton, M. G. (1953). *Cancer Research* **13**, 372–375.

Petermann, M. L., Hamilton, M. G., and Mizen, N. A. (1954). *Cancer Research* **14**, 360–366.

Petermann, M. L., Mizen, N. A., and Hamilton, M. G. (1956). *Cancer Research* **16**, 620–627.

Pileri, A., and Ledoux, L. (1957). *Biochim. et Biophys. Acta* **26**, 309–313.

Polli, E. E., and Shooter, K. V. (1958). *Biochem. J.* **69**, 398–403.

Potter, V. R. (1944). *Advances in Enzymol.* **4**, 201–256.

Potter, V. R. (1950). "Enzymes Growth and Cancer." C. C Thomas, Springfield, Illinois.

Potter, V. R. (1957). *Univ. Mich. Med. Bul.* **23**, 401–412.

Potter, V. R. (1958a). *Federation Proc.* **17**, 691–697.

Potter, V. R. (1958b). *Proc. 7th Intern. Cancer Congr. London* (abstr.), p. 12.

Potter, V. R., and Busch, H. (1950). *Cancer Research* **10**, 353–356.

Potter, R. L., Schlesinger, S., Buettner–Janusch, V., and Thompson, L. (1957). *J. Biol. Chem.* **226**, 381–394.

Potter, V. R., Brumm, A. F., and Bollum, F. J. (1958). *Proc. Am. Assoc. Cancer Research* **2**, 336.

Price, J. M., Miller, E. C., Miller, J. A., and Weber, G. M. (1949a). *Cancer Research* **9**, 96–102.

Price, J. M., Miller, E. C., Miller, J. A., and Weber, G. M. (1949b). *Cancer Research* **9**, 398–402.

Price, J. M., Miller, E. C., Miller, J. A., and Weber, G. M. (1950). *Cancer Research* **10**, 18–27.

Puck, T. T. (1957). *Spec. Publ. N.Y. Acad. Sci.* **5**, 291–298.

Randall, L. O. (1940). *Am. J. Cancer* **38**, 92–94.

Reichard, P., and Skold, O. (1958). *Biochim. et Biophys. Acta* **28**, 376–386.

Reynafarje, B., and Potter, V. R. (1957). *Cancer Research* **17**, 1112–1119.

Rondoni, P. (1955). *Advances in Cancer Research* **3**, 171–222.

Rotherham, J., and Schneider, W. C. (1958). *J. Biol. Chem.* **232**, 853–858.

Rutman, R. J., Cantarow, A., and Paschkis, K. E. (1954). *Cancer Research* **14**, 119–123.

Sanford, K. K. (1958). *Cancer Research* **18**, 747–752.

Schlenk, F. (1955). *In* "The Nucleic Acids" (E. Chargaff and J. N. Davidson, eds.), Vol. 2, pp. 309–339. Academic Press, New York.

Schlief, H., and Schmidt, G. (1955). *Naturwissenschaften* **42**, 104–105.

Schmidt, C. G. (1955). *Klin. Wochsche.* **33**, 409–419.

Schmidt, G., and Schlief, H. (1955). *Naturwissenschaften* **42**, 105–106.

Schmitz, H., Potter, V. R., and Hurlbert, R. B. (1954a). *Cancer Research* **14**, 58–65.

Schmitz, H., Potter, V. R., Hurlbert, R. B., and White, D. M. (1954b). *Cancer Research* **14**, 66–73.

Schmitz, H., Hurlbert, R. B., and Potter, V. R. (1954c). *J. Biol. Chem.* **209**, 41–54.

Schneider, W. C. (1946). *Cancer Research* **6**, 685–690.

Schneider, W. C. (1947). *Cold Spring Harbor Symposia Quant. Biol.* **12**, 169–178.

Schneider, W. C. (1955). *J. Biol. Chem.* **216**, 287–301.

Schneider, W. C. (1957). *J. Natl. Cancer Inst.* **18**, 569–577.

Schneider, W. C., and Brownell, L. W. (1957). *J. Natl. Cancer Inst.* **18**, 579–586.

Schneider, W. C., and Hogeboom, G. H. (1950). *J. Natl. Cancer Inst.* **10**, 969–975.

Schneider, W. C., and Rotherham, J. (1958). *J. Biol. Chem.* **233**, 948–953.

Schneider, W. C., Hogeboom, G. H., Shelton, E., and Striebich, M. J. (1953). *Cancer Research* **13**, 285–288.

Schweigert, B. S., Guthneck, B. T., Price, J. M., Miller, J. A., and Miller, E. C. (1949). *Proc. Soc. Exptl. Biol. Med.* **72**, 495–501.

Seabra, A., and Deutsch, H. F. (1955). *J. Biol. Chem.* **214**, 447–453.

Shack, J. (1943). *J. Natl. Cancer Inst.* **3**, 389–396.

Shelton, E., Schneider, W. C., and Striebich, M. J. (1952). *Anat. Record* **112**, 388.

Sibatani, A. (1957). *Biochim. et Biophys. Acta* **25**, 592–599.

Siebert, G. (1958). *Experientia* **14**, 448.

Siekevitz, P., and Potter, V. R. (1953). *Cancer Research* **13**, 513–520.

Siekevitz, P., and Zamecnik, P. C. (1951). *Federation Proc.* **10**, 246.

Sinsheimer, R. L. (1957). *Science* **125**, 1123–1128.

Sketol, J. A., Weiss, S., and Anderson, E. I. (1958). *J. Biol. Chem.* **233**, 936–938.

Skipper, H. E., and Bennett, L. L. Jr. (1958). *Ann. Rev. Biochem.* **27**, 137–166.

Smellie, R. M. S. (1955). In "The Nucleic Acids" (E. Chargaff and J. N. Davidson, eds.), Vol. 2, pp. 393–434. Academic Press, New York.

Sorof, S., and Cohen, P. P. (1951). *Cancer Research* **11**, 376–382.

Sorof, S., Claus, B., and Cohen, P. P. (1951a). *Cancer Research* **11**, 873–876.

Sorof, S., Cohen, P. P., Miller, E. C., and Miller, J. A. (1951b). *Cancer Research* **11**, 383–387.

Sorof, S., Young, E. M., and Ott, M. G. (1958). *Cancer Research* **18**, 33–46.

Stern, K., and Willheim, R. (1943). "The Biochemistry of Malignant Tumors." Reference Press, New York.

Striebich, M. J., Shelton, E., and Schneider, W. C. (1953). *Cancer Research* **13**, 279–284.

Swann, M. M. (1957). *Cancer Research* **17**, 727–758.

Swann, M. M. (1958). *Cancer Research* **18**, 1118–1160.

Swift, H. H. (1950). *Physiol. Zoöl.* **23**, 169–198.

Syverton, J. T. (1957). *Spec. Publ. N.Y. Acad. Sci.* **5**, 331–340.

Thomson, R. Y., and Frazer, S. C. (1954). *Exptl. Cell Research* **6**, 367–383.

Toolan, H. W. (1951). *Proc. Soc. Exptl. Biol. Med.* **78**, 540–543.

Toolan, H. W. (1954). *Cancer Research* **14**, 660–666.

Tung, T. C., and Cohen, P. P. (1950). *Cancer Research* **10**, 793–796.

Tyner, E. P., Heidelberger, C., and Le Page, G. A. (1952). *Cancer Research* **12**, 158–164.

Uzman, L. I., and Desoer, C. (1954). *Arch. Biochem. Biophys.* **48**, 63–71.

Van Vals, G. H., Van Hoeven, R. P., Bosch, L., and Emmelot, P. (1958). *Brit. J. Cancer* **12**, 448–458.

Villavicencio, M., and Barron, G. E. S. (1955). *Federation Proc.* **14**, 297.

Vogel, H. J. (1957). In "The Chemical Basis of Heredity" (W. D. McElroy and B. Glass, eds.), pp. 276–289. Johns Hopkins Press, Baltimore, Maryland.

Warburg, O. (1930). "Metabolism of Tumors." Arnold Constable, London.

Warburg, O. (1955). *Naturwissenschaften* **42**, 401–406.

Warburg, O. (1956a). *Science* **123**, 309–314.

Warburg, O. (1956b). *Science* **124**, 267–270.

Warburg, O., and Hiepler, E. (1952). *Z. Naturforsch.* **7b**, 193–194.

Warburg, O., Posener, K., and Negelein, E. (1924). *Biochem. Z.* **152**, 309–344.

Weber, G., and Cantero, A. (1957a). *Cancer Research* **17**, 995–1005.

Weber, G., and Cantero, A. (1957b). *Exptl. Cell Research* **13**, 125–131.

Weed, L. L. (1951). *Cancer Research* **11**, 470–473.

Weiler, E. (1956). *Z. Naturforsch.* **11b**, 31–38.

Weinhouse, S. (1955). *Advances in Cancer Research* **3**, 269–325.

Weinhouse, S. (1956a). *Science* **124**, 267–268.

Weinhouse, S. (1956b). *Cancer Research* **16**, 654–657.

Weinhouse, S., Millington, R. H., and Wenner, C. E. (1951). *Cancer Research* **11**, 845–850.

Weinhouse, S., Allen, A., and Millington, R. H. (1953). *Cancer Research* **13**, 367–371.

Wenner, C. E., and Weinhouse, S. (1953). *Cancer Research* **13**, 21–26.

Wenner, C. E., and Weinhouse, S. (1956). *J. Biol. Chem.* **219**, 691–704.

Wenner, C. E., Spirtes, M. A., and Weinhouse, S. (1952). *Cancer Research* **12**, 44–49.

Wenner, C. E., Hackney, J., and Herbert, J. (1957). *Proc. Am. Assoc. Cancer Research* **2**, 259.

White, J. C., Leslie, I., and Davidson, J. N. (1953). *J. Pathol. Bacteriol.* **66**, 291–306.

Wicks, L. F., and Suntzeff, V. (1942). *J. Natl. Cancer Inst.* **3**, 221–226.

Willheim, R., and Fuchs, G. (1932). *Biochem. Z.* **247**, 297–305.

Williams, A. M., and Le Page, G. A. (1958a). *Cancer Research* **18**, 548–553.

Williams, A. M., and Le Page, G. A. (1958b). *Cancer Research* **18**, 554–561.

Williams, A. M., and Le Page, G. A. (1958c). *Cancer Research* **18**, 562–568.

Williams–Ashman, H. G., and Kennedy, E. P. (1952). *Cancer Research* **12**, 415–421.

Williams–Ashman, H. G., and Lehninger, A. L. (1951). *Cancer Research* **11**, 289–290.

Wiltshire, G. H. (1953). *Brit. J. Cancer* **7**, 137–141.

Yagi, K., and Khouvine, Y. (1957). *Bull. soc. chim. biol.* **39**, 477–482.

Yarbro, C. L., and Anderson, C. E. (1954). *Federation Proc.* **13**, 326.

Yuile, C. L., Lamson, B. G., Miller, L. L., and Wipple, G. H. (1951). *J. Exptl. Med.* **93**, 539–557.

Zamecnik, P. C. (1950). *Cancer Research* **10**, 659–667.

Zamecnik, P. C., Frantz, I. D., Jr., Loftfield, R. B., and Stephenson, M. L. (1948). *J. Biol. Chem.* **175**, 299–314.

Zamecnik, P. C., Loftfield, R. B., Stephenson, M. L., and Steele, J. M. (1951). *Cancer Research* **11**, 592–602.

Zilber, L. A. (1958). *Advances in Cancer Research* **5**, 291–329.

AUTHOR INDEX

Numbers in italic indicate the pages on which the references are listed.

SUBJECT INDEX